OF THE WORLD

Brooks and Roberts Modified Van der Grinten Projection

Union of Soviet Socialist Republics

Sweden
Finland
way
E.
Ger
Poland
W.
Germany
Czech
Lech
Aus
Hung
San
Mar
Yugo.
Rom.
Italy
Bul.
Alb.
Malta
Greece
Cyprus
Turkey
Lebanon
Israel
Syria
Iraq
Iran
Afghanistan
Pakistan
Jordan
Kuwait
Bahrain
Quatar
Saudi
Arabia
United
Arab Emirates
Oman
Yemen
Yemen Arab. Rep.

Mongolia
China
North
Korea
South
Korea
Japan
Bhutan
Nepal
Bangladesh
Burma
Laos
Hong Kong
(U.K.)
China
(Taiwan)
Thailand
Vietnam
Philippines
India
Kampuchea
Brunei
Malaysia
Singapore
Sri Lanka
Maldives
I n d o n e s i a
Papua
New Guinea
Nauru
Solomon
Islands
Fiji

Libya
Egypt
Niger
Chad
Sudan
Djibouti
Nigeria
Cameroon
Cent. Af.
Rep.
Uganda
Ethiopia
Somalia
Gabon
Congo
Zaire
Rwanda
Burundi
Kenya
Cabinda
Seychelles
Tanzania
Malawi
Comoros
Angola
Zambia
Mozambique
Namibia
(S. Af.)
Zimbabwe
Madagascar
Botswana
Mauritius
Swaziland
Lesotho
South
Africa

Australia

New
Zealand

20° 40° 60° 80° 100° 120° 140° 160° 180°

Introduction to
Geography

Introduction to
Geography

Second Edition

Arthur Getis
University of Illinois, Urbana–Champaign

Judith Getis

Jerome Fellmann
University of Illinois, Urbana–Champaign

wcb
Wm. C. Brown Publishers
Dubuque, Iowa

Book Team

Editor *Edward G. Jaffe*
Developmental Editor *Nova A. Maack*
Production Editor *Barbara Rowe Day*
Designers *Mark Elliot Christianson/Tara L. Bazata*
Photo Research Editor *Shirley Charley*
Permissions Editor *Mavis M. Oeth*
Visuals Processor *Joyce E. Watters*

wcb group

Chairman of the Board *Wm. C. Brown*
President and Chief Executive Officer *Mark C. Falb*

wcb

Wm. C. Brown Publishers, College Division

President *G. Franklin Lewis*
Vice President, Editor-in-Chief *George Wm. Bergquist*
Vice President, Director of Production *Beverly Kolz*
National Sales Manager *Bob McLaughlin*
Director of Marketing *Thomas E. Doran*
Marketing Information Systems Manager *Craig S. Marty*
Marketing Manager *Matt Shaughnessy*
Executive Editor *Edward G. Jaffe*
Production Editorial Manager *Julie A. Kennedy*
Manager of Visuals and Design *Faye M. Schilling*
Manager of Design *Marilyn A. Phelps*

Cover Credits: People/Sunset/Ocean © Marc Solom/Image Bank, Chicago. Earth from Space © F. Reginato/Image Bank, Chicago.

The credits section for this book begins on page C1, and is considered an extension of the copyright page.

Contents

3

Physical Geography: Weather and Climate 85

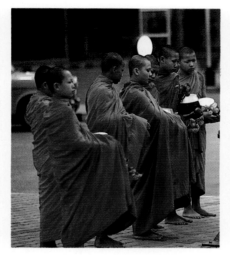

Part Three

The Locational Tradition

9

Economic Geography 293

10

Resource Geography 337

11

Urban Geography   369

Part Four

The Area Analysis Tradition

12

The Regional Concept   403

Preface

This second edition of *Introduction to Geography* retains the purpose of its predecessor: to introduce college students to the breadth and excitement of the field of geography. It retains as well the structure of the original book, with its content organized around the major research traditions of the discipline. The favorable acceptance of that earlier version encouraged us to keep its purpose and structure intact while updating its facts and making every effort to include suggestions offered by users and colleagues to improve its content.

We recognize that many students will have only a single college course and textbook in geography. Our purpose for those students is to convey concisely but clearly the nature of the field, its intellectual challenges, and the logical interconnections of its parts. Even if they take no further work in geography, we are satisfied that they will have come into contact with the richness and breadth of our discipline and gain new insights and understandings for their present and future role as informed adults. Other students may have the opportunity and interest to pursue further work in geography. For them, this text will introduce the content and scope of the subfields of geography, emphasize its unifying themes, and provide the foundation for further work in their areas of interest.

Our approach is to let the major research traditions of geography dictate the principal themes we explore. As the "Introduction" makes clear to students, the organizing traditions that have emerged through the long history of geographical thought and writing are those of earth-science, culture–environment, location, and area analysis. Each of the four parts of this book centers on one of these geographic perspectives. Within each part, except that on area analysis, are chapters devoted to the

subfields of geography, each placed with the tradition to which we think it belongs. Thus, the study of weather and climate is part of the earth-science tradition; population geography is considered under the culture–environment tradition; and urban geography is included with the locational perspective. Of course, our assignment of a topic may not seem appropriate to all users since each tradition contains many emphases and themes. Some subfields could logically be attached to more than one of the recognized traditions. The rationale for our clustering of chapters is given in the brief introductions to each part of the text. The tradition of area analysis—of regional geography—is presented in a single final chapter that draws upon the preceding traditions and themes, and is integrated with them by cross-references.

In this revision, we have tried to incorporate the most current viewpoints and approaches, and to include the data on which they are based. Essentially all of the diagrams and maps have been newly prepared or redrawn in full color for added visual impact and instructional emphasis. Almost all of the photographs are in color, new to this edition, and dramatize, demonstrate, or document important textual points. Chapter introductions, most of them new, are designed to capture students' attention, arouse their interest in the vignette itself, and hold their attention for the subject matter that follows. In addition, each chapter contains boxed inserts that further embellish ideas included in the text proper. In short, the material is designed to gain and retain student attention—the essential first step in the learning process.

An expanded "Introduction," appearing as an unnumbered chapter, prepares the student for the later substantive chapters. It introduces the field of geography as a whole, noting both its antiquity and currency, its breadth of interests, and the unifying questions, themes, and concepts that structure all geographic inquiry whatever the specialized subfield of interest. The "Introduction" also outlines for the student the organization of the book and explains the several "traditions" forming its framework.

Important to that framework, as noted, is the terminal single chapter treatment of the area analysis tradition. The case studies and examples that Chapter 12, "The Regional Concept," contains have been specially selected or newly written to illustrate the regional geographic application of the systematic themes developed by the earlier chapters. Regional understanding has always been an important motivation and justification of geography as a discipline; Chapter 12 is designed to introduce students to the diversity of regional geographic exposition. It may be read either as a separate chapter or in conjunction with the earlier material. That is, each systematic chapter contains a reference to the specific pages of Chapter 12 where a relevant regional geographic example is to be found. That referenced case study, of course, can be incorporated to demonstrate the relationships of regional and systematic geography, to show the "real world" application of geographic understandings, and to provide a springboard for further case studies as class or instructor interest may dictate. The included regional studies may also serve as models for independent student reports applying to specific cases the insights and techniques of analysis developed in the separate substantive chapters.

Learning aids at the conclusion of each chapter include a Summary; a list of Key Words together with a page reference that shows the page on which each term is defined and discussed; For Review questions designed to help students check on their understanding of the chapter material; and Suggested Readings consisting chiefly of references to recent specialized textbooks in the subject area of the chapter. We recognize that most libraries will not have a full complement of the major geographical journals nor of specialized scholarly monographs. Our references, therefore, are to easily available items that themselves contain detailed bibliographies for those students who are motivated to delve more deeply into particular subfields of geography.

At the end of the book we have placed a comprehensive Glossary of terms introduced and, as a special Appendix, a reprint of the "*1987 World Population Data Sheet*" of the Population Reference Bureau, Inc. In addition to basic demographic data and projections for countries, regions, and continents, the Appendix includes statistics on gross national product, energy consumption, and access to safe water supply. Although inevitably dated and subject to change, the Appendix statistics nonetheless will provide for some years a wealth of useful comparative data for student projects, regional and topical analyses, and study of world patterns.

A useful textbook must be flexible enough in its organization to permit an instructor to adapt it to the time and subject-matter constraints of a particular course. Although designed with a one-quarter or one-semester course in mind, this text may be used in a full-year introduction to geography when employed as a point of departure for special topics and amplifications introduced by the instructor or when supplemented by additional readings or class projects. Moreover, the chapters are reasonably self-contained and need not be assigned in the sequence here presented. The "traditions" structure may be dropped and the chapters rearranged to suit the emphases and sequences preferred by the instructor or found to be of greatest interest to the students. The format of the course should properly reflect the joint contribution of instructor and book rather than be dictated by the book alone. Further to assist the instructor in that cooperative enterprise, we have prepared an *Instructor's Manual* that highlights the main ideas of each chapter, offers topics for class discussions, and provides 50 suggested test questions for each chapter.

A number of reviewers, anonymous at the time, greatly improved the content of this book by their critical comments and suggestion. With pleasure we acknowledge the thoughtful assistance rendered by Mr. Anthony Grande, Hunter College–CUNY; Professor John Harmon, Central Connecticut State University; Dr. Paul E. Lovingood, Jr., University of South Carolina, Columbia; Professor Lisle S. Mitchell, University of South Carolina, Columbia; Professor Dale Monsebroten, Eastern Kentucky University; Professor Peter O. Muller, University of Miami; Professor Dennis Quillen, Eastern Kentucky University, Richmond; Professor Carol Rabenhorst, Catonsville Community College; Dr. Georgia Zannaras; and Dr. Charles Ziehr, East Carolina University. We are also indebted to W. D. Brooks and C. E. Roberts, Jr., of Indiana State University for the projection used for many of the maps in this book: a modified van der Grinten. We gratefully thank these and unnamed others for their help and contributions and specifically absolve them of responsibility for decisions on content and for any errors of fact or interpretation that users may detect.

Finally, we note with deep appreciation and respect the efforts of the publisher's "book team" separately named on the copyright page. In particular, our thanks go to Executive Editor Edward G. Jaffe, who brought this project to Wm. C. Brown Company; to Senior Developmental Editor Nova Maack, who labored long and carefully to bring its parts together; and to Senior Production Editor Barbara Rowe Day, who brought her keen editorial eye to the manuscript and shepherded the book to completion.

Arthur Getis
Judith Getis
Jerome Fellmann

Introduction to
Geography

Introduction

Government inspectors had issued the first of two warnings about the dam nearly a year before, but no repairs had been made. Fortunately, workers spotted the cracks early, and when the dam broke, letting loose a 20-foot (6 m) wave, the villages had been cleared and all train and highway traffic stopped. What poured through the breach was not just water, but a thick deadly brine, the impounded discharge of a major fertilizer plant. On its rush to the river, the brine swept away a railroad track, ripped up roads, smashed through a village, inundated 500 acres (200 hectares) of the nation's richest farmland, and spilled into the purest river left in the western part of the country. Worse than the immediate physical destruction were the aftereffects. Two thousand tons of fish were destroyed along with all of the vegetation that could support their replacement. Water supplies to two major and many minor cities were cut off as the formerly pure river became "brinier than the saltiest seawater," and a million tons of salt were deposited in a layer 35 feet (10.5 m) thick at the bottom of a major reservoir 300 miles (480 km) downstream.

The accident happened in 1983 in the western part of the Soviet Union and affected the Dniester River from near the Polish border all the way to the Black Sea. The details of its location are less important than the lessons contained in the event. The wall of brine, the destruction of farmland and fish, the salt layer at the lake bottom, and the scramble for alternate water supplies are dramatic evidence of the pressures humans place upon the environment of which they are a part. Radioactive waste materials and atmospheric discharges through nuclear plant accidents, deadly manufactured chemicals, accelerated erosion through unwise forestry and farming practices, the creation of deserts through overgrazing, the poisoning of soils by salts from faulty irrigation and of groundwater through deep-well injection of the liquid garbage of modern industry—all are evidence of the adverse consequences of human pressures upon natural systems. The social and economic actions of humans occur within the context of the environment and have environmental consequences too dangerous to ignore.

The interaction of human and environmental systems works both ways. People can inflict irreparable damage upon the environment and the environment can exact a frightening toll from societies that inappropriately exploit it.

November is the cyclone season in the Bay of Bengal. At the northern end of the bay lie the islands and the lowlands of the Ganges Delta—a vast fertile land that is mostly below 30 feet (9 m) in elevation, made up of old mud, new mud, and marsh. Densely settled by desperately land-needy people, this delta area is home to the majority of the population of Bangladesh. Early in November of 1970, a low-pressure weather system moved across the Malay Peninsula of Southeast Asia and gained strength in the Bay of Bengal, generating winds of nearly 150 miles (240 km) per hour. As it moved northward, the storm sucked up and drew along with it a 20-foot (6-m) wall of water. On the night of November 12, with a full moon and highest tides, the cyclone and its battering wall of water slammed into the islands and the deltaic mainland (Figure I.1). When it had passed, some of the richest rice fields in Asia were gray with the salt that ruined them, islands totally covered with paddies were left as giant sand dunes, and an estimated 500,000 persons had perished.

Should the tragedy be called the result of the blind forces of nature, or should it be seen as the logical outcome of a state of overpopulation that forced human encroachment upon lands more wisely left as the realm of river and sea?

As a discipline, geography does not attempt to make value judgments about such questions. Geography does claim, however, to be a valid and revealing approach to contemporary questions of economic, social, and ecological concern. Humans and environment in interaction; the patterns of distribution of natural phenomena affecting human use of the earth; the cultural patterns of occupance and exploitation of the physical world—these are the themes of the encompassing discipline called geography.

Roots of the Discipline

Geography literally means "description of the earth." It is a description in which people and their activities assume a central position. As a way of thinking about and analyzing the earth's surface, its physical patterns, and human occupance, geography has ancient roots. It grew out of early Greek philosophical concern with "first beginnings" and with the nature of the universe. From that broader inquiry, it was refined into a specialized investigation of the physical structure of the earth, including its terrain and its climates, and the nature and character of its contrasting inhabited portions.

Figure I.1

The swirling eye of the killer storm that struck the islands and
mainland of Bangladesh (then East Pakistan) in November 1970
is hidden in the great white mass of the low pressure system
photographed by a U.S. weather satellite. It was the small, violent
core of the storm that did the damage. Survivors reported a
deafening roar and a 50-foot (15 m) wall of water covering the
offshore islands, carrying away people, houses, animals, and
trees. As the wave moved on to the mainland, it was followed by
a violent, tearing wind. Receding water, falling back to the Bay of
Bengal, increased the destruction of the storm.

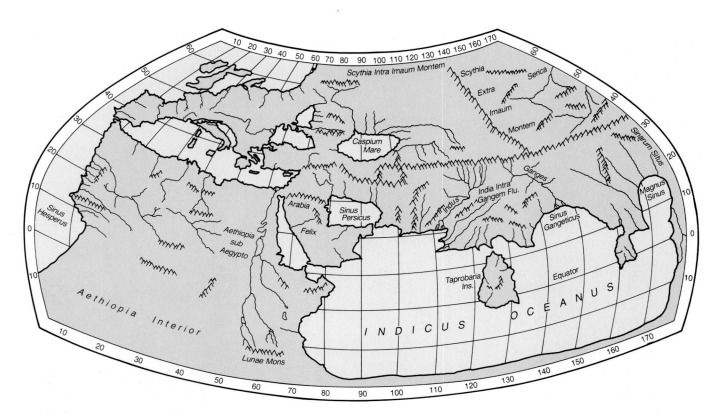

Figure I.2

The world map of the second-century A.D. Roman geographer-astronomer Ptolemy. Ptolemy adopted an already developed map grid of latitude and longitude lines based on the division of the circle into 360°, permitting a precise mathematical location for every recorded place. Unfortunately, errors of assumption and estimation rendered both map and its accompanying six-volume gazeteer inaccurate. Many variants of Ptolemy's map were published in the 15th and 16th centuries. The version shown here summarizes the extent and content of the original.

To Strabo (*c.* 64 B.C.–A.D. 20), one of the greatest Greek geographers, the task of the discipline was to "describe the known parts of the inhabited world . . . to write the assessment of the countries of the world [and] to treat the differences between countries." Even earlier, Herodotus (*c.* 484–425 B.C.) had found it necessary to devote much of his book to the lands, peoples, economies, and customs of the various parts of the Persian Empire as necessary background to his audience's understanding of the causes and course of the Persian Wars.

Geographers measured the earth, devised the global grid of latitude and longitude, and drew upon this grid surprisingly sophisticated maps of their known world

(Figure I.2). They explored the apparent latitudinal variations in climate and described in numerous works the familiar Mediterranean basin and the more remote, partly rumored lands of northern Europe, Asia, and equatorial Africa. Employing nearly modern concepts, they described river systems, explored cycles of erosion and patterns of deposition, cited the dangers of deforestation, and noted the consequences of environmental abuse. Strabo cautioned against the assumption that the nature and actions of humans were determined by their physical environment. He observed, instead, that humans were the active elements in a cultural–physical partnership.

So broad and integrated are the concerns of geography and so great the variety of facts, observations, and distributions that guide its analyses, that it has been called the mother of sciences. From its earlier concerns have sprung such specialized, independent disciplines as meteorology, climatology, cultural anthropology, geology, and a host of other fields of inquiry. From such fields of study, geographers draw the background data that contribute to their own broader investigation of human–environment systems and spatial relationships.

Basic Geographic Concepts

Although such investigations may serve broadly to define the field of geography, its nature is best understood through the questions geographers ask and the approaches they employ to answer those questions. Of a physical or cultural phenomenon, they will inquire: What is it? Where is it? How did it come to be what and where it is? Where is it in relation to other physical or social realities that affect it or are affected by it? How is it part of a functioning whole? How does its location affect people's lives?

These and similar questions derive from central themes in geography. In answering them, geographers respond by using certain fundamental concepts and terms. Together, these themes, concepts, and terms form the basic structure and vocabulary of geography. They address the fundamental truths that things are rationally organized on the earth's surface and that understanding spatial patterns is an essential starting point for understanding how humans live on the earth.

Geography is about earth places and spaces. The questions that geographers ask originate in basic observations about the nature of places and how places are similar to or different from one another. Those observations, though simply stated, are profoundly important to our comprehension of the world we occupy:

A place may be large or small.
A place has location.
A place has both physical and cultural characteristics.
The characteristics of places develop and change over time.
Places interact with other places.
Places may be generalized into regions of similarities and differences.

Size and Scale

When we say that a place may be large or small, we speak both of the nature of the place itself and of the generalizations that can be made about it. In either instance, geographers are concerned with **scale,** though we may use that term in different ways. We can, for example, study a problem—population, say, or landforms—at the local scale or on a global scale. Here, the reference is purely to the size of unit studied. More technically, scale tells us the relationship between the size of an area on a map and the actual size of the mapped area on the surface of the earth. In this sense—as Chapter 1 makes clear—scale is a feature of every map and essential to recognizing what is shown on that map.

In both senses of the word, scale implies the degree of generalization represented (Figure I.3). Geographic inquiry may be broad or narrow; it occurs at many different size-scales. Climate may be an object of study, but research and generalization focused on climates of the world will differ in degree and kind from study of the microclimates of a city. Awareness of scale is very important. In geographic work concepts, relationships, and understandings that have meaning at one scale may not be applicable when the same problem is examined at another scale.

Location

The location of a place may be described in both absolute and relative terms. **Absolute location** records a precise position on the surface of the globe, usually in terms of a mathematically based reference system. The global grid of latitude and longitude is commonly employed for absolute location. In United States property description, a reference to township, section, and range is customary in those parts of the country where that survey system exists. Absolute location is unique to each described place, is independent of any other characteristic or observation about that place, and has obvious usefulness in legal descriptions of places, in measuring the distance separating places, or in finding directions between places on the earth's surface.

When geographers remark that "location matters," however, their reference is usually not to absolute but to **relative location**—the location of a place or thing in relation to that of other places or things. Relative location expresses spatial interconnection and interdependence. It tells us that people, things, and places exist in a world of physical and cultural characteristics that differ from place to place. The attributes and potentialities of a particular locale are understandable only as that locale is seen in its spatial relationship to other places with their own attributes (Figures I.4 and 11.6). Again, the idea of scale comes into play. Depending upon the place and characteristic studied, spatial relationships may be traced at the local, regional, national, or global scales, with each scale reference to relative location adding to our understanding of place.

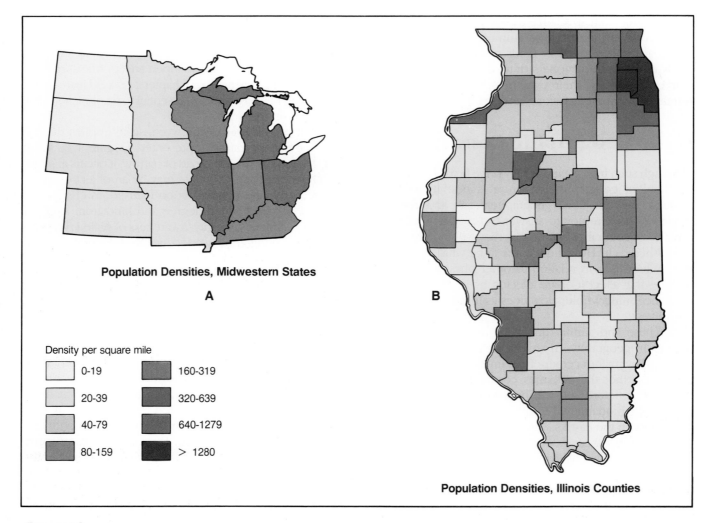

Population Densities, Midwestern States

A

Density per square mile

0-19		160-319	
20-39		320-639	
40-79		640-1279	
80-159		> 1280	

B

Population Densities, Illinois Counties

Figure I.3

"Truth" depends on one's scale of inquiry. Map A reveals that the maximum population density of Midwestern states is no more than 319 per square mile (123 per km²). From Map B, however, we see that population densities in two Illinois counties exceed 1280 people per square mile (495 per km²). Were we to reduce our scale of inquiry even further, examining individual city blocks in Chicago, we would find densities as high as 5000 people per square mile (1930 per km²).

New York City, for example, may in absolute terms be described as located at (approximately) latitude 40°43′ N (read as 40 degrees, 43 minutes north) and longitude 73°58′ W. We have a better understanding of the *meaning* of its location, however, when reference is made to its spatial relationships: to the continental interior through the Hudson–Mohawk lowland corridor; to its position on the eastern seaboard of the United States; to its role as a transportation node and multifunctional economic and administrative center of a major industrial nation of the northern hemisphere; and to its interconnections through trade routes, financial ties, and the United Nations to the international community. Within the city, we gain understanding of the locational significance of, for example, Central Park or the Lower East Side not by reference solely to the street addresses or city blocks they occupy, but by their spatial and functional relationships to the total land use, activity, and population patterns of New York City.

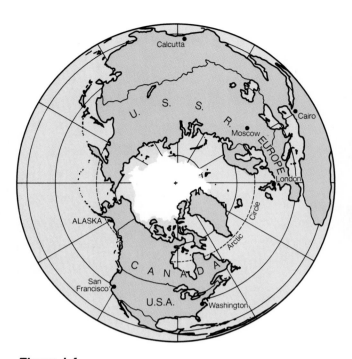

Figure I.4

The reality of *relative location* on the globe may be strikingly different from the impressions we form from flat maps. The position of the USSR with respect to North America when viewed from a polar perspective emphasizes that relative location properly viewed is important to our understanding of spatial relationships and interactions between the two world areas.

Figure I.5

These ruptured barrels from a petrochemical dump near Texas City, Texas are clear evidence of the adverse environmental impact of humans and their wastes.

Physical and Cultural Attributes

All places have individual physical and cultural attributes distinguishing them from other places and giving them character, potential, and meaning that sets them apart from other earth space. Geographers are concerned with identifying and analyzing the details of those attributes and, particularly, with recognizing the interrelationship between the physical and cultural components of area—the human–environmental interface.

The physical characteristics of a place refer to such natural aspects of a locale as its climate and soil, the presence or absence of water supplies and mineral resources, its terrain features, and the like. These attributes provide the setting within which human action occurs. They help shape—but do not dictate—how people live. The resource base, for example, is physically determined, though how resources are perceived and used is culturally conditioned. Environmental circumstances directly affect agricultural potential and reliability; indirectly, they may affect such matters as employment patterns, trade flows, population distributions, national diets, and so on. The physical environment simultaneously presents advantages and disadvantages with which humans must

deal. Thus, the danger of cyclones in the Bay of Bengal must be balanced against the agricultural bounty derived from the region's favorable terrain, soil, temperature, and moisture conditions. Physical environmental patterns and processes are explored in Chapters 2 and 3.

At the same time, by occupying a given place, people modify its environmental conditions. The existence of the American Environmental Protection Agency (and its counterparts elsewhere) is a reminder that humans are active and frequently harmful agents in the continuing spatial interplay between the cultural and physical worlds (Figure I.5). Virtually every human activity leaves its imprint on the earth's soil, water, vegetation, animal life, and other resources, and on the atmosphere common to all earth space, as Chapter 4 makes clear. The visible imprint of that human activity is called the **cultural landscape.** It, too, exists at different scales and at different levels of visibility. Differences in agricultural practices and land use between Mexico and southern California are evident in Figure I.6, while the signs, structures, and people of Los Angeles's Chinatown leave a smaller, more confined imprint within the larger cultural landscape of the metropolitan area itself.

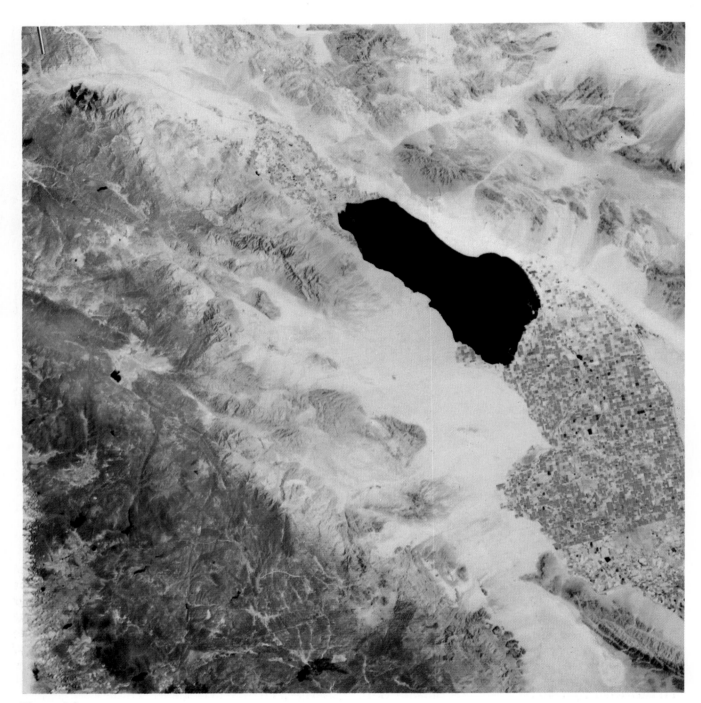

Figure I.6

This Landsat image reveals different land use patterns in
California and Mexico. Slowly move your eyes from the Salton Sea
(the dark patch near the center of the image) to the agricultural
land extending to the southeast and to the edge of the picture.
Notice how the regularity of the fields and the bright colors
(representing growing vegetation) give way to a marked break,
where irregularly shaped fields and less prosperous agriculture
are evident. Above the break is the Imperial Valley of California;
below the border is Mexico.

Figure I.7
The process of change in a cultural landscape. Before the advent
of the freeway, this portion of suburban Long Island, New York
was largely devoted to agriculture (left). The construction of the
freeway and cloverleaf interchange ramps altered nearby land use
patterns (right) to replace farming with new housing
developments and new commercial and light industrial activities.

The physical and human characteristics of places
are the keys to understanding both the simple and the
complex interactions and interconnections between
people and the environments they occupy and modify.
Those interconnections and modifications are not static
or permanent, but are subject to continual change.

Attributes of Place Are Always Changing

Characteristics of places today are the result of constantly
changing past conditions. They are the forerunners of differing human–environmental balances yet to be struck.
Geographers are concerned with place at given moments of time. But to understand fully the nature and
development of places, to appreciate the significance of
their relative locations, and to understand the interplay
of their physical and cultural characteristics, geographers
must view places as the present result of past operation
of distinctive physical and cultural processes (Figure I.7).

You will recall that one of the questions that geographers ask of a place or thing is: How did it come to be
what and where it is? This is an inquiry about process
and about becoming. The forces and events shaping the
physical and cultural environments of places today are
an important focus of geography, and are the topics of
most of the chapters of this book. To understand them
is to appreciate the changing nature of the spatial order
of our world.

Places Interact

The notion of relative location, already introduced, leads
directly to another fundamental spatial reality: places interact with other places in structured and comprehensible ways. In describing the processes and patterns of
that **spatial interaction,** geographers employ ideas of
distance, accessibility, and *connectivity.*

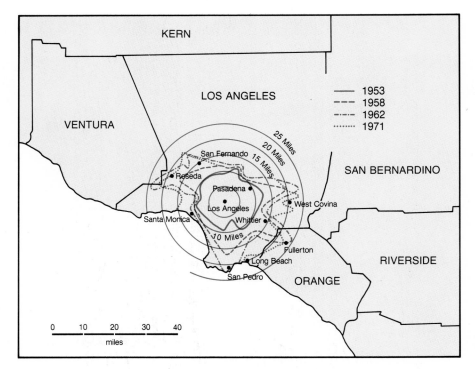

Figure I.8

Lines of equal travel time (*isochrones*) mark off different linear distances from a given starting point, depending on the condition of the route and terrain, and changes in the roads and traffic flows over time. On this map, the areas within 30 minutes of travel time from downtown Los Angeles are recorded for the period 1953 to 1971.

We may ask how far one place is from another and how that distance affects their interaction. **Distance** is a fundamental consideration in nearly everything people do. At one scale, you make decisions about where to live, eat, and shop based upon the distance you have to travel to carry out your day's activities (see Figure 8.2). At a different scale, distance isolated North America from Europe until ships (and aircraft) were developed that reduced the effective distance between the continents. From this latter example, it is evident that distance is more than a linear concept (Figure I.8). It may be measured in units other than feet or miles, and it may change as technology for bridging space and as attitudes about distance change. Whether place separation is measured in linear distance, *time-distance* or *psychological distance* (ideas explored further in Chapter 8), interaction falls off as distance increases—a statement of the idea of **distance decay**. Distance decay may be thought of as the result of a basic law of geography: everything is related to everything else, but relationships are stronger when things are near one another.

Consideration of distance implies assessment of **accessibility.** How easy or difficult is it to overcome the friction of distance? That is, how easy or difficult is it to overcome the barrier of the time or the space separation of places, and how has the changing technology of transportation altered former patterns of spatial interaction? In its turn, accessibility suggests **connectivity,** a broader concept implying all the tangible and intangible ways in which places are connected: by physical telephone lines, by street and road systems, by pipelines and sewers; by radio and TV broadcasts beamed across airwaves; and even by natural movements of wind systems and flows of ocean currents.

There is, inevitably, interchange between connected places. **Spatial diffusion** is the process of dispersion of an idea or a thing (a new consumer product or a new song, for example) from a center of origin to more distant points. The rate and extent of that diffusion are affected, again, by the distance separating the origin of the new idea or technology and other places where it is eventually adopted. Diffusion rates are also affected by

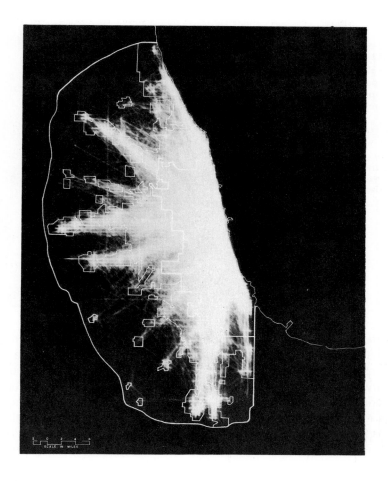

Figure I.9

The routes of the 5 million automobile trips made each day in Chicago during the late 1950s are recorded on this light-display map. The boundaries of the region of interaction that they create are clearly marked. Those boundaries (and the dynamic region they defined) were subject to change as residential neighborhoods expanded or developed, as population relocations occurred, and as the road pattern was altered over time.

such factors as population densities, means of communication, obvious advantages of the innovation, and importance or prestige of the originating node.

Geography is a study of the dynamics of spatial relationships. Movement, connection, and interaction are part of the social and economic processes that give character to places and regions (Figure I.9). Geography's study of those relationships recognizes that spatial interaction is not just an awkward necessity, but a fundamental organizing principle of the physical and social environment.

Place Similarity and Regions

The distinctive characteristics of places—physical, cultural, locational—immediately suggest to us two geographically important ideas. The first is that no two places on the surface of the earth can be *exactly* the same. Not only do they have different absolute locations, but—as in the features of the human face—the precise mix of physical and cultural characteristics of place is never exactly duplicated. Since geography is a spatial science, the inevitable uniqueness of place would seem to impose impossible problems of generalizing spatial information.

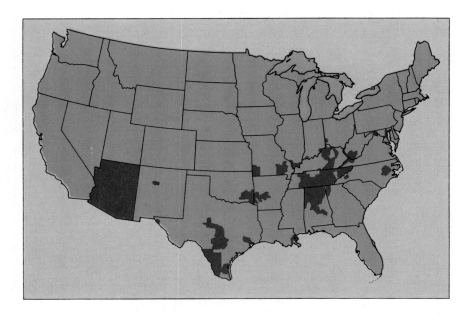

Figure I.10
Pulmonary tuberculosis mortality rates for white males—1965–1971. All spatial data may be mapped. As this example of the distribution of pulmonary tuberculosis demonstrates, mapped distributions frequently reveal regional patterns inviting analysis. Areas emphasized are those with above-average mortality rates. The question is Why?

From Thomas J. Mason, et al., *An Atlas of Mortality from Selected Diseases,* NIH Publication #81–2397, May 1981. National Institutes of Health, Bethesda, Maryland.

That this is not the case results from the second important idea: the physical and cultural features of an area and the dynamic interconnections of people and places show patterns of spatial similarity. Often the similarities are striking enough for us to conclude that spatial regularities exist. They permit us to recognize and define **regions,** earth areas that display significant elements of uniformity. Places are, therefore, both unlike and like other places, creating patterns of areal differences and coherent spatial similarity.

The problems of the historian and the geographer are similar. Each must generalize about items of study that are essentially unique. The historian creates arbitrary, but meaningful and useful, historical periods for reference and study. The "Roaring Twenties" or the "Victorian Era" are shorthand summary names for specific time spans, internally quite complex and varied, but significantly distinct from what happened before or after. The region is the geographer's equivalent of the historian's epoch: a device that segregates into component parts the complex reality of the surface of the earth. In both the time and the space need for generalization, attention is focused upon key unifying elements or similarities of the era or area selected for study. In both the historical and geographical cases, the names assigned to those times and places serve to identify the time span or region, and to convey between speaker and listener a complex set of interrelated attributes.

Regions are not "given" in nature any more than eras are given in the course of human events. Regions are devised; they are spatial summaries designed to bring order to the infinite diversity of the earth's surface. At their root, they are based upon the recognition and mapping of **spatial distributions**—the territorial occurrence of environmental, human, or organizational features selected for study. Although as many spatial distributions exist as there are imaginable physical, cultural, or connectivity elements of area to examine (Figure I.10), those that are selected for study are those that contribute to the understanding of a specific topic or problem.

Although there are as many individual regions as the objectives of spatial study and understanding demand, two generalized *types* of regions are recognized. A **formal region** is one of essential uniformity in one or a limited number of related physical or cultural features. Your home state is a formal political region within which uniformity of law and administration is found. The Columbia Plateau or the tropical rain forest are areas of uniform physical characteristics making up formal natural regions (Figure I.11).

A **functional region,** in contrast, may be visualized as a spatial system. Its parts are interdependent, and throughout its extent the functional region operates as a dynamic, organizational unit. A functional region has unity not in the sense of static content, but in the manner of its operational connectivity. It has a controlling node or core area surrounded by the total region defined by

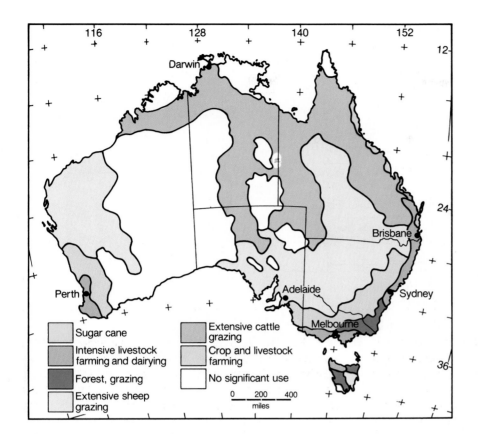

Figure I.11
This generalized land use map of Australia is composed of *formal regions* whose internal economic characteristics show essential uniformities setting them off from adjacent territories of different condition or use.

the type of control exerted. Trade areas of towns (Figure I.12), national "spheres of influence," and the territories subordinate to the financial, administrative, wholesaling, or retailing centrality exercised by such regional capitals as Chicago, Atlanta, or Minneapolis are cases in point. Further examples of formal and functional regionalism will be encountered throughout this text and make up the total content of Chapter 12.

Organization of This Book

The breadth of geographic interest and subject matter, the variety of questions that focus geographic inquiry, and the diversity of concepts and terms employed to analyze likeness and difference from place to place on the earth's surface require a simple, logical organization of topics for presentation to students new to the field. Despite its outward appearance of complex diversity, geography has a broad consistency of purpose achieved through the recognition of a limited number of distinct but closely related "traditions." William D. Pattison, who suggested this unifying viewpoint, and J. Lewis Robinson (among

others) who accepted and expanded Pattison's reasoning, found that four "traditions" were ways of clustering geographic inquiry. While not all geographic work is confined by the separate themes, one or more of the traditions is implicit in most geographical studies. The unifying themes—the four traditions within which geographers work—are listed here:

the earth-science tradition
the culture–environment tradition
the locational (or spatial) tradition
the area analysis tradition

We have employed these as the device for grouping the chapters of this book, hoping that they will help you to recognize the unitary nature of geography while appreciating the diversity of topics studied by geographers.

The **earth-science tradition** is that branch of the discipline that addresses itself to the earth as the habitat of humans. It is the tradition that in ancient Greece represented the roots of geography, the description of the

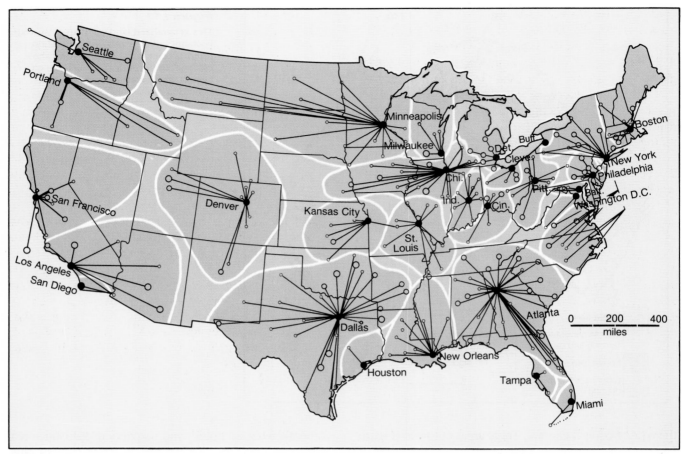

Figure I.12
Functional regions based on correspondent banking linkages
between large central cities and smaller towns are shown on this
map. The regions suggest one form of *connectivity* between
principal cities and locales beyond their own immediate
metropolitan area.

physical structure of the earth and of the natural processes that give it detailed form. In modern terms, it is
the vital environmental half of the study of human–
environment systems, which together constitute geography's subject matter. The earth-science tradition helps
the physical geographer to understand the earth as the
common heritage of humankind and to find solutions to
the increasingly complex web of pressures placed on the
earth by its demanding human occupants. By its nature
and content, the earth-science tradition forms the unifying
background against which all geographic inquiry is ultimately conducted. Consideration of the elements of the
earth-science tradition constitutes Part I of this text.

Part II concerns itself with some of the content of
the **culture–environment tradition.** Within this theme
of geography, consideration of the earth as a purely physical abstraction gives way to a primary concern with how
people *perceive* the environments they occupy. Its focus
is upon culture. The landscapes that are explored and
the spatial patterns that are central are those that are cultural in origin and expression. People in their numbers,
distributions, and diversity; in their patterns of social and
political organization; and in their spatial perceptions and
behaviors are the orienting concepts of the culture–
environment tradition. The theme is distinctive in its
thrust but tied to the earth-science tradition because
people exist, cultures emerge, and behaviors occur within
the context of the physical realities and patternings of
the earth's surface.

The **locational tradition**—or as it is sometimes called, the spatial tradition—is the subject of Part III. It is a tradition that underlies all of geographic inquiry. As Robinson suggested, if we can agree that geology is rocks, that history is time, and that sociology is people, then we can assert that geography is earth space. The locational tradition is primarily concerned with the distribution of cultural phenomena or physical items of significance to human occupance of the earth.

Part, but by no means all, of the locational tradition is concerned with distributional patterns. More central are scale, movement, and areal relationships. Map, statistical, geometrical, and systems analysis and modeling are among the research tools employed by geographers working within the locational tradition. Irrespective, however, of the analytical tools employed or the sets of phenomena studied—economic activities, resource distributions, city systems, or others—the underlying theme is the geometry, or the distribution, of the phenomenon discussed and the flows and interconnections that unite it to related physical and cultural occurrences.

A single-chapter consideration of the **area analysis tradition** constitutes Part IV and completes our survey of geography. Again, the roots of this theme may be traced to antiquity. Strabo's *Geography* was addressed to the leaders of Augustan Rome as a summary of the nature of places in their separate characters and conditions—knowledge deemed vital to the guardians of an empire. Imperial concerns may long since have vanished, but the study of regions and the recognition of their spatial uniformities and differences remain. Such uniformities and differences, of course, grow out of the structure of human–environment systems and interrelations that are the study of geography.

The identification of the four traditions of geography is not only an organizational convenience. It is also a recognition that within that diversity of subject matter called geography, unity of interest is ever preserved. The traditions, though recognizably distinctive, are intertwined and overlapping. We hope that their use as organizing themes—and their further identification in short introductions to the separate sections of this book—will help you to grasp the unity in diversity that is the essence of geographic study.

Key Words

absolute location 7	culture–environment tradition 16	formal region 14	relative location 7
accessibility 12	distance 12	functional region 14	scale 7
area analysis tradition 17	distance decay 12	isochrone 12	spatial diffusion 12
connectivity 12	earth-science tradition 15	locational tradition 17	spatial distribution 14
cultural landscape 9		regions 14	spatial interaction 11

Suggested Readings

Holt-Jensen, Arild. *Geography: Its History and Concepts.* Totowa, N.J.: Barnes and Noble Books, 1980.

James, Preston E., and Geoffrey J. Martin. *All Possible Worlds.* New York: John Wiley, 1981.

Johnston, R. J. *Geography and Geographers.* London: Methuen, 1982.

Johnston, R. J. *Philosophy and Human Geography.* London: Methuen, 1983.

Pattison, William D. "The Four Traditions of Geography," *The Journal of Geography* 63 (1964), pp. 211–16.

Robinson, J. Lewis. "A New Look at the Four Traditions of Geography," *The Journal of Geography* 75 (1976), pp. 520–30.

Wheeler, James O. "Notes on the Rise of the Area Studies Tradition in U.S. Geography, 1910–1929," *Professional Geographer* 38 (1986), pp. 53–61.

Part One

The Earth-Science Tradition

For nearly a month the mountain had rumbled, emitting puffs of steam and flashes of fire. Within the past week, on its slopes and near its base, deaths had been recorded from floods, mud slides, and falling rock. A little after 8:00 on the morning of May 8, 1902, the climax came—for volcanic Mount Pelée and the thriving port of Saint Pierre on the island of Martinique. To the roar of one of the biggest explosions the world has ever known and to the clanging of church bells aroused in swaying steeples, a fire ball of gargantuan size burst forth from the upper slope of the volcano; lava, ash, steam, and superheated air engulfed the town, and 29,933 people met their death.

More selective but—for those victimized—just as deadly was the sudden "change in weather" that struck central Illinois on December 20, 1836. Within an hour, preceded by winds gusting to 70 miles per hour, the temperature plunged from 40° F to −30° F. On a walk to the post office through the slushy snow, Mr. Lathrop of Jacksonville, Illinois, found, just as he passed the Female Academy, that "the cold wave struck me, and as I drew my feet up the ice would form on my boots until I made a track . . . more like that of a Jumbo than a No. 7 boot." Two young salesmen were found frozen to death, along with their horses; one "was partly in a kneeling position, with a tinderbox in one hand, a flint in the other, with both eyes open as though attempting to light the tinder in the box." Others died, too—inside horses that had been disemboweled and used as makeshift shelters; in fields, woods, and on roads both a short distance and an eternity from the travelers' destinations.

Fortunately, few human–environmental encounters are so tragic as these. Rather, the physical world in all its spatial variation provides the constant background against which the human drama is played. It is to that background that physical geographers, acting within the earth-science tradition of the discipline, direct their attention. Their primary concern is with the natural rather than the cultural landscape and with the interplay between the encompassing physical world and the activities of humans.

The interest of those working within the earth-science tradition of geography, therefore, is not in the physical sciences as an end in themselves, but in physical processes as they create landscapes and environments of significance to humankind. The objective is not solely to trace the physical and chemical reactions that have produced a piece of igneous rock or to describe how a glacier has scoured the rock; it is the relationship of the rock to people. Geographers are interested in what the rock can tell us about the evolution of the earth as the home of human beings or in the significance of certain types of rock for the distribution of mineral resources or of fertile soil.

The questions that physical geographers ask do not usually deal with the catastrophic, though catastrophe is an occasional element of human–earth-surface relationships. Rather, they ask questions that go to the root of understanding the earth as the home and the work place of humankind. What does the earth as our home look like? How have its components been formed? How are they changing, and what changes are likely to occur in the future through natural or human causes? How are environmental features, in their spatially distinct combinations, related to past, present, and prospective human use of the earth? These are some of the basic questions that geographers who follow the earth-science tradition attempt to answer.

The four chapters in Part I of our review of geography deal, in turn, with maps, landforms, weather and climate, and human impact upon the environment. It is easy to see why the earth's landforms, weather, and climate come within the earth-science tradition, but it may be less clear why we begin this part with a discussion of maps and end with a consideration of human impact upon the physical world.

Maps are our starting point, for it is impossible to talk of geography or to appreciate its content and lessons without understanding maps. Geography is a spatial science, and since ancient times maps have been used to represent, characterize, and interpret the nature of space. All geographers use maps as basic analytical tools, whether the maps be simple sketches made during field investigations or computer-drawn maps based on data from earth satellites. Mapmakers traditionally sought ways to represent the distribution of land and water features of the earth and to portray the character of the physical landscape. Since the 1700s, they have turned increasing attention to map rendition of cultural and political distributions and to the preparation of thematic maps. In the last few decades, geographers have begun to draw maps based on projections that distort time or space or that are intended to portray "psychological" or "perceived" reality. Whatever the state of their art or the thrust of their attention, mapmakers have been motivated to express the distribution of things central to their interests. As a spatial, distributional science, geography properly begins with a discussion of the nature of maps.

The treatment of so vast a field as landforms (Chapter 2) must be highly selective within an introductory text. The aim is to summarize the great processes by which landforms are created and to depict the general classes of features resulting from those processes without becoming overly involved in scientific reasoning and technical terms.

In Chapter 3, "Physical Geography: Weather and Climate," the major elements of the atmosphere—temperature, precipitation, and air pressure—are discussed, and their regional generalities and associations in patterns of world climates are presented. It is the study of weather and climate that adds coherence to our understanding of the earth as the home of people. Frequent reference will be made to the climatic background of human activities in later sections of our study.

1 Maps

Imagine that all of the maps in the world were destroyed at this moment. Traffic on highways would slow to a crawl. Control towers would be overwhelmed as pilots radioed airports for landing instructions; ships at sea would be forced to navigate without accurately charting their positions. Rangers would be sent to rescue hikers stranded in national parks and forests. Disputes would erupt over political boundaries; military planners would not know how to channel troop movements or where to aim artillery. People would have to rely on verbal descriptions to convey information, and no amount of verbiage can do justice to the detailed information contained on a map of even moderate complexity. Maps are tersely efficient at indicating the location of things relative to one another, and the information void created by their disappearance would have to be filled by volumes of description.

Maps are as fundamental a means of communication as the printed or spoken word or as photographs. In every age, people have produced maps, whether made out of sticks or shells, scratched on clay tablets, drawn on parchment, or printed on paper. Charlemagne even had maps made on solid plates of silver! The map is the most efficient way to portray parts of the earth's surface, to record political boundaries, and to indicate directions to travelers. In addition to their usefulness to the general populace, maps have a special significance to the geographer. In the process of studying the surface of the earth, geographers depend on maps to record, present, and aid them in analyzing the location of points, lines, or objects.

Modern scientific mapping has its roots in the 17th century, although the earth scientists of ancient Greece are justly famous for their contributions. They recognized the spherical form of the earth and developed map projections and the grid system. Much of the cartographic tradition of Greece was lost to Europe during the Middle Ages, however, and essentially had to be rediscovered. Several developments during the Renaissance gave an impetus to accurate cartography. Among these were the development of printing, the rediscovery of the work of Ptolemy and other Greeks, and the great voyages of discovery.

In addition, the rise of nationalism in many European countries made it imperative to determine and accurately portray boundaries and coastlines as well as to depict the kinds of landforms contained within the borders of a country. During the 17th century, important national surveys were undertaken in France and England. Many conventions in the way data are presented on maps had their origin in these surveys. This early concern about *physical* maps—that is, maps portraying the earth's physical features—allows us to include our discussion of maps within the part of this book exploring topics in the earth-science tradition of geography.

Knowledge of the way information is recorded on maps enables us to read and interpret them correctly. To be on guard against drawing inaccurate conclusions or to avoid being swayed by distorted or biased presentations, we must be able to understand and assess the ways in which facts are represented. Of course, all maps are necessarily distorted because of the need to portray the round earth on a flat surface, to use symbols to represent objects, to generalize, and to record features at a different size than they actually are. This distortion of reality is necessary because the map is smaller than the things it depicts and because its effective communication depends upon selective emphasis of only a portion of reality. As long as they know the limitations of the commonly used types of maps and understand what relationships are distorted, map readers may easily interpret maps correctly.

Locating Points on a Sphere: The Grid System

In order to visualize the basic system for locating points on the earth, think of the world as a sphere with no markings whatever on it. There would, of course, be no way of describing the exact location of a particular point on the sphere without establishing some system of reference. We use the **grid system,** which consists of a set of imaginary lines drawn across the face of the earth. The key reference points in that system are the North and South Poles and the equator, which are given in nature (Figure 1.1), and the prime meridian, which is agreed upon by cartographers.

The North and South Poles are the end points of the axis about which the earth spins. The line that encircles the globe halfway between the poles, and that consequently is perpendicular to the axis, is the *equator.* We can describe the location of a point in terms of its distance north or south of the equator measured as an angle at the earth's center. Because a circle contains 360 degrees, the distance between the two poles is 180 degrees and between the equator and each pole, 90 degrees. **Latitude** is the measure of distance north and south of the equator, which is itself, of course, zero degrees. As is evident in Figure 1.1, the lines of latitude, which are parallel, run east–west.

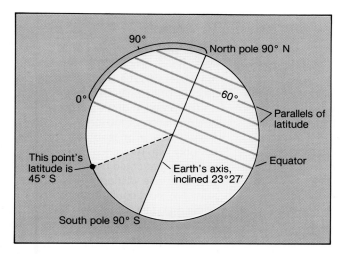

Figure 1.1

The grid system: parallels of latitude. Note that the parallels become increasingly shorter closer to the poles. On the globe, the 60th parallel is only one-half as long as the equator.

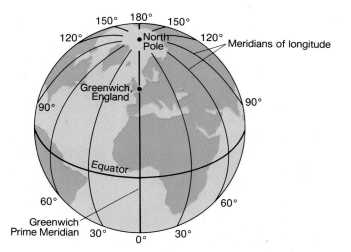

Figure 1.2

The grid system: meridians of longitude. These arbitrary but conventional lines, together with the parallels based upon the naturally given equator, constitute the globe grid. Since the meridians converge at the poles, the distance between degrees of longitude becomes shorter the farther one moves away from the equator.

The polar circumference of the earth is 24,899 miles; thus, the distance between each degree of latitude equals 24,899 ÷ 360, or about 69 miles (111 km). If the earth were a perfect sphere, all degrees of latitude would be equally long. Due to the slight flattening of the earth in polar regions, however, degrees of latitude are slightly longer near the poles (69.41 mi; 111.70 km) than near the equator (68.70 mi; 110.56 km).

To record the latitude of a place in a more precise way, degrees are divided into *minutes* and then into *seconds,* exactly like an hour of time. One minute of latitude is about 1.15 miles (1.85 km), and one second of latitude about 101 feet (31 m). The latitude of the center of Chicago, Illinois, is written 41° 51′ 50″ North. This system of subdivision of the circle is derived from the sexagesimal (base 60) numerical system of the ancient Babylonians.

Because the distance north or south of the equator is not by itself enough to locate a point in space, we need to specify a second coordinate to indicate distance east or west from an agreed-upon reference line. As a starting point for east–west measurement, cartographers in most countries use as the **prime meridian** an imaginary line passing through the Royal Observatory at Greenwich,

England. That prime meridian was selected as the zero degree longitude line by an international conference in 1884. Like all *meridians,* it is a true north–south line connecting the poles of the earth (Figure 1.2). ("True" north and south vary from magnetic north and south, the direction of the earth's magnetic poles, to which a compass needle points.) Meridians are farthest apart at the equator, come closer and closer together as latitude increases, and converge at the North and South Poles. Unlike parallels of latitude, all meridians are the same length.

Longitude is the angular distance east or west of the prime meridian. Directly opposite the prime meridian is the 180th meridian, located in the Pacific Ocean. East–west measurements range from 0° to 180°, that is, from the prime meridian to the 180th meridian in each direction. Like parallels of latitude, degrees of longitude can be subdivided into minutes and seconds. However, distance between adjacent degrees of longitude decreases away from the equator because the meridians converge at the poles. With the exception of a few Alaskan islands, all places in North and South America are in the area of west longitude; with the exception of a portion of the Chukchi Peninsula of the USSR, all places in Asia and Australia have east longitude.

Agreement on a Prime Meridian

Latitudinal location relative to the equator, a line given in nature, can be determined by measurement of the angular height of the sun above the horizon at noon on a known date. Measuring devices for this purpose (*astrolabes*) appear to have been in use by the Babylonians and the Egyptians several thousand years before the Greeks addressed the problem of constructing the globe grid and developing world maps.

The determination of longitude—the other necessity for accurate maps—was more difficult. No reference line equivalent to the equator existed, and no convenient motion of sun or stars indicated the location east or west of a naturally given baseline. This lack was merely annoying when one was traveling on land where surface distances could be measured, or on coasting vessels, from whose decks landfalls could be sighted. It was potentially disastrous when long sea voyages out of sight of land became common in the 16th and 17th centuries; even skilled navigators using dead reckoning*

could, and did, miss destinations by scores and sometimes hundreds of miles.

Innumerable proposals were made for using water wheels attached to ships' hulls and for other mechanical devices for measuring distance traveled at sea. All were rendered useless by the currents and winds acting on sailing vessels. The only way longitude could be determined was by the measurement at a given instant of the difference between the time at a known point and the time on a reference line. Since the earth rotates through 360° in 24 hours, each hour of time is the equivalent of 15° of longitude and each degree of longitude equals 4 minutes of time. The answer to the longitude problem had to be a small but extremely reliable *chronometer* (clock) dependent upon a spring-driven movement.

In 1714 an act of the English Parliament established a prize of £20,000 for any device that could determine longitude within 30 minutes of the globe grid, that is, within 2 minutes of time or 34 miles. John Harrison, to whom the prize was eventually awarded, put the chronometer of his invention to an

official sea trial on a trip from Portsmouth, England, to Jamaica in the winter of 1761–1762. It proved to be in error by only 1¼ nautical miles.

The problem of the accurate determination of the longitude of any point at sea or on land was solved, but only with reference to a prime meridian that could be established wherever one wished. For patriotic reasons, most countries located the prime meridian within their own borders, and by the 19th century over a score of different ones were in common use. Recognizing the desirability of establishing a universal prime meridian on which acceptable world maps could be based, scientists held a series of conferences as first suggested by the third International Geographical Congress meeting in Venice in 1881. Delicate negotiations finally produced agreement among 25 nations that 0° longitude would refer to the meridian passing through the Royal Observatory at Greenwich, England. This convention is now almost universally accepted, though one occasionally sees maps based on a different prime meridian.

*Dead reckoning: inferring the vessel's position without astronomical observation, for example, by estimating the course and distance traveled from a previously determined position.

Time depends on longitude. The earth is divided into 24 time zones, centered on meridians at 15° intervals. The **International Date Line,** where each new day begins, generally follows the 180th meridian. As Figure 1.3 indicates, however, the date line deviates from the meridian in some places, in order to avoid having two different dates within a country or island group.

By citing the degrees, minutes, and—if necessary—seconds of longitude and latitude, we can describe the location of any place on the earth's surface. To conclude our earlier example, Chicago is located at 41° 52′ N, 87° 40′ W. Singapore is at 1° 20′ N, 103° 57′ E (Figure 1.4).

Map Projections

The earth can be represented with reasonable accuracy only on a globe, but globes are not as convenient as flat maps to store or to use, and they cannot depict much detail. For example, if we had a large globe with a diameter of, say, 3 feet we would have to fit the details of nearly 50,000 square miles of earth surface in an area 1 inch on a side. Obviously, a globe of reasonable size cannot show the transportation system of a city or the location of very small towns and villages. In transforming a globe into a map, we cannot flatten the curved surface and keep all the properties of the original intact. So, in

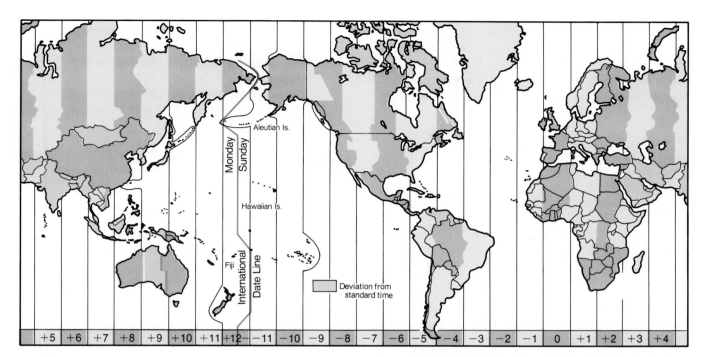

Figure 1.3

World time zones. Each time zone is about 15° wide, but variations occur to accommodate political boundaries. Thus, the International Date Line zigzags so that Siberia has the same date as the rest of the USSR and so as not to split the Aleutian Island and Fiji Island groups. New days begin at the Date Line and proceed westward, so that west of the line is always one day later than east of the line. The figures at the bottom indicate the number of hours to be added to or subtracted from Greenwich time to get local time.

drawing a map, the relationships between points are inevitably distorted in some way. To the geometer, a sphere is a *nondevelopable* surface. It can only be *projected* upon a plane surface; the term **projection** designates the method of representing a curved surface on a flat map.

Properties of Map Projections

Because no projection can be entirely accurate, the serious map user needs to know in what respects a particular map correctly reproduces, and in what respects it distorts, earth features. The four main properties of maps—area, shape, distance, and direction—are distorted in different ways and to different degrees by various projections.

Area

Some projections enable the cartographer to represent the areas of regions in correct or constant proportion to earth reality. That means that any square inch on the map represents an identical number of square miles (or of

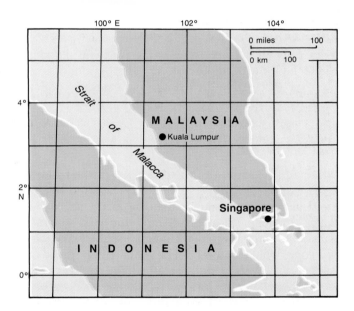

Figure 1.4

The latitude and longitude of Singapore are 1° 20′N, 103° 57′E. What are the coordinates of Kuala Lumpur?

similar units) anywhere else on the map. As a result, the shape of the portrayed area is inevitably distorted. A square on the earth, for example, may become a rectangle on the map, but that rectangle has the correct area. Such projections are called **equal-area** or **equivalent.**

Shape

Although no projection can provide correct shapes for large areas, some accurately portray the shapes of small areas by preserving correct angular relationships. That is, an angle on the globe is rendered correctly on the map. Maps that have true shapes for small areas are called **conformal.** Parallels and meridians always intersect at right angles on such maps, as they do on a globe. *A map cannot be both equivalent and conformal.*

Distance

Distance relationships are nearly always distorted on a map, but some projections do maintain true distances in one direction or along certain selected lines. Others, called **equidistant** projections, show true distance in all directions, but only from one or two central points. (See, for example, Figure 1.9.)

Direction

As is true of distances, directions between all points cannot be shown without distortion. Projections do exist, however, that enable the map user to measure correctly the directions from a single point to any other point.

Types of Projections

While all projections can be described mathematically, some can be thought of as being constructed by geometrical techniques rather than by mathematical formulas. In geometrical projections, the grid system is transformed into a geometrical figure such as a cylinder or a cone, which, in turn, can be cut and then spread out flat without any stretching or tearing. The surfaces of cylinders, cones, and planes are said to be **developable surfaces:** the first two because they can be cut and laid flat without distortion, the third because it is flat at the outset.

The selection of the surface to be developed or of the specific mathematical formula to be employed is determined by the properties of the globe grid that one elects to retain. As a reminder, **globe properties** follow:

1. **All meridians are of equal length; each is one-half the length of the equator.**
2. **All meridians converge at the poles and are true north–south lines.**
3. **All lines of latitude (parallels) are parallel to the equator and to each other.**
4. **Parallels decrease in length as one nears the poles.**
5. **Meridians and parallels intersect at right angles.**
6. **The scale on the surface of the globe is everywhere the same in every direction.**

Cylindrical Projections

The **Mercator projection** is one of the most commonly used (and misused) **cylindrical projections.** Named for its inventor, it has been mathematically adjusted to serve navigational purposes. Suppose that we roll a piece of paper around a globe, so that it is tangent to (touching) the sphere at the equator. Instead of the paper being the same height as the globe, however, it extends far beyond the poles. If we place a light source at the center of the globe, the light will project a shadow map upon the cylinder of paper. When we unroll the cylinder, the map will appear as shown in Figure 1.5.

Note the variance between the grid we have just projected and the true properties of the globe grid. The grid lines cross each other at right angles, as they do on the globe, and they are all straight north–south or east–west lines. But the meridians do not converge at the poles as they do on a globe. Instead they are equally spaced, parallel, vertical lines. Because the meridians are everywhere equally far apart, the parallels of latitude have all become the same length. The Mercator map balances the spreading of meridians toward the poles by spacing the latitude lines farther and farther apart. Mathematical tables show the cartographer how to space the parallels to balance the spreading of meridians so that the result is a conformal projection. The shapes of small areas are true, and even large regions are fairly accurately represented.

Note that the Mercator map in Figure 1.5 stops at 80° N and 70° S. A very large map would be needed to show the polar regions, and the sizes of those regions would be enormously exaggerated. The poles themselves can never be shown on a Mercator projection tangent at the equator. In fact, the enlargement of areas with increasing latitude is so great that a Mercator map should not be published without a scale of miles like that shown.

The Mercator projection has often been misused in atlases and classrooms as a general-purpose world map. Whole generations of schoolchildren have grown up convinced that the state of Alaska is nearly the size of all the lower 48 states put together. To check the distortion of this projection, compare the sizes of Greenland and Canada on Figure 1.5 with those on Figure 1.6, an equal-area projection.

The proper use of the Mercator is for navigation. In fact, it is the standard projection used by navigators because of a peculiarly useful property: a straight line drawn anywhere on the map is a line of constant compass bearing. If such a line, called a *rhumb line,* is followed,

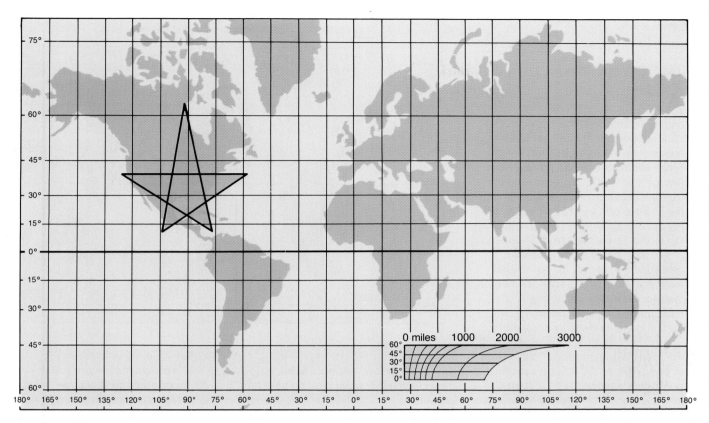

Figure 1.5

Distortion on the Mercator projection. A perfect five-pointed star was drawn on a globe, and the latitude and longitude of the points of the star were transferred to the Mercator map shown above. The manner in which the star is distorted reflects the way the projection distorts land areas. Mercator maps are usually accompanied, as here, by a diagram showing the varying scale of distance at different latitudes.

a ship's or plane's compass will show that the course is always at a constant angle with respect to geographic north. On no other projection is a rhumb line both straight and true as a direction.

Conic Projections

Of the three developable geometric forms—cylinder, cone, and plane—the cone is the closest in form to one-half of a globe. **Conic projections,** therefore, are widely used to depict hemispheres or smaller parts of the earth.

A useful projection in this category, and the easiest to understand, is the *simple conic* projection. Imagine that a cone is laid over half the globe, as in Figure 1.7, tangent to the globe at the 30th parallel. Distances are true only along the tangent circle, which is called the *standard parallel.* When the cone is developed, of course, the standard parallel becomes an arc of a circle, and all other parallels become arcs of concentric circles. With a central light source, the parallels become increasingly farther apart as they approach the pole, and distortion is accordingly exaggerated.

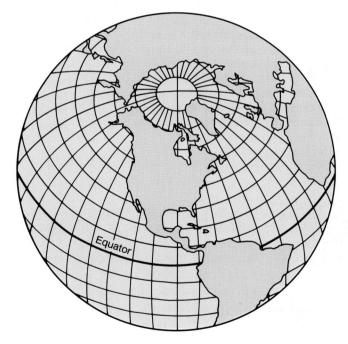

Figure 1.6

The Lambert planar equal-area projection is mathematically derived to display the property of equivalence.

The Peters Projection

Although map projections might appear to be a rather sterile, uninteresting topic, the projection shown here is creating significant controversy. Developed and promoted by a German named Arno Peters, the Peters projection purports to reflect concern for the problems of the Third World by providing a less European-centered representation of the world. In contrast to the Mercator projection, which distorts areas greatly, the Peters is an equal-area map; therefore, it is better suited than the Mercator for comparing distributions. Peters claims that his map shows the densely populated parts of the earth, the countries of the Third World, in proper proportion to one another. He has persuaded a number of agencies with a special interest in the Third World, including the World Council of Churches and several United Nations organizations, to adopt the map.

But the Peters projection is as inappropriate as Mercator's, argue two geographers (P. Porter and P. Voxland, "Distortion in Maps: The Peters Projection and Other Devilments," *Focus,* Summer 1986). Characterizing Peters's cartography

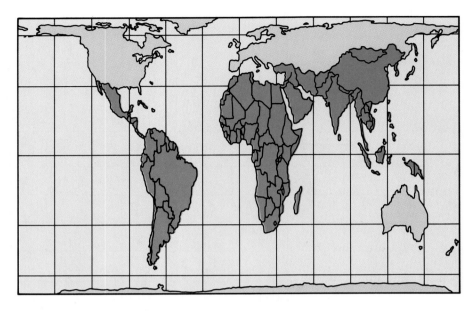

as "perverse and wrongheaded," they criticize his projection on several grounds:

1. It badly distorts shapes in the tropics and at high latitudes. Although Peters claims to support the Third World, his map preserves shapes more accurately for the developed world.
2. Distances and directions cannot be measured except under very limited conditions.
3. Peters claims his projection will meet every cartographic need, which is simply incorrect.
4. The projection is not in fact new, but identical to an equal-area projection developed by James Gall in 1885.
5. Finally, the authors point out that many projections yield an equal-area world map with less distortion of shapes than does the Peters projection.

One can lessen the amount of distortion by shortening the length of the central meridian, spacing the parallels of latitude at equal distances on that meridian, and making the 90th parallel (the pole) an arc rather than a point. Most of the conic projections that are in general use employ such mathematical adjustments. When more than one standard parallel is used, a *polyconic* projection results.

Conic projections can be adjusted to achieve desired qualities; hence, they are widely used. They are particularly suited for showing areas in the mid-latitudes that have a greater east–west than north–south extent. Both area and shape can be represented by conic projections without serious distortion (Figure 1.8). Many official map series, such as the topographic sheets and the

world aeronautical charts of the U.S. Geological Survey and the Aerospace Center of the Defense Mapping Agency, use types of conic projections. They are also used in many atlases.

Planar Projections

Planar (also called azimuthal) **projections** are constructed by placing a plane surface tangent to the globe at a single point. Although the plane may touch the globe anywhere the cartographer wishes, the polar case with the plane centered on either the North or the South Pole is easiest to visualize (Figure 1.9a).

This *equidistant* projection is useful because it can be centered anywhere, facilitating the correct measurement of distances from that point to all others. For this reason, it is often used to show air navigation routes that

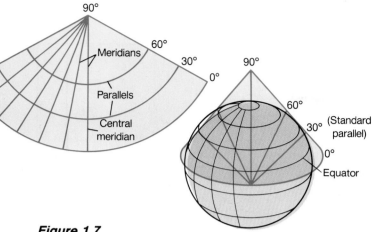

Figure 1.7
A simple conic projection with one standard parallel. Most conics are adjusted so that the parallels are spaced evenly along the central meridian.

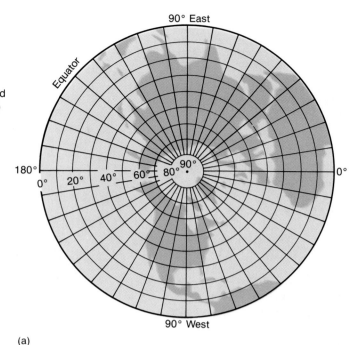

(a)

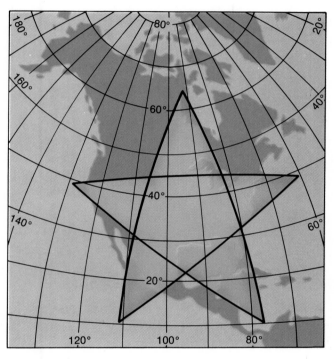

Figure 1.8
The polyconic projection. The map is produced by bringing together east–west strips from a series of cones, each tangent at a different parallel. This projection differs from the simple conic in that the parallels of latitude are not arcs of concentric circles and the meridians are curved rather than straight lines. While neither equivalent nor conformal, the projection portrays shape well. Note how closely the star resembles a perfect five-pointed star.

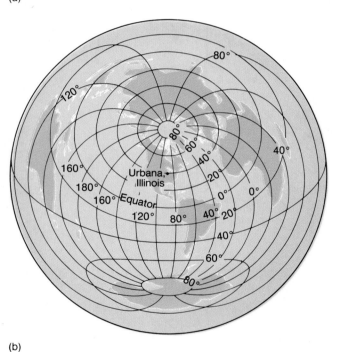

(b)

Figure 1.9
(a) The planar equidistant projection. Parallels of latitude are circles equally spaced on the meridians, which are straight lines. This projection is particularly useful because distances from the center to any other point are true. If the grid is extended to show the southern hemisphere, the South Pole is represented as a circle instead of a point. (b) A planar equidistant projection centered on Urbana, Illinois. The scale of miles applies only to distances from Urbana or on a line through it. The scale on the rim of the map, representing the antipode of Urbana, is infinitely stretched.

originate from a single place. When the plane is centered at places other than the poles, the meridians and the parallels become curiously curved, as is evident in Figure 1.9b.

Because they are particularly well suited for showing the arrangement of polar land masses, planar maps are commonly used in atlases. Depending on the particular projection used, true shape, equal area, or some compromise between them can be depicted. In addition,

Figure 1.10

The gnomonic projection is the only one on which all great circles appear as straight lines. Rhumb lines are curved. In this sense, it is the opposite of the Mercator projection, on which rhumb lines are straight and great circles are curved.

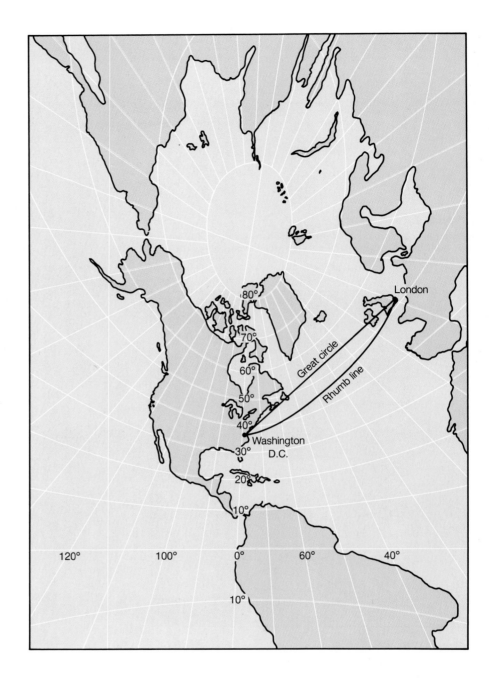

Figure 1.11

Goode's Homolosine projection, a combination of two different projections. This projection can also interrupt the continents to display the ocean areas intact.

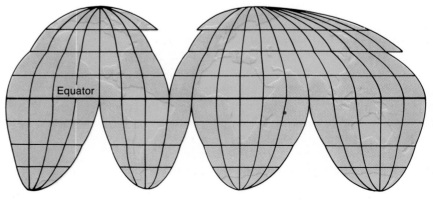

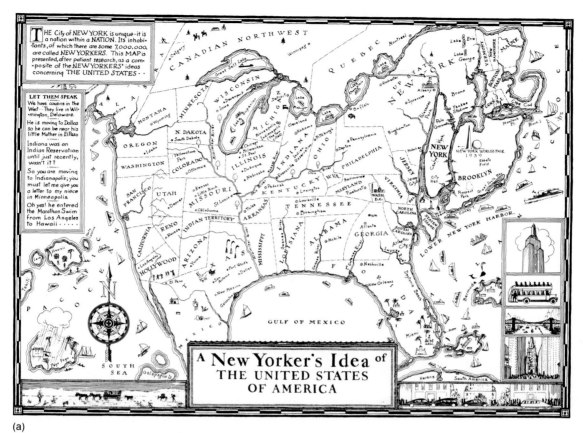

(a)

one of the planar projections is widely used for navigation and telecommunications. The *gnomonic* projection, shown in Figure 1.10, is the only one on which all *great circles* (or parts thereof) appear as straight lines. Because great circles are the shortest distances between two points, navigators need only connect the points with a straight line to find the shortest route.

Other Projections

Projections can be developed mathematically to show the world, or a portion thereof, in any shape that is desired: ovals, hearts, trapezoids, stars, and so on. One often-used projection is Goode's Homolosine. Usually shown in its interrupted form, as in Figure 1.11, it is actually a combination of two different projections. This equal-area projection also represents shapes well.

Not all maps are equal-area, conformal, or equidistant; many, such as the widely used polyconic (Figure 1.8), are compromises. In fact, some very effective projections are non-Euclidian in origin, transforming space in unconventional ways. Distances may be measured in nonlinear fashion (in terms of time, cost, number of people, or even perception), and maps that show relative space may be constructed from these data. Two examples of such transformations are shown in Figure 1.12.

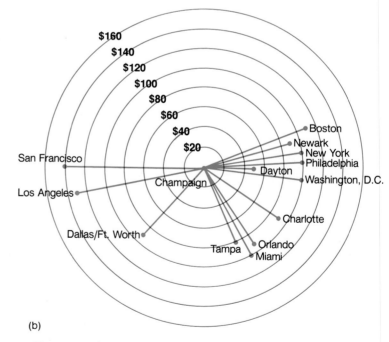

(b)

Figure 1.12

(a) A New Yorker's Idea of the United States of America portrays the country not as it actually is but as the artist conceives it. Think of it as a map based on an unknown projection. Distance, area, shape, and direction are all distorted, and a "psychological" rather than a literally true map is achieved. Such "mental maps" are discussed in Chapter 8. (Copyright Florence Thierfeldt, Milwaukee, Wisconsin) (b) Airline cost distance from Champaign, Illinois. A distance-transformation map based upon one-way coach air fares, November 1986.

Mapmakers must be conscious of the properties of the projections they use, selecting the one that best suits their purposes. If the map shows only a small area, the choice of a projection is not critical—virtually any can be used. The choice becomes more important when the area to be shown extends over a considerable longitude and latitude; then the selection of a projection depends on the purpose of the map. As we have seen, a Mercator or gnomonic projection is useful for navigation. If numerical data are being mapped, the relative sizes of the areas involved should be correct, so that one of the many equal-area projections is likely to be employed. Most atlases indicate which projection has been used for each map, thus informing the map reader of the properties of the maps and their distortions.

Selection of the map grid, determined by the projection, is the first task of the mapmaker. A second decision involves the scale at which the map is to be drawn.

Scale

The **scale** of a map is the ratio between the measurement of something on the map and the corresponding measurement on the earth. Scale is typically represented in one of three ways: verbally, graphically, or numerically as a representative fraction. A *verbal* scale, as the name implies, is given in words, such as "1 inch to 1 mile." A *graphic* scale is a line or bar marked off in map units but labeled in ground units. A graphic scale of "1 inch to 1 mile" is shown below.

A *representative fraction* (RF) scale gives two numbers, the first representing the map distance and the second the ground distance. The fraction may be written in a number of ways. There are 5280 feet in 1 mile, and 12 inches in 1 foot; 5280 times 12 equals 63,360, the number of inches in 1 mile. The fractional scale of a map at 1 inch to 1 mile can be written as follows:

$$1:63,360 \quad \text{or} \quad \frac{1}{63,360} \quad \text{or} \quad 1/63,360$$

On the simpler metric scale, 1 centimeter to 1 kilometer is 1:100,000. The units used in each part of the fractional scale are the same, thus 1:63,360 could also mean that 1 foot on the map represents 63,360 feet on the ground, or 12 miles—which is, of course, the same as 1 inch represents 1 mile. Numerical scales are the most accurate of all scale statements and can be understood in any language. Figure 1.14 is based on a fractional scale of 1:24,000.

The terms *large-scale* and *small-scale* are often applied to maps. A large-scale map, such as a plan of a city, shows a restricted area in considerable detail, and the ratio of map to ground distance is relatively large, for example, 1:600 (1 inch on the map represents 50 feet on the ground). At this scale, features such as buildings and highways that are usually represented by symbols on smaller scale maps can be drawn to scale. Small-scale maps, such as those of countries or continents, have a much smaller ratio. The scale may be 1 inch to 100 or even 1000 miles (1:6,336,000 or 1:63,360,000).

Each of the four maps in Figure 1.13 is drawn at a different scale. Although each is centered on St. Paul, notice how scale affects both (1) the area that can be shown in a square 2 inches on a side, and (2) the amount of detail that can be depicted. On Map A, at a scale of 1:24,000, about 2.6 inches represent 1 mile, so that the 2-inch square shows less than one square mile. At this scale, one can identify individual buildings, highways, rivers, and other landscape features. Map D, drawn to a scale of one to a million (1:1,000,000 or 1 inch represents almost 16 miles), shows an area of almost 1000 square miles. Now only major features, such as main highways and the location of cities, can be shown, and even the symbols used for that purpose occupy more space on the map than they should if drawn true to scale.

Small-scale maps like C and D in Figure 1.13 are said to be very *generalized*. They give a general idea of the relative locations of major features but do not permit their accurate measurement. They show significantly less detail than do large-scale maps and typically smooth out such features as coastlines, rivers, and highways.

Topographic Maps and Terrain Representation

When we speak of the topography of an area, we refer to its terrain. **Topographic maps** portray the surface features of relatively small areas, often with great accuracy (Figure 1.14). They not only show the elevations of landforms, streams, and other water bodies but also display features that people have added to the natural landscape. These may include transportation routes, buildings, and such land uses as orchards, vineyards, and cemeteries. Boundaries of all kinds, from state boundaries to field or airport limits, may be depicted on topographic maps.

The U.S. Geological Survey (USGS), the chief federal agency for topographic mapping in this country, produces several map series, each on a standard scale. A small-scale topographic series, at 1:250,000, or 1 centimeter to 2.5 kilometers, is complete for the United States. Other series are at scales of 1:125,000, 1:62,500, and 1:24,000. A single map in one of these series is called a *quadrangle*. Topographic quadrangles at the scale of

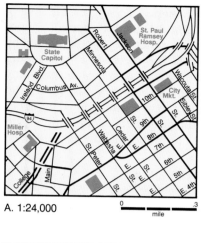

A. 1:24,000

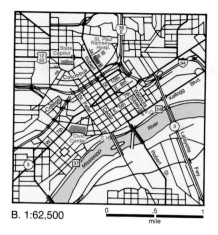

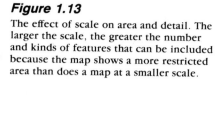

B. 1:62,500

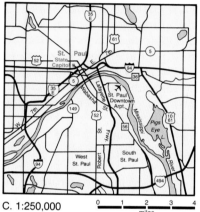

C. 1:250,000

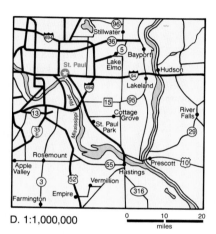

D. 1:1,000,000

Figure 1.13
The effect of scale on area and detail. The larger the scale, the greater the number and kinds of features that can be included because the map shows a more restricted area than does a map at a smaller scale.

Figure 1.14
A portion of the Santa Barbara, California, 7.5-minute series of U.S. Geological Survey topographic maps. The fractional scale is 1:24,000. The shoreline shown represents the approximate line of mean high water. The pink tint indicates built-up areas, in which only landmark buildings are shown. The light purple tint is used to indicate extensions of urban areas.

Source: U.S. Geological Survey.

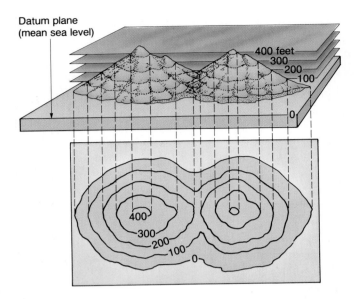

Figure 1.15
Contours drawn for an imaginary island. The intersection of the landform by a plane held parallel to sea level is a contour representing the height of the plane above sea level.

1:24,000 exist for two-thirds of the United States, excluding Alaska. As is evident from Figure 1.14, these quadrangle maps can provide detailed information about the natural and humanmade features of an area.

The principal device used to show elevation on topographic maps is the **contour line,** along which all points are of equal elevation above a datum plane, usually mean sea level. Contours are imaginary lines, perhaps best thought of as the outlines that would occur if a series of progressively higher horizontal slices were made through a vertical feature. Figure 1.15 shows the relationship of contour lines to elevation for an imaginary island.

The *contour interval* is the vertical spacing between contour lines, and it is normally stated on the map. Contour intervals of 10 and 20 feet are often used, though in relatively flat areas the interval may be only 5 feet. In mountainous areas, the spacing between contours is greater, for graphic convenience. On Figure 1.14, the contour interval is 50 feet. The more irregular the surface, usually, the greater is the number of contour lines that will need to be drawn; the steeper the slope, the closer are the contour lines rendering that slope.

Contour lines are the most accurate method of representing terrain, giving the map reader information about the elevation of any place on the map and the size, shape,

and slope of all relief features. They are not truly pictorial, however. To heighten the graphic effect of a topographic map, contours are sometimes supplemented by the use of **shaded relief.** This method of representing the three-dimensional quality of an area is illustrated by Figure 1.16. An imaginary light source, usually in the northwest, can be thought of as illuminating a model of the area, simulating the appearance of sunlight and shadows. Portions that are in the shadow are darkened on the map.

Many symbols besides contour lines appear on topographic maps. The USGS produces a sheet listing the symbols it employs (Figure 1.17), and some older maps provide legends on the reverse side. Note that in the case of running water, separate symbols are used in the legend and on the map to depict perennial (permanent) streams, intermittent streams, and springs; the location and size of rapids and falls are also indicated. There are three different symbols for dams and three more for types of bridges. On maps of cities, where it would be impossible to locate every building separately, the built-up area is denoted by special tints, and only streets and public buildings are shown.

The tremendous amount of information on topographic maps makes them useful to engineers, regional planners and land use analysts, and developers, as well as to hikers and casual users. Given such a wealth of information, the experienced map reader can make deductions about both the physical character of the area and the economic and cultural use of the land.

When topographic maps were first developed, it was necessary to obtain the data for them through fieldwork, a slow and tedious process. The technological developments that have taken place in aerial photography since the 1930s have made it possible to speed up production and greatly increase the land area represented on topographic maps. Aerial photography is only one of a number of remote sensing techniques now employed.

Remote Sensing

Although **remote sensing** is a relatively new term, the process it describes—detecting the nature of an object from a distance—has been going on for well over a century. Soon after the development of the camera, photographs were made from balloons and kites. Carrier pigeons wearing miniature cameras that took exposures automatically at set intervals were even used. The airplane, first used for mapping in the 1930s, provided a platform for the camera and the photographer, so that it was possible to take photographs from planned positions. Aerial photographs must, of course, be interpreted by using such clues as the size, shape, tone, and color of the recorded objects before maps can be made from them.

Figure 1.16
A shaded relief map of a portion of the Swiss Alps, near
Interlaken. The scale is 1:50,000. Notice how effectively shading
portrays a three-dimensional surface.

BOUNDARIES

National . . . .
State or territorial . . . .
County or equivalent . . . .
Civil township or equivalent . . . .
Incorporated-city or equivalent . . . .
Park, reservation, or monument . . . .
Small park . . . .

LAND SURVEY SYSTEMS

U.S. Public Land Survey System:

Township or range line . . . .
   Location doubtful . . . .
Section line . . . .
   Location doubtful . . . .
Found section corner; found closing corner . . . .
Witness corner; meander corner . . . .

Other land surveys:

Township or range line . . . .
Section line . . . .
Land grant or mining claim; monument . . . .
Fence line . . . .

ROADS AND RELATED FEATURES

Primary highway . . . .
Secondary highway . . . .
Light duty road . . . .
Unimproved road . . . .
Trail . . . .
Dual highway . . . .
Dual highway with median strip . . . .
Road under construction . . . .
Underpass; overpass . . . .
Bridge . . . .
Drawbridge . . . .
Tunnel . . . .

BUILDINGS AND RELATED FEATURES

Dwelling or place of employment: small; large . . .
School; church . . . .
Barn, warehouse, etc.: small; large . . . .
House omission tint . . . .
Racetrack . . . .
Airport . . . .
Landing strip . . . .
Well (other than water); windmill . . . .
Water tank: small; large . . . .
Other tank: small; large . . . .
Covered reservoir . . . .
Gaging station . . . .
Landmark object . . . .
Campground; picnic area . . . .
Cemetery: small; large . . . .

RAILROADS AND RELATED FEATURES

Standard gauge single track; station . . . .
Standard gauge multiple track . . . .
Abandoned . . . .
Under construction . . . .
Narrow gauge single track . . . .
Narrow gauge multiple track . . . .
Railroad in street . . . .
Juxtaposition . . . .
Roundhouse and turntable . . . .

TRANSMISSION LINES AND PIPELINES

Power transmission line: pole; tower . . . .
Telephone or telegraph line . . . .
Aboveground oil or gas pipeline . . . .
Underground oil or gas pipeline . . . .

CONTOURS

Topographic:

   Intermediate . . . .
   Index . . . .
   Supplementary . . . .
   Depression . . . .
   Cut; fill . . . .

Bathymetric:

   Intermediate . . . .
   Index . . . .
   Primary . . . .
   Index Primary . . . .
   Supplementary . . . .

MINES AND CAVES

Quarry or open pit mine . . . .
Gravel, sand, clay, or borrow pit . . . .
Mine tunnel or cave entrance . . . .
Prospect; mine shaft . . . .
Mine dump . . . .
Tailings . . . .

SURFACE FEATURES

Levee . . . .
Sand or mud area, dunes, or shifting sand . . . . . .
Intricate surface area . . . .
Gravel beach or glacial moraine . . . .
Tailings pond . . . .

VEGETATION

Woods . . . .
Scrub . . . .
Orchard . . . .
Vineyard . . . .
Mangrove . . . .

COASTAL FEATURES

Foreshore flat . . . .
Rock or coral reef . . . .
Rock bare or awash . . . .
Group of rocks bare or awash . . . .
Exposed wreck . . . .
Depth curve; sounding . . . .
Breakwater, pier, jetty, or wharf . . . .
Seawall . . . .

BATHYMETRIC FEATURES

Area exposed at mean low tide; sounding datum .
Channel . . . .
Offshore oil or gas: well; platform . . . .
Sunken rock . . . .

RIVERS, LAKES, AND CANALS

Intermittent stream . . . .
Intermittent river . . . .
Disappearing stream . . . .
Perennial stream . . . .
Perennial river . . . .
Small falls; small rapids . . . .
Large falls; large rapids . . . .
Masonry dam . . . .
Dam with lock . . . .
Dam carrying road . . . .
Intermittent lake or pond . . . .
Dry lake . . . .
Narrow wash . . . .
Wide wash . . . .
Canal, flume, or aqueduct with lock . . . .
Elevated aqueduct, flume, or conduit . . . .
Aqueduct tunnel . . . .
Water well; spring or seep . . . .

GLACIERS AND PERMANENT SNOWFIELDS

Contours and limits . . . .
Form lines . . . .

SUBMERGED AREAS AND BOGS

Marsh or swamp . . . .
Submerged marsh or swamp . . . .
Wooded marsh or swamp . . . .
Submerged wooded marsh or swamp . . . .
Rice field . . . .
Land subject to inundation . . . .

Figure 1.17

Standard U.S. Geological Survey topographic map symbols.

Source: U.S. Geological Survey.

Geodetic Control Data

The horizontal position of a place, specified in terms of latitude and longitude, constitutes only two-thirds of the information needed to locate it in three-dimensional space. Also needed is a vertical control point defining elevation, usually specified in terms of altitude above sea level. Together, the horizontal and vertical positions comprise *geodetic control data,* and a network of more than one million of these precisely known points covers the entire United States.

Each point is indicated by a bronze marker fixed in the ground. You may have seen some of the vertical markers, called *bench marks,* on mountain or hill tops or even on a city sidewalk. Every U.S. Geological Survey (USGS) map shows the

markers in the area covered by the map, and the USGS maintains Geodetic Control Lists containing the description, location, and elevation of each marker.

Those lists were revised in 1987 when, after 12 years effort, federal scientists completed the recalculation of the precise location of some

250,000 benchmarks across the nation. In using a satellite locating system for the first national resurvey of control points since 1927, the National Oceanic and Atmospheric Administration (NOAA) found, for example, that New York's Empire State Building is 120.5 feet northeast of where it formerly officially stood. The Washington Monument has been moved northeast by 94.5 feet, the dome of California's state Capitol in Sacramento has been relocated 300.9 feet southwest, and Seattle's Space Needle now has a position 305.26 feet west and 65.53 feet south of where maps now show it. The satellite survey provides much more accurate locations than did the old system of land measurement of distances and angles. The result will be more accurate maps and more precise navigation.

Aerial Photography

Although there is now a variety of sensing devices, aerial photography employing cameras with returned film is perhaps the most widely used remote sensing technique. Mapping from the air has certain obvious advantages over surveying from the ground, the most evident being the bird's-eye view that the cartographer obtains. Using stereoscopic devices, the cartographer can determine the exact slope and size of features such as mountains, rivers, and coastlines. Areas that are otherwise hard to survey, such as mountains and deserts, can be mapped easily from the air. Furthermore, millions of square miles can be surveyed in a very short time. Maps based on aerial photographs can be made quickly and revised easily, so that they are kept up-to-date. With aerial photography, the earth can be mapped more accurately, more completely, and more rapidly than ever before.

In 1975 the U.S. Department of the Interior adopted the National Mapping Program to improve the collection and analysis of cartographic data and to prepare maps that would assist decision makers who deal with resource and environmental problems. One of the first goals of the

program was to achieve complete coverage of the nation with orthophotographic imagery for all areas not already mapped at the scale of 1:24,000. An **orthophotomap** is a topographic map on which natural and cultural features are portrayed by color-enhanced photographic images with certain map symbols (contours, boundaries, names) added as needed for interpretation. Note in Figure 1.18 that in contrast to the conventional topographic map, the photograph is the chief means of representing information. Orthophotomaps have a variety of uses, aiding forest management, soil surveys, geological investigations, flood hazard and pollution studies, and city planning.

Standard photographic film detects reflected energy within the visible portion of the spectrum (Figure 1.19). Although they are invisible, *near-infrared* wavelengths can be recorded on special sensitized infrared film. Discerning and recording as images objects that are not visible to the human eye, infrared film has proved particularly useful for the classification of vegetation and

Figure 1.18

Topographic map (top) and orthophotomap (bottom) of a portion of the Brunswick West quadrangle in southeastern Georgia. Orthophotomaps are well suited for portraying swamp and marshlands; flat, sandy terrain; and urban areas.

Source: U.S. Geological Survey.

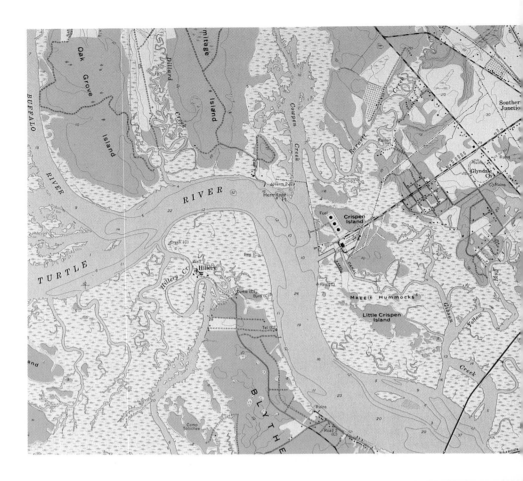

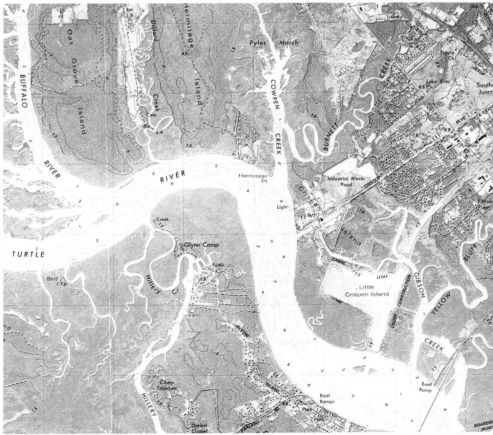

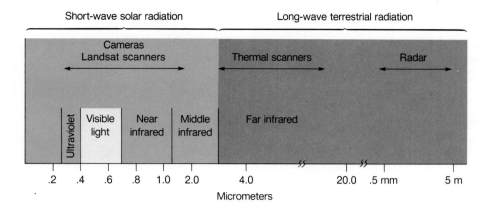

Short-wave solar radiation   Long-wave terrestrial radiation

Cameras
Landsat scanners   Thermal scanners   Radar

Ultraviolet | Visible light | Near infrared | Middle infrared | Far infrared

.2   .4   .6   .8   1.0   2.0   4.0   20.0   .5 mm   5 m

Micrometers

Figure 1.19

Wavelengths of the electromagnetic spectrum, in micrometers. One micrometer equals 1 one-millionth of a meter. The human eye can detect light in only the small portion of the spectrum that is colored yellow. Radiation from the sun has short wavelengths. Wavelengths longer than 4.0 micrometers characterize terrestrial radiation.

hydrographic features. Color-infrared photography yields what are called **false-color images**—"false" because the film does not produce an image that appears natural. For example, leaves of healthy vegetation have a high infrared reflectance and are recorded as red on color-infrared film while unhealthy or dormant vegetation appears as blue, green, or gray.

Nonphotographic Imagery

For wavelengths longer than 1.2 micrometers (a micrometer is 1 one-millionth of a meter) on the electromagnetic spectrum, sensing devices other than photographic film must be used. **Thermal scanners,** which sense the energy emitted by objects on earth, are used to produce images of thermal radiation (Figure 1.20). That is, they record the heat radiated by water bodies, clouds, and vegetation, as well as by buildings and other structures. Unlike conventional photography, thermal sensing can be employed during nighttime as well as daytime, giving it military applications. It is widely used for such purposes as detecting storms, water pollution, and the extent of forest fires.

Operating in a different band of the spectrum, **radar** systems can also be used during the day or at night. They transmit pulses of energy toward objects and sense

Figure 1.20

A midwestern utility company used a thermal-infrared scanner to measure nighttime heat loss from houses and commerical buildings.

the energy reflected back. The data are used to create images such as that shown in Figure 1.21, which was produced by radar equipment mounted on an airplane. Because radar can penetrate clouds and vegetation as well as darkness, it is particularly useful for monitoring the locations of airplanes, ships, and storm systems.

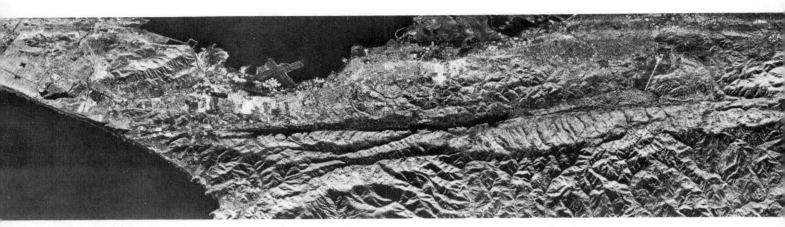

Figure 1.21
Radar image of the San Francisco peninsula. The San Andreas fault
is clearly visible.

Satellite Imagery

In the last 30 years, both manned and unmanned space-craft have supplemented the airplane as the vehicle for imaging the terrain. Concurrently, many steps have been taken to automate mapping, including the use of electronic mapping techniques, automatic plotting devices, and automatic data processing. Many images are now taken either from continuously orbiting satellites, such as those in the U.S. Landsat series and the French SPOT series, or from manned spacecraft flights, such as those of the Apollo and Gemini missions. Among the advantages of satellites are the speed of coverge and the fact that views of large regions can be obtained.

In addition, because they are equipped to record and report back to the earth information in the infrared portion of the spectrum, these satellites enable us to map the invisible. A number of federal agencies in the U.S., Japan, and the USSR have launched satellites specifically to monitor the weather. Data obtained by satellites have greatly improved weather forecasting and the tracking of major storm systems, and in the process have saved countless lives. The satellites are the source of the weather maps shown daily on television and in newspapers.

The **Landsat satellites,** first launched in 1972, take about one hour and 40 minutes to orbit the globe and can provide repetitive coverage of almost the entire globe every 18 days. Rather than recording data photographically, Landsat satellites carry scanning instruments that measure reflected light in both the visible and near-infrared portions of the spectrum. They relay electronic signals to receiving stations, where computers convert them into photo-like images that can be adjusted to fit special map projections.

A Landsat image of the Los Angeles metropolitan area appears in Figure 1.22. As it (and Figure 2.21) indicates, Landsat images can be enhanced by computers to appear as false-color images. As a result of this process, growing healthy vegetation appears in red, clear water as black, and urban areas as gray. The images have been used to produce small-scale maps ranging from 1:250,000 to 1:1,000,000. Among other things, Landsat imagery provides information on the distribution of and seasonal variation in vegetation, the state of soil moisture, snow accumulation, and drainage. It has been used to detect pollution, such as the effects of acid rain, indicate areas of potential drought or flood, and update inventories of timber and mineral deposits.

Patterns and Quantities on Maps

The study of the spatial pattern of things, whether people, cows, or traffic flows, is the essence of the locational tradition in geography, a subject explored in Part III of this book. Maps are used to record the location of these phenomena, and different kinds of techniques are employed to depict their presence or numbers at specific points, in given areas, or along lines.

EROS Data Center

The nation's only government-run center for processing data from Landsat satellites is in Sioux Falls, South Dakota. When the $50 million Earth Resources Observation Systems (EROS) complex was constructed in 1973, technology required that it be located near the center of the continent and away from interfering radio and television signals. Although current technology would permit the center to be built almost anywhere, and in fact Landsat signals are received at a dozen places outside the United States, Sioux Falls remains the EROS center's home.

One-half of the center's 350 employees are scientists with advanced degrees. They include geographers, geologists, hydrologists, meteorologists, and a host of experts in other disciplines. Their task is to process, analyze, and interpret the photography and imagery collected by high-flying aircraft, the Skylab, Apollo, and Gemini spacecraft, and Landsat satellites. Operated by the U.S. Geological Survey, EROS makes most of its millions of photographs and images of the earth's surface available to anyone through commercial distributors such as the Earth Observations Satellite Company (EOSAT) in Lanham, Maryland. Customers include U.S. intelligence agencies, foreign governments, and industry in addition to private citizens.

Figure 1.22

Landsat image of southern California. The Santa Barbara Channel appears black, the Los Angeles metropolitan region grey. The fan-shaped stain spreading into the Mojave Desert near the right-center margin is silt and other material carried down the mountains by water runoff. Analyses of Landsat images have practical applications in agriculture and forest inventory, land use classification, identification of geologic structures and associated mineral deposits, and monitoring of natural disasters.

(a)

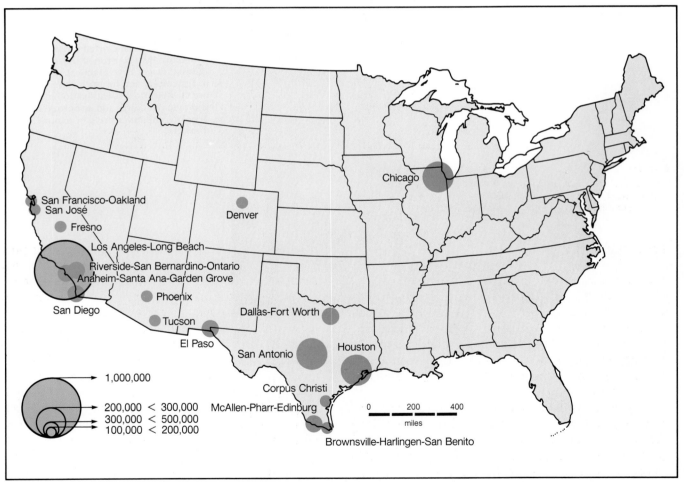

(b)

Figure 1.23

(a) Population distribution in the United States, 1980. The dot map provides a visual impression of specific locations and distributions. (Source: Geography Division, U.S. Bureau of the Census.) (b) Metropolitan areas with more than 100,000 Mexican-Americans in 1980. On this map, the area of the circle is proportional to the population. The scale in the bottom left corner aids the reader in interpreting the map.

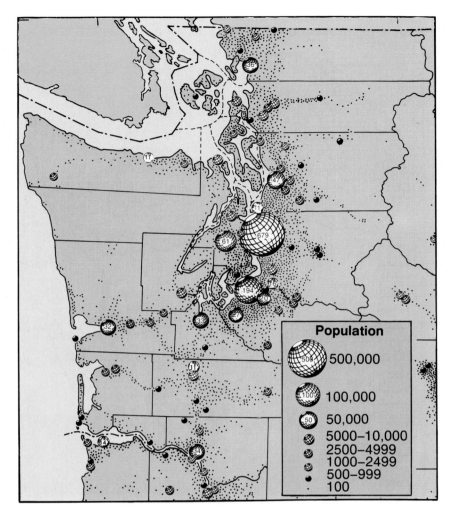

Figure 1.24

A portion of a map showing the distribution of Washington state population in 1950. The cartographer used proportional spheres to represent large urban population concentrations.

Quantities at Points

A topographic map shows the location at points of many kinds of things, such as churches, schools, and cemeteries. Each symbol counts as one occurrence. Often, however, our interest is in showing the variation in the number of some things that exist at several points, for example, the population of selected cities, or the tonnage handled at certain terminals, or the number of passengers at given airports.

There are two chief means of symbolizing such distributions, as Figure 1.23 indicates. One method is to choose a symbol, usually a dot, to represent a given quantity of the mapped item (such as 50 people) and to repeat that symbol as many times as necessary. Such a map is easily understood because the dots give the map reader a visual impression of the pattern. Sometimes pictorial symbols—for example, human figures or oil barrels—are used instead of dots.

If the range of the data is great, the cartographer may find it inconvenient to use a repeated symbol. For example, if one port handles 50 or 100 times as much tonnage as another, that many more dots would have to be placed on the map. To circumvent this problem, the cartographer can choose a second method and use proportional symbols. The size of the symbol is varied according to the quantities represented. Thus, if bars are used, they can be shorter or longer as necessary. If squares or circles are used, the *area* of the symbol ordinarily is proportional to the quantity shown (Figure 1.23b). There are occasions, however, when the range of the data is so great that even circles or squares would take up too much room on the map. In such cases, three-dimensional symbols, usually spheres or cubes, are used; their *volume* is proportional to the data. Unfortunately, many people fail to perceive the added dimension implicit in volume, and most cartographers do not recommend the use of such symbols (Figure 1.24).

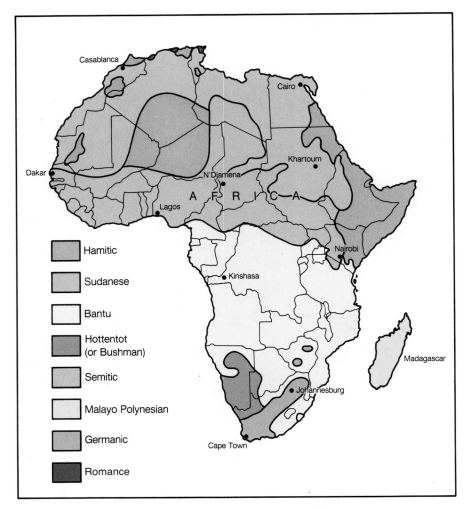

Figure 1.25

Language regions of Africa. Maps such as this one may give the false impression of uniformity within a given area, for example, that Bantu and no other languages are spoken over much of Africa. Such maps are intended to represent only the predominant language in an area.

Quantities in Area

Maps showing quantities in area fall into two general categories: those portraying the areas within which a kind of phenomenon occurs, and those showing amounts by area. Atlases contain numerous examples of the first category, such as patterns of religions, languages, political entities, vegetation, or types of rock. Normally, different colors or patterns are used for different areas, as in Figure 1.25. Among the problems involved in using such maps are that (1) they give the impression of uniformity to areas that may actually contain significant variations; (2) boundaries attain unrealistic precision and significance, implying abrupt changes between areas when, in reality, the changes may be gradual; and (3) unless colors are chosen wisely, some areas may look more important than others.

Variation in the *amount* of a phenomenon from area to area may be shown by covering subareas with different shades, colors, or patterns. Because the use of different colors or patterns may suggest qualitative rather than quantitative differences—that is, differences in kind rather than in degree—the best cartographic practice is to use a single hue and vary its value or intensity. Figure 1.26 is an example of such a map.

An interesting variation of these maps is the **cartogram,** a map simplified to present a single idea in a diagrammatic way. Railroad and subway route maps commonly take this form. Depending on the idea that the cartographer wishes to convey, the sizes and shapes of areas may be altered, distances and directions may be awry, and contiguity may or may not be preserved. In Figure 1.27 the sizes of the areas represented have been purposely distorted so that they are proportional to the population, and the map has been oriented with south at the top.

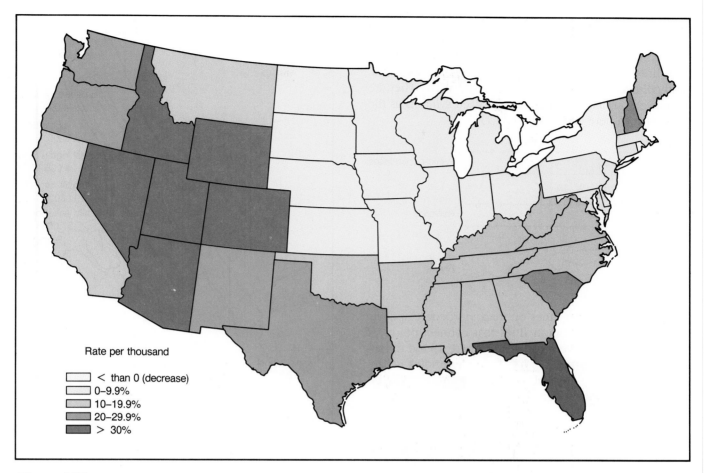

Figure 1.26

Population change by state, 1970–1980, in percent. Quantitative
variation by area is more easily visualized in map than in tabular
form.

Rate per thousand

< than 0 (decrease)
0–9.9%
10–19.9%
20–29.9%
> 30%

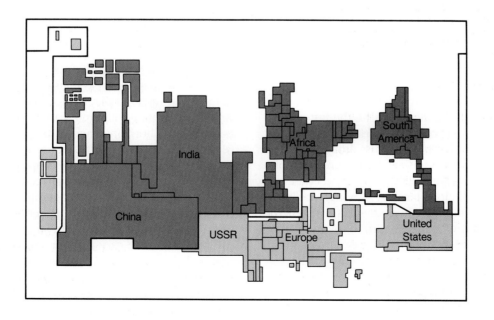

Figure 1.27

On this cartogram, the areas of countries have been made
proportional to their populations. The map has been oriented
with south at the top to make the point that northern countries
contain only a small minority of the world's people.

Lines on Maps

Some lines on maps do not have numerical significance. The lines representing rivers, political boundaries, roads, and railroads, for example, are not quantitative. They are indicated on maps by such standardized symbols as the ones shown here and in Figure 1.17.

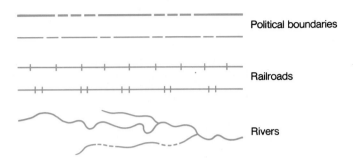

Political boundaries

Railroads

Rivers

Often, however, lines on maps do denote specific numerical values. Contour lines that connect points of equal elevation above mean sea level are a kind of **isoline,** or line of constant value. Other examples of isolines are isohyets (equal rainfall), isotherms (equal temperature), and isobars (equal barometric pressure). The implications of isolines as regional boundaries are discussed in Chapter 12.

Flow-line maps are used to portray traffic or commodity flows along a given route, usually a waterway, a highway, or a railway. The location of the route taken, the direction of movement, and the amount of traffic can all be depicted. The amount shown may be either the total or a per mile figure. In Figure 1.28, the width of the red line is proportional to the number of vehicles.

Computer-Assisted Cartography

A recent development in cartography is the use of computers to assist in making maps. Within the last two decades, computers have become an integral part of almost every stage of the cartographic process, from the collection and recording of data to the production and reproduction of maps. Computer mapping is speedier and more accurate than manual drafting. Although the initial cost of equipment is high, the investment is repaid in the more efficient production and revision of maps.

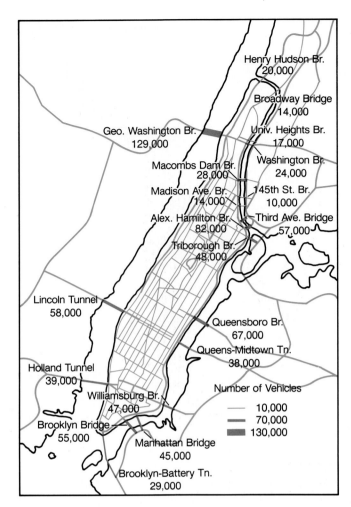

Figure 1.28

A flow-line map showing the number of vehicles entering Manhattan daily.

Data from New York City Department of Transportation, 1984.

Computers use *digital* (as opposed to *analog*) files. That is, the map image is considered to be divided into a fine-mesh grid, so that the location of any item on the map, like a dot on a graph, can be described by its position on the horizontal and vertical axes. Then the presence or absence of a particular item in any grid location is coded in the computer. The mapmaker can now draw on a growing number of such computer **data bases.** These digital records of geographic information store hundreds of millions of coordinates for physical and cultural features, census data, and so on. When a map is to be produced, the cartographer can quickly call up the desired data. Outdated data—for example, population sizes—can be erased from the file and replaced.

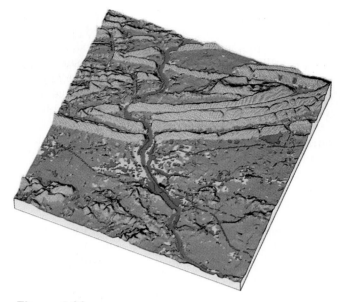

Figure 1.30
A computer-drawn map representing land use and land cover in the Harrisburg, Pennsylvania, area.

Figure 1.29
A digitizer such as this one automatically transforms and records the location of a place in terms of x and y coordinates.

The U.S. Geological Survey is digitizing categories of data from its 7.5-minute topographic quadrangles. These include latitude and longitude coordinates, political boundaries, physical features like rivers and coastlines, and place names. The 1:2,000,000-scale series of the United States already has been transformed into a digital data base. If no data base exists, cartographers can create their own by using electronic digitizing devices that transform data into digital records (Figure 1.29).

The photographic and photo-like images sensed by scanning devices on spacecraft and satellites can be broken down into a grid of digitized picture elements, or **pixels.** Each pixel has a value between 0 and 255; the brighter the pixel, the higher the numerical value. The development of ever more powerful computers has spurred the development of new techniques for the enhancement of these photo images. Edge enhancement,

linear contrast stretching, and density slicing are among the techniques that are being used to improve the contrast and resolution of the images, increase the amount of information gleaned from them, and remove the distortions caused by atmospheric haze or water reflection. Linear contrast stretching, for example, involves assigning new brightness values to the pixels on any given image, "stretching" them to fill the complete range from 0 to 255. The technique produces a dramatic rise in the contrast of the image.

Computers also permit cartographers to use sophisticated statistical measures to analyze data before they are mapped. Because they can process thousands of facts in seconds, computers are particularly useful for researchers who need to analyze many variables simultaneously and who employ such quantitative techniques as multiple regression and factor analysis. In this sense, computers make possible the production of maps that were virtually impossible to generate 20 years ago. Figure 1.30 is one example of a computer-drawn map.

Several software packages are now available for map production, particularly for commonly used projections. They enable the cartographer to produce plots of boundaries, coastlines, and grid systems at any desired scale. Some software is intended for use only on large-capacity computers, but software packages for personal computers are also available now. The computer revolution has just begun; while no one can predict all the applications computers will have to cartography, they will be significant and far-reaching.

Conclusion

In this chapter, we have not attempted to discuss all aspects of the field of cartography. Mapmaking and map design, systems of land survey, map compilation, and techniques of map reproduction are among the topics that have been deliberately omitted. Our intent has been to introduce those aspects of map study that will aid in map reading and interpretation.

A discussion of cartography is the natural entrance to the study of geography, for there is no more useful manner of examining the earth than through maps. Indeed, it may properly be said that maps are as indispensable to the geographer as words, photographs, or quantitative techniques of analysis. The number of maps contained in this book amply illustrates this point. As you read the remainder of the book, note their many different uses. For example, notice in Chapter 2 how important maps are to your understanding of the theory of continental drift; in Chapter 6, how maps aid geographers in identifying cultural regions; and in Chapter 8, how behavioral geographers use maps to record people's perceptions of space.

Geographers are not unique in their dependence on maps. People involved in the analysis and solution of many of the problems facing the world also rely on them. Environmental protection, control of pollution, conservation of natural resources, land use planning, and traffic control are just a few of the issues that call for the accurate representation of elements on the earth's surface.

Key Words

cartogram *44*	equidistant projection *26*	latitude *22*	prime meridian *23*
conformal projection *26*	false-color image *39*	longitude *23*	projection *25*
conic projection *27*	globe properties *26*	map scale *32*	radar *39*
contour line *34*	grid system *22*	Mercator projection *26*	remote sensing *34*
cylindrical projection *26*	International Date Line *24*	orthophotomap *37*	shaded relief *34*
data base *46*	isoline *46*	pixel *47*	thermal scanner *39*
developable surface *26*	Landsat satellite *40*	planar projection *28*	topographic map *32*
equal-area projection *26*			

For Review

1. What important map and globe reference purpose does the *prime meridian* serve? Is the prime, or any other meridian, determined in nature or devised by humans? How is the prime meridian designated or recognized?

2. What happens to the length of a degree of longitude as one approaches the poles? What happens to a degree of latitude between the equator and the poles?

3. From a world atlas, determine, in degrees and minutes, the locations of New York City; Moscow, USSR; Sydney, Australia; and your hometown.

4. List at least five properties of the globe grid. Examine the projections used in Figures 1.5, 1.8, and 1.9a. In what ways do each of these projections adhere to or deviate from globe grid properties?

5. Briefly make clear the differences in properties and purposes of *conformal, equivalent,* and *equidistant* projections. Give one or two examples of the kinds of map information that would best be presented on each type of projection. Give one or two examples of how misunderstandings might result from data presented on an inappropriate projection.

6. In what different ways may *map scale* be presented? Convert the following map scales into their verbal equivalents.

    1:1,000,000   1:63,360   1:12,000

7. What is the purpose of a *contour line?* What is the *contour interval* on Figure 1.14? What landscape feature is implied by closely spaced contours?

8. Examine Figure 1.14. Imagine that you are right at the beacon on Stearn's Wharf, which is 20 feet above sea level. If you looked northeast, what would you see in the foreground? In the background?

9. What kinds of data acquisition are suggested by the term *remote sensing?* Describe some ways in which different portions of the spectrum can be sensed. To what uses are remotely sensed images put?

10. The table below gives the gross national product per capita of selected European countries (in U.S. dollars, 1983). On outline maps or as cartograms, represent the data in two different ways.

Country	GNP	Country	GNP
Austria	8,904	Luxembourg	13,988
Belgium	8,243	Netherlands	9,120
Denmark	11,026	Norway	13,300
France	9,478	Portugal	2,061
W. Germany	10,672	Spain	4,049
Greece	3,544	Sweden	10,434
Ireland	4,263	Switzerland	15,390
Italy	6,208	United Kingdom	8,214

Suggested Readings

Brown, Lloyd A. *The Story of Maps.* Boston: Little, Brown, 1949; reprint ed., New York: Dover Publications, 1977.

Campbell, John. *Introductory Cartography.* Englewood Cliffs, N.J.: Prentice-Hall, 1984.

Espenshade, Edward B., Jr., and Joel Morrison, eds. *Goode's World Atlas.* 17th ed. Chicago: Rand McNally, 1986.

Lillesand, Thomas M., and Ralph W. Kiefer. *Remote Sensing and Image Interpretation.* New York: John Wiley, 1979.

Muehrcke, Phillip C. *Map Use: Reading, Analysis, Interpretation.* 2d ed. Madison, Wisc.: JP Publications, 1986.

Robinson, Arthur H.; Randall D. Sale; Joel Morrison; and Phillip C. Muehrcke. *Elements of Cartography.* 5th ed. New York: John Wiley, 1984.

2 Physical Geography: Landforms

*B*efore the morning of November 14, 1963, cartographers thought they had finished their work in Iceland. They certainly had been at it long enough—since the island's first appearance on a world map by Eratosthenes some two centuries before Christ. It was a well-mapped area early in the Middle Ages and could be set aside as one of the certainties of the North Atlantic. Then with a roar, a pillar of steam and fire, and a hail of ash, nature demanded that the cartographers go back to work. Surtsey, the Dark One—god of fire and destruction—rose from the sea 20 miles off the Icelandic coast to become new land—1 mile long, 600 feet high, 670 acres of rock, ash, and lava.

Cartographers had another job, as would have been their lot had they been around for, say, the last 100 million years. Things just won't stand still. And not only little things, like new islands, or big ones, like mountains rising and being worn low to swampy plains, but even monstrous things: continents that wander about like homeless nomads, and ocean basins that expand, contract, and split up the middle like worn-out coats. It is a fascinating story, this formation and alteration of the earth, which, from our instantaneous view, seems so eternal and unchanging.

Geologic time is long, but the forces that give shape to the land are timeless and constant. Processes of creation and destruction continually fashion the seemingly eternal structure upon which humankind lives and works. Three types of forces interact to produce those infinite local variations in the surface of the earth called *landforms*: (1) forces that push, pull, move, and raise the earth; (2) forces that scour, scrape, wash, and chip away the land; and (3) forces that deposit soil, pebbles, rocks, and sand in new areas. How long these processes have worked, how they work, and what their effects are comprise the subject of this chapter.

Geologic Time

The earth was formed about 4.7 billion years ago. Because our usual concept of time is dwarfed when we speak of billions of years, it is useful to compare the age of the earth with something that is more familiar.

Imagine that the height of the World Trade Center in New York City represents the age of the earth. The twin towers are 110 stories or 1353 feet (412 m) tall. Even the thickness of a piece of paper laid on top of the roof would be too great to represent an average person's lifetime. The height of 4.7 stories represents the 200 million years that have elapsed since the present ocean basins began to form. The first mammals made their appearance on earth 60 million years ago, or the equivalent of the height of 1.5 stories. Earth history is so long and involves so many major geologic events that scientists have divided it into a series of recognizable, distinctive stages (Figure 2.1).

At this moment, the landforms on which we live are ever so slightly being created and destroyed. The processes involved have been in operation for so long that any given location most likely was the site of an ocean and of land at a number of different times in its past. Many of the landscape features on earth today can be traced back several hundred million years. The processes responsible for building up and tearing down those features are occurring simultaneously but usually at different rates.

In the last 30 years scientists have developed a useful framework within which one can best study our constantly changing physical environment. Their work is based on the early 20th century geological studies of Alfred Wegener, one of the pioneers in the rapidly evolving science of **plate tectonics.** He believed that the present continents were once united in one supercontinent and that over many millions of years the continents broke away from each other and slowly drifted to their current positions. New evidence and new ways of rethinking old knowledge have led to the wide acceptance of the idea of moving continents by earth scientists in recent years.

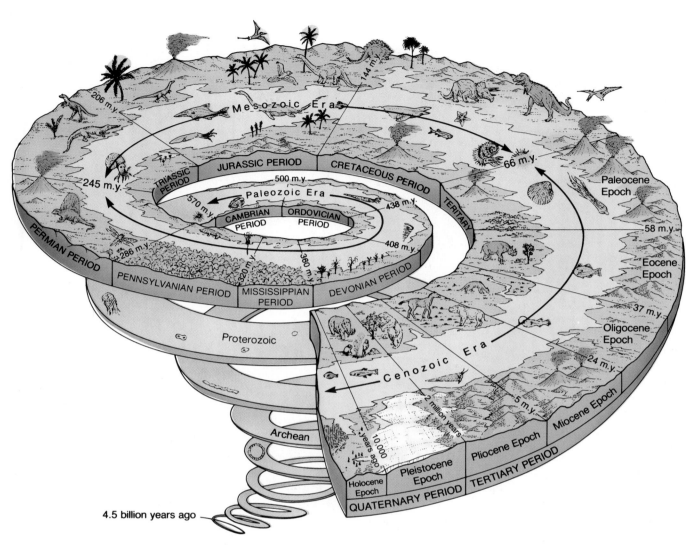

Figure 2.1

A diagrammatic history of the earth. The sketch depicts some of the known characteristics of the named geological periods.

After U.S. Geological Survey publication, "Geologic Time".

Movements of the Continents

The landforms that the cartographer maps are only the surface features of a thin cover of rock, the earth's *crust*. Above the core and the lower mantle of the earth, there is a partially molten plastic layer called the **asthenosphere** (Figure 2.2). The asthenosphere supports a thin but strong solid shell of rocks called the **lithosphere,** of which the outer, lighter portion is the earth's crust. The crust consists of one set of rocks found below the oceans and another set that makes up the continents.

The lithosphere is broken into about 10 large, rigid plates, each of which slides or drifts very slowly over the heavy semimolten asthenosphere, according to the theory of plate tectonics. A single plate may contain both oceanic and continental crust. Figure 2.3 shows that the North American plate, for example, contains the North Atlantic Ocean and most, but not all, of North America. The peninsula of Mexico—Baja California—and part of California are on the Pacific plate.

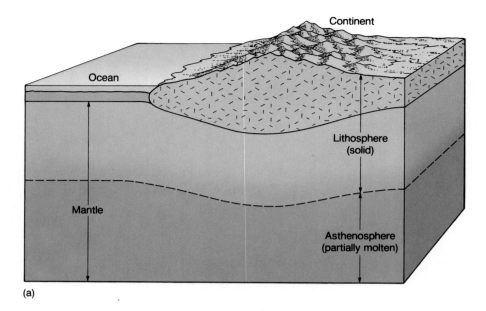

(a)

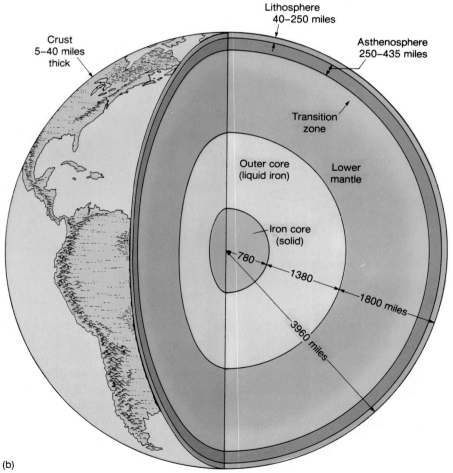

(b)

Figure 2.2

(a) The outer zones of the earth (not to scale). The *lithosphere* includes the crust and the uppermost mantle. The *asthenosphere* lies entirely within the upper mantle. (b) The very thin crust of the earth overlies a layered planetary interior. Zonation of the earth developed early in its history. Radioactive heating melted the original homogeneous planet. A dense iron core settled to the center; a surface and the remnant lower mantle—overlain by a transition zone and the asthenosphere—formed between them. Escaping gases eventually created the atmosphere and the oceans.

From John F. Dewey, "Plate Tectonics" in *Scientific American,* 1972. Copyright © 1972 Scientific American, Inc. Reprinted by permission.

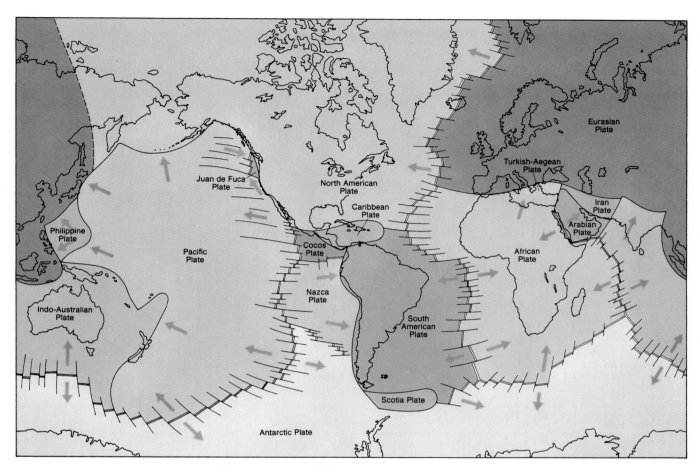

Figure 2.3
The large lithospheric plates move as separate entities and
collide. Assuming the African plate to be stationary, relative plate
movements are shown by arrows. Sea floor spreading, the
triggering mechanism, takes place along the axes of the ridges.

Scientists are not certain why the lithospheric plates
move. One reasonable theory suggests that heat and
heated material from the earth's interior rise by convection into particular crustal zones of weakness. These
zones are sources for the divergence of the plates. The
cooled materials then sink downward in subduction
zones (discussed later). In this way, the plates are thought
to be set in motion. Strong evidence indicates that about
200 million years ago the entire continental crust drifted
into one supercontinent, to which Wegener gave the
name **Pangaea** ("all earth"). Pangaea was broken into
plates as the sea floor began to spread, the major force
coming from the widening of what is now the Atlantic
Ocean (Figure 2.4).

Materials from the asthenosphere have been rising
along the Atlantic Ocean fracture, and as a result, the sea
floor has continued to spread. The Atlantic Ocean is now
4300 miles (6900 km) wide at the equator. If it widened
by a bit less than 1 inch per year, as scientists have estimated, one could calculate that the separation of the
continents did in fact begin about 200 million years ago.
Notice in Figure 2.5 how the ridge line that makes up
the axis of the ocean runs parallel to the eastern coast of
North and South America and the western coast of Europe and Africa. Scientists were led to the theory of the
continental drift of lithospheric plates by the amazing
fit of the continents.

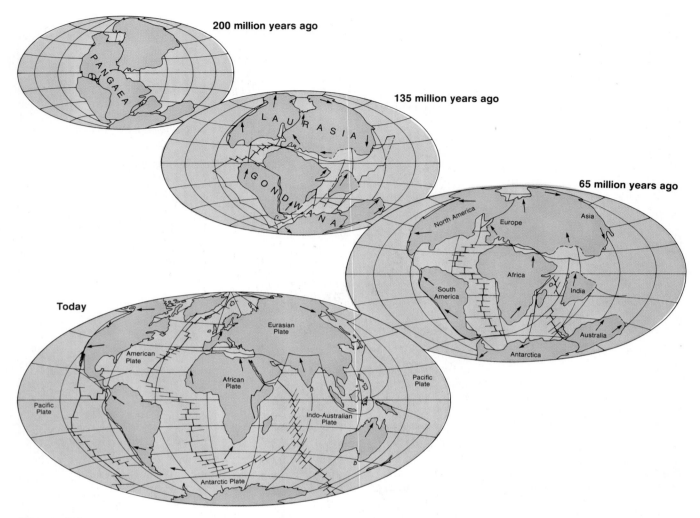

Figure 2.4
The drifting of the continents. Two hundred million years ago the continents were connected as one large land mass. After they split apart, the continents moved to their present positions. Notice how India broke away from Antarctica and collided with the Eurasian land mass. The Himalayas were formed at the zone of contact.

According to the theory, collisions occurred as the lithospheric plates moved. The pressure exerted at the intersections of plates resulted in earthquakes, which over periods of many years combined to change the shape and the features of the landforms. Figure 2.6 shows the location of near-surface earthquakes for a recent time period. Comparison with Figure 2.3 shows that the areas of greatest earthquake activity are at plate boundaries.

The famous San Andreas fault of California is a part of a long fracture separating two lithospheric plates, the North American and the Pacific. Earthquakes occur along **faults** (sharp breaks in rocks along which there is slippage) when the tension or compression at the junction become so great that only an earth movement can release the pressure. The San Andreas case is called a *transform* fault, which occurs when one plate slips past

another in a horizontal motion. Because the Atlantic Ocean is still widening at the rate of about 1 inch (2.5 cm) per year, earthquakes must occur from time to time to relieve the stress along the tension zone in the mid-Atlantic and along other fracture lines, such as the San Andreas fault.

Despite the availability of scientific knowledge about earthquake zones, the general public is peculiarly indifferent to earthquake danger. Every year there are hundreds, and sometimes thousands, of casualties from inadequate preparation for earthquakes (Figure 2.14). In some well-populated areas, the chances that damaging earthquakes will occur are very great. The distribution of earthquakes shown in Figure 2.6 points out the potential dangers to densely settled areas of Japan, the Philippines, parts of Southeast Asia, and the western rim of the Americas.

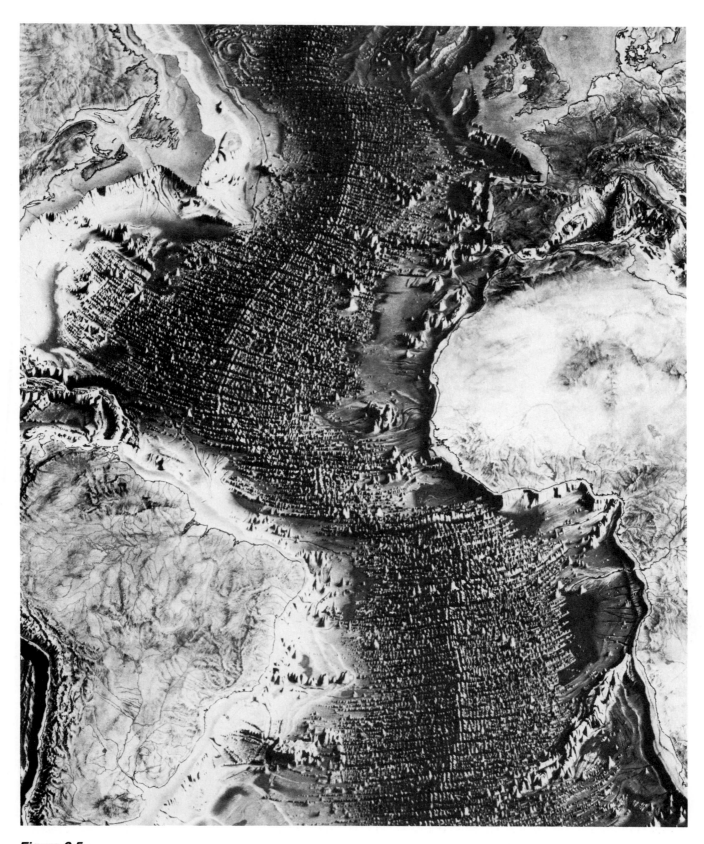

Figure 2.5
The topography of the Atlantic Ocean floor is evidence of the
dynamic forces shaping continents and ocean basins.

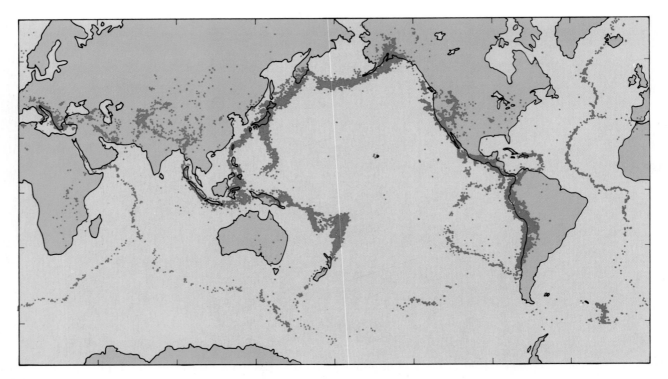

Figure 2.6
World earthquake epicenters, 1961–1967, from U.S. Coast and
Geodetic Survey.

Danger Near the San Andreas Fault

The earthquakes of July 8 and 13, 1986, in southern California reminded many that the majority of people in California live close to the San Andreas and related faults. Experts who have been monitoring seismic patterns in that area predict that a major earthquake is likely to occur along the fault before the year 2010. Great earthquakes have occurred to the east of Los Angeles about every 145 years, plus or minus several decades. The last major quake took place there in 1857, well before the city became the huge metropolis it is today.

(a)

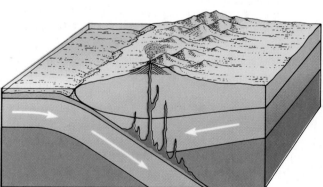

(b)

Figure 2.7

The process of *subduction*. (a) The Himalayas have been forming since the Indian subcontinent collided with Asia about 40 to 50 million years ago. (b) When lithospheric plates collide, the heavier oceanic crust is usually forced beneath the lighter continental material. Deep-sea trenches, mountain ranges, volcanoes, and earthquakes occur along the plate collision lines.

Movement of the lithospheric plates results in the formation of deep sea trenches and continental-scale mountain ranges as well as earthquakes. The continental crust is made up of lighter rocks than is the oceanic crust. Thus, where plates with different types of crust at their edges push against each other, there is a tendency for the denser oceanic crust to be forced down into the asthenosphere, causing long, deep trenches to form below the ocean. This type of collision is termed **subduction** (Figure 2.7). The edge of the overriding continental plate is uplifted to form a mountain chain that runs close to, and parallel with, the offshore trench.

Most of the Pacific Ocean is underlain by a plate that, like the others, is constantly pushing and being pushed. The continental crust on adjacent plates is being forced to rise and fracture, making an active earthquake and volcano zone of the rim of the Pacific Ocean (sometimes called the "ring of fire"). In recent years, major earthquakes and volcanic activity have occurred in Colombia, Mexico, Central America, the Pacific Northwest of the United States, Alaska, China, and the Philippines.

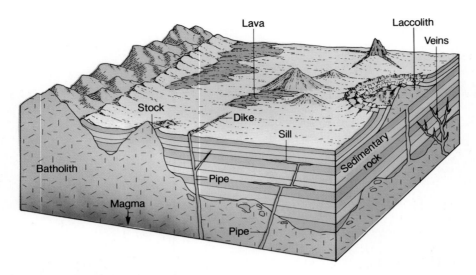

Figure 2.8
Extrusive and intrusive forms of volcanism. Lava and *ejecta* (ash and cinders) are extrusions of rock material onto the earth's surface in the form of cones or horizontal flows. *Batholiths* and *laccoliths* are irregular masses of crystalline rock that have cooled slowly below the earth's surface (intrusions).

The tremendous explosion that rocked Mount St. Helens in the state of Washington in 1980 is an example of continuing volcanic activity along the Pacific rim. Many scientists believe that there will be a damaging earthquake along the San Andreas fault in the near future.

Plate intersections are not the only locations that are susceptible to readjustments in the lithosphere. In the process of continental drift, the earth's crust has been cracked or broken in virtually thousands of places; some of the breaks are weakened to the point that they allow molten material from the asthenosphere to find its way to the surface. The molten material may explode out of a volcano or ooze out of cracks. Later in this chapter we will return to the discussion of volcanic and other forces that contribute to the shaping of landforms.

Earth Materials

The rocks of the earth's crust vary according to mineral composition. Rocks are made up of particles that contain various combinations of such common elements as oxygen, silicon, aluminum, iron, and calcium, together with less abundant elements. A particular chemical combination that has a hardness, density, and definite crystal structure of its own is called a **mineral.** Some well-known minerals are quartz, feldspar, and silica. Depending on the nature of the minerals that form them, rocks may be hard or soft, dense or open, chemically stable or unstable, and various colors. While some rocks resist decomposition, others are very easily broken down. Among the more common varieties of rock are granites, basalts, limestones, sandstones, and slates.

Although one can classify rocks according to their physical properties, the more common approach is to classify them by the way they are formed. The three main groups of rocks are igneous, sedimentary, and metamorphic.

Igneous Rocks

Igneous rocks are formed by the cooling and hardening of earth material. Weaknesses in the crust give molten material from the asthenosphere an opportunity to find its way into or onto the crust. When the molten material cools, it hardens and becomes rock. The name for underground molten material is *magma;* above ground it is *lava. Instrusive* igneous rocks were formed below ground level by the hardening of magma, while *extrusive* igneous rocks were created above nearby ground level by the hardening of lava (Figure 2.8).

The chemicals making up lava and magma are fairly uniform, but depending on the speed of cooling, different minerals form. Because it is not exposed to the coolness of the air, magma hardens slowly, allowing silicon and oxygen to unite and form quartz, a hard, dense mineral. With other components, grains of quartz combine to form the rock called *granite.*

The lava that oozes out onto the earth's surface and makes up a large part of the ocean basins contains a considerable amount of sodium or calcium aluminosilicates. These form the mineral called *feldspar* and together with the dark mineral *pyroxene* make up *basalt,* the most common rock on earth.

If instead of oozing, the lava erupts from a volcano crater, it may cool very rapidly. Some of the rocks formed in this manner contain air spaces and are light and angular, such as *pumice.* Some may be dense, even glassy, as is *obsidian.* The glassiness occurs when lava meets standing water and suddenly cools.

Sedimentary Rocks

Sedimentary rocks are composed of particles of gravel, sand, silt, and clay that were eroded from already existing rocks. Surface waters carry the sediment to oceans, marshes, lakes, or tidal basins. Compression of these materials by the weight of additional deposits on top of them, and a cementing process brought on by the chemical action of water and certain minerals, causes sedimentary rock to form.

Sedimentary rocks evolve under water in horizontal beds called *strata.* Usually one type of sediment collects in a given area. If the particles are large—for instance, the size of gravel—a gravelly rock called *conglomerate* forms. Sand particles are the ingredient for *sandstone,* while silt and clay form *shale* or *siltstone.*

Sedimentary rocks also derive from organic material, such as coral, shells, and marine skeletons. These materials settle into beds in shallow seas and congeal, forming *limestone.* If the organic material is mainly vegetation, it can develop into a sedimentary rock called *coal. Petroleum* is also a biological product, formed during the millions of years of burial by chemical reactions that transform some of the organic material into liquid and gaseous compounds. The oil and gas are light; therefore, they ooze through the pores of the surrounding rock to places where dense rocks block their upward movement.

Sedimentary rocks vary considerably in color (from coal black to chalk white), hardness, density, and resistance to chemical decomposition. Large parts of the continents contain sedimentary rocks; nearly the entire eastern half of the United States is overlain with these rocks, for example. Such formations indicate that in the geologic past, seas covered even larger proportions of the earth than they do today.

Metamorphic Rocks

Metamorphic rocks are formed from igneous and sedimentary rocks by earth forces that generate heat, pressure, or chemical reaction. The word *metamorphic* means "changed shape." The internal earth forces that cause the movement and collision of lithospheric plates may be so great that by heat and pressure, the mineral structure of a rock changes, forming new rocks. For example, under great pressure, shale, a sedimentary rock, becomes *slate,* a rock with different properties. Limestone under certain conditions may become *marble,* and granite may become *gneiss* (pronounced nice). Materials metamorphosed at great depth and exposed only after overlying surfaces have been slowly eroded away are among the oldest rocks known on earth. Like igneous and sedimentary rocks, however, their formation is a continuing process.

Rocks are the constituent ingredients of most landforms. Their strength or weakness, their permeability, and their chemical content control the way they respond to the forces that shape and reshape them. Two principal processes are at work altering rocks: the tectonic forces that tend to build landforms up and the gradational processes that wear landforms down. All rocks are part of the *rock cycle,* through which old rocks are continually being transformed into new ones by these processes. No rocks have been preserved unaltered throughout the earth's history.

Tectonic Forces

The earth's crust is altered by the constant forces resulting from plate movement. *Tectonic* (generated from within the earth) processes shaping and reshaping the earth's crust are of two types: *deformational* and *volcanic.* The great pressure acting on the plates deforms the surface by folding, twisting, breaking, or compressing rock. Volcanism is the force that transports heated material to or toward the surface of the earth. When particular places on the continents are under pressure, the changes that take place can be as simple as the bowing or cracking of rock, or as dramatic as lava exploding from the crater and sides of a Mount St. Helens.

Deformation

In the process of continental drift, pressures build in various parts of the earth's crust, and slowly, over thousands of years, the crust is transformed. By studying rock formations, geologists are able to trace the history of the development of a region. Over geologic time, most continental areas have been subjected to both tectonic and gradational activity—to building up and tearing down. They usually have a complex history of folding, faulting, and leveling. Some flat plains in existence today may hide a history of great mountain development in the past.

Folding

When the pressure caused by moving continents is great, layers of rock are forced to buckle. The result may be warping or bending the rock, and a ridge or a series of parallel ridges or **folds** may develop. If the stress is pronounced, great wavelike folds form (Figure 2.9). The folds can be thrust upward many thousands of feet and laterally for many miles. The folded ridges of the eastern United States are at present low parallel mountains 1000–3000 feet (300–900 m) above sea level, but the rock evidence suggests that the tops of the present Appalachian mountains were once valleys between 30,000-foot (9100-m) crests (Figure 2.10).

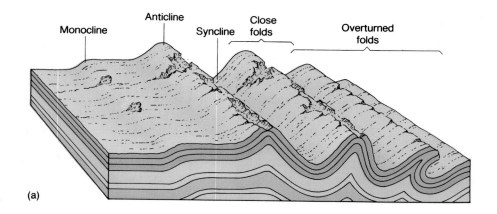

(b)

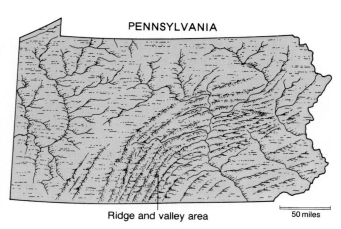

Ridge and valley area

50 miles

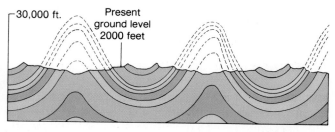

30,000 ft.

Present ground level 2000 feet

Figure 2.10

The ridge and valley region of Pennsylvania, now eroded to hill lands, is the relic of 30,000-foot (9100-m) folds that were reduced to form *synclinal* (downarched) hills and *anticlinal* (uparched) valleys. The rock in the original troughs, having been compressed, was less susceptible to erosion.

Figure 2.9

Degrees of folding vary from slight undulations of strata with little departure from the horizontal to highly compressed or overturned beds. (a) Diagram of the most common forms of folding. (b) An overturned fold in the Appalachian Mountains.

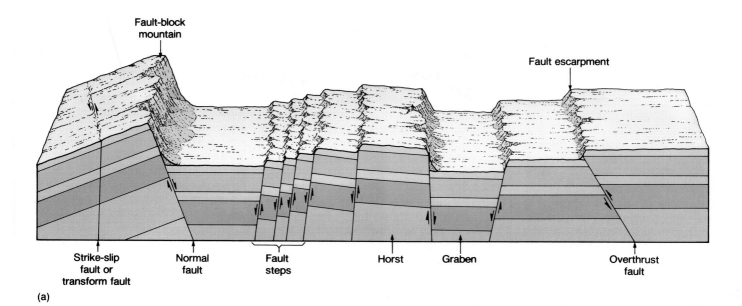

Fault-block mountain

Fault escarpment

Strike-slip fault or transform fault

Normal fault

Fault steps

Horst

Graben

Overthrust fault

(a)

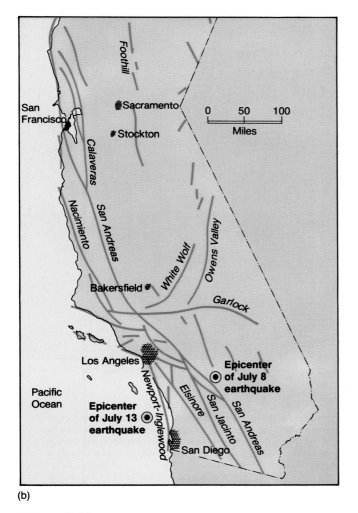

(b)

Faulting

A fault is a break or fracture in rock. The stress causing a fault results in displacement of the earth's crust along the fracture zone. Figure 2.11 depicts diagrammatic examples of fault types. There may be uplift on one side of the fault or downthrust on the other. In some cases, a steep slope known as a fault *escarpment,* which may be several hundred miles long, is formed. The stress can push one side up over the other side, or a separation away from the fault may cause the land to sink and create a *rift valley* (Figure 2.12).

Many faults are merely cracks (called *joints*) with little noticeable movement along them, but in other cases, mountains such as the Sierra Nevada of California have risen as the result of faulting. In some instances, the movement has been horizontal along the surface rather than up or down. The San Andreas transform fault mentioned earlier and pictured in Figure 2.13 is such a case.

Figure 2.11

Faults, in their great variation, are common features of mountain belts where deformation is great. (a) The different forms of faulting are categorized by the direction of movement along the plane of fracture. (b) The major faults in California with the epicenters of two 1986 quakes noted.

The Richter Scale

In 1935, C. F. Richter devised a scale of earthquake *magnitude*. An earthquake is really a form of energy expressed as wave motion passing through the surface layer of the earth. Radiating in all directions from the earthquake focus, seismic waves gradually dissipate their energy at increasing distances from the *epicenter* (the point on the earth's surface directly above the focus). On the **Richter scale,** the amount of energy released during an earthquake is estimated by measurement of ground motion. Seismographs record earthquake waves; by comparison of wave heights, the relative strength of quakes can be determined. Although Richter scale numbers run from 0 to 9, there is no absolute upper limit to earthquake severity. Presumably, nature could outdo the magnitude of the most intense earthquakes so far recorded, which reached 8.5–8.6.

Because magnitude, as opposed to intensity, can be measured accurately, the Richter scale has been widely adopted. Nevertheless, it is still only an approximation of the amount of energy released in an earthquake. In addition, the height of the seismic waves can be affected by the rock materials under the seismographic station, and some seismologists believe that the Richter scale underestimates the magnitude of major tremors.

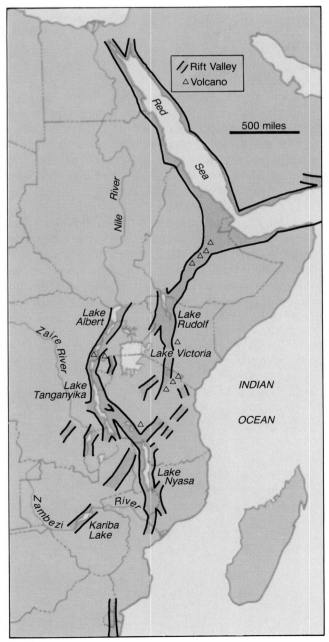

(a)

(b)

Figure 2.12

(a) Great fractures in the earth's crust resulted in the creation, through subsidence, of an extensive *rift valley* system in East Africa. The parallel faults, some reaching more than 2000 feet (600 m) below sea level, are bordered by steep walls of the adjacent plateau, which rises to 5000 feet (1500 m) above sea level and from which the structure dropped. (b) In Tanzania, the same tectonic forces have created a rift valley, one edge of which is shown here.

The Mexican Earthquake

An earthquake registering 8.1 on the Richter scale struck at the edge of the North American plate along Mexico's Pacific coast about 220 miles (350 km) from Mexico City on September 16, 1985. Pushed by the huge Nazca Plate to the south, the Cocos Plate slammed into and under the North American Plate (Figure 2.3). About 1000 times as much energy as was released by the atomic bomb at Hiroshima was expended over western Mexico. Destruction was heaviest in densely populated Mexico City. Wave after wave of violent earth shaking continued for about three minutes. The next day a strong aftershock registering 7.6 on the Richter scale added more fear and grief to the people of Mexico.

A large part of Mexico City sits on an old lake basin made up of silt, clay, sand, and gravel—a very unstable foundation for a city with tall buildings. The rock between the epicenter and Mexico City absorbed most of the quake's force, but, nonetheless, the buildings in the downtown area swayed. Fortunately, relatively few collapsed. Most were built to high standards specifically designed to resist earthquakes. Unfortunately, the 7000 buildings that did collapse killed 9000 people, caused 30,000 injuries, and left 95,000 people homeless.

A similar quake at the edge of the Cocos Plate caused additional widespread death and destruction in October 1986 in San Salvador, the capital of El Salvador. Earthquakes do their greatest damage indirectly—people's lives are jeopardized by the buildings they have erected, by landslides and mudflows that may descend on them, and by the tsunamis that ravage coastal areas.

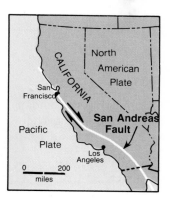

Figure 2.13

View along the San Francisco peninsula. A *transform fault,* the San Andreas marks a part of the slipping boundary between the Pacific and the North American Plates. The inset map shows the relative southward movement of the American Plate; dislocation averages about 0.4 inches (1 cm) per year.

Earthquakes

Whenever movement occurs along a fault or at some other point of weakness, an earthquake results. The greater the movement, the greater the magnitude of the earthquake. Tension builds in rock as tectonic forces are applied, and when a critical point is finally reached, the earthquake occurs and tension is reduced. The earthquake that occurred in Alaska on Good Friday in 1964 was one of the strongest known. Although the stress point of that earthquake was below ground and 75 miles from Anchorage, vibrations called *seismic waves* caused earth movement in the weak clays under the city. Sections of Anchorage literally slid downhill, and part of the business district

The Tsunami

A tsunami follows any submarine earthquake that causes fissures or cracks in the earth's surface. Water rushes in to fill the depression caused by the falling away of the ocean bottom. The water then moves outward, building in momentum and rhythm, as swells of tremendous power. The waves that hit Hawaii following the April 1, 1946, earthquake off Dutch Harbor, Alaska, were moving at approximately 400 miles (640 km) an hour, with a crest-to-crest spacing of some 80 miles (130 km).

The long swells of a tsunami are largely unnoticed in the open ocean. Only when the wave trough scrapes

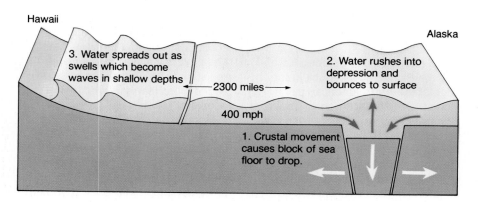

sea bottom in shallow coastal areas does the water pile up into precipitous peaks. The seismic sea waves at Hilo on the exposed northeastern part of the island of Hawaii were estimated at between 45 and 100 feet (14–30 m) in height. The

water smashed into the city, deposited 14 feet (4 m) of silt in its harbor, left fish stranded in palm trees, caused many millions of dollars in damage, and resulted in 173 deaths; many were people who had gone to the shore to see the giant waves arrive.

dropped 10 feet (3 m). Table 2.1 indicates the kinds of effects associated with earthquakes of different magnitudes, and Figure 2.14 shows various types of earthquake-induced damage.

If an earthquake occurs below an ocean, the movement can cause a **tsunami,** a large destructive sea wave. Though not noticeable on the open sea, a tsunami may become 30 or more feet (9 m) high as it approaches land, sometimes thousands of miles from the earthquake site. The Hawaiian islands now have a tsunami warning system that was developed following the devastation at Hilo in 1946 (see "The Tsunami" above).

Earthquakes occur daily in a number of places throughout the world. Most are slight and are noticeable only on *seismographs,* the instruments that record seismic waves, but from time to time there are large-scale earthquakes, like those in Guatemala and China in 1976, or the 1986 El Salvador quake that killed more than 1000 people and left 200,000 homeless. Most earthquakes take place on the rim of the Pacific (Figure 2.6), where stress from the outward-moving lithospheric plates is greatest. The Aleutian Islands of Alaska, Japan, Central America, and Indonesia experience many moderately severe earthquakes each year.

TABLE 2.1

Richter Scale of Earthquake Magnitude

Magnitude[a]	Characteristic effects of earthquakes occurring near the earth's surface[b]
0	not felt
1	not felt
2	not felt
3	felt by some
4	windows rattle
5	windows break
6	poorly constructed buildings destroyed; others damaged
7	widespread damage; steel bends
8	nearly total damage
9	total destruction

[a]Since the Richter scale is logarithmic, each increment of a whole number signifies a 10-fold increase in magnitude. Thus a magnitude 5 earthquake produces a registered effect upon the seismograph 10 times greater than a magnitude 4 earthquake, and a magnitude 6 earthquake would be 10 times 10 or 100 times greater than a 4.

[b]The damage levels of earthquakes are presented in terms of the consequences that are felt or seen in populated areas; the recorded seismic wave heights remain the same whether or not there are structures on the surface to be damaged. The actual impact of earthquakes upon humans varies not only with the severity of the quake, and such secondary effects as tsunamis or landslides, but also with the density of population in the area affected.

(a)

(b)

Figure 2.14

Two views of earthquake damage. (a) A 1964 earthquake
destroyed bridges on Seward Highway, 50 miles south of
Anchorage. (b) The Mexican earthquake of 1985 destroyed many
central Mexico City buildings.

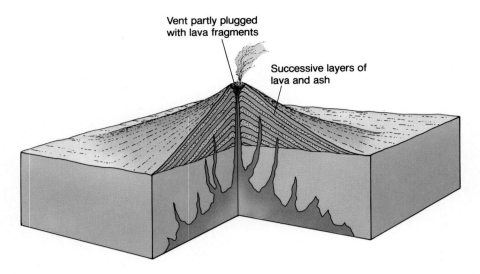

Vent partly plugged
with lava fragments

Successive layers of
lava and ash

Figure 2.15
Sudden decompression of gases contained within lavas results in
explosions of rock material to form ashes and cinders. Composite
volcanoes, such as the one diagrammed, are composed of
alternate layers of solidified lava and of ash and cinders.

Volcanism

The second tectonic force is volcano building or **volcanism.** The most likely places through which molten materials can move toward the surface are at the intersections of plates, but other fault-weakened zones are also subject to volcanic activity.

If sufficient internal pressure forces the magma upward, weaknesses in the crust, or faults, enable molten materials to reach the surface (Figure 2.15). The major volcanic belt of the world coincides with the major earthquake and fault zones. This is the zone of convergence between two plates. A second zone of volcanic activity is at diverging plate boundaries, such as in the center of the Atlantic Ocean. Molten material can either flow smoothly out of a crater or be shot into the air with great force. Some relatively quiet volcanoes have long gentle slopes indicative of smooth flow, while explosive volcanoes have steep sides. Steam and gases are constantly escaping from the nearly 300 active volcanoes in the world today. When pressure builds, a crater can become a boiling cauldron with steam, gas, lava, and ash billowing out (Figure 2.16). On Mount St. Helens in 1980, a large bulge had formed on the north slope of the mountain. An earthquake started the explosion that reduced the elevation of the mountain by over 1000 feet (300 m), completely devastating an area of about 150 square miles (400 km²) and dropping about 4 inches (10 cm) of ash on most of Washington and parts of Idaho and Montana.

In many cases the forces beneath the crust are not great enough to allow the magma to reach the surface. In these instances the magma hardens into a variety of underground formations of igneous rock that do little to affect surface landform features. However, gradational forces may erode overlying rock, so that the igneous rock, which usually is hard and resists erosion, becomes a surface feature. The Palisades, a rocky ridge facing New York City from the west, is such a landform. On other occasions, a weakness below the earth's surface may allow the growth of a mass of magma that is denied exit to the surface because of firm overlying rock. Through the pressure it exerts, however, the magmatic intrusion may still buckle, bubble, or break the surface rocks, and domes of considerable size may develop, such as the Black Hills of South Dakota (Figure 12.5).

(a)

Dogs Head

Former Summit
2950 m

New Summit
2549 m

Forsyth Gl.

Windy Pass

Timberline

(b)

(c)

Figure 2.16

(a) Mount St. Helens, Washington, before it erupted. (Mount Rainier is in the background.) (b) Aftermath of the May 18, 1980 eruption of Mount St. Helens. (c) Lava exploding from a volcano in Hawaii.

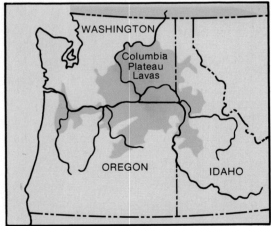

Figure 2.17
Fluid basaltic lavas created the Columbia Plateau, covering an area of 50,000 square miles (130,000 km²). Some individual flows were more than 300 feet (100 m) thick and spread up to 40 miles (60 km) from their original fissures.

Evidence from the past shows that sometimes lava has flowed through fissures or fractures without volcanoes forming. These oozing lava flows have covered large areas to great depth. The Deccan Plateau of India and the Columbia Plateau of the Pacific Northwest of the United States are examples of this type of process (Figure 2.17).

Gradational Processes

Gradational processes are responsible for the reduction of the land surface. If a land surface where a mountain once stood is now a low, flat plain, then gradational processes have been at work. The material that has been worn, scraped, or blown away is deposited in new places and, as a result, new landforms are created. In terms of geologic time, the Himalayas are a recent phenomenon; gradational processes, although active there just as they are active on all land surfaces, have not yet had time to reduce the huge mountains.

There are three kinds of gradational processes: *weathering,* the effect of the force of *gravity,* and *erosion.* Weathering processes, both mechanical and chemical, prepare bits of rock for their role in the creation of soils and for their transfer to new sites by means of gravity or erosion. The force of gravity acts to move any higher-lying rock material not held tightly in place, and the agents of running water, moving ice, wind, waves, and currents erode and carry the loose materials to other areas, where landforms are created or changed.

Mechanical Weathering

Mechanical weathering is the physical disintegration of earth materials at or near the surface. A number of processes cause mechanical weathering, the three most important being frost action, the development of salt crystals, and root action.

If the water that soaks into a rock (between particles or along joints) freezes, ice crystals grow and exert pressure on the rock. If the process is repeated—freezing, thawing, freezing, thawing, and so on—there is a tendency for the rock to begin to disintegrate. Salt crystals act similarly in dry climates, where groundwater is drawn to the surface by *capillary* action (water rising because of surface tension). This action is similar to the process in plants whereby liquid plant nutrients move upward through the stem and leaf system. Evaporation leaves behind salt crystals, which help disintegrate rocks. Roots of trees and other plants may also find their way into rock joints and, as they grow, break and disintegrate rock. These are all mechanial processes because they are physical in nature and do not alter the chemical composition of the material upon which they act.

Chemical Weathering

A number of **chemical weathering** processes cause rock to decompose rather than to disintegrate—that is, to separate into component parts by chemical reaction rather than to fragment. The three most important are oxidation, hydrolysis, and carbonation. Because each of these processes depends on the availability of water, there is less chemical weathering in dry and cold areas than in moist and warm ones. Chemical reactions are faster in the presence of moisture and heat.

Oxidation occurs when oxygen combines with rock minerals such as iron to form oxides; as a result, some rock areas in contact with the oxygen begin to decompose. Decomposition also results when water comes into contact with certain rock minerals such as aluminosilicates. The chemical change that occurs is called *hydrolysis*. When carbon dioxide gas from the atmosphere dissolves in water, a weak carbonic acid forms. The action of the acid, called *carbonation*, is particularly evident on limestone; the calcium bicarbonate salt that is created is readily dissolved and removed by ground and surface water.

Weathering, either mechanical or chemical, does not itself create distinctive landforms. Nevertheless, it acts to prepare rock particles for erosion and for the creation of soil. After the weathering process decomposes rock, the force of gravity and the erosional forces of running water, wind, and moving ice can carry the weathered material to new locations.

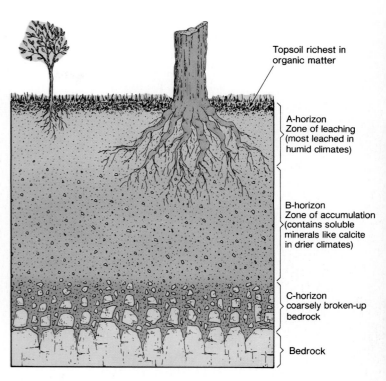

Figure 2.18
A representative soil profile, showing the various soil horizons.

Topsoil richest in organic matter

A-horizon
Zone of leaching
(most leached in humid climates)

B-horizon
Zone of accumulation
(contains soluble minerals like calcite in drier climates)

C-horizon
coarsely broken-up bedrock

Bedrock

Soils

Mechanical and chemical weathering create soil. Soil is the thin layer of fine material resting on bedrock. When digging down into the soil, one can recognize a series of layers of different colors. Soil scientists identify three sections or **soil horizons.** The top or *A-horizon* is usually the darkest; it may contain organic matter such as decayed plant leaves, twigs, and animal remains, and clays and sand grains. The second level, the *B-horizon,* is much lighter in color. It is made up largely of clay with small and large bits of minerals, and only small amounts of organic matter. The lowest level, the *C-horizon,* is the upper, broken surface of bedrock mixed with clay. The older the soil and the warmer and wetter the climate, the deeper the C-horizon will be, and the more discernible will be the three soil horizons (Figure 2.18). Soil type is much more a function of the climate of the region in which it occurs than it is of the kind of bedrock below it. Temperature and rainfall act on the minerals in conjunction with the decay of the overriding vegetation to form soils. The process of soil formation takes many hundreds and sometimes thousands of years, but duststorms and poor conservation practices by farmers and land managers can deplete an area of its crucial A-horizon in just a few years. We will return to the topic of soils in Chapter 3.

(a)

(b)

Figure 2.19
Examples of mechanical weathering. (a) Rockfall from this butte
in the Grand Canyon has begun to create a talus slope.
(b) Creeping soil has caused trees to tilt.

Gravity

The force of *gravity*—that is, the attraction of the earth's
mass for bodies at or near its surface—is constantly
pulling on all materials. Small particles or huge boulders, if not held back by bedrock or other stable material,
will fall down slopes. Spectacular acts of gravity include
avalanches and landslides, but more widespread are such
less noticeable movements as soil creep and the flow of
mud down hillsides.

Especially in dry areas, a common but very dramatic landform created by the accumulation of rock particles at the base of hills and mountains is the **talus slope**
(Figure 2.19). As pebbles, particles of rock, or even larger
stones break away from the exposed bedrock on a moun-
tainside because of weathering, they fall by force of gravity
and accumulate, producing large conelike landforms. The
larger rocks travel farther than the fine-grained sand particles, which remain near the top of the slope.

Erosional Agents and Deposition

Erosional agents such as wind and water carve already
existing landforms into new shapes. The material that has
been worn, scraped, or blown away is deposited in new
places, creating new landforms. Each erosional agent is
associated with a distinctive set of landforms.

Running Water

There is no more important erosional agent than running
water. Water, whether flowing across land surfaces or in
stream channels, plays an enormous role in wearing down
and building up landforms.

The ability of running water to erode depends on
the amount of precipitation, on the length and steepness
of the slope, and on the kind of rock and vegetative cover.
The steeper the slope, the faster the flow and, of course,

Figure 2.20
Gullying can result from poor farming techniques, including
overgrazing by livestock or many years of continuous row crops.
Surface runoff removes topsoil easily when vegetation is too thin
to protect it.

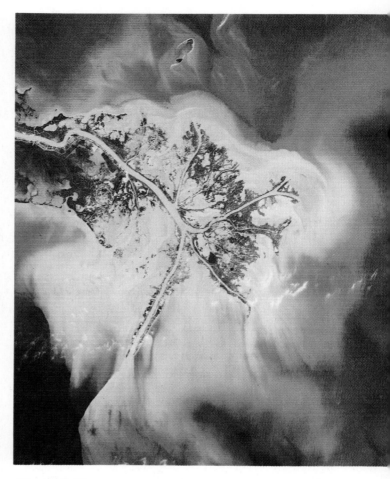

Figure 2.21
Landsat image of the delta of the Mississippi River. Notice the
ongoing deposition of silt and the effect that both river and gulf
currents have on the movement of the silt.

the more rapid the erosion. If the vegetative cover that
slows the flow of water is removed, perhaps by farming
or livestock grazing, erosion can be severe (Figure 2.20).

Even the impact force of precipitation—heavy rain
or hail—can cause erosion. After hard rain dislodges soil,
the surface becomes more compact and further precipitation fails to penetrate the soil. The result is that more
water, prevented from seeping into the ground, becomes
available for surface erosion. Soil and rock particles in
the water are carried to streams, leaving behind gullies
and small channels.

Both the force of water and the particles contained
in the stream are agents of erosion. The particles act as
abrasives, scouring the surface over which they move.
Abrasion, or wearing away, takes place when the particles strike against stream channel walls and along the
stream bed. The force of the current causes large particles such as gravel to slide along the stream bed, grinding
rock on the way. Floods and rapidly moving water are
responsible for dramatic changes in channel size and
configuration, sometimes forming new channels. In cities,
where paved surfaces cover soil that would otherwise
have absorbed or held water, runoff is accentuated so that

nearby rivers and streams rapidly increase in size and velocity after heavy rains. Flash floods and severe erosion
often result.

Small particles, such as clay and silt, are suspended
in water and constitute (together with material dissolved
in the water or dragged along the bottom) the **stream
load.** Rapidly moving floodwaters carry huge loads. As
high water or floodwater recedes and stream velocity decreases, stream sediment no longer remains suspended,
and particles begin to settle. Heavy, coarse materials drop
first; fine particles are carried farther. The decline in velocity and the resulting deposition are especially pronounced and abrupt when streams meet slowly moving
water in bays, oceans, and lakes. Silt and sand accumulate at the intersections, creating **deltas** (Figure 2.21).

A great river such as the Chang Jiang (Yangtze) in
China has a large, growing delta, but less prominent
deltas are found at the mouths of many streams. Until the
recent construction of the Aswan Dam, the huge delta of
the Nile River had continued to grow, but now the silt is
being dropped behind the dam in Lake Nasser.

(a)

(b)

Figure 2.22
(a) V-shaped valley of a rapidly downcutting stream, the Salmon River in Idaho. (b) U-shaped valley of a meandering stream, the Snake River, Wyoming, that is in the process of widening its floodplain.

In plains adjacent to streams, land is sometimes built up by the **deposition** of stream load. If the deposited material is rich, it may be a welcome and necessary part of farming activities, as historically it was in Egypt along the Nile. Should the deposition be composed of sterile sands and boulders, however, formerly fertile bottomland may be destroyed. By drowning crops or inundating inhabited areas, the floods may cause great human and financial loss. More than 900,000 lives were lost in the floods of the Huang He (Yellow River) of China in 1887.

Stream Landscapes

A somewhat misleading concept of landscape evolution conveniently places a given stream in a particular stage in its geologic history. Streams seem to have recognizable stages: *early youth, maturity,* and *old age.* The concept is misleading because each stream and its surrounding landscape have their own history of change that does not necessarily follow a path from youth to old age. A landscape is in a particular state of balance between the uplift and the erosion of land. It does not follow that rapid uplift will be followed by nicely ordered stages of erosion. Recall that uplift and erosion take place simultaneously. At a given location one force may be greater than the other at a given time, but as yet there is no way to predict the "next" stage of landscape evolution.

Perhaps the most important factor in differentiating the effect of streams on landforms is whether the recent (e.g., the last several million years) climate has tended to be humid or arid.

Stream Landscapes in Humid Areas. Perhaps weak surface material or a depression in rock allows the development of a stream channel. In its downhill run in mountainous regions, a stream may flow over precipices, forming *falls* in the process. The steep downhill gradient allows streams to flow rapidly, cutting narrow V-shaped channels in the rock (Figure 2.22a). Under these conditions, the erosional process is greatly accelerated. Over time, the stream may have worn away sufficient rock for the falls to become rapids, and the stream channel becomes incised below the height of the surrounding landforms, as can be seen in the upper reaches of the Delaware, Connecticut, and Tennessee rivers. In humid areas, the effect of stream erosion is to round landforms (Figure 2.22b), though complex landforms reflecting tectonic history and rock type may result as the regional study "The Black Hills Province" (Pages 407–8) makes evident. Streams flowing down moderate gradients tend to carve valleys that are wider than those in mountainous areas. Surrounding hills become rounded, and valleys, called **floodplains,** are U-shaped. Streams work to widen the floodplain. Their courses meander, constantly carving out new erosional channels. The channels left behind as new ones are cut become *oxbow*-shaped lakes, hundreds of which are found in the Mississippi River floodplain. In the nearly flat floodplains, the highest elevations may be the banks of the rivers, where *natural levees* are formed by the filtering and deposition of silt at river edges during flood. The lower Mississippi River is blocked from view from nearby roads by natural levees. Flood waters that breach the levees are particularly disastrous because they spread widely over the lower-lying bottom land.

Stream Landscapes in Arid Areas. A distinction must be made between the results of stream erosion in humid as opposed to arid areas. The lack of vegetation in arid regions greatly increases the erosional force of running water. Water originating in mountainous areas sometimes never reaches the sea if the channel runs through a desert. In fact, stream channels may be empty except during rainy periods, when water rushes down the

Figure 2.23

Alluvial fans, such as this one in Death Valley, California, are built where streams slow down as they flow out on the more level land at the base of the mountain slope. The abrupt change in slope and velocity greatly reduces the stream's capacity to carry its load of coarse material. Deposition occurs, choking the stream channel and diverting the flow of water. With the canyon mouth fixing the head of the alluvial fan, the stream sweeps back and forth, building and extending a broad area of deposition.

Figure 2.24

The resistant caprock of the mesa protects softer underlying strata from downward erosion. Where the caprock is removed, lateral erosion lowers the surface, leaving the mesa as an extensive and pronounced relic of the former higher-lying landscape. Here the mesa walls have a sloping accumulation of talus.

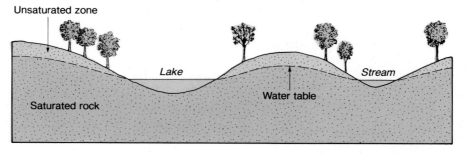

Figure 2.25

The groundwater table generally follows surface contours, but in subdued fashion. Water flows slowly through the saturated rock, emerging at earth depressions that are lower than the level of the water table. During a drought, the table is lowered and the stream channel becomes dry.

hillsides to collect and form temporary lakes called *playas*. In the process, **alluvium** (sand and mud) builds up in the lakes and at lower elevations, and *alluvial fans* are formed along hillsides (Figure 2.23). The fan is produced by the scattering of silt, sand, and gravel outward as the stream reaches the lowlands at the base of the slope it traverses. If the process has gone on over a particularly long time, alluvial deposits may bury the eroded mountain masses. In desert regions in Nevada, Arizona, and California, it is not unusual to observe partially buried mountains poking through alluvium.

Because the streams in arid areas have only a temporary existence, their erosional power is less certain than that of the permanent streams of humid areas. In some instances, they may barely mark the landscape; in other cases, swiftly moving water may carve deep, straight-sided **arroyos.** Water may rush onto an alluvial plain in a complicated pattern resembling a multistrand braid, leaving in its wake an alluvial fan. The channels resulting from this rush of water are called *washes*. The erosional power of unrestricted running water in arid regions is dramatically illustrated by the steep-walled configuration of **buttes** and **mesas** (large buttes) such as those shown in Figure 2.24. These are remnants of a larger plateau, most of which have been eroded by water rushing over the former plateau.

Groundwater

Some of the water that falls as rain and snow sinks underground into the pores and cracks in rocks and soil—not in the form of an underground pond or lake, but simply as very wet subsurface material. When it accumulates, a zone of saturation forms. The upper level of this zone is the **water table;** below it, the soils and rocks are saturated with water (Figure 2.25). Groundwater

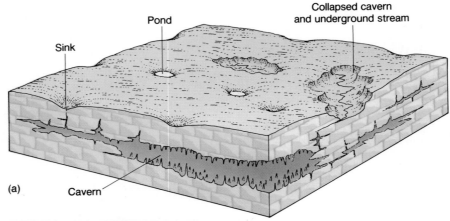

(a)

Sink

Pond

Collapsed cavern
and underground stream

Cavern

(b)

Figure 2.26

Limestone erodes easily in the presence of water. (a) *Karst* topography, such as that shown here, occurs in humid areas where limestone in flat beds is at the surface. (b) This satellite photo of central Florida shows the many round lakes formed in the sinkholes of a karst landscape.

moves constantly but very slowly (only a few feet or inches a day). Most remains underground, seeking the lowest level; however, when the surface of the land dips below the water table, ponds, lakes, and marshes form. Some water finds its way to the surface by capillary action in the ground or in vegetation. Groundwater, particularly when combined with carbon dioxide, dissolves soluble materials by a chemical process called *solution.*

Although groundwater tends to decompose many types of rocks, its effect on limestone is most spectacular. Many of the great caves of the world have been created by the underground movement of water through limestone regions. Water sinking through the overlying rock leaves carbonate deposits as it drips. The deposits hang from the cave roofs *(stalactites)* and build upward from the cave floors *(stalagmites).* In some areas, the uneven effect of groundwater erosion of limestone leaves a landscape pockmarked by a series of *sinkholes,* surface depressions in an area of plentiful caverns. **Karst topography** refers to a large limestone region marked by sinkholes, caverns, and underground streams (Figure 2.26). This type of topography gets its name from a region near the Adriatic Sea at the Italy–Yugoslavia border. Central Florida, a karst area, has suffered considerable damage from the creation and widening of sinkholes. The Mammoth Cave region of Kentucky, another karst area, has tens of miles of interconnected limestone caves.

Glaciers

Another agent of erosion and deposition is glaciers. Although they are much less extensive today, glaciers covered a large part of the earth's land area as recently as 8000 to 15,000 years ago, during the *Pleistocene* geologic epoch. Many landforms were created by the erosional or depositional effects of glaciers.

Glaciers form only in very cold places with short or nonexistent summers, where annual snowfall exceeds annual snowmelt and evaporation. The weight of the snow causes it to compact at the base and form ice. When the snowfall reaches a thickness of several hundred feet, the ice at the bottom becomes like plastic and begins to move slowly. A **glacier,** then, is a large body of ice moving slowly down a slope or spreading outward on a land surface (Figure 2.27). Some glaciers appear to be stationary simply because the melting and evaporation at the glacier's edge equals the speed of the ice advance. Glaciers can, however, move as much as several feet per day.

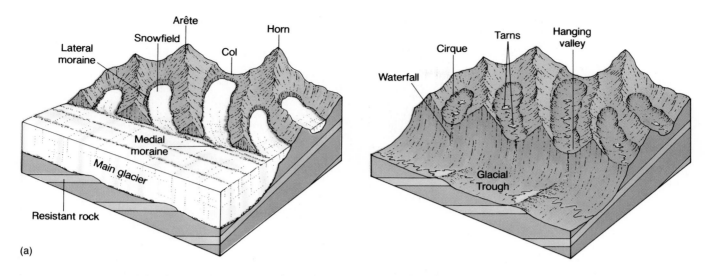

(a)

(b)

Figure 2.27

The evolution of alpine glacial landforms. (a) Frost shattering
and ice movement carve *cirques,* the irregular bottom of which
may contain lakes (*tarns*) after the glacier melts. Where cirque
walls adjoin from opposite sides, knife-like ridges called *arêtes*
are formed, interrupted by overeroded passes or *cols.* The
intersection of three or more arêtes creates a pointed peak, or
horn. Rock debris falling from cirque walls is carried along by
the moving ice. *Lateral moraines* form between the ice and the
valley walls; *medial moraines* mark the union of such debris
where two valley glaciers join. (b) Tarns in glacial cirques.

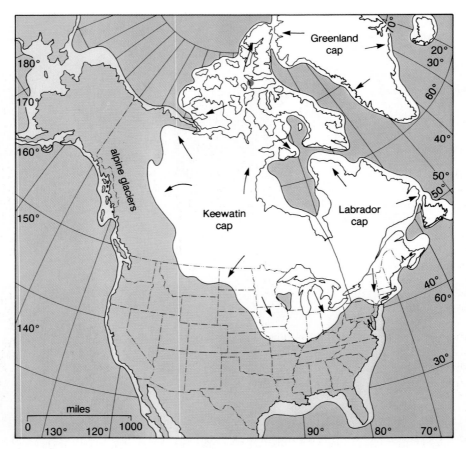

Figure 2.28

Farthest extent of glaciation in the northern hemisphere. Separate
centers of snow accumulation and ice formation developed.
Large lakes were created between the western mountains of
North America and the advancing ice front; to the south, huge
rivers carried away glacial meltwaters. Since so much water was
trapped as ice on the land, sea levels were lowered and
continental margins were extended.

Most theories of the formation of the great Pleistocene continental glaciers concern climatic cooling of
the earth. Perhaps some combination of the following
theories explains the evolution of glaciers. The first theory
attributes the ice ages to periods when there may have
been excessive amounts of volcanic dust in the atmosphere. The argument is that the dust, by reducing the
amount of solar energy reaching the earth, effectively
lowered temperatures at the surface. A second theory attributes the ice ages to known changes in the shape, the
tilt, and the seasonal positions of the earth's orbit around
the sun over the last million years. Such changes alter
the amount of solar radiation received by the earth and
its distribution over the earth. A recent theory suggests
that when large continental plates drift over polar regions, temperatures on earth become more extreme, and
as a result induce the development of glaciers. This
theory, of course, cannot explain the most recent ice ages.

Today, continental-size glaciers exist only on Antarctica and Greenland, but mountain glaciers are found
in many parts of the world. About 10% of the earth's land
area is under ice. During the most recent advance of ice,
the continental ice of Greenland was part of an enormous glacier that covered nearly all of Canada (Figure
2.28) and the northernmost portions of the United States.
The giant glacier reached thicknesses of 10,000 feet (its
depth in Greenland today), enveloping entire mountain
systems.

The weight of glaciers breaks up underlying rock
and prepares the rock for transportation by the moving
mass of ice. Consequently, glaciers alter landforms by
weathering and erosion. Glaciers scour the land as they
move, leaving surface scratches on the rocks that remain.
Much of eastern Canada has been scoured by glaciers that
left little soil but many ice-gouged lakes and streams. The
erosional forms created by glacial scourings have a variety of names. A *glacial trough* is a deep, U-shaped valley
visible only after the glacier has receded (Figure 2.27).
If the valley is today below sea level, as in Norway or
British Columbia, **fiords,** or arms of the sea, are formed.

Permafrost

In 1577, on his second voyage to the New World in search of the Northwest Passage, Sir Martin Frobisher reported finding ground in the far north that was frozen to depths of "four or five fathoms, even in summer," and that the frozen condition "so combineth the stones together that scarcely instruments with great force can unknit them." Permanently frozen ground, now termed **permafrost,** underlies perhaps a fifth of the earth's land surface. In the lands surrounding the Arctic Ocean, its maximum thickness has been reported in thousands of feet.

For almost 300 years after Frobisher's discovery, little attention was paid to this frost phenomenon. But in the 19th and 20th centuries, during the building of the Trans-Siberian railroad, the construction of buildings during the gold rush in Alaska and the Yukon Territory, and

the development of the oil pipeline that now connects Prudhoe Bay in northern Alaska with Valdez in southern Alaska, attention has focused on the unique nature of permafrost.

Uncontrolled construction activities thaw permafrost. The ground then becomes unstable and susceptible to soil movement and landslides, ground subsidence, and frost heaving. Scientists have found that only special construction techniques causing the least possible disturbance of the frozen ground will

avoid thawing and the resulting effects on the buildings. The Alaska pipeline was built above the earth's surface in order to reduce the likelihood that the relatively warm oil would disturb the permafrost and thus destroy the pipeline. It was necessary to preserve the insulating value of the ground surface as much as possible, and to this end the vegetation mat was not removed. In addition, a coarse gravel fill was added along the surface below the pipeline.

Glaciers create landforms when they deposit the debris they have transported. These deposits, called **glacial till,** consist of rocks, pebbles, and silt. As the great tongues of ice move forward, debris accumulates in parts of the glacier. The ice that scours valley walls and the ice at the tip of the advancing tongue are particularly filled with debris (Figure 2.29). As a glacier melts, it leaves behind hills of glacial till, called **moraines.**

Many other landforms have been formed by glaciers. The most important is the **outwash plain,** a gently sloping area in front of a melting glacier. The melting along a broad front sends thousands of small streams running out from the glacier in braided fashion, streams that deposit neatly stratified glacial till. Outwash plains, which are essentially great alluvial fans, cover a wide area and provide rich new parent material for soil formation. Most of the midwestern part of the United States owes some of its soil fertility to relatively recent glacial deposition.

Figure 2.29
Arêtes and lateral and medial moraines.

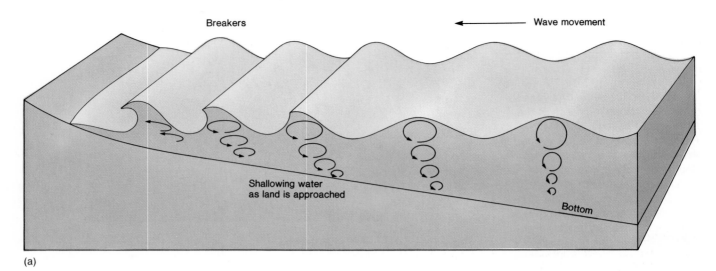

Breakers

Wave movement

Shallowing water
as land is approached

Bottom

(a)

(b)

Figure 2.30

Formation of waves and breakers. (a) As the offshore swell
approaches the gently sloping beach bottom, sharp-crested waves
form, build up to a steep wall of water, and break forward in
plunging surf. (b) Evenly spaced breakers form as successive
waves touch bottom along a regularly sloping shore.

Before the end of the most recent ice age, there
were at least three previous glacial advances during the
roughly 2 million years of the Pleistocene period. Firm
evidence is not available on whether we have emerged
from the cycle of ice advance and retreat. For the first half
of this century, the world's glaciers were melting faster
than they were building up, but factors concerning the
earth's changing temperature, which are discussed in
Chapter 3, must be considered before it is possible to
assess the likelihood of a new ice advance. Current trends
are not clear.

Waves and Currents

While glacial action is intermittent in earth history, ocean
waves breaking on continental coasts and on islands is
unceasing and causes considerable change in coastal
landforms. As waves reach the shallow water close to
shore, they are forced to become higher until a breaker
is formed (Figure 2.30). The uprush of water not only
carries sand for deposition but also erodes the landforms
at the coast, while the backwash carries the eroded material away. This type of action results in different kinds
of landforms, depending on conditions.

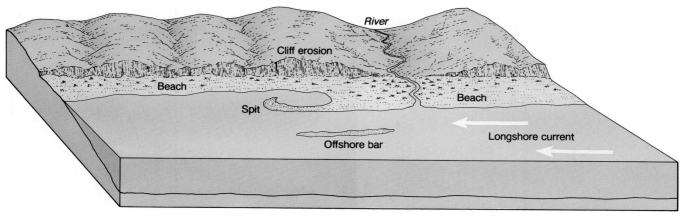

Figure 2.31
The cliffs behind the shore are eroded by waves during storms and high water. Sediment from the cliff and the river forms the beach deposit; the longshore current moves some sediment downcurrent to form a *spit*. Offshore *sandbars* are created from material removed from the beach and deposited by retreating waves.

If the land at the coast is well above sea level, the wave action causes cliffs to form. The cliffs then erode at a rate dependent on the resistance of the rock to the constant assault from the salt water. During storms, a great deal of power is released by the forward thrust of the waves, and much weathering and erosion take place. Landslides are a hazard during coastal storms; they occur particularly in areas of weak sedimentary rock or glacial till.

Beaches are formed by the deposition of sand grains contained in the water. The sand originates from the vast amount of coastal erosion and from streams (Figure 2.31). **Longshore currents,** which move roughly parallel to the shore, transport the sand, forming beaches and *sandspits*. The more sheltered the area, the better the chance for a beach to build. The backwash of waves, however, takes sand away from the beaches if there is no longshore current. As a result, *sandbars* develop a short distance away from the shoreline. If the sandbars become large enough, they eventually close off the shore, creating a new coastline that encloses *lagoons* or *inlets*. *Salt marshes* very often develop in and around these areas. Figure 2.32 shows one kind of area partially carved by waves.

Coral reefs, made not from sand but from coral organisms growing in shallow tropical water, are formed by the secretion of lime in the presence of warm water and sunlight. Reefs, consisting of millions of colorful skeletons, develop short distances offshore. Off the coast of Australia lies the most famous coral reef, the Great Barrier Reef. **Atolls,** found in the south Pacific, are nearly circular reefs formed in shallow water.

Figure 2.32
Sea cliffs, headlands, embayments, and offshore erosional remnants are typical of cliffed shorelines such as this one along the coast of Oregon. Beaches occur only near the mouths of rivers or in embayments.

Figure 2.33
The prevailing wind from the left has given these transverse dunes a characteristic gentle windward slope and a steep, irregular leeward slope.

Wind

In humid areas, the effect of wind is confined mainly to sandy beach areas, but in dry climates, wind is a powerful agent of weathering, erosion, and deposition. Here the limited vegetation leaves exposed particles of sand, clay, and silt subject to movement by wind. Thus many of the sculptured features found in dry areas result from mechanical weathering, that is, from the abrasive action of sand and dust particles as they are blown against rock surfaces. Sand and dust storms in a drought-stricken farm area may make it unusable for agriculture. Inhabitants of Oklahoma, Texas, and Colorado suffered greatly in the 1930s when their farmlands became the "dust bowl" of the United States.

Several types of landforms are produced by wind-driven sand (Figure 2.33). Although sandy deserts are much less common than gravelly deserts, their characteristic landforms are better known. Most of the Sahara,

the Gobi, and the western United States deserts are covered not with sand but with rocks, pebbles, and gravel. Each also has a small portion (and the Saudi Arabian desert, a large area) covered with sand blown by wind into a series of waves or **dunes.** Unless vegetation stabilizes them, the dunes move as sand is blown from their windward faces onto and over their crests. One of the most distinctive sand desert dunes is the crescent-shaped *barchan.* Along seacoasts and inland lakeshores, in both wet and dry climates, wind can create sand ridges that may reach a height of 300 feet. Sometimes coastal communities and farmlands are threatened or destroyed by moving sand.

Another kind of wind-deposited material, silty in texture and pale yellow or buff in color, is called **loess.** Usually encountered in mid-latitude westerly wind belts,

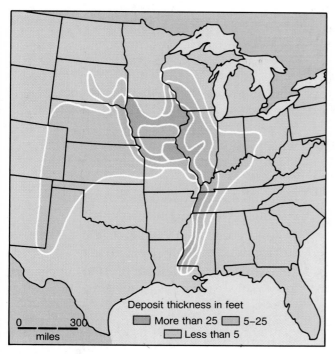

Figure 2.34

Location of wind-blown silt deposits, including *loess,* in the United States. The thicker layers, found in the upper Mississippi valley, are associated with the wind movement of glacial debris. Farther west, in the Great Plains, wind-deposited materials are sandy in texture, not loessal.

it covers extensive areas in the United States (Figure 2.34), central Europe, central Asia, and Argentina. It has its greatest development in northern China, where it covers hundreds of thousands of square miles, often to depths of more than 100 feet (30 m). The wind-borne origin of loess is confirmed by its typical occurrence downwind from extensive desert areas, though major deposits are assumed to have resulted from wind erosion of nonvegetated sediment deposited by meltwater from retreating glaciers. Because rich soils usually form from loess deposits, if climatic circumstances are appropriate, these areas are among the most productive agricultural lands in the world.

Conclusion

Landforms are the most visible elements on the stage upon which the drama of human life is played. We have seen them to be a setting that is continuously resculpted by the forces that gave them shape. In this chapter we emphasized those factors that are responsible for landforms, but only by implication did we indicate how humans interact with them. A fundamental knowledge of landforms and climate—the earth and the sky—is necessary to understand how we cope with or are affected by the natural world in which we live. Landforms and the migratory plates on which they develop are both favorably and adversely affected by the actions of humans. A few examples of how humans have changed the face of the earth are the construction of great dams that alter river gradients and impinge upon the processes of erosion; the filling of tidal marshes and the dredging of reefs that destroy the natural coastlines; and the conversion of the North African Sahel region, a fertile grassland, into a desert following human disturbance of its vegetation. These processes are not remote, they are the very stuff of life.

The landscape affecting and affected by human action is more than landforms, moving continents, and occasional earthquakes. Except at times of natural disasters, these elements of the physical world are usually quiet, accepted background. More immediately affecting our lives are the great patterns of climate that help define the limits of the economically possible; the daily changes of weather that affect the success of a picnic or a crop yield; and the patterns of vegetation and soils that are related to the realms of climate but used, altered, and perhaps destroyed by humans. In the next chapter, we turn our attention to these elements of the natural environment as the necessary prelude to the consideration of the human side of the human/earth-surface equation of geography.

Key Words

alluvium 75	faults 56	longshore current 81	Richter scale 64
arroyo 75	fiord 78	mechanical	sedimentary rocks 61
atoll 81	floodplain 74	weathering 71	soil horizons 71
asthenosphere 53	folds 61	mesa 75	stream load 73
butte 75	glacial till 79	metamorphic rocks 61	subduction 59
chemical weathering 71	glacier 76	mineral 60	talus slope 72
continental drift 55	gradational processes 70	moraine 79	tsunami 66
delta 73	igneous rocks 60	outwash plain 79	volcanism 68
deposition 74	karst topography 76	Pangaea 55	water table 75
dunes 82	lithosphere 53	permafrost 79	
erosional agents 72	loess 82	plate tectonics 52	

For Review

1. What evidence makes the theory of plate tectonics plausible?

2. What is the meaning of and the name of the process that occurs when two plates collide?

3. In what ways may rocks be classified? List three classes of rocks according to their origin. In what ways can they be distinguished from one another?

4. Explain how gradation and volcanism create new landforms.

5. Draw a diagram showing the varieties of ways faults may occur.

6. With what earth movements are earthquakes associated? What is a tsunami, and how does it develop?

7. What is the distinction between mechanical and chemical weathering? Is weathering responsible for landform creation? In what way do glaciers engage in mechanical weathering?

8. Explain the origin of the various landforms one usually finds in desert environments.

9. How do glaciers form? What landscape characteristics are associated with glacial erosion? With glacial deposition?

10. What landform features can you identify on Figure 1.14? What were the agents of their creation?

11. What theories are offered to explain the cooling of the earth's temperature?

12. How is groundwater erosion different from surface water erosion?

13. How are the processes that bring about change due to waves and currents related to the processes that bring about change by the force of wind?

14. What processes account for the landform features of the area in which you live?

Suggested Readings

Butzer, K. W. *Geomorphology from the Earth*. New York: Harper and Row, 1976.

Gabler, R., et al. *Essentials of Physical Geography*. 2d ed. Philadelphia: Saunders Publishing Co., 1982.

Goudie, A. S. *Environmental Change*. New York: Oxford University Press, 1977.

Marsh, William M. *Earthscape: A Physical Geography*. New York: Wiley, 1987.

Muller, Robert A., and Theodore M. Oberlander. *Physical Geography Today*. 3d ed. New York: Random House, 1984.

Press, Frank, and Raymond Siever. *Earth*. 3d ed. San Francisco: Freeman, 1982.

Progress in Physical Geography. A journal containing essays on new directions in the study of the physical environment. Cambridge University Press, New York.

Ritter, Dale F. *Process Geomorphology*. 2d ed. Dubuque, Iowa: Wm. C. Brown, 1986.

Strahler, Arthur N., and Alan H. Strahler. *Elements of Physical Geography*. 3d ed. New York: Wiley, 1987.

3 Physical Geography: Weather and Climate

Like so many before, it first appeared on a hot, humid day in early June 1985 as a tiny blip on Indian radar screens. This one was recognized as a small circulation pattern developing in the Bay of Bengal south of the crowded country of Bangladesh. By late afternoon it grew large enough to be called a tropical depression, the possible start of a major storm system. The Indian meteorologists warned the Bangladesh government that a killer storm, a hurricane (called a typhoon or cyclone in Asia), was headed for the densely settled islands at the mouth of the Ganges River (Figure 3.1). Hourly warnings were broadcast on radio and television that all coastal residents must seek shelter immediately.

Unfortunately, most of the people of Bangladesh were too poor to own radios, telephones, or television and did not hear the warning. By late that night, what was to become the most damaging storm in Bangladesh since 1970 came ashore. The afflicted area, an already-flat landscape, was stripped of all huts, houses, and other buildings. People were swept away by the waves. At week's end, it was estimated that about 20,000 people had died, thousands were injured, and a quarter million were left homeless. Those who survived had clung to floating bamboo rooftops or pieces of driftwood. Many eventually died because supplies and medicine could not be transported to people marooned on islands; the storm had destroyed roads and bridges. There was no shelter, no clothing, no food.

The power of hurricanes is concentrated in a narrow path. Whether such meterological events occur in Asia or in North America, they do great damage. The lives of all of those in the paths of storms are affected.

Tropical storms are an extreme type of weather. Most people are "weather watchers": they watch the TV forecasts with great interest and plan their lives around weather events. In this chapter, we review that subsection of physical geography concerned with weather and climate. It deals with normal, patterned phenomena out of which such an abnormality as the Bangladesh hurricane occasionally emerges.

A weather forecaster describes current conditions for a limited area, such as a metropolitan area, and predicts tomorrow's and perhaps long-range weather conditions. If the elements that make up the **weather,** such as temperature, wind, and precipitation, are recorded at specified moments in time, such as every hour, an inventory of weather conditions can be developed. By finding trends in data that have been gathered over an extended period of time, we can speak about typical conditions. These characteristic circumstances describe the **climate** of a region. Weather is a moment's view of the lower atmosphere, while climate is a description of typical weather conditions in an area or at a place over a long period of time. Geographers analyze the differences in weather and climate from place to place in order to understand how climatic elements affect human occupance of the earth.

In geography, we are particularly interested in the physical environment that surrounds us. That is why the **troposphere,** the lowest layer of the earth's atmosphere, attracts our attention. This layer, extending about 6 miles (10 km) above the ground, contains virtually all of the air, clouds, and precipitation of the earth.

In this chapter, we try to answer the questions that are usually raised about the characteristics of the lower atmosphere. By discussing the answers to these questions from the standpoint of averages or average variations, we attempt to give a view of the earth's climatic differences as background important for understanding the way people use the land. Climate is a key to understanding, in a broad way, the distribution of world population. People have great difficulty living in areas that are, on the average, very cold, very hot, very dry, or very wet. They are also negatively affected by huge storms or flooding. In this chapter, we discuss the elements that constitute weather conditions, then we describe the various climates of the earth.

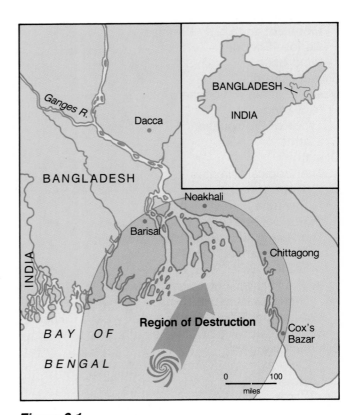

Figure 3.1

Path of the 1985 hurricane (cyclone, typhoon) in the Bay of Bengal.

Air Temperature

Perhaps the most fundamental question about weather is Why do temperatures vary from place to place? The answer requires a discussion of a number of concepts that help to explain the way heat accumulates on the earth's surface.

Energy from the sun, called *solar energy*, is transformed into heat, primarily at the earth's surface and secondarily in the atmosphere. Not every part of the earth or its overlying atmosphere receives the same amount of solar energy. At any given place, the amount of available incoming solar radiation, or **insolation,** depends on the intensity and duration of radiation from the sun. These are determined by (1) the angle at which the sun's rays strike the earth, and (2) the number of daylight hours. These two fundamental factors plus the following five modifying variables determine the temperature at any given location: (1) the amount of water vapor in the air, (2) the degree of cloud cover, (3) the nature of the surface of the earth, (4) the elevation above sea level, and (5) the degree of air movement. Let us briefly look at the circumstances that bring these factors into play.

Earth Inclination

The earth, as Figure 3.2 indicates, spins on an axis that is not perpendicular to a line connecting the center of the earth and the sun. Rather, the earth's axis is tilted about 23.5° away from the perpendicular; every 24 hours the earth rotates once on that axis, as shown in Figure 3.3. While rotating, the earth is slowly revolving around the sun in a nearly circular annual orbit. The axis of the earth, that is, the imaginary line connecting the North Pole to the South Pole, always remains in the same position. This is to say that there is parallelism of the earth's axis during rotation and revolution (Figure 3.4). If the earth were not tilted from the perpendicular, the solar energy received *at a given latitude* would not vary during the course of the year. The rays of the sun would strike the equator most directly, and as the distance away from the equator became greater, the rays would strike the earth at ever-increasing angles, diminishing the intensity of the energy and giving climates a latitudinal standardization (Figures 3.5 and 3.6).

Because of the inclination, however, the location of highest incidence of incoming solar energy varies during the course of the year. When the northern hemisphere is tilted directly toward the sun, the vertical rays

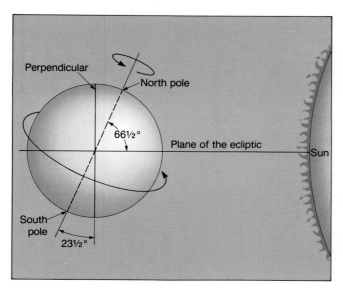

Figure 3.2

The earth spins on an axis tilted about 23½° from the perpendicular or 66½° from the *plane of the ecliptic,* an imaginary plane that contains the lines connecting the center of the earth at all times of the year with the center of the sun.

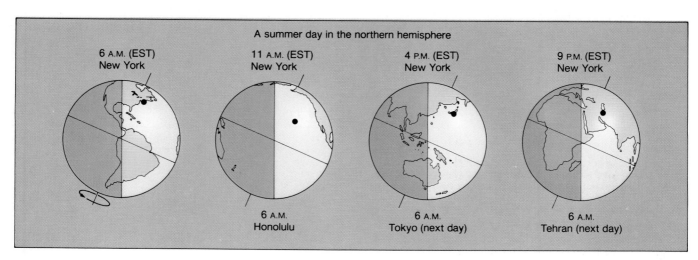

Figure 3.3

The process of the 24-hour rotation of the earth on its axis.

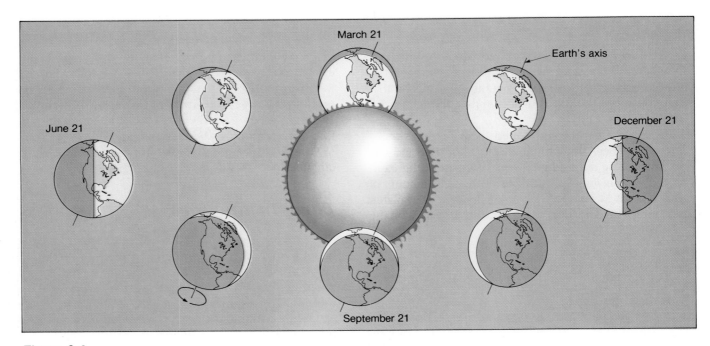

Figure 3.4
The process of the yearly revolution of the earth about the sun.

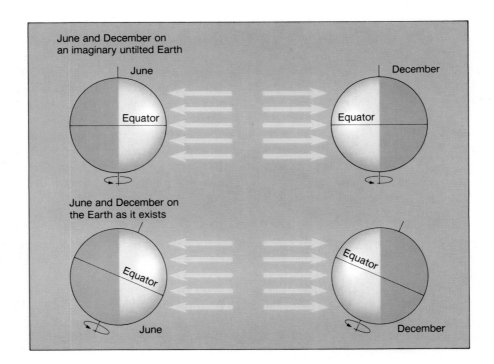

Figure 3.5
Notice in the lower diagram that as the earth revolves, the north polar area in June is bathed in sunshine for 24 hours, while the south polar areas are dark. The most intense of the sun's rays are felt north of the equator in June and south of the equator in December. None of this would be true if the earth's axis were not tilted.

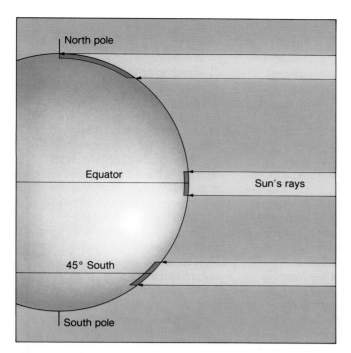

Figure 3.6
Three equal, imaginary rays from the sun are shown striking the earth at different latitudes at the time of the equinox. As distance increases away from the equator, the rays become more diffused, showing how the sun's intensity is diluted in the high latitudes.

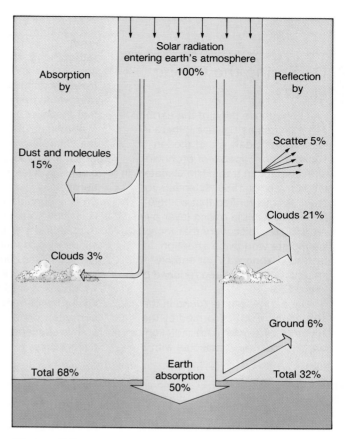

Figure 3.7
Consider the incoming solar radiation as 100%. The portion that is absorbed into the earth (50%) is eventually released to the atmosphere and then reradiated into space. Notice that the outgoing radiation is equal to 100%, showing that there is an energy balance on the earth.

are felt as far north as 23.5° N latitude. This position of the earth occurs about June 21, the summer *solstice* for the northern hemisphere and the winter solstice for the southern. About December 21, when the vertical rays of the sun strike near 23.5° S latitude, it is the beginning of summer in the southern hemisphere and the onset of winter in the northern. During the rest of the year, the position of the earth relative to the sun results in direct rays migrating from about 23.5° N to 23.5° S and back again. About March 21 and September 21 (the spring and autumn *equinoxes*), the vertical rays of the sun strike the equator.

The tilt of the earth also means that the length of the day and night varies during the year. One-half of the earth is always illuminated by the sun, but only at the equator is it light for 12 hours each day of the year. As distance away from the equator becomes greater, the hours of daylight or darkness increase, depending on whether the direct rays of the sun are north or south of the equator. In the summer, daylight increases to the maximum of 24 hours in the summer polar region, and during the same period, night-time finally reaches 24 hours in length in the other polar region.

Because of the 24-hour daylight, it would seem that much solar energy should be available in the summer polar region, but this is not the case. The angle of the sun is so narrow (the sun is so low in the sky) that the solar energy is spread over a wide surface. By contrast, the combination of relatively long days and sun angles

close to 90° makes an enormous amount of energy available to areas in the neighborhood of 15° to 30° north and south latitude during each hemisphere's summer.

Reflection and Reradiation

Much of the insolation potentially receivable is, in fact, sent back to outer space or diffused in the troposphere in a process known as **reflection.** Clouds, which are dense concentrations of suspended tiny water or ice particles, reflect a great deal of energy. Light-colored surfaces, especially snow cover, also reflect large amounts of solar energy.

Energy is lost through reradiation as well as reflection. In the **reradiation** process, the earth acts as a communicator of energy. The shortwave energy that is absorbed into the land and water is returned to the atmosphere in the form of longwave terrestrial radiation (Figure 3.7). On a clear night, when no clouds can block or diffuse the movement, temperatures become lower and lower as the earth reradiates as heat the energy it has received and stored during the course of the day.

Depleting the Ozone Layer

Although we think of the earth's climates as "givens," there is increasing evidence that modern societies are capable of producing effects that can transform climates. In the summer of 1986, scientists for the first time verified that a "hole" had formed in the ozone layer over Antarctica, a discovery that received immediate worldwide attention. In fact, the ozone was not entirely absent, but it had been reduced by some 40%.

The ozone layer is found in the *stratosphere* and is formed by the action of ultraviolet (UV) radiation on oxygen creating a molecule with 3 rather than 2 atoms. It acts as a shield to protect the troposphere from the damaging effects of the UV rays, but current evidence indicates that the layer is being depleted by synthetic gases. The chief culprits are chlorofluorocarbons (CFCs), which are used as coolants, cleansing agents, and propellants for aerosols.

Why should the hole in the ozone layer have occurred over Antarctica? In most parts of the world, horizontal winds tend to keep chemicals in the air well mixed. But circulation patterns are such that the freezing whirlpool of air over Antarctica in winter is not penetrated by air currents from the hotter continents, so it is susceptible to chemical disturbance.

The depletion of the ozone layer is expected to have a number of adverse consequences. Among other things, the Environmental Protection Agency has warned, the increased exposure of UV radiation could be responsible for millions of cases of skin cancer, dysfunctions of the human immune system, and damage to crops and aquatic life. In addition, CFCs are now thought to be playing a major role in creating the "greenhouse" effect, the name given to the heating of the earth's atmosphere caused by the accumulation of gases that prevent the reradiation of heat from earth back into space. Possible effects of a doubling of greenhouse gases and a global warming trend are discussed on Page 140.

Some kinds of earth surface material, especially water, store solar energy more effectively than others. Because water is transparent, solar rays can penetrate a great distance below its surface. If the water body has currents, the heat is distributed even more effectively. On the other hand, land surfaces are opaque, and so all of the energy received from the sun is concentrated at the surface. Land, having more heat available at the surface, reradiates its energy faster than does water. Air is heated by this process of reradiation from the earth and not directly by energy from the sun passing through it. Thus, because land heats and cools much more rapidly than water, the extremes of hot and cold temperatures recorded on earth occur on land and not on the sea.

Temperatures are moderated by the presence of large bodies of water near land areas. Note in Figure 3.8 that coastal areas have lower summer temperatures and higher winter temperatures than places at the same distance from the equator that are not seacoasts. Land areas affected by the moderating influences of water are considered *marine* environments; those areas not affected by nearby water are *continental* environments.

Temperatures vary in a cyclical way over the 24 hours of a day. In the course of a day, as incoming solar energy exceeds the energy lost through reflection and reradiation, temperatures begin to rise. The ground stores some heat, and temperatures continue to rise until the angle of the sun becomes so low that the energy received no longer exceeds that lost by the reflection and reradiation processes. Not all of the heat is lost during the night, but long nights appreciably deplete the stored energy.

The Lapse Rate

We might think that as we moved vertically away from the earth toward the sun, temperatures would increase; such is not true within the troposphere. The earth is a body that absorbs and reradiates heat; therefore temperatures are usually warmest at the earth's surface and lower as elevation increases. Note on Figure 3.9 that this temperature **lapse rate** (the rate of change of temperature with altitude in the troposphere) averages about 3.5° F per 1000 feet (6.4° C per 1000 m). For example, the difference in elevation between Denver and Pikes Peak is about 9000 feet (2700 m), which would normally result in a 32° F (17° C) difference in temperature. Jet planes flying at an altitude of 30,000 feet (9100 m) are moving through air about 100° F (58° C) colder than sea level temperatures.

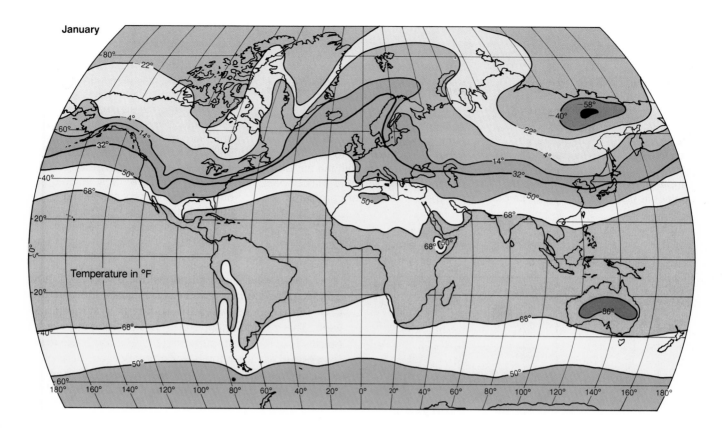

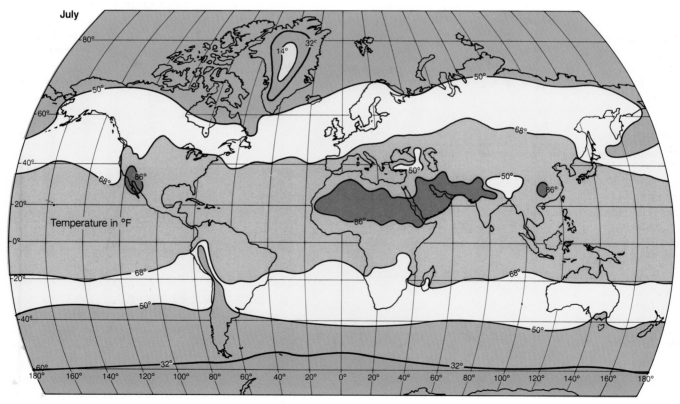

Figure 3.8

At a given latitude, water areas are warmer than land areas in
winter and cooler in summer. *Isotherms* are lines of equal
temperature.

The Donora Tragedy

A heavy fog settled over the valley town of Donora, Pennsylvania, in late October 1948. Stagnant, moisture-filled air was trapped by surrounding hills and by a temperature inversion that held the cooler air, gradually filling with smoke and fumes from the town's zinc works, against the ground under a lid of lighter, warmer upper air. For 5 days the smog increased in concentration; the sulfur dioxide emitted from the zinc works continually converted to deadly sulfur trioxide by contact with the air. Both old and young, with and without past histories of respiratory problems, reported to doctors and hospitals that they had difficulty in breathing and unbearable chest pains. Before rains washed the air clean nearly a week after the smog buildup, 20 were dead and hundreds hospitalized. A normally harmless, water-saturated inversion had been converted to deadly poison by a tragic union of natural weather processes and human activity.

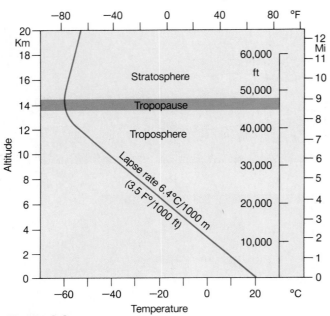

Figure 3.9

The temperature lapse rate under normal conditions.

The normal lapse rate does not always hold, however. Rapid reradiation sometimes causes temperatures to be higher above the earth's surface than at the surface itself. This particular condition, in which air at lower altitudes is cooler than air aloft, is called a **temperature inversion.** An inversion is important because of its effect on air movement. Warm air at the surface, which would normally rise, may be blocked by the still warmer air of the temperature inversion. Thus the surface air is trapped, and if it is filled with automobile exhaust emissions or smoke, a serious smog condition may develop (see Figure 3.10). Because of the nearby mountains, Los Angeles (Figure 3.11) often experiences temperature inversions, causing the sunlight to be reduced to a dull haze.

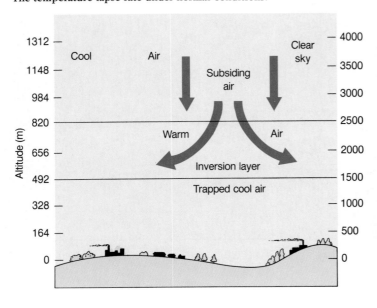

Figure 3.10

Temperature inversion. Subsiding air has forced up the temperature at a height of several hundred feet.

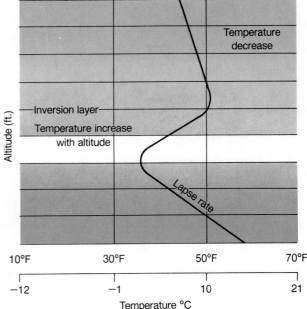

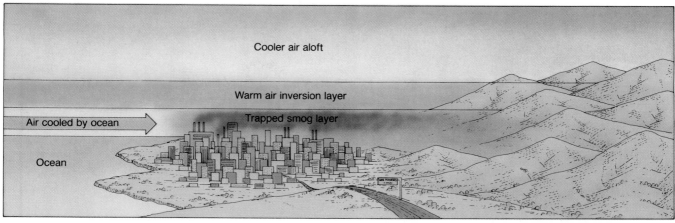

(a)

Figure 3.11

Smog in the Los Angeles area. (a) Below the inversion layer, stagnant air holds increasing amounts of pollutants, caused mainly by automobile exhausts. (b) Afternoon smog in Los Angeles.

Air Pressure and Winds

A second fundamental question about weather and climate concerns **air pressure**. How do differences in air pressure from place to place affect weather conditions? The answer requires that we first explain why there are differences in air pressure.

Air is a gaseous substance whose weight affects air pressure. If it were possible to carve out 1 cubic inch of air at the surface of the earth at sea level and weigh it along with all the other cubic inches of air above it, under normal conditions the total weight would come to about 14.7 pounds (6.67 kg). Actually this is not very heavy when you consider the dimensions of the 1-inch column of air: 1 inch by 1 inch by about 6 miles, or about 220 cubic feet (6.2 m³). If you weighed the air 3 miles (4.8 km) above the earth's surface, however, it would weigh considerably less than 14.7 pounds because there is correspondingly less air above it. So, air is heavier (denser) and air pressure is higher close to the earth's surface. It is a physical law that for equal amounts of cold and hot air, the cold air is heavier. That is why hot-air balloons, filled with lighter (less dense) air, can rise into the atmosphere. A cold morning is characterized by relatively heavy air, but as afternoon temperatures rise, the air becomes lighter.

As early as the 17th century, it was discovered that when a column of mercury is contained within a tube, normal atmospheric pressure at sea level is sufficient to balance the weight of the mercury column to a height of 29.92 inches (76 cm). *Barometers* of varying types are used to record changes in air pressure, and barometric readings in inches (or centimeters) of mercury are a

(b)

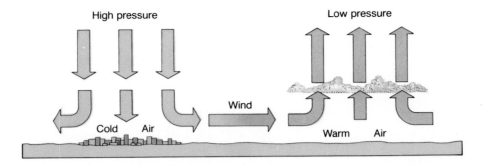

Figure 3.12
If the distance between centers of high pressure and of low
pressure is short, the wind is strong. If the pressure differences
are great, winds are again strong. The strongest winds are
produced by extreme pressure differences over very short
distances.

normal part, along with recorded temperatures, of every
weather report. Since air pressure at a given location
changes as surface heating lightens air, barometers record a drop in atmospheric pressure in warming air and
a rise in pressure when air is cooling.

In order to visualize the effect of air movements on
weather, it is useful to think of air as a liquid made up
of two fluids of different densities (representing light air
and heavy air), such as water and gasoline. If the fluids
are introduced into a tank at the same time, the lighter
liquid will move to the top. This upward movement represents the vertical motion of air. The heavier liquid
spreads out horizontally over the bottom of the tank, so
that it is the same thickness everywhere. This flow represents the horizontal movement of the air, or the wind.
Air attempts to achieve an equilibrium by evening out
the pressure imbalances that result from the heating and
cooling processes just discussed (see Figure 3.13). Air
races from heavy air locations (cold air) to light air locations (warm air). Thus the greater the differences in
air pressure between places, the stronger the wind.

Pressure Gradient Force

Because of differences in the nature of the earth's surface—water, snow cover, dark green forests, cities, and
so on—and the other factors mentioned that affect energy receipt and retention, zones of high and low air
pressure develop. Sometimes these high- and low-pressure zones cover entire continents, but usually they are
considerably smaller—several hundred miles wide—and
within these regions small differences are noted over
short distances. When pressure differences exist between areas, a **pressure gradient** is formed.

In order to balance out the pressure differences that
have developed, air from the heavier high-pressure areas
flows to the low-pressure zones. The heavy air stays close
to the earth's surface as it moves, producing winds, and

helps to speed the movement of warm air upward. The
velocity, or speed, of the wind is in direct proportion to
pressure differences. As depicted in Figure 3.12, winds
are caused by pressure differences that induce airflow
from points of high pressure to points of low pressure.
If distances between high and low pressure are short,
pressure gradients are steep and wind velocities are great.
The equalization process results in more gentle air
movements when different pressure zones are far apart.

The Convection System

As you know, the temperature of the room you are in is
lower near the floor than near the ceiling because warm
air rises and cool air descends. The circulatory motion
of descending cool air and ascending warm air is known
as **convection.** A convectional wind system results from
the horizontal inflow of air replacing warm, rising air.

A good example of a convectional system is land
and sea breezes. Close to a large body of water, the differential daytime heating between land and water is great.
As a result, the warmer air over the land rises vertically,
to be replaced by cooler air from over the sea. At night,
just the opposite occurs. The water is now warmer than
the land, which has reradiated much of its heat, and the
result is a land breeze toward the sea. These two winds
make seashore locations in warm climates particularly
comfortable.

Mountain and Valley Breezes

Gravitational force causes the heavy cool air that accumulates over snow in mountainous areas to descend into
lower valley locations (Figure 3.13). Consequently, valleys can become much colder than the slopes, and there
is a temperature inversion. Slopes are the preferred sites
for agriculture in mountainous regions because the cold
air from the **mountain breezes** can cause freezing conditions in the valleys. In densely settled narrow valleys

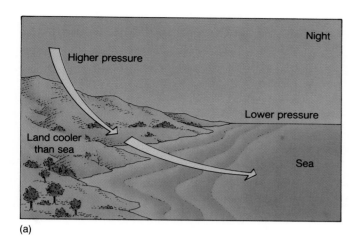

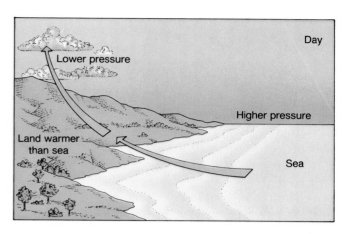

(a)

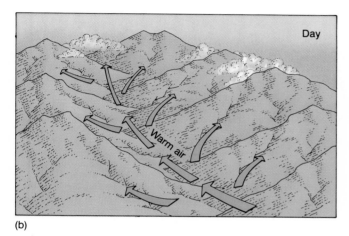

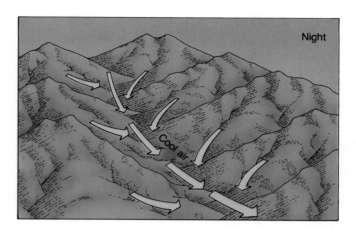

(b)

Figure 3.13
Convectional wind effects due to differential heating and cooling.
(a) Land and sea breezes. (b) Mountain and valley breezes.

where industry is concentrated, air pollution can become particularly dangerous. Mountain breezes usually occur during the night; **valley breezes**—caused by warm air moving up slopes in mountainous regions—are usually a daytime phenomenon. The canyons of southern California are the scene of strong mountain and valley breezes, and in the dry season they become dangerous areas for the windswept spread of brush and forest fires.

Coriolis Effect

In the process of moving from high to low pressure, the wind appears to veer toward the right in the northern hemisphere and toward the left in the southern hemisphere, no matter what the compass direction of the path. This apparent deflection is called the **Coriolis effect**. Were it not for this effect, winds would move in exactly the direction specified by the pressure gradient.

To illustrate the impact of the Coriolis effect upon the winds, think of a line of ice skaters holding hands while skating in a circle, with one of the skaters nearest the center of the circle. This skater turns slowly, while the one at the outside of the circle must skate very rapidly in order to keep the line straight. In a similar way, the equatorial regions of the earth are rotating at a much faster rate than the areas around the poles. Next, suppose that the center skater threw a ball directly toward the skater at the end of the line. By the time the ball arrived, it would pass behind the outside skater. If the skaters are going in a counterclockwise direction, as the earth appears to be moving when viewed from the position of the North Pole, the ball seems to the person at the North Pole to pass to the right of the outside skater. If the skaters are going in a clockwise direction, as the earth appears to be moving viewed from the South Pole position, the deflection is to the left. Because air (like the ball) is not firmly attached to the earth, it, too, will appear to be deflected. The air maintains its direction of movement, but the earth's surface moves out from under it. Since the position of the air is measured relative to the earth's surface, the air appears to have diverged from its straight path.

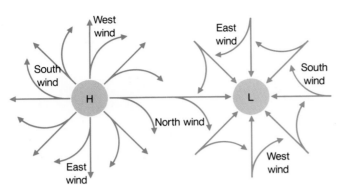

Figure 3.14

The impact of the Coriolis effect on flowing air. The straight arrows indicate the paths winds would follow flowing out of an area of high (H) pressure, or into one of low (L) pressure, were they to follow the paths dictated by pressure differentials. The curved arrows represent the apparent deflecting effect of the Coriolis force. Wind direction—indicated on the diagram for selected curved arrows—is always given by the direction *from* which the wind is coming.

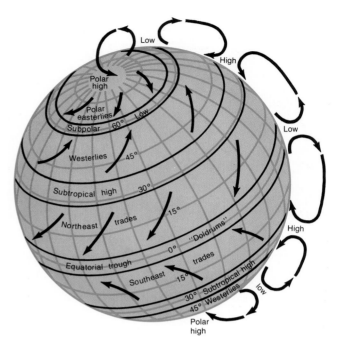

Figure 3.15

The planetary wind and pressure belts as they would develop on an earth of homogeneous surface. The named high and low pressure belts represent surface pressure conditions; the named wind belts are prevailing surface wind movements responding to pressure gradients and the Coriolis effect. In the upper atmosphere the conditions are reversed; descending air, creating surface high pressure, represents the contraction of the air column through cooling and the creation of lower pressure aloft. Air rising from surface flows increases upper air pressure at those zones. The upper atmospheric air flows shown on the diagram respond to pressure contrasts at higher altitudes. Land and water contrasts on the actual earth, particularly evident in the northern hemisphere, create complex distortions of this simplified pattern.

The Coriolis effect and the pressure gradient force produce spirals rather than simple straight-line patterns of wind (Figure 3.14). The spiral of wind is the basic form of the many storms that are so very important to the earth's air circulation system. These storm patterns are discussed later in the chapter.

Wind Belts

Equatorial areas of the earth are zones of low pressure. Intense solar heating in these areas is responsible for a convectional effect. Note in Figure 3.15 how the warm air rises and tends to move away from the *equatorial low pressure* in both a northerly and a southerly direction. As the equatorial air rises, it cools and eventually becomes heavy. Finally, the heavy air falls, forming zones of high pressure. These areas of *subtropical high pressure* are located at about 30° N and 30° S of the equator.

When this cooled air reaches the earth's surface, it, too, moves in both a northerly and a southerly direction. The Coriolis effect, however, modifies the wind direction and creates, in the northern hemisphere, the belts of winds called the *northeast trades* in the tropics and the *westerlies* (really the southwesterlies) in the middle latitudes. The names refer to the direction from which the winds come. Most of the United States lies within the belt of westerlies; that is, the air usually move across the country from southwest to northeast. There is also a series of cells of ascending air over the oceans to the north of the westerlies called the *subpolar low*. These areas tend to be cool and rainy. The *polar easterlies* connect the subpolar low areas to the *polar high*.

These belts move in unison as the vertical rays of the sun change position with the seasons. For example, equatorial low conditions move to the area just north of the equator in the northern hemisphere summer and just south of the equator in the southern hemisphere summer. The general planetary air-circulation pattern is modified by local wind conditions like sea, land, and mountain breezes.

The strongest flow of upper-air winds are the **jet streams.** These streams of rapidly flowing air—100 to 200 miles (150 to 300 km) per hour—move in a west-to-east direction in the northern hemisphere around the earth in an undulating pattern, first north, and then several hundred miles to the east, bending south. These undulations control the flow of air masses on the earth's surface. The more stable the wave pattern, the greater the repetitiveness of current conditions. The pattern of movement sometimes dies out and then reforms perhaps a month or two later. The winter jet streams are stronger than the summer ones.

Nowhere does the manner in which the seasonal shift takes place have such a profound effect on humanity as in densely populated southern and eastern Asia. The wind, which comes from the southwest in summer in India, reaches the landmass after picking up a great deal of moisture over the warm Indian Ocean. As it crosses the coast mountains and then the foothills of the Himalayas, the monsoon rains begin. **Monsoon** means "seasonal winds." Southern and eastern Asia's farm economy, particularly the rice crop, is totally dependent

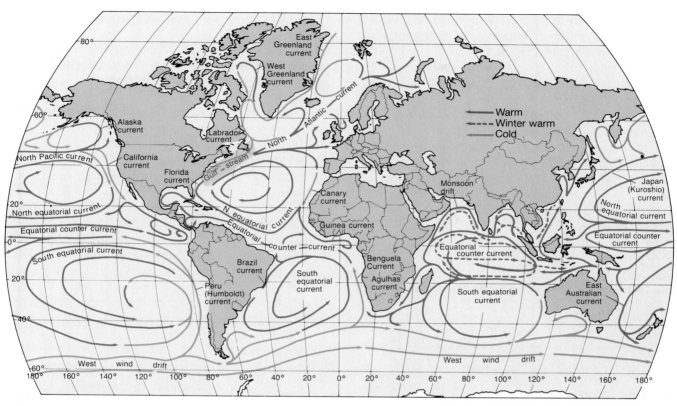

Figure 3.16

The principal surface ocean currents of the world. Notice how the warm waters of the Gulf of Mexico, the Caribbean, and the tropical Atlantic Ocean drift to northern Europe.

on summer monsoon rainwater. If, for any of several possible reasons, the wind shift is late or the rainfall is significantly more or less than optimum, crop failure may result. The unusual prolongation of the summer monsoon rains in 1978 caused disastrous flooding and crop and life loss in eastern India and Southeast Asia. The transition to the dry northeast winter winds occurs gradually across the region. First, the dry winds become noticeable in the north of India in September, and by January, most of the subcontinent is dry. Then, beginning in March in the south, the yearly cycle repeats itself.

Ocean Currents

Surface ocean currents correspond roughly to wind direction patterns because the winds of the world set the ocean currents in motion. In addition, just as differences in air pressure cause wind movements, so differences in the density of water help to force water to move. When water evaporates, the salt and other minerals that will not evaporate are left behind, making the water denser. High-density water exists in areas of high pressure, where descending dry air readily picks up moisture. In areas of low pressure, where rainfall is plentiful, the ocean water is of low density. Wind direction (including the Coriolis effect) and the differences in density cause water to move in wide paths from one part of the ocean to another.

There is an important difference between surface air movements and surface water movements. Landmasses are barriers to water movement, deflecting currents and sometimes forcing them to move in a direction opposite to the main current; air, on the other hand, moves freely over both land and water.

The shape of an ocean basin also has an important effect on ocean current patterns. For example, the North Pacific current, which moves from west to east, strikes the west coast of Canada and the United States and then is forced to move both north and south (although the major movement is the cold ocean current that moves south along the coast of California). In the Atlantic Ocean, however, the current is deflected in a northeasterly direction by the shape of the coast (Nova Scotia and Newfoundland jut far out into the Atlantic), and then it moves freely across to and past the British Isles and Norway, all the way to the extreme northwest coast of the USSR (Figure 3.16). This massive movement of warm water to northerly lands, called the **North Atlantic drift,** has enormous significance to the inhabitants of those areas. Without it, northern Europe would be much colder than it is.

Ocean currents affect the temperature and the precipitation on the land areas adjacent to the ocean. A cold ocean current near a land area robs the air above the current of its potential moisture supply, thus denying moisture to nearby land. Coastal deserts of the world usually border cold ocean currents. On the other hand, warm ocean currents—such as that off India—bring moisture to the adjacent land area, especially when the prevailing winds are landward.

Our earlier question concerning the ways in which differences in air pressure affect weather conditions is now answered in terms of the movement of warm and cool air over various surfaces at different times of the year and different times of the day. A more complete answer about the causes of different types of weather conditions, however, requires an explanation of the likelihood of places to receive precipitation, because rainfall and wind patterns are highly related.

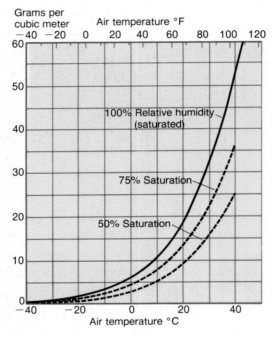

Figure 3.17

The water-carrying capacity of air and relative humidity. The actual water in the air (water vapor) divided by the water-carrying capacity (×100) equals the relative humidity. The solid line represents the maximum water-carrying capacity of air at different temperatures.

Moisture in the Atmosphere

Air contains water vapor (what we feel as humidity), which is the source of all precipitation. **Precipitation** is water in any form (rain, sleet, snow, hail) deposited on the earth's surface. Ascending air can easily expand because there is less pressure on it. When the heat of the lower air spreads out through a larger volume in the troposphere, the mass of air becomes cooler. Cool air is less able than warm air to hold water vapor (Figure 3.17). The air is said to be *saturated* when its contained water vapor condenses; that is, changes from a gas to a liquid, and forms droplets around fine particles, called *condensation nuclei*. These particles, mostly dust, pollen, smoke, and salt crystals, are nearly always available. At first, the tiny water droplets are usually too light to fall. When there are so many droplets that they coalesce into drops and become too heavy to remain suspended in air, they fall as rain. If temperatures are below the freezing point, the water vapor changes to ice crystals instead of water droplets. When enough ice crystals are present, snow is formed (Figure 3.18).

Rain droplets or ice crystals in large numbers form clouds, which are supported by light upward movements of air. The form and altitude of clouds depend on the amount of water vapor in the air, the temperature, and the wind movement. Descending air in high-pressure zones usually yields cloudless or nearly cloudless skies. Whenever warm, moist air rises, clouds form. The most dramatic cloud formation is probably the *cumulonimbus* (Figure 3.19). This is the anvil-head cloud that often accompanies heavy rain. Low, gray *stratus* clouds appear more often in cooler seasons than in the warmer months. The very high, wispy *cirrus* clouds that may appear in all seasons are made entirely of ice crystals.

The amount of water vapor in the air compared to the amount when condensation would begin at a given temperature is called the air's **relative humidity.** The warmer the air, the greater the amount of water vapor it can contain. If the relative humidity is 100%, the air is completely saturated with water vapor. A value of 60% on

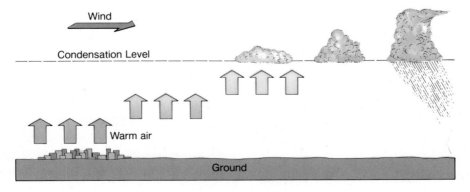

Figure 3.18

As warm air rises, it cools. As it cools, its water vapor condenses and clouds form. If the water content of the air is greater than the air's capacity to retain moisture, some form of precipitation occurs.

(a)

(b)

(c)

Figure 3.19
Cloud types: (a) cumulonimbus, (b) stratus, (c) cirrus.

El Niño

El Niño, Spanish for The Child, is a term coined years ago by fishermen who noticed that the usually cool waters off the Peruvian coast were considerably warmer every three or four years around Christmastime. The fish catch was reduced significantly during those periods, and, if the fishermen were able to make the scientific associations that present-day oceanographers and climatologists can, they would recognize that a host of other effects follow El Niño. In 1982–83 an unusually severe and extended period for El Niño caused enormous damage and many deaths. Parts of South America were ravaged by floods while droughts baked Australia and parts of Asia. Hurricanes (typhoons) that rarely strike Tahiti hit the island six times during that period.

The winds that usually blow from east to west over the central Pacific Ocean, from the cold ocean current to the warm waters of eastern Asia, slow or even reverse during periods of El Niño. The warm water builds up along the west coast of South America, and its effects are felt around the world. For example, in 1982–83, huge downpours, coastal erosion, and flooding caused tens of millions of dollars of damage in southern California.

The phenomenon occurs every two to seven years, but at different degrees of intensity. For example, an El Niño occurred in 1986, but the modest warm water buildup did not cause extraordinary effects.

The El Niño condition is an example of the interaction of atmospheric pressure and ocean temperature. The atmospheric and oceanic states reinforce each other. Under normal conditions, the contrast in temperatures across the ocean helps drive the winds, which in turn keep pushing water to the west, maintaining the contrast in water temperature. But in some years there is what is called a *southern oscillation.* Atmospheric pressure rises near Australia, the wind falters, and El Niño is created off the coast of South America.

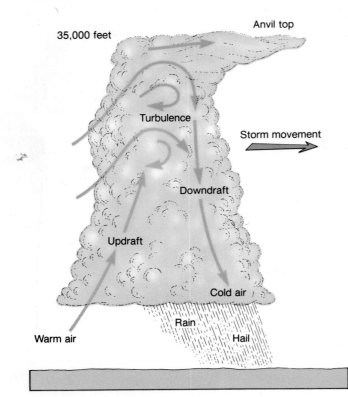

Figure 3.20
When warm air laden with moisture rises, a cumulonimbus cloud may develop and convectional precipitation occur. The turbulence within the system creates a downdraft of cold upper-altitude air.

a hot day means that the air is extremely humid and very uncomfortable. A 60% reading on a cold day, however, tells us that although the air contains relatively large amounts of water vapor, it holds in absolute terms much less than on the hot, muggy day. This example demonstrates the point that relative humidity is meaningful only if we keep the air temperature in mind.

Dew or frost on the ground in the morning means that during the night the temperature dropped to the level at which condensation took place (Figure 3.17). The critical temperature for condensation is the **dew point.** Foggy or cloudy conditions on the earth's surface imply that the dew point has been reached and that the relative humidity is 100%.

Types of Precipitation

When large masses of air rise, precipitation may take place. It can be one of three types: (1) convectional, (2) orographic, or (3) cyclonic or frontal precipitation.

Convectional precipitation results from rising, heated, moisture-laden air. As the air rises, it cools; when its dew point is reached, condensation and precipitation occur (Figure 3.20). This process is typical of summer storms or showers in tropical and continental climates. Usually the ground is heated during the morning and the early afternoon. The warm air that accumulates begins to rise, and first cumulus clouds then cumulonimbus clouds develop. Finally, there is lightning and thunder, and heavy rainfall, which may affect each part of the ground for only a brief period since the storm is moving. It is not unusual for these convectional storms to occur in late afternoon or early evening.

If quickly rising air currents violently circulate the air within the cloud, ice crystals may form near the top of the cloud. These ice crystals may begin to fall when they are large enough, but a new updraft can force them back up, with a subsequent enlargement of the pieces of

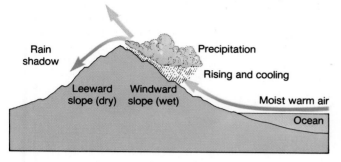

Figure 3.21

Orographic precipitation. Surface winds may be raised to higher elevations by hills or mountains lying in their paths. If such orographically lifted air is sufficiently cooled, precipitation occurs. Descending air on the leeward side of the upland barrier becomes warmer, its capacity to retain moisture is increased, and water absorption rather than release takes place.

ice. This process may occur again and again, until the ice can no longer be sustained by the updrafts and pieces of ice fall to the ground as *hail.*

Orographic precipitation occurs as warm air is forced to rise because hills or mountains stand in the way of moisture-laden winds (Figure 3.21). This type of precipitation is typical of areas where mountains and hills are situated close to oceans or large lakes. Saturated air from over the water blows onshore, rising as the land rises. Again, the process of cooling, condensation, and precipitation takes place. The *windward* side (the side toward which the wind blows) of the hills and mountains may receive a great deal of precipitation. The other side, called the *leeward* side, and the adjoining regions downwind are very often dry. The air that passes over the mountains or hills descends and warms, and as we have seen, descending air does not produce precipitation and warming air absorbs moisture from surfaces over which it passes. A graphic depiction of the great differences in rainfall over very short distances is shown on the map of the state of Washington (Figure 3.22).

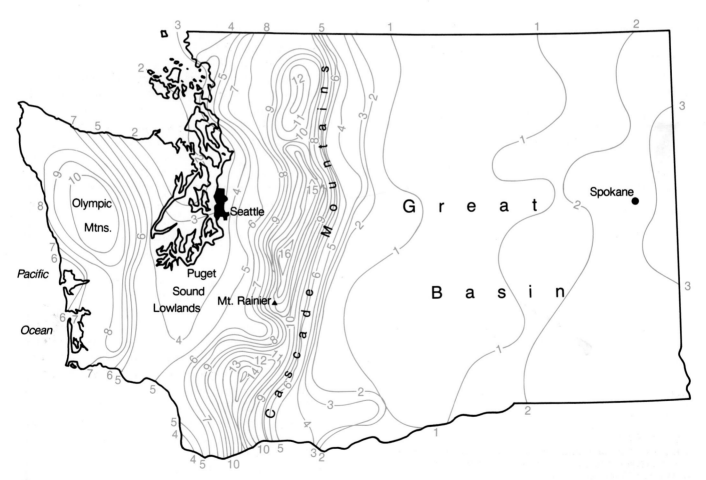

Figure 3.22

Precipitation in inches for the month of November 1985 for the state of Washington. The moisture-laden Pacific air is first forced up over the 5000–7000-foot Olympic Mountains; then it descends into the Puget Sound Lowlands; then up over the 9000–14,000-foot Cascade Mountains; and finally down into the Great Basin of eastern Washington.

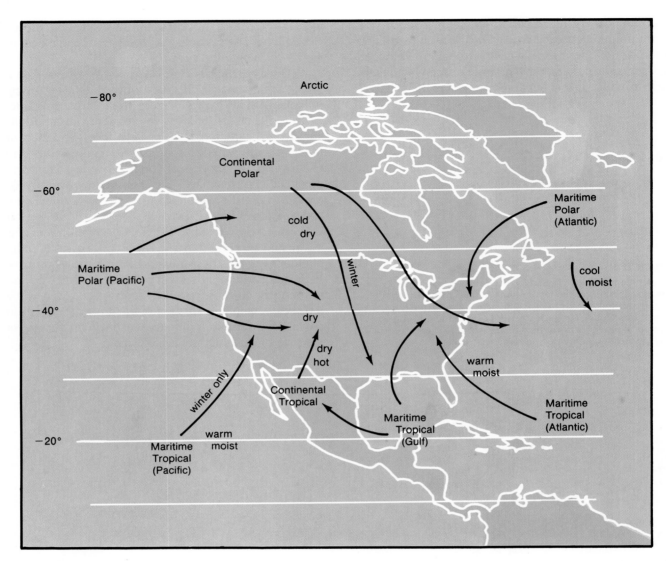

Figure 3.23

Source regions for air masses in North America. The United States and Canada, lying between major contrasting air-mass source regions, are subject to numerous storms and changes of weather.

Source: After Haynes, U.S. Department of Commerce.

Cyclonic or **frontal precipitation,** the third type, is common to the middle latitudes, where cool and warm air masses meet. It also occurs in the tropics, though less frequently, as the originator of hurricanes and typhoons. The most extreme form of cyclonic storm is the tornado, which is much more a windstorm than a rainstorm. In order to understand cyclonic or frontal precipitation, one must first visualize the nature of air masses and the way cyclones develop.

Air masses are large bodies of air with similar temperature and humidity characteristics throughout; they form over a **source region.** Source regions include large areas of uniform surface and relatively consistent temperatures, such as the cold land areas of northern Canada, the north central part of the Soviet Union, or the warm tropical water areas in any of the oceans close to the equator. Source regions for North America are shown in Figure 3.23. During a period of a few days or a week, an air mass may form in a source region. For example, in northern Canada, in the fall, when snow has already covered the vast subarctic landscape, cold, heavy, dry air develops over the frozen land surface. (A further discussion of air masses as regional entities may be found in Chapter 12, Pages 408–10.)

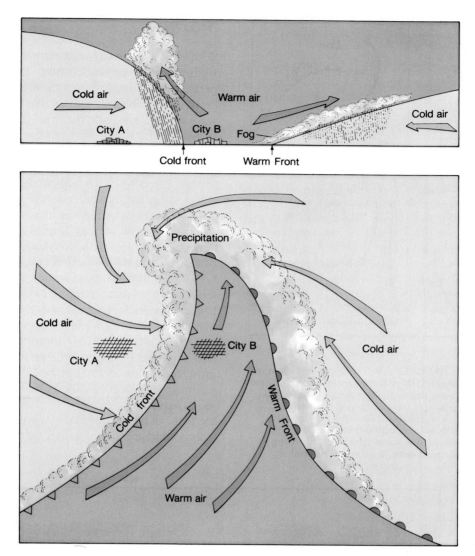

Figure 3.24

In this diagram, the cold front has recently passed over city A and is heading in the direction of city B. The warm front is moving away from city B.

After developing, this continental polar air mass begins to move toward the lighter, warmer air to the south. The leading edge of the tongue of air is called a **front.** The front in this case separates the cold, dry air from whatever air is found in its path. If a warm, moist air mass is in the path of the polar air mass, the heavier cold air hugs the ground and forces the lighter air up above it. The rising moist air condenses, and frontal precipitation occurs. On the other hand, if it is warm air that is on the move, it slides over the cold air pushing the cold air back, and again, precipitation occurs as the warm air is forced upwards. In the first case, when cold air moves toward warm air, cumulonimbus clouds form and precipitation along the *squall line* is brief and heavy. As the front passes, temperatures drop appreciably, the sky clears, and the air becomes noticeably drier. In the second case, when warm air is moving over cold air, steel-gray nimbostratus (*nimbo* means "rain") clouds form, and the precipitation is steady and long-lasting. As the front passes, warm, muggy air becomes characteristic of the area. Figure 3.24 summarizes the movement of fronts.

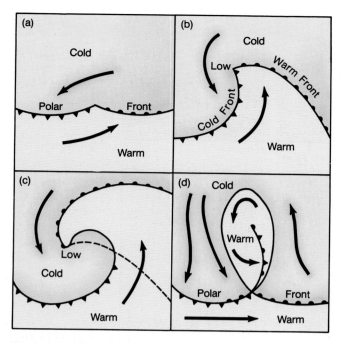

Figure 3.25

When wedges of warm and cold air develop along a low-pressure trough in the middle latitudes, there is the possibility of cyclonic storm formation. (a) A wave begins to form along the polar front. (b) In the northern hemisphere, cold air begins to turn in a southerly direction, while warm air moves north. The meeting lines of these unlike air masses are called *fronts*. Cold air overtakes (c) and isolates (d) the warm air pocket, removing it from its energy and moisture source. The cyclonic storm dissipates as the polar front is reestablished.

Storms

When two air masses come into contact, there is the possibility that storms will develop. If the contrasts in temperature and humidity are sufficiently great, or if the wind directions of the two masses that touch are opposite, a wave might develop in the front (Figure 3.25). Once established, the waves may enlarge and become wedge-like. One wedge is the cooler air moving along the surface, and the other wedge is the warmer air moving up over the cold air. In both wedges, the warm air rises, and a low-pressure center forms. Considerable precipitation is accompanied by counterclockwise winds around the low-pressure area. A large system of air circulation centered on a region of low atmospheric pressure like this is called a mid-latitude **cyclone.**

A cyclone may be a weak storm or one of great intensity. The tropical storm is likely to begin over warm tropical waters, far enough from the equator for the Coriolis effect to be significant. Here a wave that is not associated with a front may form. If conditions are such that a wedge develops, an intense tropical storm may grow, fed by the energy embodied in the rising moist, warm air. This storm is called a **hurricane** in the Atlantic region and a *typhoon* in the Pacific. Figure 3.26 shows the paths of hurricanes in the Atlantic. The winds of these storms whirl around the center, called the *eye*. But within

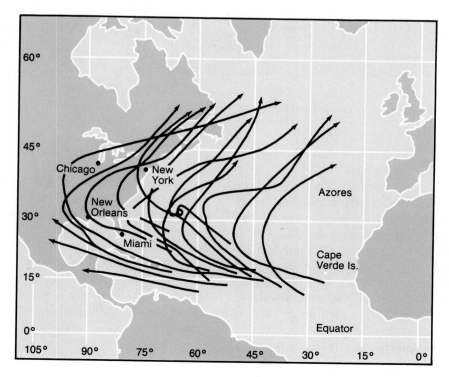

Figure 3.26

Tracks of typical hurricanes. In the United States, the most vulnerable areas are the Gulf coast of Florida; Cape Hatteras, North Carolina; Long Island, New York; and Cape Cod, Massachusetts.

Source: After U.S. Navy Oceanographic Office.

the eye of the storm, there are gentle breezes and relatively clear skies. Over land, these storms lose their warm-water energy source and subside quickly. If they move far into the colder northern waters, they are pushed or blocked by other air masses and also lose their energy source and abate.

The most violent storm of all—but also the smallest—is the **tornado** (Figure 3.27). A typical tornado is less than 100 feet in diameter. Tornadoes are spawned in the huge cumulonimbus clouds that sometimes travel in advance of a cold front along a squall line. The central part of the United States in spring or fall, when adjacent air masses contrast the most, is the scene of many of these funnel-shaped killer clouds. Although winds are estimated to reach 500 miles (800 km) per hour, these storms are small and usually travel on the ground for less than a mile, so that they affect only limited areas.

Climate

We have traced some of the causes of weather changes that occur as air from high-pressure zones flows toward low-pressure areas, fronts pass and waves develop, dew points are reached, and sea breezes arise. As we have noted, in some parts of the world these changes occur more rapidly and more often than in other parts.

Weather conditions from day to day can be explained by the principles we have examined. However, one cannot understand the effect of the weather elements—temperature, precipitation, air pressure, and winds—unless one is conscious of the earth's surface features. Weather forecasters in each location on earth must deal with the elements of weather in the context of the local environment.

Figure 3.27
Tornado. In the United States, these violent storms occur most frequently in the central and south central part of the country (especially in Oklahoma, Kansas, and North Texas), where cold polar air very often meets warm, moist gulf air.

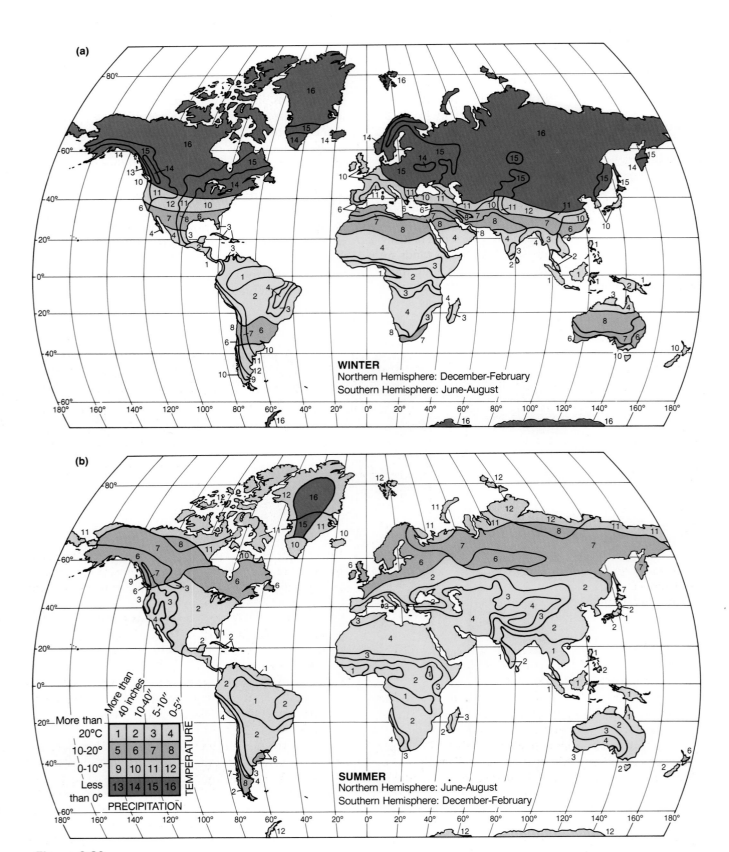

Figure 3.28

These maps combine temperature and precipitation data to display seasonal variations in the basic components of climate. In reading the maps, remember that the summer climate shown in (b) represents the summer season for both the northern and southern hemispheres. This means that June, July, and August data were used for the north latitudinal part of the world, and December, January, and February data were used for the south latitudes. The result is a summer view of the world—one of nearly universally warm temperatures and large amounts of precipitation. The winter view (a) shows a world of greatly varying temperatures and small amounts of precipitation.

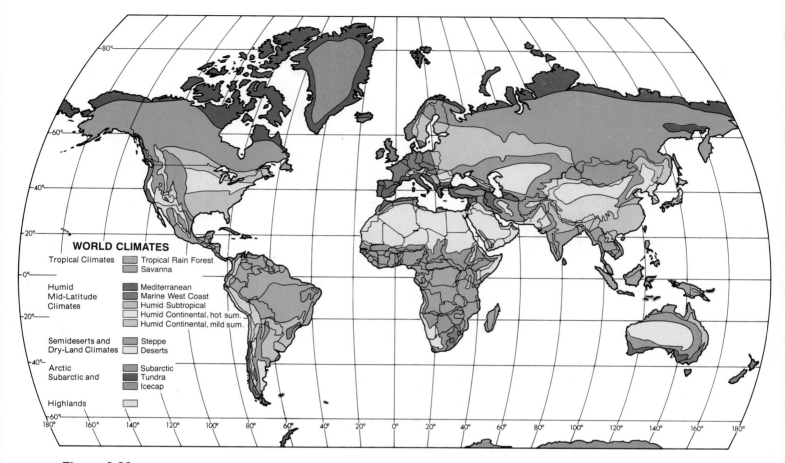

WORLD CLIMATES

Tropical Climates
- Tropical Rain Forest
- Savanna

Humid Mid-Latitude Climates
- Mediterranean
- Marine West Coast
- Humid Subtropical
- Humid Continental, hot sum.
- Humid Continental, mild sum.

Semideserts and Dry-Land Climates
- Steppe
- Deserts

Arctic Subarctic and
- Subarctic
- Tundra
- Icecap

Highlands

Figure 3.29
Climates of the world.

The complexities of daily weather conditions may be summarized by statements about climate. The climate of an area is a generalization based on daily and seasonal weather conditions. Are the summers warm on the average? Is there a tendency toward heavy snow in the winter? Are the winds normally from the southeast? Are the climatic averages typical of daily weather conditions, or are the variations from day to day or from week to week so great that one should speak of average variations rather than just averages? These are the questions we must ask in order to form an intelligent description of the differences in conditions from place to place.

The two most important elements that differentiate weather conditions are temperature and precipitation. While air pressure is also an important weather element, differences in air pressure are hardly noticeable without the use of a barometer. So, we may regard warm, moderate, cold, and very cold temperatures as characteristic of a place or a region. In addition, high, moderate, and low precipitation are good indicators of the degree of humidity or aridity. We shall take these two scales, define the terms more precisely, and map the areas of the world having various combinations of temperatures and precipitation.

Because extreme seasonal changes do occur, two global climatic maps are shown in Figure 3.28, one for winters (a) and one for summers (b). Bear in mind that a summer map of the world is a combination of the climates of the northern hemisphere in July, August, and September and of the southern hemisphere in January, February, and March because the seasons in the two hemispheres are reversed.

Figure 3.29 depicts the various climates of the world. It is based on the type of information presented in Figure 3.28. Note that in rugged mountain areas, we use the symbol M to denote that climate at any one location may differ appreciably from nearby areas depending on such factors as elevation, latitude, southward or northward facing slopes, exposure to moisture laden winds, and so on.

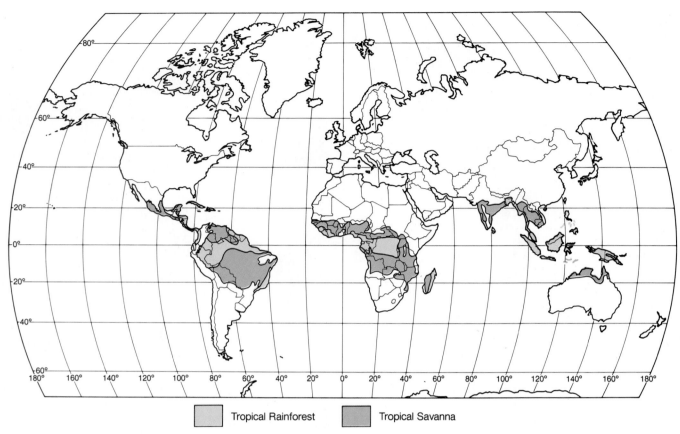

Tropical Rainforest   Tropical Savanna

Figure 3.30

The location of humid tropical climates.

The following paragraphs briefly describe the major climates and indicate the important characteristics of their typifying *natural vegetation,* that is, the plant life that would exist in each area if humans did not directly interfere in the growth process. Natural vegetation, little of which remains today in densely settled portions of the globe, has close interrelationships with soils, landforms, groundwater, elevation, and other features of habitat. The tie between **biochores**—the major structural subdivisions of natural vegetation—and climate is particularly close. Indeed, the earliest maps of the world climatic zones were based not on statistical variation in recorded temperatures, pressures, precipitation, and other measures of the atmosphere, but on observed variation in vegetative regions. Even the descriptive names of the biochores—*forest, savanna, grassland, desert*—are climatic in their implications. The discussion of climates, therefore, benefits from a visualization of regional differences by reference to the type of mature plant communities that, in nature, developed within those climates.

Since soil characteristics, discussed briefly in Chapter 2, are mainly a creation of overlying vegetation (which becomes humus), available moisture, and temperature operating over long periods of time, mention will be made of the soils typically found in each climate region. We will identify the scientific names for each of the major soil types. Soils, however, vary considerably over short distances depending on such factors as the underlying rock material, the slope of the land, and the availability of transported materials such as wind-blown silt (loess) and alluvium deposited by flooding rivers and streams.

The world's major climate regions are briefly discussed in the paragraphs that follow. The numbers following the section headings below refer to the keys on Figure 3.28. The first figure represents typical winter conditions, and the second is for summer. Each represents idealized conditions.

The Tropical Climates

Tropical climates are generally associated with the earth areas lying between the farthest north and farthest south lines of vertical rays of the sun—the *Tropic of Cancer* and the *Tropic of Capricorn* (Figure 3.30).

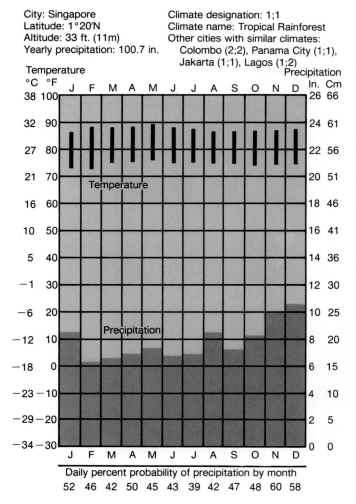

City: Singapore
Latitude: 1°20'N
Altitude: 33 ft. (11m)
Yearly precipitation: 100.7 in.

Climate designation: 1;1
Climate name: Tropical Rainforest
Other cities with similar climates:
　Colombo (2;2), Panama City (1;1),
　Jakarta (1;1), Lagos (1;2)

Daily percent probability of precipitation by month
52　46　42　50　45　43　39　42　47　48　60　58

Figure 3.31
This and succeeding climate charts (*climagraphs*) show average daily high and low temperatures for each month, the average precipitation for each month, and the probability of precipitation on any particular day in a designated month. For Singapore, the average daily high temperature in August is 87° F (30.5° C), the low is 75° F (24° C). The rainfall for the month, on average, is 8.4 inches (21 cm), and on a given day in August, there is a 42% chance of rainfall.

Tropical Rain Forest (1;1)

The areas that straddle the equator are located generally within the equatorial low-pressure zone. They have warm, wet climates in both winter and summer (Figure 3.31). Their rainfall usually comes in the form of daily convectional thunderstorms. Although most days are sunny and hot, by afternoon cumulonimbus clouds form and convectional rain falls.

　　The natural vegetation is **tropical rain forest**, which is still present but declining rapidly in such areas as the Amazon Basin of South America and the Zaire River basin of Africa. Tall, dense forests of broadleaf trees and heavy vines predominate. Among the hundreds of species of trees that are found here, there are dark woods, light woods, spongy softwoods such as balsa, and hardwoods such as teak and mahogany (Figure 3.32). Rain

Figure 3.32
Tropical rain forest. The vegetation is characterized by tall, broad-leaf, hardwood trees and vines.

forests also extend some distance away from the equator along coasts where prevailing winds supply a constant source of moisture to coastal uplands and where the orographic effect provides enough precipitation for heavy vegetation to develop.

　　The typical soils of these regions are called *oxisols.* They are red, yellow, and yellowish-brown just below a darkened surface layer indicative of extreme chemical weathering. As a result of the weathering, most of the soil nutrients necessary for cultivated plants are washed away by the heavy rains. Only with large inputs of fertilizers can the soils be made to sustain continued agricultural use.

(a)

(b)

Figure 3.33
The parklike landscapes of grasses and trees, characteristic of the
(a) drier (Kenya) and (b) wetter (Ivory Coast) tropical savanna.

Savanna (3;1)

As the sun's vertical rays move away from the equator in
summer, the equatorial low-pressure zone follows the
sun's path. Thus the areas to the north and south of the
rain forest are wet in the summer months, although still
hot, but are dry the remainder of the year because the
moist equatorial low has been replaced by the dry air of
the subtropical highs. These areas are known as **savanna** lands because of the kind of natural vegetation
found there. Although the natural vegetation of these
areas appears to have been a form of scrub forest, the
savanna is now a grassland with widely dispersed trees.

The possibility of more forested cover has been dis-
couraged by periodic clearing by burning carried out by
local agriculturalists and hunters. Sometimes savannas
seem to have been purposely designed, for they have a
parklike look (Figure 3.33). The East African region of
Kenya and Tanzania contains well-known grasslands and
fire-resisting species of trees where large animals like gi-
raffes, lions, and elephants roam. The *campos* and *llanos*
of South America are other huge savanna areas.

The wetter portions of the savanna lands are often
underlain by *ultisols,* soils that develop in warm, wet-dry
regions beneath forest vegetation. These soils are poor
in nutrients for cultivated plants but respond well when

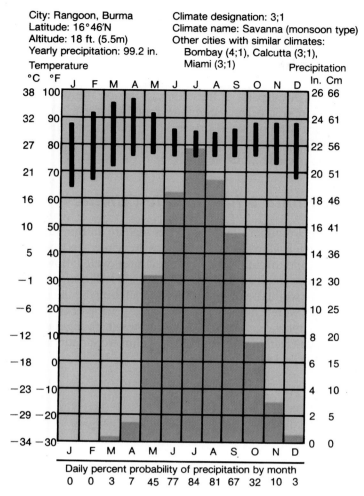

City: Rangoon, Burma
Latitude: 16°46′N
Altitude: 18 ft. (5.5m)
Yearly precipitation: 99.2 in.

Climate designation: 3;1
Climate name: Savanna (monsoon type)
Other cities with similar climates:
Bombay (4;1), Calcutta (3;1),
Miami (3;1)

Daily percent probability of precipitation by month
0   0   3   7   45   77   84   81   67   32   10   3

Figure 3.34
Climagraph for Rangoon, Burma. (See Figure 3.31 for
explanation.)

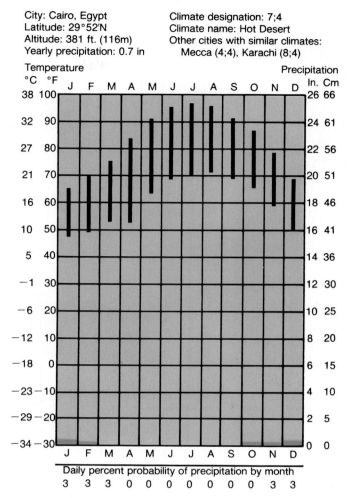

City: Cairo, Egypt
Latitude: 29°52′N
Altitude: 381 ft. (116m)
Yearly precipitation: 0.7 in

Climate designation: 7;4
Climate name: Hot Desert
Other cities with similar climates:
Mecca (4;4), Karachi (8;4)

Daily percent probability of precipitation by month
3   3   3   0   0   0   0   0   0   0   3   3

Figure 3.35
Climagraph for Cairo, Egypt. (See Figure 3.31 for explanation.)

lime and fertilizers are applied. In the drier parts of the savanna the characteristic soils are *vertisols,* which form under grasses in warm climates. When it rains, the surface becomes plastic-like, and some of the soil slides into the cracks that have formed in the dry season. Consequently, they are difficult to till and are used productively as grazing land.

There are usually no marked breaks between one kind of climate and another; instead there are zones of transition. These transitional zones are typical of plains and plateaus; they are not so gradual in mountainous regions. Between the tropical rain forest and the savanna, there are forests that are less dense than the rain forest itself.

A special case needs mentioning. In Asia, when summer monsoon winds carry water-laden air to the mainland, far more rain falls on the hills, mountains, and adjacent plains than in the savanna (notice the pattern of precipitation in Figure 3.34). As a result, the vegetation is much denser, even though the winters are dry.

Jungle growth and large forests are the natural vegetation. Much of this vegetation, however, no longer exists because people have been using the land for rice and tea production for many generations.

The most common soils in the monsoon lands are ultisols, but in valleys where alluvium has been deposited by flooding rivers, a nutrient-rich soil type called *inceptisols* develops.

Hot Deserts (7;4)

As one continues away from the equator, eventually the grasses shorten, and desert shrubs become evident on the poleward side of the savannas. Here we approach the belt of subtropical high pressure that brings considerable sunshine and hot summer weather but very little precipitation. Note the almost total lack of rainfall in Figure 3.35. The precipitation that does fall is of the convectional variety, but it is sporadic. As conditions become

drier, there are fewer and fewer drought-resistant shrubs, and in some areas only gravelly and sandy deserts exist, as suggested on Figure 3.36. The great hot deserts of the world, such as the Sahara, the Arabian, the Australian, and the Kalahari, are all the products of high-pressure zones. Often the driest parts of these deserts are on the west coast, where cold ocean currents are found (Figure 3.28). Except for salt-encrusted or sand-covered deserts, the soils, called *aridisols,* respond well to cultivation when irrigation is made available.

The Humid Mid-Latitude Climates

There are a number of climate types that are all humid, that is, not having desert conditions in winter and/or summer, and having winter temperatures well below those of the tropical climates (Figure 3.37). These climate types would be neatly defined, paralleling the lines of latitude, were it not for mountain ranges, warm or cold ocean currents, and particularly land-water configurations. These factors cause the greatest climatic variations in the middle latitudes.

Figure 3.36
Devoid of stabilizing vegetation, desert sands are constantly rearranged in complex dune formations.

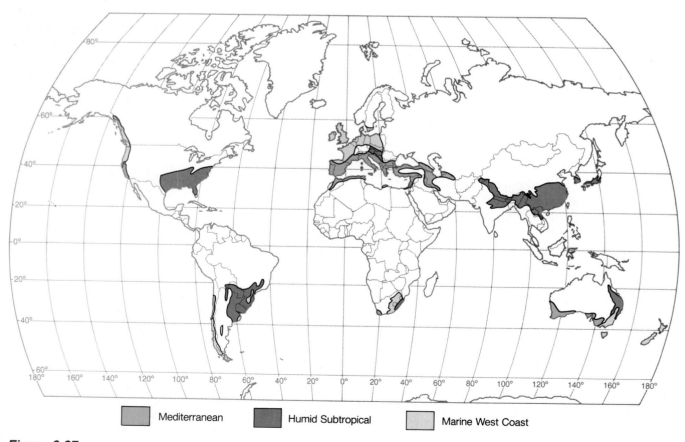

Mediterranean   Humid Subtropical   Marine West Coast

Figure 3.37
The location of humid mid-latitude climates. Humid continental climate areas are not shown.

City: Rome, Italy
Latitude: 41°48'N
Altitude: 377 ft. (115m)
Yearly precipitation: 33.3 in.

Climate designation: 6;3
Climate name: Mediterranean
Other cities with similar climates:
Athens (6;3), Los Angeles (6;4),
Valparaiso (6;4)

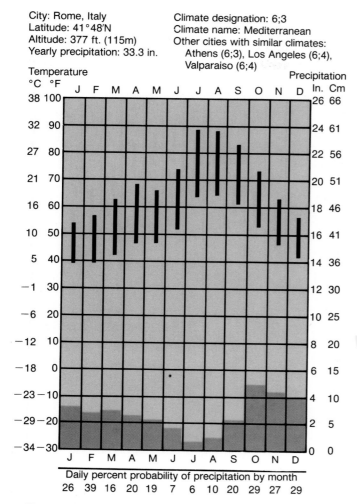

Figure 3.38

Climagraph for Rome, Italy. (See Figure 3.31 for explanation.)

Mediterranean Climate (6;3)

Mid-latitude winds are generally from the west in both the northern and the southern hemispheres, and a significant amount of the precipitation is generated by frontal systems. Thus, it is important to know whether the water is cold or warm near land areas. Several climatic zones are noticeable in the middle latitudes. They are all marked by warm summer temperatures except in those areas cooled by westerly winds from the ocean. To the poleward side of the hot deserts, there is a transition zone between the subtropical high and the moist westerlies zones. Here, cyclonic storms bring rainfall only in the winter, when the westerlies shift toward the equator. The summers are dry and hot as the subtropical highs shift slightly poleward (Figure 3.38), and the winters are not cold. These conditions describe the **Mediterranean climate,** which is often found on west coasts of continents in middle latitudes. Southern California, the Mediterranean area itself, western Australia, the tip of South Africa, and central Chile in South America are characterized by such a climate. In these areas, where there is enough precipitation, shrubs and small deciduous trees (trees that lose their leaves once a year) such as scrub oak grow (Figure 3.39).

The Mediterranean climate area—long and densely settled in southern Europe, the Near East, and North Africa—has more moisture and a greater variety of vegetation and soil types than is found in the desert. Clear, dry air predominates, and winters are relatively short and mild. Plants and flowers grow all year around. Even though the summers are hot, the nights are usually cool and clear. Much of the vegetation of this area is now in the form of crops.

Figure 3.39

Typical vegetation in an area with a Mediterranean climate. Trees such as scrub oak are short as shown in this photograph of coastal California.

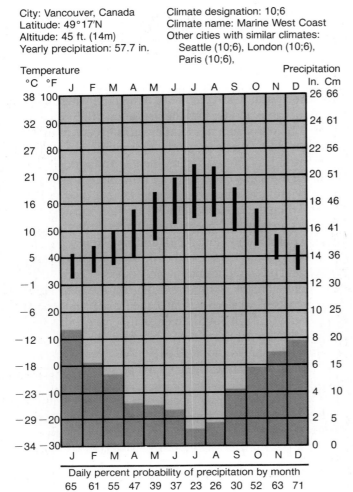

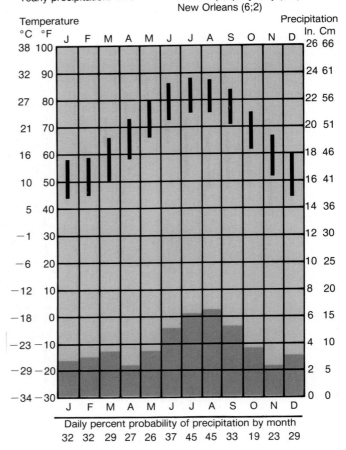

City: Vancouver, Canada
Latitude: 49°17'N
Altitude: 45 ft. (14m)
Yearly precipitation: 57.7 in.

Climate designation: 10;6
Climate name: Marine West Coast
Other cities with similar climates:
    Seattle (10;6), London (10;6),
    Paris (10;6),

Daily percent probability of precipitation by month
65 61 55 47 39 37 23 26 30 52 63 71

Figure 3.40
Climagraph for Vancouver, Canada. (See Figure 3.31 for explanation.)

City: Charleston, S.C.
Latitude: 32°46'N
Altitude: 16 ft. (5m)
Yearly precipitation: 47.5 in.

Climate designation: 6;2
Climate name: Humid Subtropical
Other cities with similar climates:
    Canton (6;2), Sydney (6;2),
    New Orleans (6;2)

Daily percent probability of precipitation by month
32 32 29 27 26 37 45 45 33 19 23 29

Figure 3.41
Climagraph for Charleston, South Carolina. (See Figure 3.31 for explanation.)

Marine West Coast Climate (10;6)

Closer to the poles, but still within the westerly wind belt, are areas of **marine west coast climate.** Here, cyclonic storms play a relatively large role. In the winter, more rain falls and cooler temperatures prevail than in the Mediterranean zones. (Compare the patterns in Figures 3.38 and 3.40.) Little rain falls in the summer in the transitional zone just poleward of the Mediterranean climate, but farther toward the poles, rainfall increases appreciably in summer and even more in winter. Marine winds from the west moderate both summer and winter temperatures. Thus, summers are pleasantly cool, and winters, though cold, do not normally produce freezing temperatures.

This climate exists somewhere on all continental west coasts that reach sufficient latitude, but it affects only relatively small land areas in all but one region. Because northern Europe has no great mountain belt to thwart the west-to-east flow of moist air, the marine west coast climate stretches well across the continent to Poland. At that point, cyclonic storms originating in the Arctic regions begin to be felt. Northern Europe's moderate climate, then, owes its existence to a lack of mountains and to a relatively warm ocean current whose influence is felt for about 1000 miles, from Ireland to central Europe.

The orographic effect from the mountains in areas such as northwestern United States, western Canada, and southern Chile produces enormous amounts of precipitation, very often in the form of snow, on the windward side. Vast coniferous forests (needle-leaf trees, such as pine, spruce, and fir) cover the mountains' lower elevations. Because the mountains prevent moist air from continuing to the leeward side, mid-latitude deserts are found to the east of these marine west coast areas.

The main soils of marine west coast areas are called *spodosols*. They are strongly acidic and low in plant nutrients as a result of the acidic humus that develops from the fallen needles of the coniferous trees. Fertilizers must be used for farming in order to neutralize the acidity.

The Humid Subtropical Climate (6;2)

On the east coasts of continents, the transition is from the equatorial climate to the **humid subtropical climate.** Here, convectional summer showers and winter cyclonic storms are the sources of precipitation. This climate has hot, moist summers and moderate, moist winters (Figure 3.41). In the fall, on occasion, hurricanes born in tropical waters strike the coastal areas.

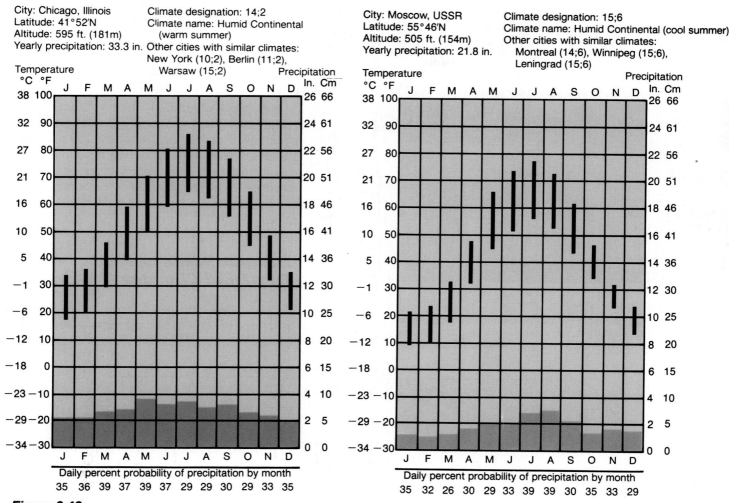

City: Chicago, Illinois
Latitude: 41°52'N
Altitude: 595 ft. (181m)
Yearly precipitation: 33.3 in.

Climate designation: 14;2
Climate name: Humid Continental
(warm summer)
Other cities with similar climates:
New York (10;2), Berlin (11;2),
Warsaw (15;2)

Daily percent probability of precipitation by month
35 36 39 37 39 37 29 29 30 29 33 35

Figure 3.42
Climagraph for Chicago, Illinois. (See Figure 3.31 for explanation.)

City: Moscow, USSR
Latitude: 55°46'N
Altitude: 505 ft. (154m)
Yearly precipitation: 21.8 in.

Climate designation: 15;6
Climate name: Humid Continental (cool summer)
Other cities with similar climates:
Montreal (14;6), Winnipeg (15;6),
Leningrad (15;6)

Daily percent probability of precipitation by month
35 32 26 30 29 33 39 39 30 35 33 29

Figure 3.43
Climagraph for Moscow, USSR. (See Figure 3.31 for explanation.)

The generally even distribution of rainfall allows for forests of deciduous, hardwood trees such as oak and maple, whose leaves turn orange and red before falling in autumn. In addition, conifers become mixed with the deciduous trees as a second-growth forest. Southern Brazil, the southeastern United States, and southern China all have the humid subtropical climate.

The transition poleward to the continental climates is accompanied by increasingly colder winters and shorter summers. In this direction as well, cyclonic storms become responsible for more rainfall than convectional showers, but at this point, the region can no longer be described as humid subtropical; rather it is described as *humid continental.*

Beneath the deciduous forests of both the humid subtropical and the humid continental climates are soils called *alfisols.* The A-horizon is generally gray-brown in color. The soils, usually rich in plant nutrients derived from the strongly basic humus created by the fallen leaves of deciduous trees, retain moisture during the hot days of summer, allowing for productive agriculture.

Humid Continental Climate (10;2)

The air masses that drift toward the equator from their origin close to the poles and the air masses that drift toward the poles from the tropics produce frontal precipitation. Whenever warmer air or marine air is in the way of the cold continental air masses, or vice versa, frontal storms develop. We speak of the climates that these air masses influence as **humid continental.** Figures 3.42 and 3.43 show the range and the dominance of winter temperatures within this climatic type. The continental climate may be contrasted to the marine west coast climates: the former has prevailing winds from the land, the latter from the sea. Coniferous forests become more plentiful in the direction of the poles, until temperatures become so low that trees are denied an adequate growing

Figure 3.44
In the extensive region of east central Canada and the area around Moscow, USSR, the summers are long and warm enough to support a dense coniferous forest. Farther north, growth is less luxuriant.

season and are stunted (Figure 3.44). Along with the coniferous forests are the infertile *spodosols.* The transition from the very cold air masses of the winter to the occasional convectional storms of the summer means that there are four distinct seasons.

Three huge areas of the world are characterized by a humid continental climate: (1) the north and central United States and southern Canada, (2) most of the European portion of the Soviet Union, and (3) northern China. Because there are no land areas at a comparable latitude in the southern hemisphere, this climate is not represented there. In fact, the only nonmountain cold climate in the southern hemisphere is the polar climate of Antarctica.

Mid-Latitude Semideserts and Dry-Land Climates (10;4)

The locations of these climates is shown in Figure 3.45. In the interior of the continents, behind mountains that block the west winds, or in lands far from the reaches of

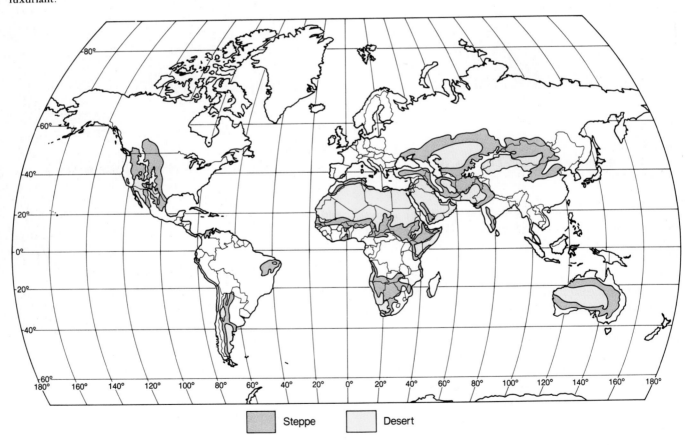

Steppe   Desert

Figure 3.45
The location of mid-latitude dry climates.

moist tropical air, there are extensive regions of *semidesert* conditions. Figure 3.46 illustrates typical temperature and precipitation patterns in these areas. Occasionally a summer convectional storm or a frontal system that still contains some moisture brings rainfall. These moderately dry lands are called **steppes.** The natural vegetation is grass, although desert shrubs are found in the drier portions of the steppes (Figure 3.47). Although rain is not plentiful, the soils are rich because the grasses return nutrients to the soil. The soils, called *mollisols,* have a dark brown to black A-horizon, and are among the most naturally fertile soils in the world. As a result, humans have made the steppes of the United States, Canada, the Soviet Union, and China among the most productive agricultural regions of the world. The steppes are also known for their hot, dry summers and for biting winter winds that sometimes bring blizzards.

Figure 3.47
Desert shrubs in the mid-latitude drylands of northern Mexico.

City: Tehran, Iran
Latitude: 35°41′N
Altitude: 4002 ft. (1220m)
Yearly precipitation: 10.1 in.

Climate designation: 10;4
Climate name: Midlatitude Dryland
Other cities with similar climates:
    Salt Lake City (12;4), Ankara (10;4)

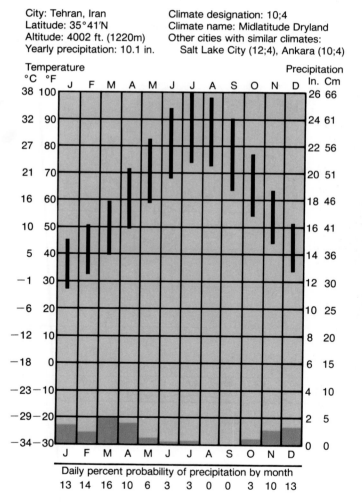

Daily percent probability of precipitation by month
13 14 16 10 6 3 3 0 0 3 10 13

Figure 3.46
Climagraph for Tehran, Iran. (See Figure 3.31 for explanation.)

Subarctic and Arctic Climates (16;7)

Toward the north and in the interior of the North American and Eurasian land masses, colder and colder temperatures prevail (Figures 3.48 and 3.49). Trees become stunted, and eventually only mosses and other cool-weather plants of the type shown in Figure 3.50 will grow. Because long, cold winters predominate, the ground is frozen most of the year. Very often the word **tundra** is used to describe the northern zone beyond the forested subarctic regions. A few cool summer months, with an abundant supply of mosquitoes, break up the monotony of extremely cold but not very snowy conditions. Strong easterly winds blow the snow, which, combined with ice fogs and little winter sunlight, contributes to a very bleak climate indeed. Alaska, northern Canada, and the northern USSR are covered with the stunted trees of the subarctic climate or by the bleak, treeless expanse of the tundra. Antarctica and Greenland, however, are icy deserts.

Soils in these vast arctic regions are varied, but perhaps most characteristic are *histosols*. These soils tend to be peat or muck, consisting of plant remains that accumulate in water. In forested areas, they will tend more toward spodosols. They form in areas of poor drainage both in the arctic and the middle latitudes.

Summary These rough sketches of climatic conditions throughout the world give us the basic patterns of the larger regions. On any given day, conditions may be quite different from those discussed or mapped, but it is the physical climatological processes in general that concern us.

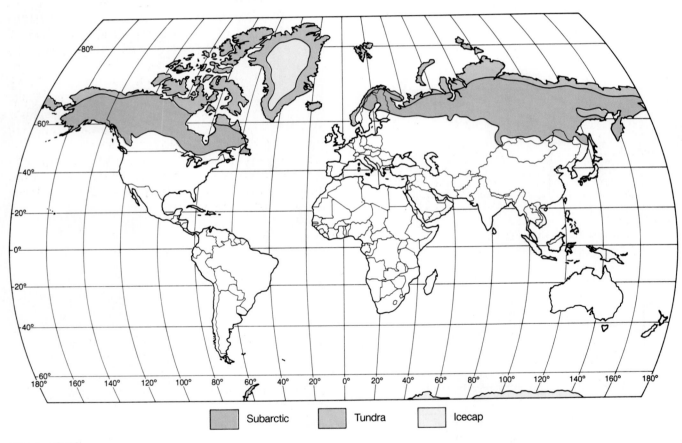

Subarctic   Tundra   Icecap

Figure 3.48
The location of arctic and subarctic climates.

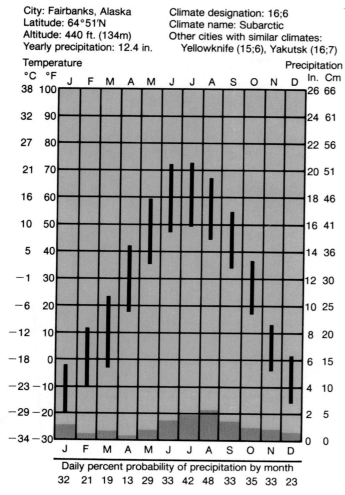

City: Fairbanks, Alaska
Latitude: 64°51′N
Altitude: 440 ft. (134m)
Yearly precipitation: 12.4 in.

Climate designation: 16;6
Climate name: Subarctic
Other cities with similar climates:
    Yellowknife (15;6), Yakutsk (16;7)

Daily percent probability of precipitation by month
32   21   19   13   29   33   42   48   33   35   33   23

Figure 3.49
Climagraph for Fairbanks, Alaska. (See Figure 3.31 for explanation.)

Climatic Change

It has been stressed that climates are only averages of perhaps greatly varying day-to-day changes. The global variation in precipitation from year to year is shown in Figure 3.51. Temperatures are less changeable than precipitation on a year-to-year basis, but they vary, too. How can we account for these variations? Scientists in research stations all over the world are investigating this question. The data they use range from daily temperature and precipitation records to calculations of the position of the earth relative to the sun. Because day-to-day records for most places go back only 50 to 100 years, scientists look for additional information in rock formations, the chemical composition of earth materials, evidence of past vegetation, and astronomical changes.

The fact that a glacier covered most of Canada and about one-third of the United States about 20,000 years ago as well as at three other times in the last 1 million years, has raised the question of periodic changes in climate. By analyzing layers of fossil microorganisms on the ocean floor, scientists have determined that for the last 400,000 years, there have been climatic cycles of 100,000 years' duration. Some scientists believe that this 100,000-year cycle corresponds to the cyclical changes in the earth's orbit around the sun, and that such changes are responsible for the succession of the ice ages. When the orbit is nearly circular, the earth experiences relatively cold temperatures, and when it is elliptical, as it is now, the earth is exposed to more total solar radiation and thus experiences warmer temperatures.

Figure 3.50
Tundra vegetation in the arctic.

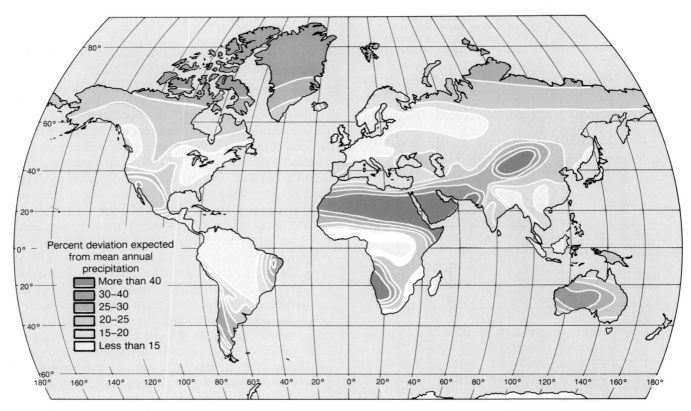

Figure 3.51

The world pattern of precipitation variability. Note that regions of low total precipitation tend to have high variability. In general, the lower the amount of long-term average annual precipitation, the lower is the probability that the "average" will be recorded in any single year.

Our Inconstant Climates

In recent years archaeologists and historians have found evidence that ancient seats of power—Sumeria in the Middle East, Mycenae in southern Greece, or Mali in Africa—may have fallen not to barbarians but to unfavorable alterations in the climates under which they came to power. Between A.D. 1550 and 1850, a "little ice age" descended on the northern hemisphere. Arctic ice expanded, glaciers advanced, and drier areas of the earth were desiccated. Crop failures were common; Iceland lost one-quarter of its population between 1753 and 1759, and 1816 was the "year

without a summer" in New England, with snow in June and frost in July.

A pronounced warming trend began about 1890 and peaked by 1945. During that period, the margin of agriculture was extended northward, the pattern of commercial fishing shifted poleward, and the reliability of crop yields increased. The equable climate, plus medical advances, presumably did much to encourage accelerating growth in the world's population.

Now, ample evidence tells us that climate is again changing. From the 1940s to the 1970s, the mean temperature of the globe declined; the growing season in England became two weeks shorter; disastrous droughts occurred in

Africa and Asia; unexpected freezes altered crop patterns in Latin America.

The recent global cooling trend altered atmospheric circulation and rainfall. The *circumpolar vortex,* high-altitude winds circling the poles from west to east, moved nearer the equator than in previous decades. Its stability occasionally prevented the expected northward movement of seasonal monsoon winds in the northern hemisphere and widened and relocated desert conditions associated with the subtropical high. In recent years, however, scientists have uncovered substantial evidence that indicates that the cooling trend may be more than offset by the effects of increased carbon dioxide in the atmosphere (see Chapter 4).

The tilt of the earth relative to the orbital plane follows another cycle of 42,000 years. The tilt varies from 22.1° to 24.5°. A low tilt position—that is, a more perpendicular position of the earth—is accompanied by periods of colder climate. Cooler summers are thought to be critical in the formation of ice sheets.

Even though the last ice age ended about 11,000 years ago and the earth is now in one of its warmest periods, scientific evidence suggests that there will be substantially more ice on earth 3000 years from now. Climate can change more abruptly than this time scale suggests, however. Relatively small changes in upper-air wind movements can change a climate significantly in decades. Today, polar air is more dominant than tropical air. Changes in the amounts of precipitation or in the reliability of precipitation, as well as temperature, can have an enormous impact on agriculture and patterns of human settlement. In India, monsoon winds have fluctuated greatly in intensity recently, and in northern Africa, summer rains have periodically failed for a number of years.

Great volcanic eruptions that spread dust particles around the world may also produce climatic change because the dust blocks the sun's rays to some extent. For three years after the volcano Krakatoa near Java erupted in 1883, there was a noticeable decline in temperatures throughout the world.

Finally, some scientists are concerned about the effect of human civilization on climatic change. There is evidence that industrial and automobile emissions are changing the gaseous mixture in the atmosphere. These changes are among the topics discussed in Chapter 4, dealing with environmental concerns.

Conclusion

In this chapter we introduced various concepts and terms that are used to understand weather and climate. We identified solar energy as the great generator of the main weather elements of temperature, moisture, and atmospheric pressure. Spatial variations in these elements are caused both by the earth's broad physical characteristics, like greater solar radiation at the equator than at the poles, and by local physical characteristics, such as the effect of water bodies or nearby mountains on local weather conditions.

Climate regions help us to simplify the complexities that arise from such special conditions as monsoon winds in Asia or cold ocean currents off the west coast of South America. In just a few descriptive sentences, it is possible to identify the essence of the wide variations in weather conditions that may exist at a place over the course of a year. When one says that Seattle is in a region of marine west coast climate, for example, images of the average daily conditions communicate to us not only the facts of weather at a place but also the reasons for conditions being as they are. In addition, knowledge of climate tells us about the conditions within which one carries out life's daily tasks.

Weather and climate are among the building blocks that help us to better understand the role of humans on the surface of the earth. Although in the following chapters attention will be devoted mainly to the characteristics of human cultural landscapes, one should keep in mind that the physical landscape significantly affects human behavior.

Key Words

air mass *102*
air pressure *93*
biochore *108*
climate *86*
convection *94*
convectional
  precipitation *100*
Coriolis effect *95*
cyclone *104*
cyclonic or frontal
  precipitation *102*
dew point *100*

El Niño *100*
front *103*
humid continental
  climate *115*
humid subtropical
  climate *114*
hurricane *104*
insolation *87*
jet stream *96*
lapse rate *90*
marine west coast
  climate *114*

Mediterranean
  climate *113*
monsoon *96*
mountain breeze *94*
North Atlantic drift *97*
orographic
  precipitation *101*
precipitation *98*
pressure gradient *94*
reflection *89*
relative humidity *98*
reradiation *89*

savanna *110*
source region *102*
steppe *117*
temperature
  inversion *92*
tornado *105*
tropical rain forest *109*
troposphere *86*
tundra *118*
valley breeze *95*
weather *86*

For Review

1. What is the difference between *weather* and *climate?*

2. What determines the amount of insolation received at a given point? Does all potentially receivable solar energy actually reach the earth? If not, why?

3. How is the atmosphere heated? What is the *lapse rate,* and what does it indicate about the atmospheric heat source? Describe a *temperature inversion.*

4. What is the relationship between atmospheric pressure and surface temperatures? What is a *pressure gradient,* and of what concern is it in weather forecasting?

5. In what ways do land and water areas respond differently to equal insolation? How are these responses related to atmospheric temperatures and pressures?

6. Draw and label a diagram of the planetary wind and pressure system. Account for the occurrence and character of each wind and pressure belt. Why are the belts latitudinally ordered?

7. How is *relative humidity* affected by changes in air temperatures? What is the relative humidity at the *dew point?*

8. What are the three types of large-scale precipitation? How does each occur?

9. Describe the development of a cyclonic storm, showing how it is related to air masses and fronts.

10. What factors were chiefly responsible for today's weather?

11. Summarize the distinguishing temperature, moisture, vegetation, and soil characteristics of each type of climate.

12. What is the climate of Tokyo, London, São Paulo, Leningrad, Bangkok?

Suggested Readings

Bryson, Reid A., and Thomas J. Murray. *Climates of Hunger: Mankind and the World's Changing Weather.* Madison, Wis.: University of Wisconsin Press, 1977.

Cole, Franklyn W. *Introduction to Meteorology.* New York: John Wiley, 1980.

Hays, J. D., J. Imbrie, and N. J. Schackleton. "Variations in the Earth's Orbit: Pacemaker of the Ice Ages," *Science* 194 (December 10, 1976), pp. 1121–32.

Henderson-Sellers, Ann, and Peter J. Robinson. *Contemporary Climatology.* New York: John Wiley, 1986.

Lydolph, Paul E. *Weather and Climate.* Totowa, N.J.: Rowman and Allanheld, 1985.

Muller, R. A., and Theodore M. Oberlander. *Physical Geography Today.* 3d ed. New York: Random House, 1984.

Oliver, John E., and John J. Hidore. *Climatology: An Introduction.* Columbus, Ohio: Charles E. Merrill, 1984.

Stewart, George R. *Storm.* New York: Random House, 1941.

Strahler, Arthur N. *Physical Geography.* 4th ed. New York: John Wiley, 1975.

Trewartha, Glenn T., and Lyle H. Horn. *An Introduction to Climate.* 5th ed. Madison, Wis.: University of Wisconsin Press, 1981.

Weatherwise. A periodical issued six times a year by Weatherwise, Inc., 230 Nassau St., Princeton, N.J. 08540.

4 Human Impact on the Environment

*T*imes Beach, Missouri, used to be a tight-knit, working-class community along the Meramec River southwest of St. Louis. Now the people are gone. Empty houses and stores, abandoned cars, and rusting refrigerators stand silent. "Thanks for coming," says the sign on the Easy Living Laundromat to nonexistent customers. Wildflowers bloom on overgrown lawns. Flies buzz, squirrels chase each other up trees, and an occasional coyote prowls the streets. Were they allowed, visitors might think they were seeing an uncannily realistic Hollywood stage set. Instead, the ghost town is a symbol of what happens when hazardous wastes contaminate a community.

The trouble began in the summer of 1971, when Times Beach hired a man to spread oil on its 10 miles of dirt roads to keep down the dust. Unfortunately, his chief occupation was removing used oil and other waste from a downstate chemical factory, and it was a mixture of these products that he sprayed all over town that summer and the next. The tens of thousands of gallons of purple sludge were contaminated with dioxin, the most toxic substance ever manufactured.

Effects of the dioxin exposure appeared almost immediately in two riding arenas that had been sprayed. Within days, hundreds of birds and small animals died, kittens were stillborn, and many horses became ill and died. Then residents began reporting a variety of medical disorders, including miscarriages, seizure disorders, liver impairment, and kidney cancer. Not until 1982 did the Environmental Protection Agency (EPA), alerted to high levels of dioxin in wastes stored at the chemical plant, make a thorough investigation of Times Beach. They found levels of the dioxin compound TCDD as high as 300 parts per billion in some of the soil samples; 1 part per billion is the maximum concentration deemed safe. The EPA purchased every piece of property in town and ordered its evacuation.

The story of Times Beach is one of many that could have been selected to illustrate how people can affect the quality of the water, air, and soil on which their existence depends.

Terrestrial features and ocean basins, elements of weather and characteristics of climate, flora, and fauna comprise the building blocks of that complex mosaic called the **environment.** It is an overworked word that means the totality of things that in any way may affect an organism. Humans exist within a natural environment—the sum of the physical world—that they have modified by their individual and collective actions. Forests have been cleared, grasslands plowed, dams built, and cities constructed. On the natural environment, then, has been erected a cultural environment, modifying, altering, or destroying the balance of nature that existed before human impact was expressed. This chapter is concerned with the interrelation between humans and the natural environment that they have so greatly altered and endangered.

Since the beginnings of agriculture, humans have changed the face of the earth, have distorted delicate balances and interplays of nature, and, in the process, have both enhanced and endangered the societies and the economies that they have erected. The essentials of the natural balance and the ways in which humans have altered it are not only our topics here but are also matters of social concern that rank among the principal domestic and international issues of our times. As we shall see, the fuels we consume, the raw materials we use, the products we create, and the wastes we discard all contribute to the harmful alteration of the **biosphere** or **ecosphere**—the thin film of air, water, and earth within which we live.

Ecosystems

This biosphere is composed of three interrelated parts: (1) the *atmosphere,* some 6–7 miles (9.5–11.25 km) thick; (2) surface and subsurface waters in oceans, rivers, lakes, glaciers, or groundwater (*hydrosphere*)—much of it locked in ice or earth and not immediately available for use; and (3) the upper reaches of the earth's crust (*lithosphere*), a few thousand feet at most, containing the soils that support plant life, the minerals that plants and animals require to exist, and the fossil fuels and ores that humans exploit. The biosphere is an intricately interlocked system, containing all that is needed for life, all that is available for life to use, and, presumably, all that ever will be available. The ingredients of the thin ecosphere must be, and are, constantly recycled and renewed in nature: plants purify the air; the air helps to purify the water; the water and the minerals are used by plants and animals and are returned for reuse.

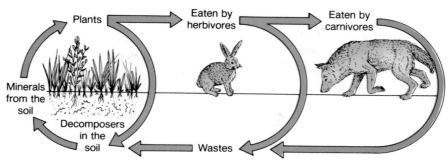

Figure 4.1

A simplified example of an ecosystem illustrating the interdependence of organisms and the physical environment.

The biosphere, therefore, consists of two inextricable components: (1) a nonliving outside (solar) energy source and requisite chemicals, and (2) a living world of plants and animals. In turn, the biosphere may be subdivided into specific **ecosystems,** which are made up of the population of organisms (*biomes*) existing together in a particular area and of the necessary base environment of energy, air, water, soil, and chemicals. Figure 4.1 depicts one kind of ecosystem and in crude form suggests the most important principle concerning all ecosystems: everything is interconnected. Any intrusion or interruption in the balance that has been naturally achieved inevitably results in dislocations and generally unexpected and perhaps undesirable effects elsewhere in the system. Each organism occupies a specific and essential *niche,* or place, within an ecosystem. In the energy exchange system, each organism plays a definite role; individual organisms survive because of other organisms that also live in that environment. The problem lies not in recognizing the niches but in anticipating the chain of causation and the readjustments of the system consequent on disturbing the occupants of a particular niche.

A case study of the Everglades ecosystem may be found in Chapter 12, pages 411–12.

Food Chains and Renewal Cycles

Life depends on the energy and nutrients flowing through an ecosystem. The transfer of energy from one organism to another is one link in a **food chain,** defined as the sequence of such transfers within a biome. Some food chains have only two links, as when human beings eat rice; most have only three or four. Because the ecosystem in nature is in a continuous cycle of integrated operation, there is no "start" or "end" to a food chain; there are, simply, nutritional transfer stages in which each lower level in the food chain transfers part of its contained energy to the next-higher-level consumer.

The *decomposers* pictured in Figure 4.1 are essential in maintaining food chains and the cycle of life. They cause the disintegration of organic matter—animal carcasses and droppings, dead vegetation, paper, and so on. In the process of decomposition, the chemical nature of the material is changed, and the nutrients contained within it become available for reuse by plants or animals. *Nutrients,* the minerals and other elements that organisms need for growth, are never destroyed; they keep moving from living to nonliving things and back again. Our bodies contain nutrients that were once part of other organisms, perhaps a hare, a hawk, or an oak tree.

The idea of a cycle is important in furthering our understanding of the natural renewal of the elements essential to life, which include carbon, oxygen, hydrogen, and phosphorus. Because the supply of these materials is fixed, they must be continuously recycled through food chains. Figures 4.2 and 4.3 show two important cycles, the phosphorus cycle and the hydrologic cycle.

Phosphorus is an element essential to the survival of many species of plants and animals. It is an important component of DNA and of bones and teeth. Plants absorb phosphate from the soil; other organisms acquire it by eating plants. It is returned to the soil through animal excretions or through the decay of plants and animals. When phosphate that has been used as an agricultural fertilizer is carried to rivers and eventually to the sea through runoff, only a small percentage is returned to the land through fish catches and the excretions (guano) of sea birds, which eat fish. Most of it remains in deep ocean waters until geologic processes cause uplifting of the ocean floor.

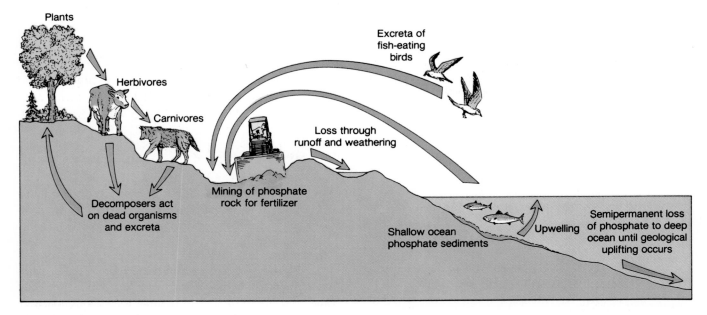

Figure 4.2

The phosphorous cycle. The natural phosphorus cycle is
exceedingly slow and has been upset by human activity.
Currently, phosphorous is being washed into the sea faster than it
is being returned to the land, primarily because it is being mined
for use as a fertilizer.

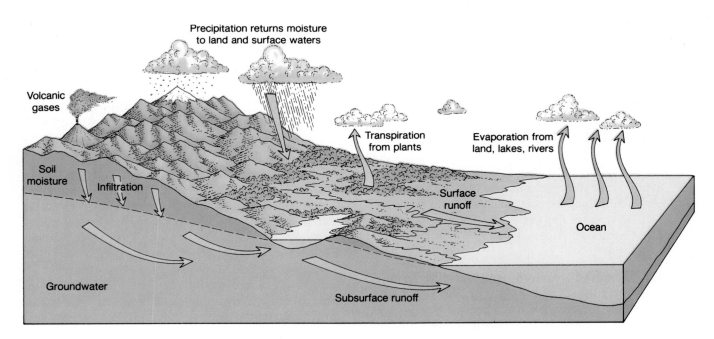

Figure 4.3

The hydrologic cycle. The sun provides energy for the
evaporation of fresh and ocean water. The water is held as vapor
until the air becomes supersaturated. Atmospheric moisture is
returned to the earth's surface as solid or liquid precipitation to
complete the cycle. Because precipitation is not uniformly
distributed, moisture is not necessarily returned to areas in the
same quantity as it has evaporated from them. The continents
receive more water than they lose; the excess returns to the seas
as surface water or groundwater. A global water balance,
however, is always maintained.

Food chains, systems of energy transfer, and chemical element recycling are all in natural balance and sequence in undisturbed biomes. Humans, however, the one universally dominant species, have had a long record of altering the flow of these cycles. The impact of humans on ecosystems—small at first, with low population size, energy consumption, and technological levels—has increased so rapidly and pervasively as to present us with widely recognized and varied ecological crises. Some of the effects of humans on the natural environment are the topic of the remainder of this chapter.

Impact on Water

The supply of water is constant; the system by which it is continuously circulated through the biosphere is called the **hydrologic cycle** (Figure 4.3). In that cycle, water may change form and composition, but under natural environmental circumstances, it is marvelously purified in the recycling process and is again made available with appropriate properties to the ecosystems of the earth. *Evaporation* and *transpiration* (the emission of water vapor from plants) are the mechanisms by which water is redistributed. Water vapor collects in clouds, condenses, and then falls again to earth. There it is reevaporated and retranspired, only to fall once more as precipitation.

Humans have a discernible impact on the hydrologic cycle by affecting the speed of water recirculation and by hastening the return of water to the sea. Massive withdrawals of water from surface supplies by thermal power plants or by industry for cooling purposes can abbreviate the normal cycle by prematurely converting water from a liquid to its vaporous state. Comparable accelerated vaporization occurs through evaporation from reservoirs, particularly in arid regions. The amount of usable water available at some point in the cycle is thereby reduced.

Because the need for water will continue to grow as the world population increases, many scientists fear a coming water crisis. Agriculture currently accounts for almost 80% of all the water used by people, most of it for irrigation. The Food and Agriculture Organization (FAO) of the United Nations has estimated that the amount of water used for irrigation will have to double to feed all those who will be alive in the year 2000. Regional water shortages will affect developed as well as less developed countries. For example, agriculture in large parts of Kansas, Nebraska, Colorado, and Texas draws on the water stored in a vast underground formation called the Ogallala Aquifer. (An *aquifer* is water-bearing, porous rock lying between impermeable layers.) The water bed is less than 200 feet thick in part of the area; yet, because large amounts are withdrawn for irrigated agriculture, the water table falls from 2 to 5 feet each year. Preliminary studies indicate that if present rates of pumping continue, as much as 40% of the irrigated acreage in portions of the Great or High Plains States will be lost by the year 2020 (Figure 4.4).

Urbanization and industrialization also place growing demands on water supplies. Some 150 tons (40,000 gallons) of water are used to produce a ton of steel; 250 tons are needed to make a ton of paper. As more and more countries industrialize, pressure on the water supply will increase.

People's dependence on water has long led to efforts to control its supply. Such manipulation has altered the quantity and the quality of water in rivers and streams.

Modification of Rivers and Streams

To prevent flooding, to regulate the water supply for agriculture and urban settlements, or to generate power, people have for thousands of years manipulated rivers by constructing dams, canals, and reservoirs. Although they generally have achieved their purposes, these structures can have unintended environmental consequences. These include reduction in the sediment load downstream, followed by a reduction in the amount of the nutrients available for crops and fish; an increase in the salinity of the soil; and subsidence (see page 144 for a discussion of subsidence).

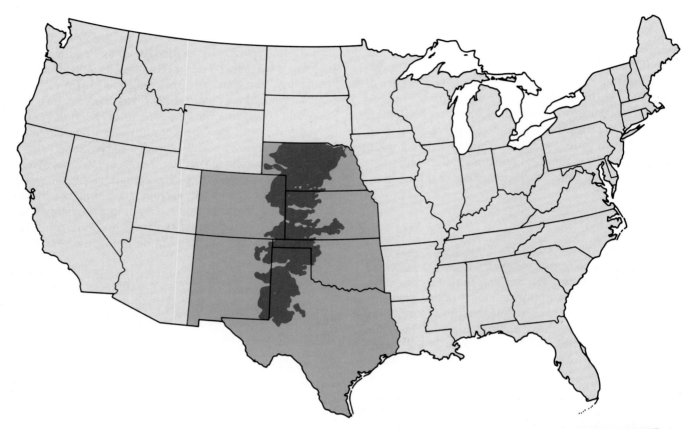

Figure 4.4

The giant Ogallala aquifer, the nation's biggest underground water supply. Some 20 million acres (8 million hectares) of land are now watered by pumping from the aquifer, which supports about one-quarter of the country's cotton crop and a great deal of its corn and wheat. About 40% of the nation's cattle are fed grain grown in the High Plains region. Approximately 300 gallons of water in the field are needed to grow enough wheat for one loaf of bread, 4200 gallons to produce one pound of beef (one-half Kg) for the table—figures that help explain why the aquifer is being depleted faster than it can be replenished by nature.

(above) Source: High Plains Association, Six State High Plains—Ogallala Aquifer Regional Resources Study, A Report to the U.S. Department of Commerce and the High Plains Study Council, March 1982. (at right) Source: After U.S. Geological Survey, 1982 Annual Report.

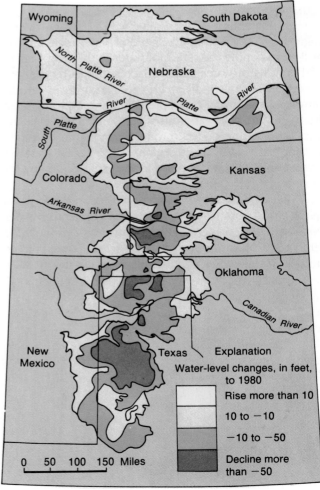

Explanation

Water-level changes, in feet, to 1980

Rise more than 10

10 to −10

−10 to −50

Decline more than −50

0   50   100   150 Miles

Hydropower Negatives

Generation of hydroelectric power in developing nations has the potential to bring those areas plentiful energy for economic growth. The great tropical river systems have a sizable percentage of the world's undeveloped power potential. Excluding the Amazon itself, the tributary rivers of the Amazon Basin of South America have a total energy potential of 100,000 megawatts, the equivalent of oil wells eternally producing 5 million barrels a day. The Zaire of Africa and Salween of Asia are other rain forest rivers with great power potential.

Tropical rain forests have proved to be a particularly difficult environment for dam builders, however, and the dams often carry a heavy ecological price. The creation of Brokopondo in Suriname in 1964 marked the first large reservoir in a rain forest locale. Without being cleared of their potentially valuable timber, 570 square miles (1480 km²) of dense tropical forest disappeared underwater. As the trees decomposed, producing hydrogen sulfide, an intolerable stench polluted the atmosphere for scores of miles downwind. For more than two years, employees at the dam wore gas masks at work. Decomposition of vegetation produced acids that corroded the dam's cooling system, leading to costly continuing repairs and upkeep.

Water hyacinth spreads rapidly in tropical impoundments, its growth hastened by the rich nutrients released by tree decomposition. Within one year of its completion, a 50-square-mile (130 km²) blanket of the weed was afloat on Lake Brokopondo, and after another year almost half the reservoir was covered. Another 170 square miles (440 km²) were claimed by a floating fern, *Ceratopteris*. Identical problems have plagued other tropical power projects.

Impacts on humans are as great and as serious as those on the environment. In many nations, settlement is concentrated along the very rivers most promising for power development. The Aswan High Dam in Egypt displaced 80,000 residents; the Lake Volta project in Ghana forced the relocation of 75,000.

Tropical reservoirs have expanded the breeding areas for snails that spread schistosomiasis, a disease causing potentially fatal damage to liver, kidneys, and other organs. Malaria, too, typically increases following river impoundments. The Tucuruí Dam on the Tocantins River in the Amazon Basin of Brazil has flooded more than 760 square miles (2000 km²) of forest. During periods of reservoir drawdown, up to 340 square miles (880 km²) of ideal mosquito breeding area will be exposed.

Finally, the expense, the human disruption, and the environmental damage all may be in vain. Deforestation of river banks and clearing of vegetation for permanent agriculture may result in accelerated erosion, rapid sedimentation of reservoirs, and drastic reduction of electrical generating capacity. The Ambuklao Reservoir in the Philippines, built with an expected pay-back period of 60 years, now appears certain to silt up in half that time. The Anchicaya Reservoir in Colombia was almost totally filled with silt within 10 years of its completion.

Channelization, another method of modifying river flow, is the construction of embankments and dikes and the straightening, widening, and/or deepening of channels to control floodwaters or to improve navigation. Many of the great rivers of the world, including the Nile and the Huang He, are lined by embankment systems. Like dams, these systems can have unforeseen consequences. They reduce the natural storage of floodwaters, can aggravate flood peaks downstream, and can cause excessive erosion. Some of the possible ecological effects of channelization are shown in Figure 4.5.

Channelization and dam construction are deliberate attempts to modify river regimes, but other types of human action also affect river flow. *Urbanization,* for example, has significant hydrologic impacts, including a lowering of the water table, pollution, and increased flood runoff. Likewise, *deforestation* (the removal of forest cover) increases runoff, promotes flash floods, lowers the water table, and hastens erosion. Nevertheless, the primary adverse human impact on water is felt in the area of water quality. People withdraw water from lakes, rivers, or underground deposits to use for drinking, bathing, agriculture, industry, and many other purposes. Although the water that is withdrawn returns to the water cycle, it is not always returned in the same condition as at the time of withdrawal. Water, like other segments of the ecosystem, is subject to serious problems of pollution.

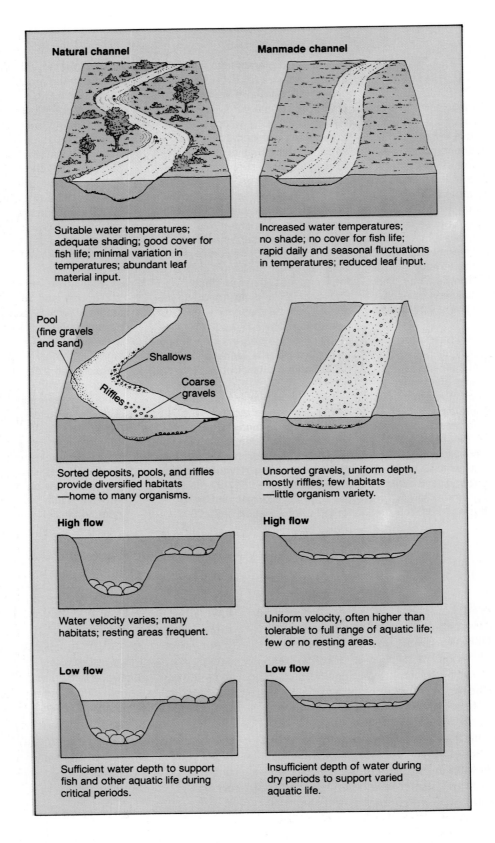

Natural channel

Suitable water temperatures; adequate shading; good cover for fish life; minimal variation in temperatures; abundant leaf material input.

Pool (fine gravels and sand)
Shallows
Riffles
Coarse gravels

Sorted deposits, pools, and riffles provide diversified habitats —home to many organisms.

High flow

Water velocity varies; many habitats; resting areas frequent.

Low flow

Sufficient water depth to support fish and other aquatic life during critical periods.

Manmade channel

Increased water temperatures; no shade; no cover for fish life; rapid daily and seasonal fluctuations in temperatures; reduced leaf input.

Unsorted gravels, uniform depth, mostly riffles; few habitats —little organism variety.

High flow

Uniform velocity, often higher than tolerable to full range of aquatic life; few or no resting areas.

Low flow

Insufficient depth of water during dry periods to support varied aquatic life.

Figure 4.5
Comparison of a natural channel with a channelized stream. Some of the possible effects of channelization are depicted. Dredging and filling (the deposition of dredged materials to create new land) are other ways in which people modify the configuration of water bodies.

TABLE 4.1
Sources and Types of Water Pollutants

Source of waste	Kind of pollutant
Municipalities and residences	Human wastes, detergents, garbage, trash
Urban drainage	Suspended sediment, fertilizers, pesticides, road salt
Agriculture	
Crop production	Fertilizers, herbicides, pesticides, erosion sediment
Irrigation return flow	Mineral salts and erosion sediment
Industry	The widest possible range of pollutants, including biodegradable wastes in the paper and food-processing industries, heat discharge in a variety of activities, nondegradable wastes in the chemical and iron and steel industries, and radioactivity.
Mining	Acids, sediment, metal wastes, culm (coal dust)
Electric power production	Heat and nuclear wastes
Recreation and navigation	Human wastes, garbage, fuel wastes

Reproduced from *Environmental Issues* by Lawrence G. Hines, by permission of W. W. Norton & Co., Inc., copyright © 1973 by W. W. Norton & Co., Inc.

The Meaning of Water Pollution

As a general definition, **environmental pollution** by humans means the introduction into the biosphere of wastes that, because of their volume or their composition or both, cannot be readily disposed of by natural recycling processes. In the case of water, the central idea is that pollution exists when water composition has been so modified by the presence of one or more substances that either it cannot be used for a specific purpose or it is less suitable for that use than it was in its natural state. Pollution is brought about by the discharge into water of substances that cause unfavorable changes in its chemical or physical nature or in the quantity and quality of the organisms living in the water. Pollution is a relative term. Water that is not suitable for drinking may be completely satisfactory for cleaning streets. Water that is too polluted for fish may provide an acceptable environment for certain water plants.

Human activity is not the only cause of water pollution. Leaves that fall from trees and decay, animal wastes, oil seepages, and other natural phenomena may affect water quality. There are natural processes, however, to take care of such pollution. Organisms in water are able to degrade, assimilate, and disperse such substances in the amounts in which they naturally occur. Only in rare instances do natural pollutants overwhelm the cleansing abilities of the recipient waters. What is happening now is that the quantities of wastes discharged by humans often exceed the ability of a given body of water to purify itself. In addition, humans are introducing pollutants such as metals or inorganic substances that cannot be broken down at all by natural

mechanisms, or that take a very long time to break down. Table 4.1 shows the kinds of water pollutants associated with different sources.

As long as there are people on earth, there will be pollution. Thus the problem is one not of eliminating pollution but of controlling it. This is particularly true in the technologically advanced countries. The more developed and affluent the country, the more resources it uses and the more it pollutes.

Chemical Pollution by Agriculture

Agriculture is a chief contributor of excess nutrients to water bodies. Pollution occurs when nitrates and phosphates that have been used in fertilizers and that are present in animal manure drain into streams and rivers, eventually accumulating in ponds, lakes, and estuaries. The nutrients hasten the process of **eutrophication,** or the enrichment of waters by nutrients. Eutrophication occurs naturally when nutrients in the surrounding area are washed into the water, but when the sources of enrichment are artificial, as is true of commercial fertilizers, the body of water may become overloaded with nutrients. The end result may be an oxygen deficiency in the water. Scientists have estimated that as many as one third of the medium and large-sized lakes in the United States have been affected by accelerated eutrophication. Symptoms of a eutrophic lake are prolific weed growth, large masses of algae, fish kills, rapid accumulation of sediments on the lake bottom, and water that has a foul taste and odor.

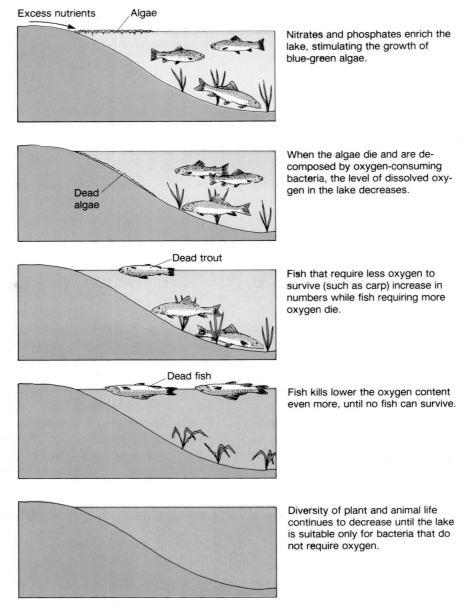

Figure 4.6

Eutrophication is hastened by artificial sources of nutrients. Although eutrophication is primarily a result of agricultural activities, additional sources of nitrates in both surface and underground water supplies are urban drainage, industrial wastewater, and septic tanks.

Figure 4.6 illustrates one form that overfertilization of a water body can take. Algae and other plants are stimulated to grow abundantly. When they die, the level of dissolved oxygen in the water decreases, primarily because of the bacteria acting on the dead and decomposing vegetation. Fish and plants that cannot tolerate the poorly oxygenated water are eliminated. In addition to being potentially lethal for fish, eutrophication affects the suitability of water for drinking and bathing, because excess nutrients have been shown to pose a health hazard to humans.

The herbicides and pesticides used in agriculture are another source of the chemical pollution of water bodies. Runoff from farms where such *biocides* have been applied contaminates both ground and surface waters. One of the problems connected with the use of biocides is that the long-term effects of such usage are not always immediately known. DDT, for example, was used for many years before people discovered its effect on birds, fish, and water plant life. One of the dilemmas that we are facing is that we have not yet discovered how to balance our need for increased agricultural yields in the short run—and the manipulation of the environment that it entails—with what is in our best interest in the long run.

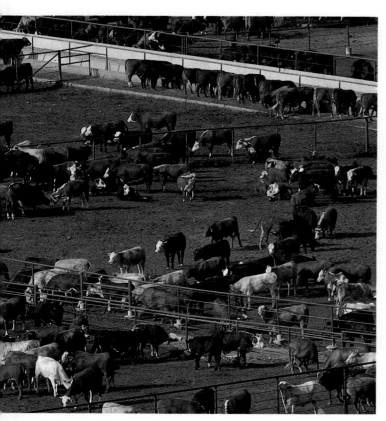

Figure 4.7
A feedlot near San Angelo, Texas. The sanitary disposal of organic wastes generated by such concentrations of animals is a problem only recently addressed by environmental protection agencies.

A final agricultural source of chemical pollution is animal wastes, especially in countries where animals are raised intensively. This is a problem particularly in feedlots, where animals are crowded together at maximum densities to be fattened before slaughter. Large feedlots, such as the one pictured in Figure 4.7, may produce as much waste as would a large city. It is estimated that animal wastes in the United States total about 1.5 billion tons per year, with feedlots generating about half of the total. If not treated properly, the manure pollutes both soil and water with infectious agents and excess nutrients.

Chemical Pollution by Other Sources

Worldwide, agriculture probably contributes more to water pollution than does any other single activity, but in developed countries, industry is equally culpable. In the United States, about half the water used daily is used by industry. Many industries dump organic and inorganic wastes into bodies of water. These may be acids, highly toxic minerals such as mercury or arsenic, or, in the case of petroleum refineries, toxic organic chemicals.

Organisms not adapted to living in water that is thus contaminated die; the water may become unsuitable for domestic use or irrigation; or the wastes may reenter the food chain, with deleterious effects on humans. A particularly unfortunate example brought to the attention of the public in 1953 was of mercury poisoning in Japan. A chemical plant in Minamata Bay that used mercury chloride in its manufacturing process discharged the waste mercury into the bay, where it settled with the mud. Fish that fed on organisms in the mud absorbed the mercury and concentrated it; the fish were in turn eaten by humans. Over 100 people died, or suffered deformity or other permanent disability. Many people fear that the discharge of radioactive isotopes from nuclear power plants may lead to similar contamination of the water supply.

More recently, attention in this country has focused on **polychlorinated biphenyls (PCBs),** a family of related chemicals used in electrical devices, paints, and plastics. During the manufacturing process, companies have dumped PCBs into rivers, from which they have entered the food chain. Several states have banned commercial fishing in lakes and rivers where fish have higher levels of PCBs than are considered safe. A rock bass containing 355 parts per million of PCBs was caught in the Hudson River; had that fish been eaten, the consumer would have taken in one third the lifetime PCB limit as set by the U.S. Food and Drug Administration. Although not all of the effects of PCBs on human health are known, skin eruptions, excessive eye discharges, and possibly cancer of the liver are thought to be linked to them. In 1977, the Environmental Protection Agency banned the direct discharge of PCBs into United States waters.

Other industries that contribute to the chemical pollution of water are the petroleum industry, the nuclear industry, and mining. Oceans, for example, are becoming increasingly contaminated by oil. The greatest single source of petroleum pollution in open ocean regions is the discharge of tank flushings and ballast from tanker holds; other sources include tanker collisions, wrecks, and explosions, and seepage from offshore oil installations. Recent studies indicate that the Gulf of Mexico, the site of extensive offshore drilling, is among the most seriously polluted major bodies of water in the world.

The nuclear industry has caused some water pollution when radioactive material has seeped from the tanks in which the wastes have been buried, either at sea or underground. Surface mining for coal, iron, copper, and other substances contributes to contamination of the water supply through the wastes generated. Rainwater reacts with the wastes, and dissolved minerals seep into nearby water bodies. The exact chemical changes produced depend on the composition of the coal or ore slag heaps, and the reaction of the minerals with sediments or river water.

Figure 4.8
Cooling towers at Three Mile Island nuclear plant near Harrisburg, Pennsylvania. Heated wastewaters are often significantly warmer than the waters into which they are discharged, disrupting the growth, reproduction, and migration of fish populations.

Finally, a host of pollutants derives from the activities associated with urbanization. The use of detergents has increased the phosphorus content of rivers, and salt (used for deicing roads) increases the chloride content of runoff. Water runoff from urban areas contains contaminants from garbage, animal droppings, litter, vehicle drippings, and the like. Sewage can also be a major pollutant, depending on how well it is treated before being discharged. In general, this is less a problem in countries such as the United States than in countries where such treatment is either not practiced or is not thorough. This country is not without such sources of water pollution, however. Many communities still don't have adequate sewage-treatment facilities. The aged sewer system that serves Boston and 42 nearby communities dumps millions of gallons of raw sewage from some 2 million people into Boston Harbor annually. The city of Houston, Texas, among many others, has been allowed to dump raw, untreated human waste into nearby water bodies for years—waste containing viruses responsible for polio, hepatitis, spinal meningitis, and other diseases.

Because the sources of pollution are so varied, the water supply in any single area is often affected by diverse contaminants; this diversity complicates the problem of controlling water quality.

Contaminated drinking-water wells have been found in more than half of the states. Hundreds of wells in the New York metropolitan area have been closed in recent years because of chemical contamination; thousands more may be closed in coming years. Chemicals have reached the groundwater by seeping into aquifers from landfills, from ruptured gasoline and fuel-oil storage tanks, from septic tanks, and from fields sprayed with pesticides and herbicides. The pollution of aquifers is particularly troublesome because, unlike surface waters, groundwater lacks natural cleansing properties; it can remain contaminated for centuries.

Thermal Pollution

Many industrial processes require the use of water as a coolant (Figure 4.8). **Thermal pollution** occurs when water that has been heated is returned to the environment and has adverse effects on the plants and animals in the water body. Many plants and fish cannot survive changes of even a few degrees in the water temperature. They either die or migrate; the species that depend on them for food must also either die or migrate. Thus the food chain has been disrupted. In addition, the higher the temperature of the water, the less oxygen it contains, which means that only lower-order plants and animals can survive.

Controlling Water Pollution

In recent years, concern over increased levels of pollution has brought about major improvements in the quality of some surface waters, both in the United States and abroad. The federal government in 1972 took the lead in regulating water pollution with the enactment of the Clean Water Act. Its objective was "to restore and maintain the chemical, physical, and biological integrity of the nation's waters." Congress established uniform nationwide controls for each category of major polluting industry and directed the government to pay most of the cost of new sewage-treatment plants. Since 1972 such plants have been built to serve over 80 million Americans, and industries have spent billions of dollars to comply with the Clean Water Act by reducing organic waste discharges.

The gains have been impressive. Many rivers and lakes that were ecologically dead or dying are now thriving. Once dumping grounds for all kinds of human and industrial waste, the Hudson, Potomac, Cuyahoga, and Trinity Rivers are cleaner, more inviting, and more productive than before and they now support fishing, swimming, and recreational boating. Similarly, Seattle's Lake Washington and the Great Lakes are healthier than they were two decades ago. Recently, authorities have announced ambitious plans to clean up the waters of Chesapeake Bay, the nation's largest estuary, and to undo much of the damage that has been inflicted on Florida's Everglades by improving the water quality of the Kissimmee River and Lake Okeechobee.

Environmental awareness in other countries has also prompted legislation and action. For example, the River Thames in southern England, which had become seriously contaminated by the dumping of sewage and industrial wastes, is now cleaner than it has been in centuries. The enforcement of stringent pollution-control standards has halted the downward trend in quality. Algae and seaweed, fish and wildfowl have returned to the river in abundance. Even the Mediterranean Sea is on its way to gradual recovery. When the 18 countries that border the sea signed the Convention for the Protection of the Mediterranean Sea Against Pollution in 1976, all coastal cities dumped their untreated sewage into the sea, tankers spewed oily wastes into it, and tons upon tons of phosphorus, detergents, lead, and other substances contaminated the waters. Now many cities have built or are building sewage-treatment plants, ships are prohibited from indiscriminate dumping, and national governments are beginning to enforce control of pollution from land-based sources.

Such gains should not mislead us. While some of the most severe problems have been attacked, serious pollution still plagues about one-fourth of the nation's rivers and streams, lakes and reservoirs. The solution to water pollution lies in the effective treatment of municipal and industrial wastes; the regulation of chemical runoff from agriculture, mining, and forestry; and the development of less polluting technologies. Although pollution-control projects are expensive, the long-term costs of pollution are even higher.

Impact on Air and Climate

The **troposphere,** the thin layer of air just above the earth's surface, contains all the air that we breathe. Every day thousands of tons of pollutants are discharged into the air by cars and incinerators, factories and airplanes. Air that contains substances in sufficient concentrations to have a harmful effect on living things is polluted.

Air Pollutants

Truly clean air has probably never existed. Just as there are natural sources of water pollution, so there are substances that pollute the air without the aid of humans. Ash from volcanic eruptions, marsh gases, smoke from forest fires, and wind-blown dust are natural souces of air pollution.

Normally these pollutants are of low volume and they are widely dispersed throughout the atmosphere. On occasion, a major volcanic eruption may produce so much dust that the atmosphere is temporarily altered. The eruption in 1883 of a volcano on the island of Krakatoa, west of Java, produced dust that lingered in the area for over a year. In general, however, the natural sources of air pollution do not have a significant, long-term effect on air, which, like water, is able to cleanse itself.

Far more important than naturally occurring pollutants are the substances that people cause to be discharged into the air. These pollutants result primarily from burning fossil fuels (coal, gas, and oil) and other materials. Fossil fuels are burned in power plants that generate electricity, in many industrial plants, in home furnaces, and in cars, trucks, buses, and airplanes. Scientists estimate that about three-quarters of all air pollutants come from burning fossil fuels. Industrial processes other than fuel burning, incinerating solid

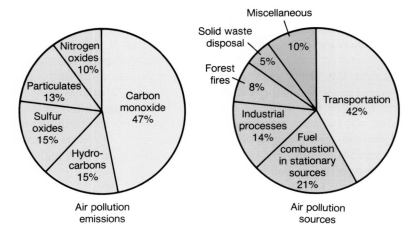

Figure 4.9

Air pollution emissions in the United States by weight. If pollution from vehicles were eliminated, emissions would be reduced by 42%.

Source: National Air Pollution Control Administration, U.S. Department of Health and Human Services.

wastes, forest and agricultural fires, and the evaporation of solvents account for most of the remaining pollutants. Figure 4.9 depicts the major sources and types of air pollutants.

Perhaps the best known air pollutants and the most dangerous to health are *sulfur oxides,* which come primarily from burning fossil fuels. They cause industrial smog, which is associated with respiratory ailments in humans. When unusual weather conditions cause the smog to remain over an area for several days, the result may be a disaster. For example, in 1952, London, England, experienced an acute episode of such pollution, as a result of which 4000 more people died than normal.

Sulfur oxides are also largely responsible for a phenomenon popularly known as acid rain, although acid precipitation is a more precise description. **Acid rain** is a term for pollutants, chiefly oxides of sulfur and nitrogen, that are created by burning fossil fuels and that change chemically as they are transported through the atmosphere and fall back to earth as acidic rain, snow, fog, or dust. When sulfur dioxide is absorbed into water vapor in the atmosphere, it becomes sulfuric acid. Sulfur dioxide contributes about two-thirds of the acids in the rain; about one-third come from *nitrogen oxides,* transformed into nitric acid in the atmosphere. When washed out of the air by rain, snow, or fog, the acids change the *pH factor* (the measure of acidity/alkalinity on a scale of 0 to 14) of the soil and water, setting off a chain of chemical and biological reactions. The average pH of normal rainfall is 5.6, slightly acidic, but acid rainfalls with a pH of 2.4—approximately the acidity of vinegar and lemon juice—have been recorded (Figure 4.10). Acid rain also coats the ground with particles of toxic heavy metals such as cadmium and lead.

The effects of acid precipitation on water bodies, forests, wildlife, and buildings are deadly. Acid rains have been linked to the disappearance of fish in thousands of streams and lakes in New England, Canada, and Scandinavia, and to a decline in fish populations in other lakes. In parts of North America, northern and western Europe, the USSR, China, and Japan, atmospheric acid has rendered the soil less fertile, damaging crops and reducing rates of forest growth. One can see its corrosive effects on marble and limestone sculptures and buildings and on metals such as iron and bronze (Figure 4.11).

Many of the fuel-burning processes already mentioned discharge solid particles of soot and dust, sulfate and fluoride, and asbestos and metallic particles (berylium, arsenic, cadmium, and lead) into the air. Other *particulates* are formed in the air from gases emitted during combustion. Human production of particulates accounts for a significant proportion of the total volume, although natural sources of particulate pollution exist.

Large and medium-sized particles, those greater than 1 micron in diameter, are heavy and tend to wash out of the air rapidly, discoloring buildings, clothing, and cars. The technology is available to collect most of these particles in industrial and power plants before they are discharged into the air. The emission of very fine particulates is more difficult to control. Because they are light, they can remain suspended in the air for days or even weeks. Many are known to have an adverse effect on human health; lung cancers, for example, have been linked to the inhalation of asbestos fibers. The possibility also exists that the increase in particulates in the atmosphere may have a long-term effect on the climate by reducing the amount of solar radiation that reaches the earth's surface.

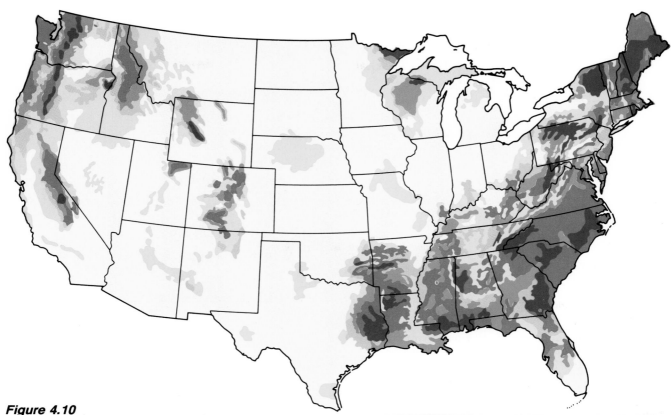

Figure 4.10

Acid precipitation and surface water. The light-colored regions indicate where surface waters are most alkaline and therefore most resistant to damage from acid rain. The darker the shading, the less alkaline are the waters. Because the waters and soil in the darkest areas cannot naturally buffer, or neutralize, the acids deposited by rain or snow, they are the most susceptible to acidification.

Many people fear that the discharge of radioactive pollutants from nuclear power plants and nuclear weapons may lead to the contamination of both air and water, a problem of particular concern because of the long time it takes for some isotopes to decay. The *half-life* of an isotope is the time required for one-half of the atomic nuclei to decay. After one half-life has elapsed, half of the nuclei remain; after another half-life, half of the remainder (one-quarter of the original nuclei) remain, and so on. The half-life of the isotope indicates whether the problem posed by a particular radioactive pollutant is a short- or a long-term one. The half-life of uranium-238, which is used in breeder reactors, is 4.5 billion years. Other isotopes have half-lives measured in fractions of a second.

Figure 4.11

The destructive effect of acid rain may be seen on this limestone statuary at the cathedral in Reims, France. Sulfur emissions could be reduced by restricting the burning of high-sulfur coal, washing coal to remove the sulfur before burning, and fitting plants with gas-stack scrubbers.

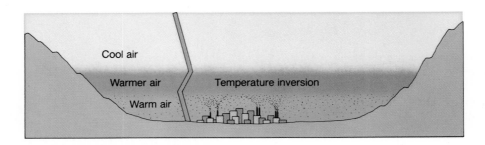

Figure 4.12

Under normal conditions (top), air temperature decreases as altitude increases. During a temperature inversion (bottom), there is a mass of warmer air between surface and upper-altitude air. A temperature inversion intensifies the effect of air pollution; instead of being dispersed, the pollutants are held close to the ground.

Figure 4.13

The brown cloud that hovers over Denver when temperature inversions keep air pollutants from dispersing. Denver frequently has the worst carbon monoxide level in the country from mid-November to mid-January, when cold winter air over the city is trapped by warm, still air at higher altitudes. The city has recently embarked on a campaign to reduce air pollution.

Factors Affecting Air Pollution

Many factors affect the type and the degree of air pollution found at a given place. Those over which people have relatively little control are climate, weather, wind patterns, and topography. These determine whether pollutants will be blown away or whether a buildup is likely to occur. Thus a city on a plain is less likely to experience a buildup than is a city in a valley.

Unusual weather can alter the normal patterns of pollutant dispersal. A **temperature inversion** magnifies the effects of air pollution (Figure 4.12). Under normal circumstances, air temperature decreases away from the earth's surface. If, however, there is a stationary layer of warm, dry air over a region, the normal rising and cooling of air from below is prevented.

As described in Chapter 3, the air becomes stagnant during an inversion. Pollutants accumulate in the lowest layer instead of being blown away, so that the air becomes more and more contaminated. Normally inversions last for only a few hours, although certain areas experience them much of the time. Temperature inversions occur often in Los Angeles in the fall and Denver in the winter (Figure 4.13). If an inversion lingers long enough—for example, several days—it can contribute to the accumulation of air pollutants to levels that seriously affect human health. (See "The Donora Tragedy," Page 92.)

The Tall Smokestack Paradox

The dramatic increase in acid precipitation in recent decades is partly the result of an effort to curb air pollution. Written before concern about acid rain became widespread, the U.S. Clean Air Act of 1970 restricts the deposit of specific pollutants over the surrounding countryside and sets standards only for ground-level air quality.

Most of the millions of tons of sulfur and nitrogen oxides that are released into the atmosphere each year come from "stationary source" fuel combustion, primarily coal- and oil-burning power plants. In order to keep air in local communities clean enough to meet the air-quality standards, industries have since the early 1970s been building ever-taller smokestacks that discharge sulfur dioxide and other pollutants into the upper atmosphere. Stacks 1000 feet (305 m) high are now a common sight at utility plants and factories; previously, stacks 200–300 feet (60–90 m) high were the norm. The situation is not unlike disposing of garbage by throwing it over your backyard fence. It still comes down, but not in your yard. Ironically, the farther and higher the noxious emissions go, the longer they have to combine with other atmospheric components and moisture and form

acids; thus, the taller stacks have directly aggravated the acid rain problem.

Recognizing the problem that the stacks had created, the Environmental Protection Agency in June 1985 issued rules discouraging the use of tall smokestacks to disperse emissions. Plants now must seek ways to reduce rather than simply disperse emissions, for example by installing scrubbers on their stacks or using low-sulfur coal.

Some areas are particularly likely to suffer from **photochemical smog,** which is created when oxides of nitrogen react with the oxygen present in water vapor in the air to form nitrogen dioxide. In the presence of sunlight, nitrogen dioxide reacts with hydrocarbons from automobile exhausts and industry to form new compounds, such as **ozone.** A key ingredient of smog, ozone causes coughing and breathing problems, especially among asthmatics. Warm, dry weather and stagnant air promote ozone (ironically, the same chemical that in the upper atmosphere shields the earth from the damaging effects of ultraviolet radiation). Los Angeles, Salt Lake City, and Denver have the kinds of climate and topography that favor the formation of photochemical smog.

The air pollutants generated in one place may have their most serious effect in areas hundreds of miles away. Thus the worst effects of the photochemical air pollution that originates in New York City are felt in Connecticut and parts of Massachusetts. The chemical reaction that produces ozone takes a few hours, and by that time air currents have carried the pollutants away from the city.

Figure 4.14
Researchers are trying to determine the long-term effect of the cloud cover produced by contrails from jet aircraft.

In a similar fashion, New York City is the recipient of pollutants produced in other places. Much of the acid rain that affects New England is thought to originate in the coal-fired power plants along the lower Great Lakes and in the Ohio Valley that use extremely high smokestacks to disperse sulfurous emissions. And the coal-based industries in the Soviet Union and Europe produce sulfate, carbon, and other pollutants that are transported by air currents to the land north of the Arctic Circle, where they result in a contamination known as **Arctic haze.** Discovered 25 years ago, Arctic haze is more than just aesthetically troubling. It has the potential to affect climates worldwide. Particles of soot both in the air and on the ground absorb the sun's radiation, effectively holding heat in the atmosphere. Conceivably, this could cause ice to melt and hinder ice formation in the sea, altering weather patterns.

Other factors that affect the type and the degree of air pollution at a given place are the levels of urbanization and industrialization. Population densities, traffic densities, the type and density of industries, and home-heating practices all help to determine the kinds of substances discharged into the air at a single point. In general, the more urbanized and industrialized a place is,

the more responsible it is for pollution. The United States may contribute as much as one third of the world's air pollution, a figure roughly equivalent to the proportion of the world's fossil fuel and mineral resources consumed in this country.

Effects of Air Pollution

In recent years, researchers have begun to realize that air pollution has an effect on the major factors that control the temperature of the earth's surface, but how much of an effect is not yet known. The pollution processes that send particulates into the upper atmosphere, as well as the cloud cover produced by contrails from jet planes (Figure 4.14), have an **"icebox" effect** on the earth. The particulates and the clouds reflect incoming sunlight back into space before it reaches the earth. The result is a cooler earth than normal.

On the other side of the heat balance sheet is the **"greenhouse" effect.** The burning of fossil fuels adds carbon dioxide (CO_2) to the atmosphere. CO_2 slows down the reradiation of heat from earth back into space,

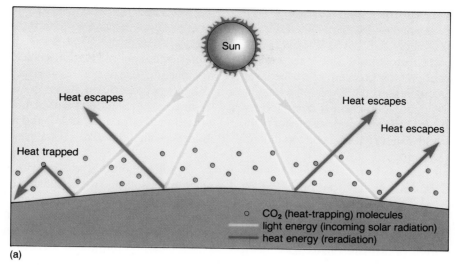

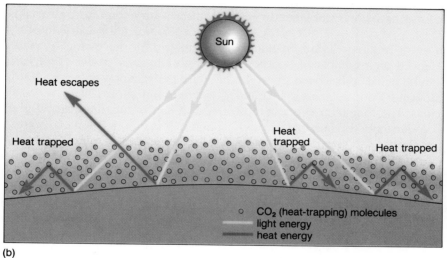

Figure 4.15

When there is a low level of carbon dioxide in the air, as in (a), incoming solar radiation strikes the earth's surface, heating it up, and the earth radiates the energy back into space as infrared light (heat). The "greenhouse" effect, depicted in (b), is the result of billions of tons of CO_2 that the burning of fossil fuels releases into the air each year. The CO_2 molecules absorb some of the reradiated energy, deflecting it groundward and preventing it from escaping from the atmosphere.

acting as glass in a greenhouse does to heat up the climate (Figure 4.15). The more CO_2 there is in the atmosphere, the higher the temperatures on earth should be, and increasing industrial activity over the next one or two centuries is likely to raise the carbon dioxide content of the atmosphere many times.

It should be noted that there is considerable scientific controversy over this presumed consequence of rising carbon-dioxide levels; some authorities suggest that a net cooling—not warming—of the earth's atmosphere results from increases in its CO_2 content. But many scientists believe that doubling the concentration of CO_2, which could occur within 50 years, would raise surface temperatures in the middle latitudes by 3–5° F (1.7–2.8° C), and those in polar regions as much as 15° F

(8.3°C). Even such seemingly small increases in temperature, combined with the increased amount of CO_2, would affect patterns of precipitation, agriculture, and ecological balance. Some analysts suggest that a global warming trend could conceivably result in one or two centuries' time in a melt of glacial ice, which would slowly raise ocean levels and inundate coastal areas.

Also contributing to global warming are emissions from a family of synthetic chemicals developed in 1931 and known as **chlorofluorocarbons (CFCs).** As discussed in Chapter 3, the CFCs are eating away at the earth's stratospheric **ozone layer,** which shields all forms of life on earth from the sun's harmful ultraviolet (UV) radiation. Commercially valuable, CFCs are used as coolants for refrigerators and air conditioners, as aerosol spray propellants, and as a component in foam packaging material. Because CFC emissions disperse rapidly throughout the atmosphere, the problem of ozone-layer depletion is global. As increased UV levels penetrate to the lower atmosphere, ozone depletion is likely to produce an increase in skin cancers and respiratory ailments, reduce crop and fishery yields, and, as noted, contribute to global warming.

The effect of air pollution on plant life has been well documented. Chemicals washed out of the air by rain and snow contaminate soil and water. Acid rains and smog have caused extensive crop damage and the reduction of yields in some areas; for example, citrus groves in the vicinity of Florida phosphate plants have been devastated. Long-term changes in climate induced by air pollution would, of course, have important secondary effects on vegetation.

The short-term effect of air pollution on human health has been evidenced on a number of occasions. Highly polluted air can be lethal, as has been shown by disasters in London, England; Donora, Pennsylvania; and a number of other cities. The long-term effect of most pollutants is less well understood. It is possible that constant exposure to low levels of polluted air damages lungs, increasing the incidence of such respiratory ailments as pneumonia, emphysema, and asthma. Increases in certain cancers and in heart disease may also turn out to be related to air pollution.

Impact on Land and Soils

People have affected the earth wherever they have lived. Whatever we do, or have done in the past, to satisfy our basic needs has had an impact on the landscape. To provide food, clothing, shelter, transportation, and defense, we have cleared the land and replanted it, rechanneled waterways, and built roads, fortresses, and cities. We have mined the earth's resources, logged entire forests, terraced mountainsides, even reclaimed land from the sea. The nature of the changes made in any single area depends on what was there to begin with and how people have used the land. Some of these changes are examined in the following sections.

Landforms Produced by Excavation

Although we tend to think of landforms as "givens," created by natural processes over millions of years, people have played and continue to play a significant role in shaping the physical landscape. Some features are created deliberately, others unknowingly or indirectly. Pits, ponds, ridges and trenches, subsidence depressions, canals, and reservoirs are the chief landform features resulting from excavations. Some date back to Neolithic times, when people dug into chalk pits to obtain flint for toolmaking. Excavation has had its greatest impact within the last two centuries, however, as earth-moving operations have been undertaken for mining; for building construction and for agriculture; and for the construction of transport facilities like railways, ship canals, and highways.

Surface mining, which involves the removal of vegetation, topsoil, and rocks from the earth's surface in order to get at the resources underneath, has perhaps had the greatest environmental impact. Open-pit mining and strip mining are the methods most commonly used.

Open-pit mining is used primarily to obtain iron and copper, sand, gravel, and stone. As Figure 4.16 indicates, an enormous pit remains after the mining has been completed because most of the material has been removed for processing. *Strip* mining is being increasingly employed in this country as a source of coal; more coal per year now comes from strip mines than from underground mines. Phosphate is also mined in this way. A trench is dug, the material is excavated, and another trench is dug, the soil and waste rock being deposited in the empty first trench, and so on. Unless reclamation is practiced, the result is a ridged landscape.

Landscapes marred by vast open pits or unevenly filled trenches are one of the most visible results of surface mining. Thousands of square miles of land have been affected, with the prospect of thousands more to come as the amount of surface mining increases. Damage to the aesthetic value of an area is not the only liability of surface mining. If the area is large, wildlife habitats are disrupted, and surface and subsurface drainage patterns are disturbed. In the United States in recent years, concern over the effect of strip mining has prompted federal and state legislation to increase regulation and to stop the worst abuses. Strip-mining companies are now expected to restore mined land to its original contours and to replant vegetation.

Landforms Produced by Dumping

Excavation in one area often leads to the creation, via dumping, of landforms nearby. Both surface and subsurface mining produce tons of waste and enormous spoil piles. In fact, in terms of tonnage, mining is the single greatest contributor to solid wastes, with about 2 billion tons per year left to be disposed of in this country alone. The normal custom is to dump waste rocks and mill tailings in huge heaps near the mine sites. Unfortunately this practice has secondary effects on the environment.

Figure 4.16
Aerial view (above) of the Bingham Canyon open-pit copper mine in Utah. The pit is over 1500 feet (450 m) deep; the operations cover more than 1000 acres (405 hectares). About 150 square miles (390 km²) of land surface (at right) in the United States are lost each year to the strip mining of coal and other resources.

Carried by wind and water, dust from the wastes pollutes the air, and dissolved minerals pollute nearby water sources. Occasionally the wastes cause greater damage, as happened in Wales in 1966, when slag heaps from the coal mines slid onto the village of Aberfan, burying over 140 schoolchildren. Such tragedies call attention to the need for less potentially destructive ways of disposing of mine wastes.

Another example of the combined effect of excavating and filling on the landscape is the agricultural terrace characteristic of parts of Asia. In order to retain water and increase the amount of arable land, terraces are cut into the slopes of hills and mountains (Figure 6.1). Low

Figure 4.17
Tell Hesi, northeast of Gaza, Israel. Here and elsewhere in the Middle East, the debris of millennia of human settlement gradually raised the level of the land surface, producing such *tells,* or occupation mounds. In some cases, the striking landforms may rise hundreds of feet above the surrounding plains.

walls protect the patches of level land. The *tells* of the Middle East, pictured in Figure 4.17, are another type of landform produced by the accumulation of waste material.

Human impact on land has been particularly strong in areas where land and water meet. Dredging and filling operations undertaken for purposes of water control create landscape features like embankments and dikes. In many places, the actual shape of the shoreline has been altered, as builders in need of additional land have dumped solid wastes into landfills. In the Netherlands, millions of acres of land have been reclaimed from the sea by the building of dikes to enclose polders and canals to drain them. Farming practices in river valleys have had significant effects on deltas; for example, increased sedimentation has often extended the area of land into the sea.

Formation of Surface Depressions

The extraction of material from beneath the ground can lead to **subsidence,** the settling or sinking of a portion of the land surface. Many of the world's great cities are sinking because of the removal of *fluids* (groundwater, oil, and gas) from beneath them. Cities threatened by such subsidence are located on unconsolidated sediments (New Orleans, Bangkok), coastal marshes (Venice, Tokyo), or lake beds (Mexico City). When the fluids are removed, the sediments compact and the land surface sinks. Because many of the cities are on coasts or estuaries and are often only a few feet above sea level, subsidence makes them more vulnerable to flooding from the sea.

Groundwater abstraction has created serious subsidence in many places. Recent evidence indicates that the withdrawal of trillions of gallons of water from a 4500-square-mile (11,650-km²) area of Arizona has resulted in widespread ground subsidence and the formation of more than a hundred earth fissures, jagged ground cracks that can be as much as 9 miles (14.5 km) long and 400 feet (120 m) deep.

The removal of *solids* (such as coal, salt, and gold) by underground mining may result in the collapse of land over the mine. *Sinkholes* or *pits* (circular, steep-walled depressions) and *sags* (larger and shallower depressions) are two types of landscape features produced by such collapse. If surface drainage patterns are disrupted, subsidence *lakes* may form in the depressions. Subsidence has become a more serious problem as towns and cities have expanded over mined-out areas.

As one might expect, subsidence damages structures built on the land, including buildings, roads, and sewage lines. A dramatic example occurred in Los Angeles in 1963, when subsidence caused the dam at the Baldwin Hills Reservoir to crack; in less than 2 hours, the water emptied into the city, resulting in millions of dollars worth of property damage (Figure 4.18). The withdrawal of groundwater from beneath Mexico City has led to severe though differential subsidence. One of the reasons the 1985 earthquake in that city was so damaging was that subsidence had weakened building structures.

Agriculture and Soils

A human activity that affects both the landscape and the nature of the soil is agriculture. By design or by accident, people have brought about many changes in the physical, chemical, and biochemical nature of the soil and altered its structure, fertility, and drainage characteristics. The exact nature of the changes in any area depends on past practices as well as on the original nature of the land.

Two forms of abuse of the world's soil that have noticeably diminished its amount and productivity are depletion and erosion. **Soil depletion** simply means that the soil has lost some or all of its vital nutrients through crop removal. The soil has been "mined," crop yields drop, plants are weakened, and their nutritive value is reduced. Plant nutrients can be replaced by fertilization. What is more difficult to repair in damaged soil is the deterioration in physical structure resulting from depletion, particularly of the organic matter (humus) essential to hold water in the soil.

Figure 4.18
Subsidence caused cracking and emptying of the Baldwin Hills
Reservoir of Los Angeles in December 1963.

Figure 4.19

A drought-stricken area in Burkina Faso, one of the countries in the Sahel, where desertification has been accelerated by both climate and human activity. The cultivation of marginal land, overgrazing by livestock, and climatic change during the 1970s and 1980s have led to the destruction of native vegetation, to erosion, and to the expansion of the deserts of the world.

The depletion of humus, deforestation, the plowing of hillsides, overgrazing, and other acts of land resource exploitation accelerate **erosion** by preparing the thin layer of productive soil for the erosive forces of wind and running water. Through the processes of **desertification,** erosion has turned vast areas into useless dust bowls. As much as 5% of the world's deserts may have been caused not by climate but by human abuse of the land. Desertification is particularly severe in arid and semiarid lands, where population pressures impel millions of people to use marginal lands for cultivation, for animal grazing, or as a source of fuel wood (Figure 4.19).

People have known for decades that the Sahara is spreading southward (Figure 4.20); less well-known is the fact that hundreds of thousands of acres of land become desert each year on the Sahara's northern edge, as well as in India, the Middle East, and eastern Africa. It has been estimated that on a worldwide basis, desertification claims about 14 million acres (5.7 million hectares) per year—an area about half the size of Ohio.

The harmful effects of accelerated erosion are not limited to Africa and Asia. Indeed, in recent years, soil erosion in the United States has been at an all-time high (Figure 4.21). Wind and water are blowing and washing soil off pasturelands in the Great Plains, ranches in Texas, and farms in the Southeast. The nation's croplands lose

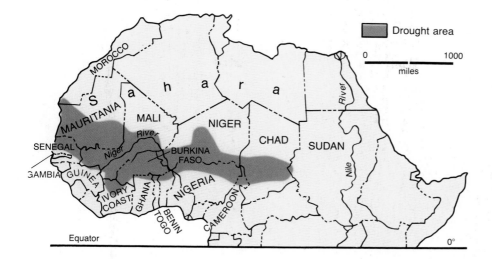

Drought area

0   1000
miles

Equator   0°

Figure 4.20

The expansion of the fringes of the Sahara has been well documented, and because of tragic recurring droughts in the Sahel (Mauritania, Senegal, Gambia, Mali, Burkina Faso, Niger, and Chad) beginning in 1968, it has been called sharply to world attention. In the last 50 years, some 250,000 square miles (650,000 km²) have been added to the southern Sahara, areas reduced to eroded, stony wasteland and to drifting sand. Elsewhere in Africa, and particularly in the Middle East and India, similar human-induced extensions of arid landscapes—eroded by wind, scarred by infrequent runoff, or subject to moving dunes—give vivid and catastrophic evidence of cultural impact on landform processes.

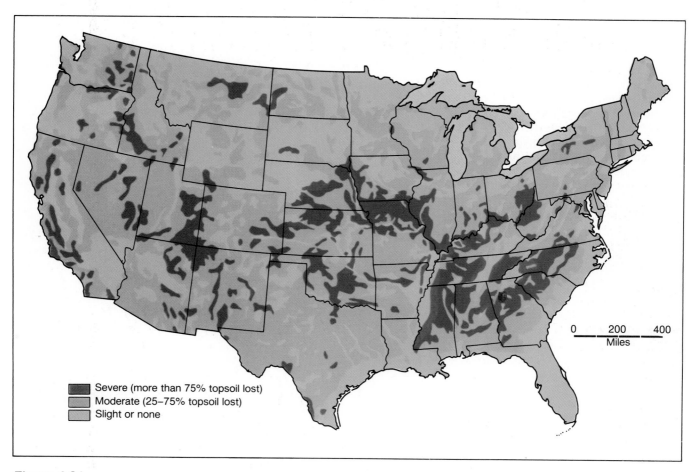

Severe (more than 75% topsoil lost)
Moderate (25–75% topsoil lost)
Slight or none

0   200   400
Miles

Figure 4.21

Soil erosion in the United States. Although many activities (mining, construction, and urbanization) contribute to erosion, agriculture and deforestation are particularly significant. In recent years, erosion has been most severe not in the southwestern Dust Bowl but in the moist, rolling-hill regions of western Mississippi, western Tennessee, and Missouri. Each of these areas loses about 10 tons of topsoil per acre of cropland annually. However, soil resource stress and depletion affect all parts of the nation.

Data from USDA Soil Conservation Service.

Soil Conservation

Economic conditions of the past decade have contributed to the current national high rate of soil erosion, the worst since the Dust Bowl of the 1930s. Federal tax laws and the high farmland values of the 1970s encouraged farmers to plow virgin grasslands and to tear down windbreaks to increase their yields. The secretary of agriculture exhorted farmers to plant all of their land, "from fencerow to fencerow," to produce more grain for export. Land was converted from cattle grazing to corn and soybean production as livestock prices declined.

When prices of both land and agricultural products declined in the 1980s, farmers were impelled to produce as much as they could in order to meet their debts and make any profit at all. To maintain or increase their productivity, many neglected conservation practices, plowing under marginal lands and using fields for the same crops every year.

Techniques to reduce erosion by holding the soils in place are well-known. They include contour plowing, terracing, strip cropping and crop rotation, erecting windbreaks and water diversion channels, and no-till farming (allowing crop residue such as cut corn stalks to remain on the soil surface throughout the winter). In addition, farmers can be paid to idle marginal, highly erodible land. Only by employing such practices can we maintain the long-term productivity of soil—the resource base upon which all depend.

No-till farming

Strip cropping

almost 3 billion tons of soil per year to erosion, an average annual loss of some 7 tons per acre; in many areas, the average is 15–20 tons per acre. Recent studies have shown that of the roughly 413 million acres (167 million hectares) of land that are intensively cropped in this country, over one-third are losing topsoil faster than it can be replaced naturally (it can take up to a century to replace an inch of topsoil). In parts of Illinois and Iowa, where the topsoil was once a foot deep, less than half of it remains. Every hour about 40,000 tons of topsoil wash into the Mississippi River.

Like most processes, erosion has secondary effects. Streams and reservoirs experience accelerated siltation when farm-soil loss increases. The topsoil contains millions of tons of agricultural chemicals, making erosion-borne silt one of the biggest pollutants of the country's water system. Urban water supplies are reduced, and more farmland is removed from cultivation as new and larger reservoirs are constructed. Flood danger is increased as bottomlands fill with silt, and the costs of maintaining navigation channels grow.

(a)

Figure 4.22

(a) The walking catfish. About a foot (30 cm) long, the walking catfish can hop out of water and creep along the ground by arching its neck and dragging its tail forward. It can remain out of water for more than 24 hours. Introduced into this country by a tropical fish farm in Florida, the catfish are now found over several thousand square miles in the state and have become an environmental menace. (b) Whitefish with lamprey attached. Ironically, the cleanup of Wisconsin's Green Bay and Fox River poses a threat that lamprey, which like clear water, may move into the Winnebago lakes, endangering the sturgeon population.

(b)

Irrigation, a component of agriculture for thousands of years, has led to excessive salinity of the soil in many parts of the world, particularly in arid and semiarid areas. These regions are naturally salty because evapotranspiration exceeds precipitation. Because irrigation water tends to move slowly, and thus to evaporate more rapidly, the salts carried in the water are absorbed into the ground. In time, increased salinity affects the productivity of the land. In Iran and Iraq, thousands of once-fertile acres have had to be abandoned; it has been estimated that over 25% of the irrigated areas of Pakistan, Syria, and Egypt are affected by such **salinization.** In this country, the 250-mile-long (400 km) San Joaquin Valley of California is experiencing a similar buildup of harmful salty residues. Because the valley is a major supplier of fruits and vegetables to the rest of the country, the conversion of over 1 million acres (405,000 hectares) to a barren salt flat would have an impact well beyond California's borders.

Impact on Plants and Animals

Introduction of New Species

People have modified plant and animal life on the earth through several different processess. One is by deliberately or inadvertently introducing a plant or an animal into an area where it did not previously exist. If conditions in the new area favor the species, it may multiply rapidly, often with damaging and unforeseen consequences.

The rabbit, for example, was purposely introduced into Australia in 1859. The original dozen pairs multiplied to a population in the thousands in only a few years and, despite programs of control, to an estimated one billion a hundred years later. Inasmuch as 5 rabbits eat about as much as 1 sheep, a national problem had been created. Much of the grassland on which sheep could graze had been denuded by the rabbits.

Goats were imported to the island of St. Helena by the Portuguese, who could not have foreseen that in time the goats would completely destroy the indigenous vegetation. The sea lamprey, a parasite that attaches itself to a fish and sucks on its blood until the fish dies, found its way into the Great Lakes when the Welland Canal was deepened in 1932 (Figure 4.22). Within a few years, catches of lake trout and whitefish declined precipitously, all but putting an end to these fish as commercial crops. Sparrows and starlings, deliberately introduced into the United States in the last century, have largely displaced native songbirds such as wrens and bluebirds in many areas.

Plants as well as animals can alter vegetative patterns. The Asiatic chestnut blight for instance, has destroyed most of the native American chestnut trees in the United States, trees with significant commercial as well as aesthetic value. The cause was the importation of some chestnut trees from China to the United States. They carried a fungus fatal to the American chestnut tree but not to the Asiatic variety, which is largely immune to it. The water hyacinth entered this country in 1884, when bulbs were given as souvenirs at an exposition in New Orleans.

(a)

(b)

Figure 4.23

(a) A crippled calf. Some livestock disorders are caused by ingestion of agricultural biocides with feed. (b) A gray fox with a freshly killed rabbit. Much killing of wildlife to protect crops and livestock is simplistic and even detrimental. The killing of foxes and coyotes, which occasionally eat young sheep, has resulted in an increase in the rabbit population on western ranges. Rabbits eat grasses that the sheep need for forage.

Before the end of the century, the spread of the plant had become a matter of concern. In one growing season, a single plant can produce over 60,000 offshoots. To the consternation of those who sail and fish, swamps, lakes, and canals in the southern United States have become clogged with hyacinths, which also affect fish and plankton.

These are just a few of the many examples that illustrate an often ignored ecological truth: plant and animal life are so, interrelated that when people introduce a new species to a region, whether by choice or by chance, there may be unforeseen and far-reaching consequences.

Destruction of Plants and Animals

Another way in which humans have affected plants and animals is through deliberate destruction. When people clear an area to plant crops, spread defoliants in a war, or use herbicides, they are destroying vegetation. Herbicides are chemical weed killers designed to increase agricultural yields in the short run, but they may have adverse consequences in the long run (Figure 4.23a), as may the deliberate elimination of a presumed destructive animal species (Figure 4.23b).

Whenever we modify the vegetation in an area, we affect plant and animal populations and soils. Thus soil in the tropics exposed by defoliants to wind and sun may be completely altered structurally and chemically, and become less productive. If herbicides kill microorganisms in the soil, the soil ecology is affected. We are learning that the food chain is part of an interlocking web; to touch one strand of the web is to set off reverberations in the rest.

Humans have caused the extinction of some plants and animals and have placed many other species in jeopardy. We have overhunted and overfished, for food, fur, or sport. Jaguars, buffalo, passenger pigeons, whales, and hundreds of other animals have been thoughtlessly exploited. Around the world, about 20,000 species of flowering plants—10% of the world's flora—are endangered or threatened with extinction. In the United States, 15% of the approximately 400 native mammals are in the Rare and Endangered Species list maintained by the national government and updated yearly.

We have also destroyed animals by disrupting the habitats in which they live. One of the main causes of extinction has been the loss or alteration of habitats for wildlife. By clearing forest land, draining wetlands, extending farmland, and building cities, people modify or destroy the habitats in which animals have lived. Often the animals can simply move elsewhere, but not always. Tidal marshlands have been subjected to dredging and filling for residential and industrial development. The loss of such areas reduces the essential habitat of waterfowl, fish, crustaceans, and mollusks. Many waterfowl breed and feed in coastal marshes and use them for rest during

Timely Reprieves

As evidence of our ability to inadvertently destroy species has mounted, attempts have been made to take corrective action. DDT was banned in this country in 1971, and the use of many other chlorinated hydrocarbons has been restricted. Most new pesticides are now designed to kill specific pests rather than a broad range and to dissipate quickly after use. The bald eagle, prairie falcon, gyrfalcon, merlin, and northern goshawk—five of America's rarest birds of prey—have made dramatic recoveries over the last decade, chiefly because of the reduction in use of DDT and other persistent pesticides.

There are other examples of species saved from extinction. Declared endangered only 20 years ago, after decades of intensive hunting, the alligator has reappeared across the South, from Texas to the Carolinas. The wild turkey, which had all but disappeared from northeastern states by the time of the Civil War—a victim of both hunting and habitat destruction—has recently been successfully reintroduced into the area. And although commercial hunting has reduced populations of blue, humpback, right, and bowhead whales to levels that may be beyond restoration, other species of whales may be saved by the total moratorium on commercial whaling that the International Whaling Commission imposed in 1985.

long migrations. The whooping crane has been virtually eliminated in this country, partly because the marshes in which it lived were drained; its comeback, sought by breeding programs in the United States and Canada, is still uncertain.

Many people fear that as countries in Africa and South America become more industrialized and more urbanized and expand their areas under cultivation, there will be an increasingly negative impact on wildlife. It is already known that in Africa, wild animals are vanishing fast, the victims of habitat destruction, hunting and poaching, starvation, and disease. Of a population of 18,000 in 1969, only 500 rhinoceros remained in Kenya by the middle 1980s. Farther south, in Botswana, 250,000 antelope and zebra died in a decade—disoriented by fences erected to protect cattle. Elephant herds have disappeared from Burundi, and the elephant population dropped from an estimated 371,000 to 150,000 in just 5 years in Zaire. As selected animal species decline in numbers, balances among species are upset and entire ecosystems are disrupted.

Perhaps the worst ecological disaster now facing humankind is the destruction of the world's rain forests. Found primarily in Central and South America, the Zaire River basin of Africa, and Southeast Asia, tropical rain forests are the most biologically diverse places on earth, but some 40,000 square miles (100,000 km²)—an area larger than Indiana—are being destroyed every year. Corporate cattle and timber operations are responsible for some of the loss, but the major cause of **deforestation** is clearing the land for crops. Unhappily, soil erosion and the loss of fuel supplies associated with deforestation create hardships for already poor people. In addition, it is estimated that at current rates, 10%–20% of the earth's animal species will disappear by the year 2000.

Poisoning and Contamination

Humans have also affected plant and animal life by poisoning or contamination. In the last several years, we have become acutely conscious of the effect of insecticides, rodenticides, and herbicides, known collectively as *biocides*. The best-known and most widely used has been DDT, although there are now thousands of compounds in use.

DDT was first used during World War II to kill insects that carried diseases such as malaria and yellow fever. In the years following the war, tons of DDT and other biocides were used, sometimes to combat disease, sometimes to increase agricultural yields. Insects, after all, can destroy a significant percentage of a given crop, either when it is in the field or after it has been harvested.

In the last ten years, some of the side effects of these biocides have been well enough documented for us to question their indiscriminate use. Once used, a biocide settles into the soil, where it may remain or may be washed into a body of water. In either case, it is absorbed by organisms living in the soil or the mud. Through a process known as **biological magnification,** the biocide accumulates and is concentrated at progressively higher levels in the food chain (Figure 4.24). By remaining as residue in fatty tissues, a very small amount of a biocide produces unexpected effects, with predators accumulating larger amounts than their prey.

Thus the concentration of DDT in the mud at the bottom of a lake may be on the order of 0.01 parts per million (ppm); small crustaceans eating algae or other organisms may exhibit concentrations of DDT of 0.5 ppm;

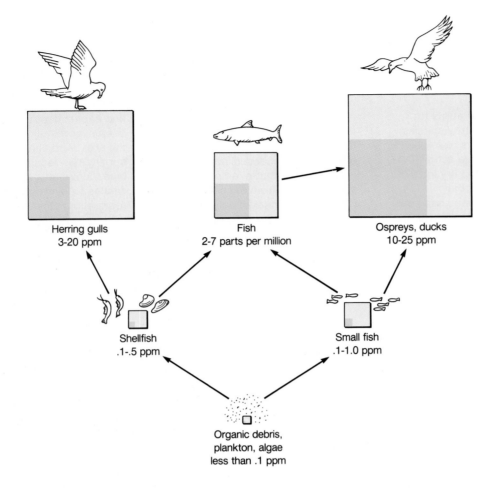

Herring gulls
3-20 ppm

Fish
2-7 parts per million

Ospreys, ducks
10-25 ppm

Shellfish
.1-.5 ppm

Small fish
.1-1.0 ppm

Organic debris,
plankton, algae
less than .1 ppm

Figure 4.24

A simplified example of biological magnification. Although the
level of DDT in the water and mud may be low, the impact on
organisms at the top of the food chain can be significant. In this
example, birds at the top of the chain have concentrations of
residues as much as 250 times greater than the concentration in
the water. Radioisotopes such as strontium-90 and cesium-137
undergo magnification in the food chain just as insecticides do.

fish feeding on them may show 4 ppm; and at the highest
level of that food chain, gulls or other birds that feed on
those fish may have concentrations of 20 ppm. The higher
the level of an organism in the food chain, the greater
the concentration of DDT will be, and that concentration
may be lethal. Robins and other small birds die when
they eat earthworms that have ingested DDT that has settled into the earth after being sprayed on trees.

DDT has also been shown to cause a decrease in
the thickness of the eggshells of some of the larger birds,
causing a greater number of eggs to break than normally
would. Peregrine falcons, bald eagles, and brown peli-
cans were among the birds brought to the edge of extinction by this disruption of the reproductive process.
(See "Timely Reprieves," Page 29.)

DDT is so long-lived and has been so widely used
that it is present everywhere. Even penguins and cor-
morants in the Antarctic, hundreds of miles from the
nearest point of DDT use, have detectable levels of the
insecticide. In the United States, DDT and other biocides are present in most foods, including human milk.

Although the use of DDT has declined as its effects
have become apparent, other chlorinated hydrocarbon
compounds have been developed and are in wide use.
Indeed, the use of pesticides in this country has more
than doubled in the last quarter century. More than 2 billion pounds (900 million kg) containing more than 600
active ingredients are applied each year, most of them
inadequately tested for properties that cause mutations,
cancer, or other adverse health effects.

PCBs have, like DDT, entered the food chain on a
worldwide basis. They, too, are long-lived, accumulate in
fatty tissues, and display biological magnification in the
food chain. As is true of DDT, the long-term effect of low
concentrations of PCBs in humans is unknown.

Biocides may exacerbate the problem their use is
designed to eradicate. By altering the natural processes
that determine which insects in a population will survive, we have inadvertently spurred the development of
resistant species. If all but 5% of the mosquito population in an area are killed by an insecticide, the ones that
survive are the most resistant individuals, and they are

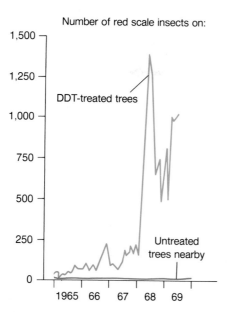

Number of red scale insects on:

Figure 4.25

Using DDT to destroy sap-sucking red scale insects on lemon trees had the opposite effect, actually increasing the numbers of the pest. Spraying an infested crop kills, perhaps, 90% of the pests but also the insects that eat the pests. With their food abundant and their predators rare, the remaining pests recover faster than their enemies, whose prey is now scarce and harder to locate.

Source: "Ecology," Charles J. Krebs, as reprinted in *The Economist*, April 26, 1986, p. 93.

the ones that will produce the succeeding generations (Figure 4.25). Within the last decade, the World Health Organization has abandoned its malaria eradication program because mosquitoes developed resistance to virtually all applicable insecticides, and malarial parasites developed resistance to antimalarial drugs.

There are now insects whose total resistance to certain pesticides has led some scientists to conclude that the entire process of insecticide development may be self-defeating. Despite the enormous growth in the use of pesticides, crop loss to insect and weed pests has actually grown. According to Department of Agriculture figures, 32% of crops were lost to pests in 1945; 40 years later, such losses had increased to 37%.

Pesticides have been known to increase problems by destroying the natural enemies of the intended target, leaving it to breed unchallenged. Examples include the tobacco budworm and the brown planthopper, which were relatively minor pests before intensive crop-spraying destroyed rival pests. The budworm severely cut cotton production in Mexico and Texas in the 1960s; the brown planthopper continues to affect rice growing in Asia.

Many environmental scientist advocate *integrated pest control,* or using a combination of traditional chemical pesticides with other techniques of pest management. These techniques include breeding parasites to attack specific pests, growing resistant rather than high-yielding varieties of plants, and planting crops so that they miss the high season of pests. With an integrated pest-management system, farmers' outlays of expensive chemicals would be lower and they would be less in danger of experiencing the near-total loss of a crop.

Solid Waste Disposal

Modern technologies and the societies that have developed them produce enormous amounts of **solid wastes** of which disposition must be made. The rubbish heaps of past cultures (Figure 4.17) suggest that humankind has always been faced with the problem of ridding itself of materials no longer needed. The problem for advanced societies, with their ever-greater variety, amount, and durability of refuse, is even more serious: how to dispose of the solid wastes produced by residential, commercial, and industrial processes. Although these account for much less tonnage than the wastes produced by mining or agriculture, they are everywhere, a problem with which each individual and each municipality must deal. The amount of waste has increased rapidly in recent years, to its current level in the United States of about 1 ton per person per year.

Municipal Waste

The wastes that communities must somehow dispose of include newspapers and beer cans, toothpaste tubes and old television sets, broken refrigerators and rusted cars. Although such ordinary household trash does not meet the governmental designation of **hazardous waste**— defined as discarded material that may pose a substantial threat to human health or the environment when improperly stored or disposed of—it is hazardous nonetheless. Products containing toxic chemicals include paint and paint remover, furniture polishes, bleaches, deodorizers, used motor oil, and garden weed killers and pesticides.

Every method of disposing of such wastes has its own impact on the environment. Loading wastes onto barges and dumping them in the sea, long a practice for coastal communities, inevitably pollutes the ocean. Open dumps on the land are a menace to public health, for they harbor disease-carrying rats and insects. When combustibles are burned, chemicals and particulates are discharged into the air. In extreme cases, gases and chemicals may be a health hazard. For example, when polyvinyl chloride (PVC), a component of plastic, is burned, hazardous hydrogen chloride is released.

Beginning in the 1960s, communities were encouraged to dispose of their wastes in what was considered a more environmentally sound manner: the *sanitary landfill.* "Sanitary" has turned out to be a deceptive word.

Figure 4.26
Storage tanks under construction in Hanford, Washington. Built to contain high-level radioactive wastes, the tanks are shown before they were encased in concrete and buried underground.

Even if no commercial or industrial waste has been dumped at the site, toxic heavy metals and carcinogens may seep through the soil, contaminating the groundwater. Heavy metals are leached from batteries and old electrical parts, and vinyl chlorides come from the plastic in household products.

Hazardous Waste

The EPA has classified more than 400 substances as hazardous, posing a threat to human health or the environment. Industrial wastes have grown steadily more toxic; currently about 10% of industrial waste materials are considered hazardous.

Every organization that either uses or produces radioactive materials generates *low-level waste*. Nuclear power plants produce about half the total low-level waste in the form of used resins, filter sludges, lubricating oils, and detergent wastes. Industries that manufacture radiopharmaceuticals, smoke alarms, radium watch dials, and other consumer goods produce low-level waste consisting of machinery parts, plastics, and organic solvents.

High-level waste is nuclear waste with a relatively high level of radioactivity. It consists primarily of spent power reactor fuel assemblies—termed "civilian waste"—and waste produced as a by-product of the manufacture of nuclear weapons, or "military waste."

No satisfactory method for disposing of hazardous waste has yet been devised. Some wastes have been sealed in protective tanks and dumped at sea. However, the tanks may be moved from the original dumping site by strong currents and may be crushed by water pressure, causing leakage of the wastes. Over 47,000 drums of radioactive wastes were dumped into the ocean 35 miles (56 km) west of San Francisco in the period 1946–70. Recent investigations have shown that some of the drums have been crushed, contaminating sediment in the area with plutonium. Even without such physical damage, the life expectancy of the containers must be presumed to be far shorter than the half-life of their radioactive contents.

Other radioactive wastes have been placed in tanks and buried in the earth. By 1976, 75 million gallons of liquid high-level waste and 51 million cubic feet of low-level waste were stored at nine locations in the United States, one of which is shown in Figure 4.26. At least one of these storage areas, that at Hanford, Washington, has experienced leakages, with seepage of one-half million gallons of high-level waste into the surrounding soil.

Another method of waste disposal, used for chemicals as well as for radioactive wastes, has been to inject them into deep steel- or concrete-lined wells. Because underground injection poses a threat of groundwater contamination and may contribute to earth tremors, the injection of wastes into or above strata that contain aquifers is being phased out.

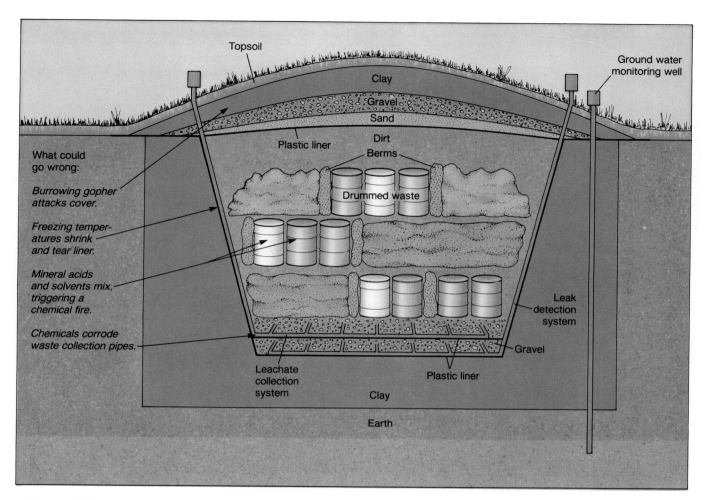

Figure 4.27

The closest thing to what experts consider a secure landfill for
hazardous waste is a pit 60 feet (18 m) deep, sealed in a plastic
liner to keep contaminated liquids from seeping into the soil.
Toxic waste is placed in drums and covered with dirt. Leachate,
the liquid mixture of wastes that seep to the bottom of the pit,
passes into perforated collection pipes and is pumped to the
surface for treatment. But landfills fail easily. Bulldozers can tear
the plastic liner and leachate disintegrate it. The waste may crush
collection pipes, debris clog the perforations. If erosion breaches
the landfill's protective cover, rain may mix with waste,
overloading the collection system or causing the landfill to
overflow.

The most common method of disposal of low-level
waste is in landfills, often the local municipal dump,
where the waste chemicals may leach through the soil
and into the groundwater. By EPA estimates, the United
States contains at least 25,000 legal and illegal dumps with
hazardous waste; as many as 2000 are deemed potential
ecological disasters. Some scientists contend that a safe
dump exists only on paper, that even following the federal government's suggested guidelines for the best possible hazardous waste landfill will yield an insecure one
(Figure 4.27).

Solid waste will never cease to be a problem, but
its impact on the environment can be lessened by reducing the volume of waste that is generated, reducing
the production of toxic residues, halting irresponsible
dumping, and finding ways to reuse the resources that it
contains. Several methods of reuse are being explored,
to different degrees in different countries. The conversion of the wastes into fuels or into compost, a low-grade
fertilizer, is feasible if there is a market for those products. Likewise many materials in the waste, excluding
plastics or textiles, can be recycled and probably will be
once the costs are shown to be economical. Until then,
current methods of waste disposal will continue to pollute soil, air, and water.

Superfund

As disasters like that at New York's Love Canal made it clear that hazardous wastes were a significant environmental menace, posing a direct threat to public health and the environment, the U.S. Congress reacted by enacting the Comprehensive Environmental Response, Compensation and Liability Act. Generally called the **"Superfund" law,** the goals of the program it established were primarily (1) to identify and analyze all uncontrolled and abandoned hazardous waste sites, and (2) to assure that remedial action was taken at sites that pose a grave danger.

Some 25,000 hazardous waste sites are potential cleanup targets. Of those, the EPA had by the close of 1986 put 888 on the *National Priorities List* (NPL) of sites that present a chronic or long-term threat, indicating that they are candidates for remedial action. Time and funding permitting, samples of air, earth, and water will be taken at these sites to determine the degree and nature of the contamination, and available remedies (such as incineration, treatment for safe disposal) will then be considered. Typically, these studies take 2–3 years, and only after they are completed is remedial action undertaken.

Because of concern that only six sites had been cleaned and deleted from the NPL in the first 6 years of the program, the Superfund bill that passed Congress in late 1986 required EPA administrators to initiate at least 375 cleanups during the next 5 years. It should be noted that there would still remain almost 25,000 abandoned and uncontrolled sites in the country. And the bill did not address the roughly 100,000 municipal or industrial landfills that all evidence indicates are far from safe, or the estimated 180,000 pits, ponds, and lagoons filled with liquid wastes.

Conclusion

The intricately interconnected systems of the ecosphere—the atmosphere, the hydrosphere, the lithosphere, and their contained biomes—have been subjected to profound and frequently unwittingly destructive alteration by humans. The effects of human impact on the environment are complex and are never isolated. An external action that impinges on any part of the web of nature inevitably triggers chain reactions, the ultimate impacts of which appear never to be fully anticipated.

Human disruption of the fragile structure of the environment is growing at an increasing rate. World population expansion and the ecological pressures that it implies, increases in per capita energy consumption, and the creation of ever more grandiose schemes to alter nature to satisfy felt developmental needs are among the obvious causes of that disruption and are indications of the human conviction that we alone stand above the imperatives of nature. Other species may perish in a changing environment; only humans have sought to divorce themselves from it.

Of course, it is now accepted as an abstract truism that humans are part of the natural environment and depend, literally, for their lives on the water, air, food, and energy resources that the biosphere contains. But for many, it is an acceptance outside the realm of daily concern. Ever more forceful, however, is the evidence that humans cannot manipulate, distort, pollute, or destroy any part of the ecosystem without diminishing its quality or disrupting its structure. The increasing frequency of actual environmental danger and disaster has converted the abstraction to a present reality.

Key Words

acid rain *136*
Arctic haze *140*
biological
  magnification *151*
biosphere *124*
channelization *129*
chlorofluorocarbons
  (CFCs) *141*
deforestation *151*

desertification *146*
ecosphere *124*
ecosystem *125*
environment *124*
environmental
  pollution *131*
erosion *146*
eutrophication *131*
food chain *125*

"greenhouse" effect *140*
hazardous waste *153*
hydrologic cycle *127*
"icebox" effect *140*
ozone *139*
ozone layer *141*
photochemical smog *139*
polychlorinated
  biphenyls (PCBs) *133*

salinization *149*
soil depletion *144*
solid waste *153*
subsidence *144*
"Superfund" law *156*
temperature
  inversion *138*
thermal pollution *134*
troposphere *135*

For Review

1. Sketch and label a diagram of the *biosphere.* Briefly indicate the content of its component parts. Is that content permanent and unchanging? Explain.

2. What is the relationship of the concepts of *ecosystem, biome,* and *niche?* How does each add to our understanding of the "web of nature"?

3. Draw a diagram of or briefly describe the *hydrologic cycle.* In what ways do people have an impact on that cycle? What effect does urbanization have on it?

4. Is the concept of pollution meaningful if humans are absent? Is all pollution the result of human action? When can we say that pollution of a part of the biosphere has occurred?

5. Describe the chief sources of water pollution. What steps have been taken here and abroad to control such pollution?

6. What effects has the increasing use of fossil fuels over the past 200 years had on the environment? What is *acid rain,* and where is it a problem? What factors affect the type and degree of air pollution found at a place? What is the relationship of *ozone* to *photochemical smog?*

7. What causes the "icebox" and "greenhouse" effects? How might they affect the environment?

8. What kinds of landforms have been produced by excavation? By dumping? What are the chief causes and effects of subsidence?

9. How have people diminished the amount and the productivity of soil? What types of areas are particularly subject to desertification? How may soil erosion be controlled?

10. What are the chief ways in which humans have affected plants and animals? Define *biological magnification* and indicate its implications for humans.

11. What methods do communities use to dispose of solid waste? How does the government define *hazardous waste,* and how is it disposed of? What ecological problems does solid waste disposal present?

Suggested Readings

Dasman, Raymond F. *Environmental Conservation.* 5th ed. New York: John Wiley, 1984.

Di Silvestro, Roger L., ed. *Audubon Wildlife Report 1986.* Covelo, Calif.: Island Press, 1986.

Epstein, S. S., L. O. Brown, and C. Pope. *Hazardous Waste in America.* San Francisco: Sierra Club Books, 1982.

Global 2000 Report to the President. A Report Prepared by the Council on Environmental Quality and the Department of State, Gerald O. Barney, Study Director. Harmondsworth, Middlesex, England: Peregrine Books, 1982.

Goudie, Andrew. *The Human Impact: Man's Role in Environmental Change.* Cambridge, Mass.: MIT Press, 1982.

Greenland, David. *Guidelines for Modern Resource Management.* Columbus, Ohio: Charles E. Merrill, 1983.

Miller, G. Tyler. *Living in the Environment.* 3d ed. Belmont, Calif.: Wadsworth, 1982.

ReVelle, Penelope, and Charles ReVelle. *The Environment.* 2d ed. Boston: Willard Grant, 1984.

Southwick, Charles H. *Global Ecology.* Sunderland, Mass.: Sinauer Associates, 1985.

The State of the Environment 1985. The Organization for Economic Cooperation and Development. Paris, France, 1986.

Strahler, Arthur N., and Alan H. Strahler. *Geography and Man's Environment.* New York: John Wiley, 1977.

Part Two

The Culture—Environment Tradition

T he Crow country. The Great Spirit put it exactly in the right place; while you are in it, you fare well; whenever you get out of it, whichever way you travel, you fare worse. . . . The Crow country is in exactly the right place. It has snowy mountains and sunny plains; all kinds of climates and good things for every season. When the summer heats scorch the prairies, you can draw up under the mountains, where the air is sweet and cool. . . . In the autumn when your horses are fat and strong from the mountain pastures, you can go down on the plains and hunt the buffalo or trap beaver on the streams. And when winter comes on, you can take shelter in the woody bottoms along the rivers.

The Crow country is exactly in the right place. Everything good is found there. There is no country like the Crow country.

Such was the opinion of Arapoosh, chief of the Crows, speaking of the Big Horn basin country of Wyoming in the early 19th century. In the 1860s, Captain Raynolds reported to the secretary of war that the basin was "repelling in all its characteristics, surrounded on all sides by mountain ridges [and presenting] but few agricultural advantages."

In the four chapters of Part I, our primary concern was the physical landscape. But while the physical environment may be described by process and data, it takes on human meaning only through the filter of culture. Arapoosh and Raynolds viewed the same landscape, but from the standpoints of their separate cultures and conditionings. Culture is like a piece of tinted glass, affecting and distorting our view of the earth. Culture conditions the way people think about the land, the way they use and alter the land, and the way they interact with one another upon the land.

Such conditioning is the focus of the culture-environment tradition of geography, a tradition still concerned with the landscape but not in the physical science sense of Part I of this book. In Part II, humans are the focus. Of course, the physical environment is always in our minds as we develop the notion of cultural difference and reality. Our landscapes, however, take on an added dimension and become human rather than purely physical.

The four chapters in this portion of our study, therefore, concentrate upon the "people" portion of geography's environment–culture–people relationship. In Chapter 5, "Population Geography," we start with the basics of human populations in their numbers, compositions, distributions, growth trends, and the pressures they exert on the resources of the lands they occupy. These quantitative and distributional aspects of peoples are important current concerns, as frequent popular reference to a "population explosion," public debate about legal and undocumented immigration, and speculation about population growth and food availability attest. More fundamentally, of course, numbers and locations of people are the essential background to all other understandings of human geography.

People are, however, more than numbers in a global counting parlor. They are, separately, individuals who think, react, and behave in response to the physical and social environments they occupy. In turn, those thoughts, actions, and responses are strongly conditioned by the standards and structures of the cultures to which the individuals belong. The world is a mosaic of culture groups and human landscapes that invite geographic study. Chapter 6, "Cultural Diversity," introduces that study by examining the components and subsystems of culture, the ways in which culture changes, and the key variables in defining a culture and in producing variations in culture from place to place.

One of those variables, politics, is explored in Chapter 7, "Political Geography." Political systems and processes strongly influence the form and distribution of many elements of culture. Economic and transportation systems coincide with national boundaries. Political regulations, whether detailed zoning codes or broad environmental protection laws, have a marked effect on the cultural landscape. In some countries, even such apparently nonpolitical matters as religion, literature, music, and the fine arts may be affected by governmental support of certain forms of expression and rejection and prohibition of others. In an increasingly interrelated world community, the international aspects of political association create supranational patterns of culture distinct from traditional linguistic, religious, or ethnic affiliations.

Patterns of human spatial behavior and the factors that account for the manner in which people use space are the subject of Chapter 8, "Behavioral Geography." The way people view the environment is important, for human actions are guided as much by how things are perceived to be arranged and structured as by the objective reality of their location and content. The individual and group behaviors of humans create the cultural landscapes—those of production, resource use, and urban settlement—that are the topics of Part III, "The Location Tradition." Chapter 8, therefore, may be seen as an entity in itself and as a bridge to other subfields of the discipline.

Our theme in Part II, however, is people and the collective and personal cultural landscapes they create or envision. Let us begin with people themselves before we move on to consider their cultures and behaviors.

5 Population Geography

Zero, possibly even negative [population] growth" was the 1972 slogan proposed by the prime minister of Singapore, an island nation in Southeast Asia. His country's population, which stood at 1 million at the end of the Second World War, had doubled by the mid-1960s. In support, the government decreed "Boy or girl, two is enough," and refused maternity leaves and access to health insurance for third or subsequent births. Abortion and sterilization were legalized, and children born fourth or later in a family were to be discriminated against in school admissions policies. In response, birth rates by the mid-1980s fell to below the level necessary to replace the population and abortions were terminating more than one-third of all pregnancies.

"At least two. Better three. Four if you can afford it" was the national slogan proposed by that same prime minister in 1986, reflecting fears that the stringencies of the earlier campaign had gone too far. From concern that overpopulation would doom the country to perpetual Third World poverty, Prime Minister Lee Kuan Yew was moved to worry that population limitation would deprive it of the growth potential and national strength implicit in a youthful, educated work force adequate to replace and support the present aging population.

The policy reversal in Singapore reflects an inflexible population reality: the structure of the present determines the shape of the future. The size, characteristics, and growth trends of today's populations help shape the well-being of peoples whose numbers, nationalities, and distributions are yet to be determined.

A respected geographer once commented that "population is the point of reference from which all other elements [of geography] are observed." Certainly, it may fairly be said that a knowledge of humankind in all its diversity demands special emphasis in our study of the earth's surface and of the manner in which people occupy and use that surface. The numbers, age, and sex distribution of people; patterns and trends in their fertility and mortality; their density of settlement and rate of growth—all these aspects of population affect and are affected by the social, political, and economic organization of a society. Through them we begin to understand how the people in a given area live, how they may interact with one another, how they use the land, what pressure there is on resources, and what the future may bring. These are topics that underlie the essentially *human* component of the geographer's "human–earth's surface" equation.

Population geography provides the background tools and understandings of those topics. It is concerned with the number, composition, and distribution of human beings in relation to variations in the conditions of earth space. It differs from *demography*, the systematic study of human population, in its concern with *spatial* analysis—the relationship of numbers to area. In seeking to describe and account for the uneven distribution of people over the earth's surface, population geographers study the temporal and spatial variation of the components of population change. Birth and death rates, fertility rates, and growth rates are among the data they examine as they study the spatial associations of population with both physical landscape elements and cultural geographic characteristics. Population geographers are also concerned with the significance of spatial differences and of current population trends in relation to the kind and quality of life that people lead—that is, the relationship of societies to such matters as type of economic development, level of living, supply of food and other resources, and conditions of health and well-being.

These are concerns thrust into prominence in recent years by a "population explosion." Dramatic increases in human numbers have been experienced in many parts of the globe. Those increases and the association between them and their supporting earth resources are fundamental expressions of the human–environmental relationships that are the substance of geographic inquiry.

Population Growth

Sometime during the summer of 1987, a human birth raised the earth's population to 5 billion persons. In 1961 there were about 3 billion. That is, over the quarter century between those two years, the world's population grew on average by some 80 million persons annually, or some 220,000 per day. United Nations estimates, which have proven reliable in the past, predict a year 2000 population of just over 6 billion (the present century began with fewer than 2 billion) inhabitants. Eventually, most analysts assume, world population will stabilize at between 8 and 14 billion persons, with the majority of the growth occurring in nations now considered "less developed" (Figure 5.1). We will return to these projections and the difficulties inherent in making them later in this chapter.

Just what is implied by numbers in the millions and billions? With what can we equate the 1987 population of Botswana in Africa (a little over 1 million) or of China (about 1 billion)? Unless we have some grasp of their scale and meaning, our understanding of the data and data manipulations of the population geographer can at best be superficial. It is difficult to appreciate how vast a number is 1 million or 1 billion, and how great the distinction between them is. An example or two offered by the Population Reference Bureau may help you to visualize their immensity and implications.

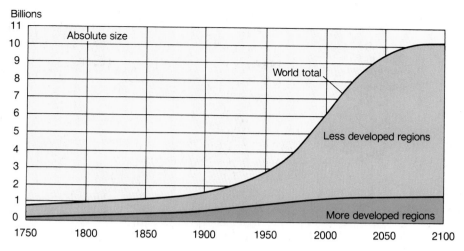

Figure 5.1

Population growth, 1750–2100. After two centuries of slow growth, world population began explosive expansion after World War II. World Bank 1984 medium projections are for a global population of about 10 billion for the first half of the 21st century. Nearly all of the growth will be in countries now considered "less developed."

Source: Adapted from Merrick, Thomas W., with PRB staff, "World Population in Transition," *Population Bulletin,* 41, no. 2, Washington, D.C.: Population Reference Bureau, Inc., 1986, p. 4.

A 1-inch stack of U.S. paper currency contains 233 bills. If you had a *million* dollars in thousand dollar bills, they would make a stack 4.3 inches high. If you had a *billion* dollars in thousand dollar bills, your pile of money would reach 357 feet in height—about the length of a football field.

You had lived a *million* seconds when you were 11.6 days old. You won't be a *billion* seconds old until you are 31.7 years of age.

The supersonic airplane, the Concorde, could theoretically circle the globe in only 18.5 hours at its cruising speed of 1340 miles per hour. It would take 31 days for a passenger to journey a *million* miles on the Concorde, while a trip of a *billion* miles would last 85 years.

The implications of the present numbers and of the potential increases in population are of vital current social, political, and—above all—ecological concern. Population numbers were much smaller some 11,000 years ago when continental glaciers began their retreat, people spread to formerly unoccupied portions of the globe, and human experimentation with food sources initiated the Agricultural Revolution. The 5 or 10 million people who then constituted all of humanity obviously had considerable potential to expand their numbers. In retrospect, we see that the natural resource base of the earth had a population-supporting capacity far in excess of the pressures exerted upon it by early hunting and gathering groups.

Some would maintain that despite present numbers or even those we can reasonably anticipate for the future, the adaptive and exploitive ingenuity of humans is in no danger of being taxed. Others, however, compare the earth to a self-contained "spaceship" and declare with chilling conviction that a finite vessel cannot bear an ever-increasing number of passengers. They point to recurring problems of malnutrition and starvation (though these are realistically more a matter of failures of distribution than of inability to produce enough foodstuffs worldwide). They cite dangerous conditions of air and water pollution, the nearing exhaustion of mineral resources and of fossil fuels, and other evidence of strains on world resources as foretelling the discernible outer limits of population growth.

Why are we suddenly confronted with what seems to many an insoluble problem—the apparently unending tendency of humankind to increase in numbers? On a worldwide basis, populations grow only one way: the number of births in a given period exceeds the number of deaths. Ignoring for the moment regional population changes resulting from migration, we can conclude that the observed and projected dramatic increases in population must result from the failure of natural controls to limit the number of births or to increase the number of deaths, or from the success of human ingenuity in circumventing such controls when they exist. The implications of such considerations will become clearer after we define some important terms in the study of population geography and explore their significance.

Some Population Definitions

Demographers employ a wide range of measures of population composition and trends, though all of their calculations start with a count of events: of individuals in the population; of births, deaths, marriages, and so on. To those basic counts are brought refinements that make the figures more meaningful and useful in population analysis. Among them are **rates** and **cohort measures.**

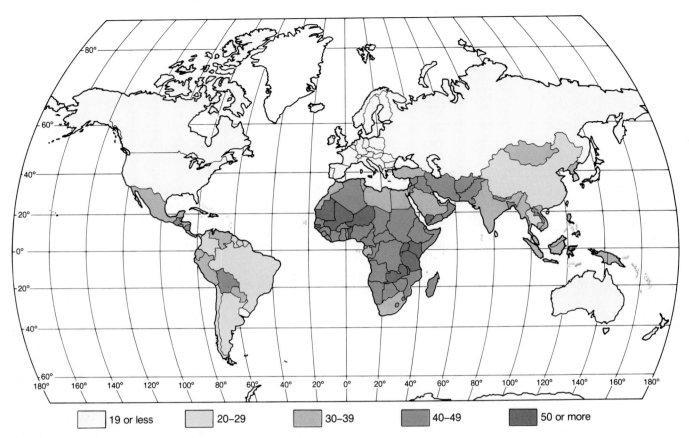

| | 19 or less | | 20–29 | | 30–39 | | 40–49 | | 50 or more |

Figure 5.2

This map of the world pattern of crude birth rates suggests a degree of precision that is misleading in the absence of reliable, universal registration of births. It serves, however, as a generally useful summary of comparative reproduction patterns if class divisions are not taken too literally.

Source: Data from Population Reference Bureau, Inc.

Rates simply record the frequency of occurrence of an event during a given time frame for a designated population—for example, the marriage rate as the number of marriages performed per 1000 population in the United States last year. Cohort measures refer data to a population group unified by a specified common characteristic—such as, the age cohort of 1–5 years, or the college class of 1990. Basic 1987 counts and rates useful in the analysis of world population and population trends have been reprinted with the permission of the Population Reference Bureau, Inc., as an appendix to this book. Examination of them will document the discussion to follow.

Birth Rates

The **crude birth rate** (CBR), often referred to simply as the "birth rate," is the annual number of live births per 1000 population (both male and female). A country with a population of 2 million and with 40,000 births a year would have a birth rate of 20 per 1000.

$$\frac{40,000}{2,000,000} = 20 \text{ per } 1000$$

The birth rate of a nation is, of course, strongly influenced by the age and sex structure of that nation, by the customs and family size expectations of its inhabitants, and by its adopted population policies. Since these conditions vary widely, recorded national birth rates vary—from, in 1987, a high of 53 per 1000 in the eastern African nations of Malawi and Rwanda to the low of 10 per 1000 in Denmark, Italy, and West Germany in western Europe. Although birth rates greater than 30 per 1000 are considered *high,* more than half of the world's people live in countries with rates that are that high or higher (Figure 5.2). In these countries, the population is predominantly agricultural and rural, and a high proportion of the female population is young; they are found chiefly in Africa, southern Asia, and Latin America.

Birth rates of less than 20 per 1000 are reckoned *low* and are characteristic of industrialized, urbanized nations. Most European countries, the Soviet Union, Anglo-America, Japan, Australia, and New Zealand have such rates as, importantly, do a few developing nations

China's Way

"Every stomach comes with two hands attached" was Chairman Mao's explanation for his permissive attitude toward population growth. An ever-larger population is "a good thing" he announced in 1965 when the birth rate was 37 per 1000. At the start of Mao's rule in 1949, China had an estimated 540 million stomachs. At his death in 1976, population had risen to 852 million, the consequence of a drop in infant mortality rates, longer life spans, and—during the 1950s—the government's failure to recognize the impact of unchecked birth rates upon the economy and upon the livelihood and well-being of the population.

During the 1970s, when it was realized that population growth was consuming more than half of the annual increase in gross national product, China introduced a well-publicized campaign advocating the "two-child family" and providing services, including abortions, supporting that program. In

response, China's growth rate dropped by 1975 to 15.7 per 1000, down from 28.5 per 1000 just ten years earlier. Mao's death cleared the way for an even more forceful program of birth restriction. "One couple, one child" became the slogan of a new population control drive launched in 1979. "Husband and wife," stated the new constitution of 1982, "have a duty to practice family planning."

The "one child" program was consistently pushed, its message carried by billboards, newspapers, radio, and television, and its achievement assured in a tightly controlled society by both incentives and penalties. Late marriages were encouraged. Free contraceptives, abortions, and sterilizations were provided. Cash awards and free medical care were promised to parents who agreed to limit their families to just a single child. Penalties for second births, including fines equal to 15% of family income for 7 years, were levied. At the campaign's height in 1983, the government ordered the sterilization

of either husband or wife for couples with more than one child. Social pressures in the form of public criticism and even, reportedly, forced abortions helped guarantee compliance with the birth limitation goals of the campaign. Tragically, infanticide—particularly the exposure or murder of female babies—was a reported means both of conforming to a one-child limit and of increasing the chances that the one child would be male.

By 1986, China's growth rate had fallen to 1%, far below the 2.4% registered among the rest of the world's less developed nations. Its birth rate stood at 18 per 1000, and its year 2000 population was projected by Chinese observers to be no more than 1.2 billion, the goal set at the outset of family planning in 1971. The 1987 jump in the birth rate to 21 reflected a relaxation of stringent controls, an increase in numbers of young people of child-bearing ages, and changes in the age at marriage.

---

that have adopted stringent family planning programs. As recently as 1986, China was in this category, but recent relaxation of family size controls have resulted in a jump in its birth rates (see "China's Way"). *Transitional* birth rates (between 20 and 30 per 1000) characterize some, mainly smaller, "developing" countries.

As the recent population histories of Singapore, recounted at the opening of this chapter, and of China indicate, birth rates are subject to change. The low birth rates of European countries and of some of the areas that they colonized are usually ascribed to industrialization, urbanization and, in recent years, maturing populations. While stringent family planning policies in China rapidly reduced the birth rate from over 33 per 1000 in 1970 to 18 per 1000 in 1986, Japan experienced a 15-point decline in the decade 1948–1958 with little governmental intervention.

The stage of economic development is closely related to variations in birth rates among nations, although rigorous testing of this relationship proves it to be imperfect (Figure 5.2). As a group, the more developed nations of the world showed a crude birth rate of 15 per 1000 in 1987; less developed nations (excluding China) registered 36 per 1000. Religious and political beliefs can

also affect the rates. The convictions of many Roman Catholics and Muslims that their religion forbids the use of artificial birth control techniques often lead to high birth rates among believers (but dominantly Catholic Italy shows Europe's lowest birth rate). Similarly, some governments of both western and eastern Europe—concerned about birth rates that are too low to sustain present population levels—subsidize births in an attempt to raise those rates. Regional variations in percentage contributions to world population growth are summarized in Figure 5.3.

Fertility Rates

Crude birth rates may display such regional variability because of variations in age and sex composition or disparities in births among the reproductive age, rather than total, population. A more accurate statement than the birth rate of the amount of reproduction in the population is the **total fertility rate** (TFR) (Figure 5.4). This is the average number of children a woman will bear,

Figure 5.3

Percent contributions to world population growth, by region, 1975–1985. Birth-rate changes are altering the world pattern of population increase. Between 1965 and 1975, China's contribution to world growth was two and a half times that of Africa; between 1975 and 1985, Africa's numerical growth was 1.3 times that of China. India's share of world population growth between 1965 and 1975 was 16%, far below its 21% contribution in the next 10 years.

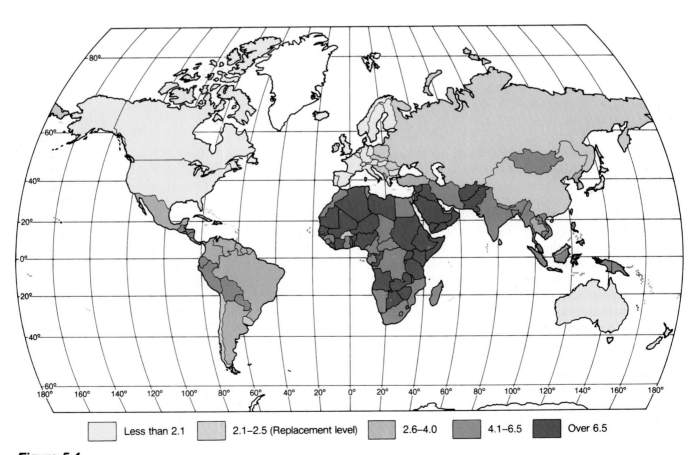

Figure 5.4

Total fertility rate. TFR indicates the average number of children that would be born to each woman if, during her child-bearing years, she bore children at the current year's rate for women that age. Since the total fertility rate is age-adjusted, two nations with identical birth rates may have quite different fertility rates and therefore different prospects for growth. Depending upon mortality conditions, a total fertility rate of 2.1 to 2.5 children per family is considered the "replacement level," at which a population will eventually stop growing.

Source: Data from Population Reference Bureau, Inc.

TABLE 5.1

Fertility Change in Selected Less Developed Nations and Regions

Region or country	Total fertility rate		
	Early 1960s	Early 1980s	Percent change
East Asia	5.3	2.3	−56.6
China	5.9	2.1	−64.4
South Asia	6.1	4.6	−24.6
Afghanistan	7.0	6.5	−7.1
Bangladesh	6.7	6.2	−7.5
India	5.8	4.5	−22.4
Nepal	5.9	6.3	+6.8
Thailand	6.4	3.5	−45.3
Africa	6.7	6.4	−4.5
Egypt	7.1	5.3	−25.4
Kenya	8.2	8.0	−2.4
Nigeria	6.9	6.3	−8.7
Latin America	5.9	4.1	−30.5
Brazil	6.2	4.1	−34.4
Guatemala	6.8	5.8	−14.7
Mexico	6.7	4.4	−34.3

Source: Thomas W. Merrick, with PRB staff, "World Population in Transition," *Population Bulletin* 41, no. 2 (Washington, D.C.: Population Reference Bureau, Inc., 1986).

assuming that current birth rates remain constant throughout her child-bearing years. The fertility rate minimizes the effects of fluctuation in the population structure and is thus a more reliable figure for regional comparative and predictive purposes than the crude birth rate.

A total fertility rate of 2.1 is necessary just to replace present population. On a worldwide basis, the TFR for the mid–1980s was 3.7. The more developed countries recorded a 2.0 rate, while less developed nations (excluding China) had a collective TFR of 5.0 (1984–1987). Despite the rate disparity between the two classes of nations, the fertility rates for many less developed nations have declined a third or more since the early 1960s (Table 5.1). Substantial reductions occurred in several Asian and many Latin American nations; China's drop from a TFR of 5.8 births per woman in 1970 to 2.4 in 1987 was both impressive and demographically significant because of China's status as the world's foremost nation in population. However, fertility rates are still well over 6 per woman in most of Africa and in many smaller Asian nations.

Death Rates

The **crude death rate** (CDR), also called the **mortality rate,** is calculated in the same way as the crude birth rate: the annual number of events per 1000 population. In general, the death rate, like the birth rate, varies with national levels of economic development. Highest rates (over 20 per 1000) are found in the less developed countries, particularly in Africa, and the lowest (less than 10) in developed countries, though the correlation is far from exact (Figure 5.5). Dramatic reductions in death rates occurred in the years following World War II as antibiotics, vaccinations, and pesticides to treat diseases and control disease carriers were disseminated to almost all parts of the world. Distinctions between more developed and less developed nations in mortality were reduced, but by no means erased, as death rates in many tropical, newly independent countries and in isolated rural areas of developed nations fell to a quarter or less of what they had been as late as the 1940s.

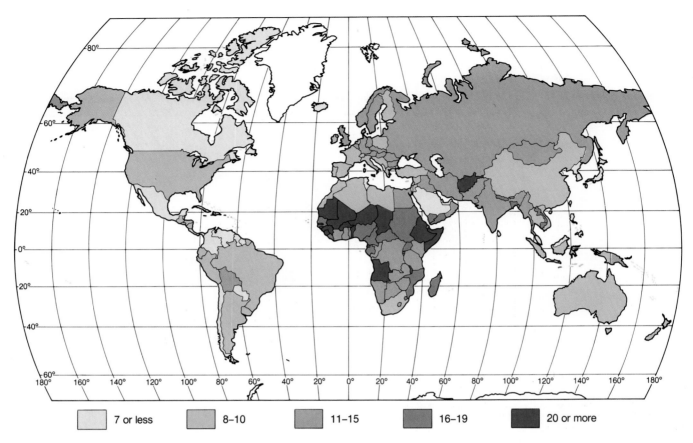

| | 7 or less | | 8–10 | | 11–15 | | 16–19 | | 20 or more |

Figure 5.5

Crude death rates show less worldwide variability than do the
birth rates on Figure 5.2, the result of widespread availability of
at least minimal health protection measures and a generally
youthful population in the developing nations, where death rates
are frequently lower than in "old-age" Europe.

Source: Data from Population Reference Bureau, Inc.

Like crude birth rates, death rates are meaningful
for comparative purposes only when identically structured populations are studied. Countries with a high proportion of elderly people, such as Austria or West
Germany, would be expected to have higher death rates
than those with a high proportion of young people like
Spain, assuming equality in other national conditions affecting health and longevity. The pronounced youthful-
ness of populations in developing nations is an important
factor in the recently reduced mortality rates of those
areas.

To overcome that lack of comparability, death rates
can be calculated for specific age groups. The **infant
mortality rate,** for example, is the ratio of deaths of infants aged 1 year or under per 1000 live births:

$$\frac{\text{deaths, age 1 year or less}}{\text{1000 live births}}$$

Infant mortality rates are significant because it is at these
ages that the greatest declines in mortality have occurred, largely as a result of the increased availability of
health services. The drop in infant mortality accounts for
a large part of the decline in the general death rate in
the last few decades, for mortality during the first year of
life is usually greater than in any other year.

Two centuries ago, it was not uncommon for 200–
300 infants per 1000 to die in their first year. Even today,
despite significant declines in those rates over the last
half century in many individual nations (Figure 5.6),
striking world regional and national contrasts remain. For
all of Africa, infant mortality rates approach 115 per 1000;
individual African nations (for example, Gambia, Mali,
Sierra Leone—all in West Africa) show rates near or above
170. In contrast, infant mortality rates in North America
and western Europe are in the 9–11 range.

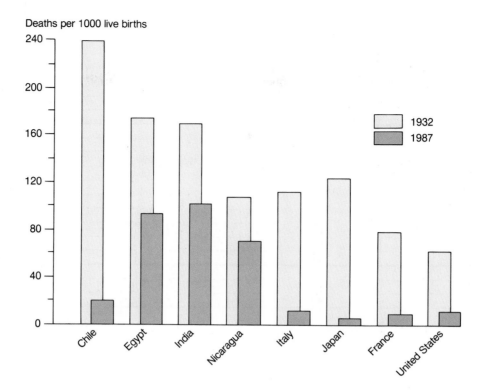

Figure 5.6

Infant mortality rates for selected countries. Dramatic declines in the rate have occurred in all countries, a result of international programs of health care delivery aimed at infants and children in developing nations. Nevertheless, the decreases have been proportionately greatest in the urbanized, industrialized nations, where sanitation, safe water, and quality health care are more widely available.

Source: Data from U.S. Bureau of the Census and Population Reference Bureau, Inc.

The time and space variations in infant mortality and in crude death rates are obviously not due to differences in the innate ability of different national groups to survive. They are essentially the result of differential access to the health maintenance technologies known and widely adopted since the 1940s, but not yet universally available. Penicillin and other antibiotics, as well as DDT and other pesticides, protect, preserve, and prolong lives that, before their introduction, would have been lost. Modern medicine and sanitation have increased life expectancy and have altered age-old relationships between birth and death rates. The availability and employment of these modern methods, however, have varied regionally, and the developed countries have been most able to benefit from them. In such underdeveloped and impoverished areas as much of sub-Saharan Africa, the chief causes of death are those no longer of concern in more developed lands: infectious diseases such as tuberculosis and malaria; intestinal infections; and, among infants and children especially, malnutrition and dehydration from diarrhea.

Population Pyramids

Another means of comparing populations is through the **population pyramid,** a graphic device that represents a population's age and sex composition. The term "pyramid" is descriptive of the diagram's design for many countries in the 1800s, when the display was created: a broad base of younger age groups and a progressive narrowing toward the apex as older populations were thinned by death. Now, many different shapes are encountered, each reflecting a different population history (Figure 5.7). Data for several generations of people are grouped, highlighting the impact of "baby booms," wars, birth rate reductions, and external migrations. A rapidly growing country such as Mexico has most people in the lowest age cohorts; the percent in older age groups declines successively, yielding a pyramid with markedly sloping sides. Female life expectancy is, typically, reduced in older cohorts of less developed countries, so that for Mexico the proportion of females in older age

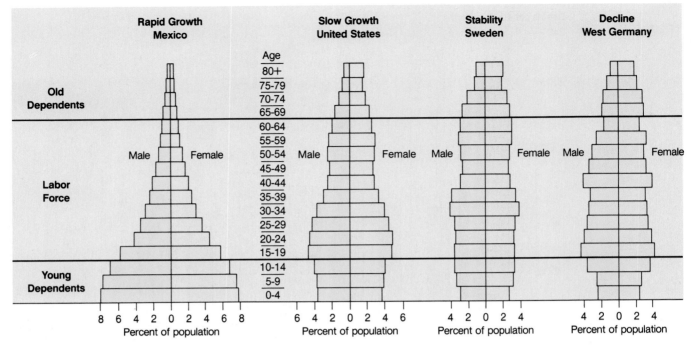

Figure 5.7

Four patterns of population change. These diagrams show that population "pyramids" assume many shapes. The age distribution of national populations both records the present and foretells the future. In countries like Mexico, social costs related to the young are important, and economic expansion is vital to provide employment for new entrants in the labor force. West Germany's negative growth means a future with fewer workers to support a growing demand for social services for the elderly.

groups is lower than in, for example, the United States or Sweden. In the latter country, a wealthy nation with a very slow rate of growth, the population is nearly equally divided among the age groups, giving a "pyramid" with almost vertical sides. Among older cohorts, as West Germany shows, there may be an imbalance between men and women because of the greater life expectancy of the latter.

The population pyramid provides a quickly visualized demographic picture of immediate practical and predictive value. For example, the percent of a nation's population in each age group strongly influences demand for goods and services within that national economy. A country with a high proportion of young has a high demand for educational facilities and certain types of health delivery services; additionally, of course, a large percent of the population is too young to be employed (Figure 5.8). On the other hand, a population with a high percent of elderly people also requires medical goods and services specific to that age group; these people must be supported by a smaller proportion of workers. The **dependency ratio** is a simple measure of the number of dependents, old or young, that each 100 people in the productive years (usually 20–64) must support. Population pyramids give quick visual evidence of that ratio.

Over the next several decades, its population pyramid tells us, the United States will likely become more and more like Sweden in demographic structure. The proportion of elderly will increase, the proportion of young people will decrease, and the median age of the population will rise. Social services and costs related to an aging population will become a greater national concern. Mexico, on the other hand, faces continuing problems: expansion of educational facilities, training of teachers, creation of jobs, and provision of health services required by a young and growing population.

Natural Increase

Knowledge of the sex and age distributions also enables nations to forecast their future population levels, though the reliability of projections decreases with increasing length of forecast (Figure 5.9). Thus, a country with a high proportion of young people will experience a high rate of natural increase unless there is a very high mortality rate among infants and juveniles, or fertility and birth rates change materially. The **rate of natural increase** of a population is derived by subtracting the crude death rate from the crude birth rate. *Natural* means that increases or decreases due to migration are not included.

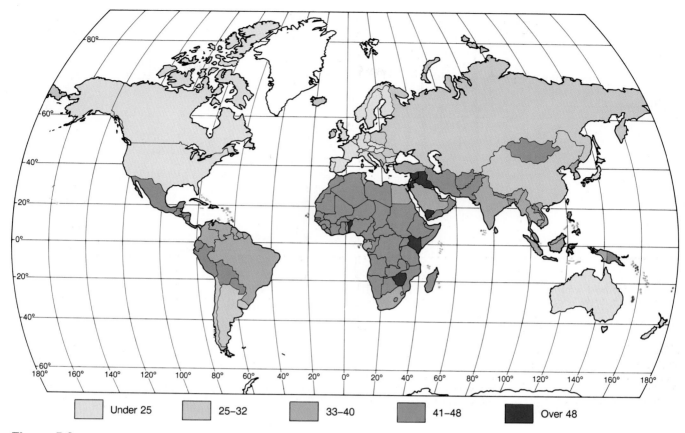

| | Under 25 | | 25–32 | | 33–40 | | 41–48 | | Over 48 |

Figure 5.8

Percent of population under 15 years of age. A high
proportion of a national population under age 15 increases the
dependency ratio of that nation and promises future population
growth as the youthful cohorts enter child-bearing years.

Source: Data from Population Reference Bureau, Inc.

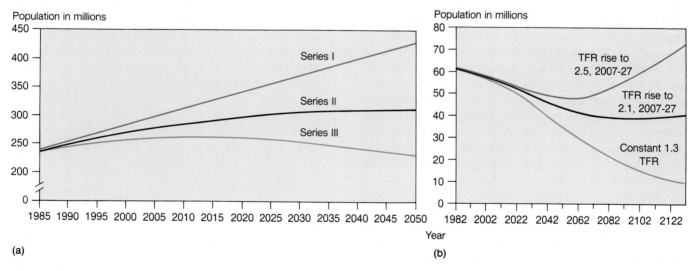

(a)

(b)

Figure 5.9

**Possible population futures: United States and West
Germany.** (a) United States population projections to year 2050.
Population projections often prove inaccurate because birth and
death rates and the number of immigrants are constantly
changing. The middle series projection assumes middle levels of
fertility, mortality, and immigration. Depending on the
assumptions, U.S. population in 2050 might range from 231
million to 429 million. (b) The West German projections are
based solely upon varying assumptions about that nation's total
fertility rate, among the world's lowest at 1.3 births per woman.
Should that rate remain constant, population will drop to just
under 10 million by 2132. Even if the total population rate rose
to the "replacement level" of 2.1 children during the years
2007–2027 (the middle projection), the population decrease
would still amount to a one-third reduction from the 61 million
of 1987.

Source: U.S. Bureau of the Census and Population Reference Bureau, Inc.

TABLE 5.2
Doubling Time in Years at Different Rates of Increase

Annual percentage increase	Doubling time (years)
0.5	140
1.0	70
2.0	35
3.0	24
4.0	17
5.0	14
10.0	7

TABLE 5.3
Population Growth Yielded by a 2% Rate of Increase

Year	Population
0	1000
5	1104
10	1219
15	1345
20	1485
25	1640
30	1810
35	2000

TABLE 5.4
Population Growth and Approximate Doubling Times Since A.D. 1

Year	Estimated population	Doubling time (years)
1	250 million	
1650	500 million	1650
1850	1 billion	200
1930	2 billion	80
1975	4 billion	45
?	8 billion	?

If a country had a birth rate of 22 per 1000 and a death rate of 12 per 1000 for a given year, the rate of natural increase would be 10 per 1000. This rate is usually expressed as a percentage, that is, as a rate per 100 rather than per 1000. In the example given, the annual increase would be 1%.

Doubling Times

The rate of increase can be related to the time it takes for a population to double. Table 5.2 shows that it would take 70 years for a population with a rate of increase of 1% (approximately the rate of growth of the Soviet Union in the middle 1980s) to double; a 2% rate of increase means that the population will double in only 35 years. How could adding only 20 persons per 1000 cause a population to grow so quickly? The principle is the same as that used to compound interest in a bank. Table 5.3 shows the number yielded by a 2% rate of increase at the end of successive 5-year periods.

For the world as a whole, the rates of increase have risen over the span of human history. Therefore, the **doubling time** has decreased. Note in Table 5.4 how the population of the world has doubled in successively shorter periods of time. It may reach 9.4 billion during the first half of the 21st century if the present rate of growth continues (Figure 5.1). In countries with high rates of increase (Figure 5.10), the doubling time is less than the 40 years projected for the world as a whole (at growth rates displayed in 1987). Should world fertility rates decline, population doubling time will correspondingly increase.

Here, then, lies the answer to the question posed earlier. Even small annual additions accumulate to large total increments because we are dealing with geometric or exponential (1, 2, 4, 8) rather than arithmetic (1, 2, 3, 4) growth. The ever-increasing base population has reached such a size that each additional doubling results in an astronomical increase in the total. One observer, G. Tyler Miller, has offered a graphic illustration of the inevitable consequences of such doubling, or J-curve, growth. He asks that we mentally fold in half a page of this book. We have, of course, doubled its thickness. We can—mentally, though certainly not physically—continue the doubling process several times without great effect, but 12 such doublings would yield a thickness of 1 foot (30.5 cm) and 20 doublings would result in a thickness of nearly 260 feet (79 m). From then on, the results of further doubling are astounding. Doubling the page only about 52 times, Miller reports, would give a thickness reaching from the earth to the sun. Rounding the bend on the J-curve, which world population has done (Figure 5.11), poses problems and has implications for human occupance of the earth of a vastly greater order of magnitude than ever faced before.

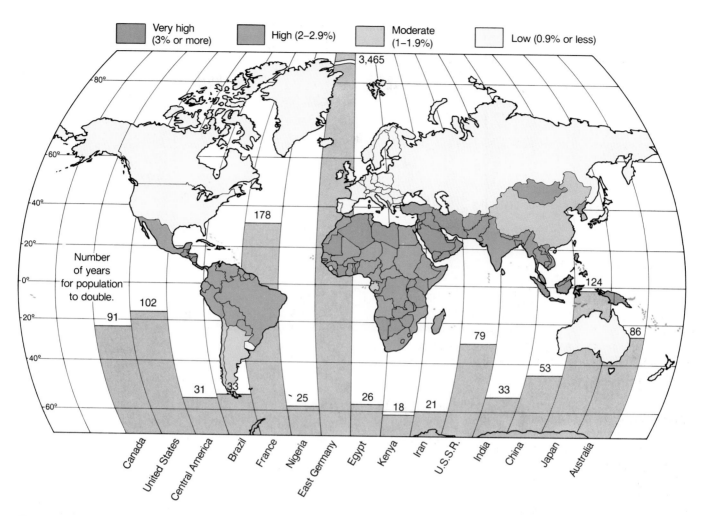

Figure 5.10

Annual rates of natural increase. While the world's 1987 rate of natural increase (1.7%) would mean a doubling of population in 40 years, many individual continents and countries deviate widely from the global average rate of growth and have vastly different doubling times. Africa as a whole has the highest rates of increase, followed by Latin America and Asia. China's rate of natural increase has dropped since the early 1970s to below the world average. Europe and North America are prominent among the low-growth areas, with some individual countries of Western Europe actually declining in population size.

Source: Data from Population Reference Bureau, Inc.

The Demographic Cycle

The theoretical consequence of exponential population growth cannot be realized. A braking mechanism, the substantial lowering of birth rates through what is known as the **demographic transition,** may result in the voluntary control of totally unregulated population growth. Should that fail, involuntary controls of an unpleasant nature will be set in motion.

As long as births only slightly exceed deaths, even when the rates of both are high, the population will grow only slowly. This was the case for most of human history until about A.D. 1750 (Figure 5.11). Demographers think that it took from approximately A.D. 1 to A.D. 1650 for the population to increase from 250 million to 500 million, a doubling time of more than a millennium and a half. Growth was not steady, of course. There were periods of regional expansion that were usually offset by sometimes catastrophic decline. Wars, famine, and other disasters took heavy tolls. For example, the bubonic plague (the Black Death), which swept across Europe in the 14th century, is estimated to have killed over one-third of the population of that continent. High birth and high, but

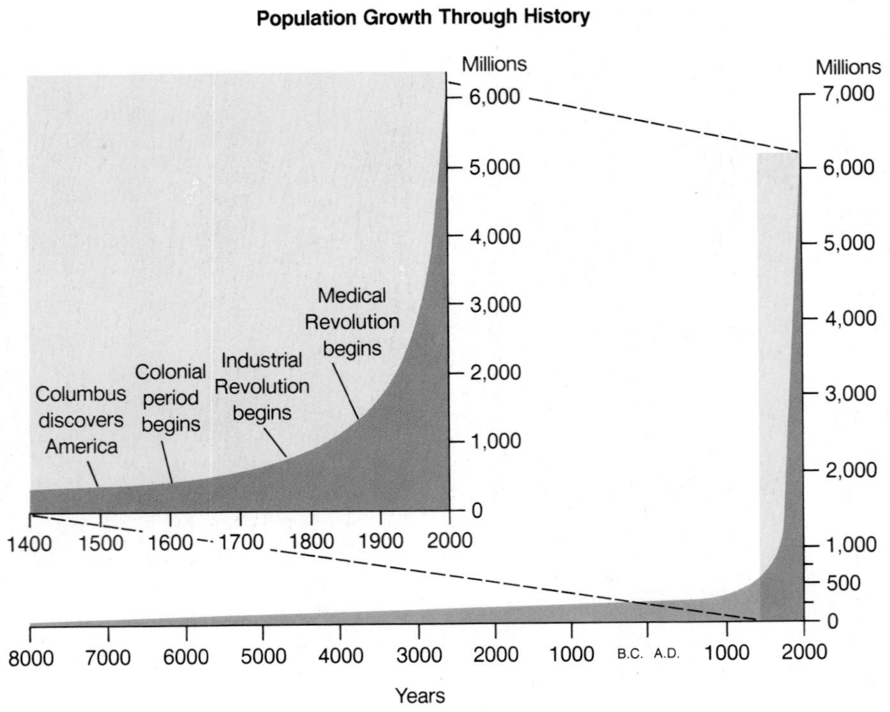

Population Growth Through History

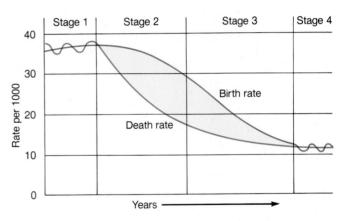

Figure 5.11

World population growth, 8000 B.C. to A.D. 2000. Notice that
the bend in the J-curve begins about the mid-1700s, when
industrialization started to provide new means to support the
population growth made possible by revolutionary changes in
agriculture and food supply. Improvements in medical science
and nutrition served to reduce death rates around 1900 in the
industrializing nations.

fluctuating, death rates represent the *first stage* of the **demographic cycle** (Figure 5.12). That first stage is no
longer found in any country. In 1987, the highest death
rates (nearly 30 per 1000) were in Gambia and Sierra
Leone, in West Africa, but the birth rates in those countries were even higher, almost 50 per 1000.

The Western Experience

The demographic transition model was developed to
summarize the population history of western Europe.
That area entered a *second stage* of the cycle with the
industrialization that began about 1750. Its effects—declining death rates accompanied by continuing high birth
rates—have been dispersed worldwide even without
universal conversion to an industrial economy. Rapidly
rising populations during the second demographic stage
result from dramatic increases in life expectancy. That,
in turn, reflects falling death rates due to advances in
medical and sanitation practices, improved mechanisms
for foodstuff storage and distribution, a rising per capita
income, and the urbanization that provides the environment in which sanitary, medical, and food distributional
improvements are concentrated (Figure 5.13). Birth rates

Figure 5.12

Stages in the demographic cycle. During the first stage, birth
and death rates are both high, and population grows slowly.
When the death rate drops and the birth rate remains high, there
is a rapid increase in numbers. During the third stage, birth rates
decline and population growth is less rapid. The fourth stage is
marked by low birth and death rates and, consequently, by a low
rate of natural increase.

do not fall as soon as death rates; ingrained cultural patterns change more slowly than technologies. In many societies, large families are considered advantageous.
Children contribute to the family by starting to work at
an early age and by supporting their parents in old age.

Many countries in Latin America and parts of
southern and southeastern Asia display the characteristics of this second stage in the population cycle. Pakistan,
with a birth rate of 44 and a death rate of 15, or Bolivia,
with respective rates of 40 and 14 (1987 estimates), are
typical. The annual rates of increase of such countries are

Figure 5.13

Vienna, Austria in the 1870s. A modernizing Europe experienced improved living conditions and declining death rates during the 19th century.

near 30 per 1000; their populations will double in about 25 years. Such rates, of course, do not mean that the full impact of the Industrial Revolution has been worldwide; they do mean that the underdeveloped societies have been beneficiaries of the life preservation techniques associated with it.

The *third stage* follows when birth rates decline as people begin to control family size. The advantages that having many children bring in an agrarian society are not so evident in urbanized, industrialized cultures. In fact, children may be viewed as economic liabilities rather than assets. When the birth rate falls, and death rates remain low, the population size begins to level off. Chile, Sri Lanka, and Thailand are among the many nations now displaying the low death and transitional birth rates of the third stage.

The *demographic transition*—the development stages during which high birth and death rates are replaced by low birth and death rates—ends with the *fourth* and final stage of the demographic cycle. Essentially all European countries, Canada, Australia, and Japan are among the 30 or so nations that have entered this phase. Because it is characterized by very low birth and death rates, it yields very slight percent increases in population. Population doubling times may be as long as a thousand or more years if those present low birth rates continue. In the case of a few countries—Denmark, West Germany, and Hungary in 1987, for example—death rates have begun to exceed birth rates, and populations are declining (see "Europe's Population Dilemma").

The population cycle described by the demographic transition model recounts the experience of northwestern European countries—primarily England and Wales—as they went from rural, agrarian societies to urban-industrial ones. The demographic history of what are today's rich, developed nations may not fully reflect the prospects of contemporary developing countries. In Europe, church and municipal records—some dating from the 16th century—show that people married late or not at all; in England before the Industrial Revolution as many as half of all women in the 15–50 age cohort were unmarried. Infant mortality was high, life expectancy low. With the coming of industrialization in the 18th and 19th centuries, immediate factory wages instead of long apprenticeship programs permitted earlier marriage and more children. Since improvements in sanitation and health came only slowly, death rates remained high. Around 1800, 25% of Swedish infants died before their first birthday. Population growth rates remained below 1% per year in France throughout the 19th century.

Beginning about 1860, first death rates and then birth rates began their dramatic decline. The mortality revolution came first. Many formerly fatal epidemic diseases had become endemic—that is, they were essentially continual within a population. As people developed partial immunities, mortalities associated with the diseases decreased. Improvements in animal husbandry, crop rotation and other agricultural practices, and new

Europe's Population Dilemma

Although international dismay may be expressed over rising world populations, some European nations are facing an opposite domestic concern. Europe's population is older than that of any other continent, and for many of its nations, population is stagnating or declining. The fertility rates of 21 of the continent's 30 nations in 1987 were below the replacement level of 2.1 per 1000, and 2 were at that level. None of the large countries of western Europe is at present replacing its population through natural increase. At a fertility rate of 1.3, West Germany stands at the bottom of the international reproduction scale, and Denmark and Italy are almost equally infertile. Not surprisingly, West Germany also has the oldest population in the world, with a small proportion of young and a large share of middle-aged and retired persons in its population pyramid. Most of eastern Europe has the same problem of declining population growth, reduced work-age cohorts, and an aging citizenry. Europe's population will begin to decline in the 1990s.

Spatially, the continent's fertility is peripheral. Catholic Ireland and Poland have fertility rates a little above the replacement level; only Albania and Turkey, where high Muslim reproduction rates are the rule, are far above the replacement fertility point. The Soviet Union, with a fertility rate of 2.4, owes much of its growth to its Central Asian Muslim, not its European, population. "In demographic terms," France's prime minister has remarked, "Europe is vanishing."

The national social and economic consequences of population stability or reduction are not always perceived by those who advocate **zero population growth**—a condition achieved when births plus immigration equal deaths plus emigration. An exact equation of births and deaths means an increasing proportion of older citizens, fewer young people, and a rise in the median age of the population. Actual population decline, now the common European condition, exaggerates those consequences. Already, schools are closing and universities being cut back in the face of permanently reduced demand. By the end of the century, the unemployment of the

1980s may be replaced by a shortage of workers, particularly skilled workers. Governments will have to provide pensions and social services for the one-quarter of their citizens older than 60 and pay for them by taxes on a diminishing work force. Already in West Germany there are four pensioners for every ten workers; by 2030, the numbers will be equal.

France, which grants substantial payments and services to parents to help in child rearing, is the only West European nation with an official probirth policy. East Europe is much more active in encouraging growth. Three children per family has become the adopted goal of several eastern bloc countries that fear for national survival if present trends continue. Parental subsidies, housing priorities, and generous maternity leaves at full salaries are among the inducements for multiple-child families. In Romania, contraceptives are banned and abortions permitted only on doctor's orders in the hope that birth rates can be increased. In general, despite some short-term successes, Europeans of both east and west have ignored both blandishments and restrictions. The consequences will only gradually emerge over the next several decades.

---

foodstuffs (the potato was an early example) from overseas colonies raised the level of health and disease resistance of the European population in general. At the same time, sewage systems and sanitary water supplies became common in larger cities, and general levels of hygiene improved everywhere. Death rates fell dramatically, as did birth rates as European societies began to redefine downward the traditional concepts of ideal family size. In cities, child labor laws and mandatory schooling meant that children became a burden, not a contribution, to family economies. As "poor-relief" legislation and other forms of public welfare substituted for family support structures, the insurance value of children declined. Family consumption patterns altered as the Industrial Revolution made more widely available goods that served consumption desires, not just basic living needs, and children hindered rather than aided the

achievement of the age's promise of social mobility and life-style improvement. Perhaps most important, and by some measures preceding and independent of the implications of the Industrial Revolution, were changes in the status of women and in their spreading conviction that control over childbearing was within their power and to their benefit.

The Developing World

Many aspects of the population history of Europe, the basis of the demographic transition model, appeared not to be directly and immediately applicable to the developing world of the middle and late 20th century. Death rates could be, and were, lowered quickly and dramatically worldwide, but their reduction followed a different pattern. In Europe, health improvements were gradual,

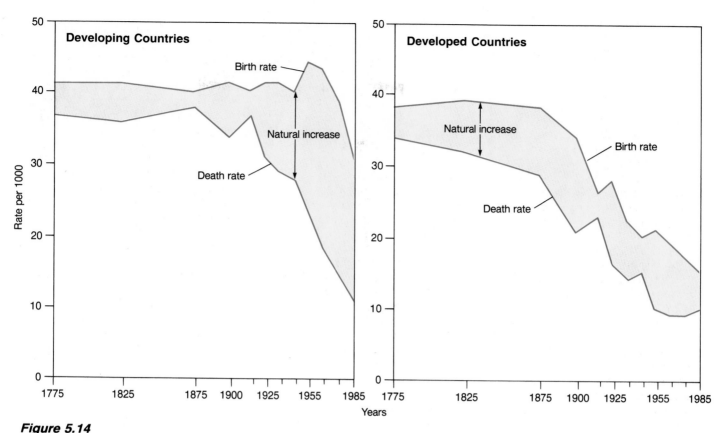

Figure 5.14

World birth and death rates. The "population explosion" after World War II (1939–1945) reflected the effects of drastically reduced death rates in developing nations without compensating reductions in births. Mortality declines in European and Western nations had been more gradual, with birth-rate reductions keeping pace.

Source: Adapted from *World Population: Toward the Next Century*, 1985, Population Reference Bureau, Inc.

resulting from a composite of slowly changing economic and social conditions. In the developing nations, comparable reductions seemed to happen overnight through the introduction of Western technologies of medicine and public health: insecticides, antibiotics, sanitation, immunization, infant and child health care, and eradication of smallpox. Frequently introduced under the auspices of such international agencies as the World Health Organization, these imported technologies and treatments accomplished in a few years what it took Europe 50 or 100 years to experience. Sri Lanka, for example, sprayed extensively with DDT to combat malaria; life expectancy jumped from 44 years in 1946 to 60 only 8 years later. With similar public health programs, India also experienced a steady reduction in its death rate after 1947. Simultaneously, and with comparable international sponsorship, food aid cut the death toll of developing nations during drought and other disasters. The dramatic decline in mortality, which emerged only gradually throughout the European world but occurred so rapidly in contemporary developing nations, has been the most fundamental demographic change in human history.

Decreases in death rates and increases in life expectancies were almost immediately reflected in the population structure of developing nations. Balancing reductions in birth rates have been harder to achieve and depend less upon supplied technology and assistance than they do upon social acceptance of the idea of fewer children and smaller families (Figure 5.14). Fertility falls, simply put, as incomes increase, health improves, and (particularly, female) educational levels rise—all conditions that make parents desire fewer children. But not all women or cultures want smaller families. Surveys in several African nations indicate a continuing preference among women for large families of between six and nine children. The cultural bias favoring many offspring still evident in much of Africa may eventually change spontaneously or coercively, as it has in Singapore, China, Japan, and many other nations.

For much of the developing world, however, the differentials between birth and death rates remain great. The resulting rapid population growth often hinders economic development because of the strain it puts on

Figure 5.15

Fleeing war, repression, and famine, millions of people in developing nations, like these Ethiopians, have become refugees from their homelands.

the economy. When population doubles each generation, as it must at the mid–1980s fertility rates in more than 100 nations, it is hard to achieve social and economic progress; just keeping up with the new mouths to feed is a major accomplishment.

Population Relocations

The rapid increase in population that accompanied Europe's industrialization was alleviated by the emigration of millions of people to North America, Australia, New Zealand, and other areas. Most of that movement was voluntary, as people sought economic opportunity or, for some, religious or political freedoms they had not enjoyed in their native lands. In the century from 1820 to 1920, some 50 million Europeans migrated to Anglo-America alone. Emigration for economic reasons is still important in the modern world but is increasingly restricted by national policies attempting to limit by quotas and border control the freer movement of people of earlier periods. The debate about the flow of illegal migrants from Mexico into the United States during the 1970s and 1980s reflected a variety of national concerns, with competition for jobs and drain on social services most frequently cited as reasons for controlling cross-border movements. Nigeria in West Africa and Switzerland in Europe, both nations that had received "guest workers" at times of domestic labor shortages, forcibly

expelled them during economic recessions of the 1980s. Migration as an aspect of behavioral geography is discussed further in Chapter 8.

Not only economics but also politics induce large-scale cross-border movements. Refugees from war and political turmoil or repression numbered 12 million in 1987, according to United Nations figures. These are the officially designated refugees; unofficially, the total might have reached 20 million or more. In the past, refugees sought asylum primarily in Europe and other developed areas. More recently, the flight of people is primarily from developing countries to other developing regions. About 80% of the officially recorded refugees have found asylum in Southeast Asia, southern and eastern Africa, the Near East and south Asia, and South and Central America. Africa alone housed nearly 3 million refugees, including more than 1 million Ethiopians (Figure 5.15); in Somalia, one out of eight persons was a refugee in 1986. The largest refugee population in the world at that date was the 3.5 million Afghans who sought safety in Pakistan (2.5 million) and Iran (1 million). Indo-Chinese refugees (including the "boat people") and uprooted or fleeing populations from strife-ridden sections of Central America were other sizable groups seeking foreign shelter.

Numerically, these refugees represent massive dislocations of populations. Proportional to total world population, however, they are insignificant. Such contemporary migrations can have no similar impact to the filling of near-empty continents by European immigrants

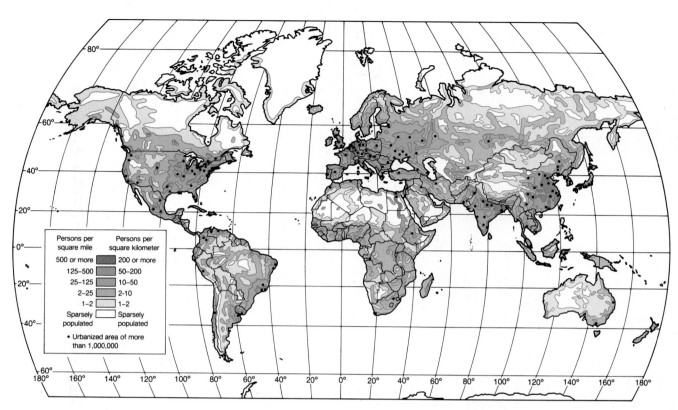

Figure 5.16
World population density.

and their descendants in past centuries, or by the gradual spread of human populations over the habitable earth in prehistory. The spatial impact of even catastrophic population events is reduced as total human numbers in a crowded world reach into the billions.

World Population Distribution

The millions and billions of people of our discussion are not uniformly distributed over the surface of the earth. The most striking feature of the world population distribution map (Figure 5.16) is the very unevenness of the pattern. Some land areas are nearly uninhabited, others are sparsely settled, and yet others contain dense agglomerations of people. More than half the world's people are found—unevenly concentrated, to be sure—in rural areas, but more than 40% are urbanites, and a constantly growing proportion are residents of very large cities of 1 million or more. Earth regions of apparently very similar physical makeup show quite different population numbers and densities, perhaps the result of differently timed settlement or of settlement by different cultural groups. Had North America been settled by Chinese instead of Europeans, for example, it is likely

that its western sections would be far more densely settled than they now are. Europe, inhabited thousands of years before North America, contains about twice the population of the United States on less land.

Certain generalizing conclusions can be drawn from the uneven but far from irrational distribution of population shown on Figure 5.16. First, about 90% of all people live north of the equator, where is also found some 80% of the world's land area. Second, a large majority of the world's inhabitants occupy only a small part of its land surface. Over half the people live on about 5% of the land; almost nine-tenths reside on less than 20%. Third, people congregate in lowland areas; their numbers decrease sharply with increases in elevation. Temperature, length of growing season, slope and erosion problems, even oxygen reductions at very high altitudes, all appear to limit the habitability of upper-elevation locations. One estimate is that between 50% and 60% of all people live below 650 feet (200 m), a zone containing less than 30% of total land area; nearly 80% reside below 1650 feet (500 m).

Fourth, although low-lying areas are preferred settlement locations, not all such areas are equally favored. Continental margins have attracted densest settlement; about two-thirds of world population is concentrated

Figure 5.17
Terracing of hillsides is one device to extend a naturally limited productive area. The technique is effectively used on the densely settled Indonesian island of Bali.

within 300 miles (480 km) of the ocean, much of it on alluvial lowlands and river valleys. Latitude, aridity, and elevation, however, limit the attractiveness of many seafront locations. Low temperatures and infertile soils of extensive coastal lowlands of higher latitudes of the northern hemisphere have restricted settlement there despite the low elevations of the Arctic coastal plain and the advantages of ocean proximity. Mountainous or desert coasts are sparsely occupied at any latitude, and some tropical lowlands and river valleys that are marshy, forested, and disease-infested are unevenly settled.

Within the sections of the world generally conducive to settlement, four areas contain great clusters of population: East Asia, South Asia, Europe, and northeastern United States–southeastern Canada. The *East Asia* zone, which includes Japan, China, Taiwan, and South Korea, is areally the largest cluster. The four nations forming it contain 25% of all people on earth, and China alone accounts for one in five of the world's inhabitants.

The *South Asian* cluster is composed primarily of countries associated with the Indian subcontinent—Bangladesh, India, Pakistan, and the island nation of Sri Lanka—though some might add to it the Southeast Asian nations of Burma, Kampuchea (Cambodia), and Thailand. The four core countries alone account for another one-fifth, 20%, of the world's 5 billion inhabitants. The South and the East Asian concentrations are, thus, home to nearly one-half of the world's people. *Europe*—southern, western, and eastern through much of European USSR—is the third extensive world population concentration, with another 13% of its inhabitants. Much smaller in extent and total numbers is the cluster in *northeastern United States and adjacent Canada*. Other such smaller, but pronounced, concentrations are found around the globe: on the island of Java in Indonesia, along the Nile River in Egypt, and in discontinuous pockets in Africa and Latin America. (A study of "Population Patterns of Latin America" is included in Chapter 12, Page 413.)

The term **ecumene** is applied to permanently inhabited areas of the earth's surface. The ancient Greeks used the word, derived from their verb "to inhabit," to describe their known world between what they believed to be the unpopulated searing southern equatorial and permanently frozen northern polar reaches of the earth. Clearly, natural conditions are less restrictive than Greek geographers believed, and both ancient and modern technologies have rendered habitable areas that natural conditions make forbidding. Irrigation, terracing, diking, and draining are among the methods devised to extend the ecumene locally and at least temporarily (Figure 5.17).

At the world scale, the ancient observation of habitability appears remarkably astute. The **nonecumene,** the uninhabited or very sparsely occupied zone, does include the permanent icecaps of the Far North and Antarctica and large segments of the tundra and coniferous forest (taiga) of northern Asia and North America. But the nonecumene is not continuous as the ancients supposed (Figure 5.16). It is discontinuously encountered in all portions of the globe and includes parts of the tropical rain forests of equatorial zones, mid-latitude deserts of both northern and southern hemispheres, and high mountain areas. Even parts of these unoccupied or sparsely occupied districts have localized dense settlement nodes or zones based on irrigation agriculture, mining and industrial activities, and the like. Perhaps the most anomalous case of settlement in the nonecumene world is that of the dense population in the Andes Mountains of South America and the plateau of Mexico. Here, Native Americans found temperate conditions away from the dry coast regions and the inhospitable wet Amazon Basin; the fertile high basins have served a large population for more than a thousand years. Even with these locally important exceptions, the nonecumene portion of the earth is extensive. Perhaps 35%–40% of all the world's land surface is barren and without significant settlement. This is, admittedly, a smaller proportion of the earth than would have qualified as uninhabited in ancient times or even during the last century. Humans have, since the end of the Ice Ages some 11,000 years ago, steadily expanded their areas of settlement.

Population Density

Margins of habitation could only be extended, of course, as humans learned to support themselves from the resources of new settlement areas. The numbers that could be sustained in old or new habitation zones were, and are, related to the resource potential of those areas and the cultural levels and technologies possessed by the occupying populations. We express the relationship between number of inhabitants and the area they occupy as **population density.**

Density figures are useful if sometimes misleading representations of regional variations of human distribution. The **crude density** or **arithmetic density** of population is the commonest and least satisfying expression of that variation. It is the calculation of the number of people per unit area of land, usually within the boundaries of a political entity. It is an easily reckoned figure; all that is required is information on total population and total area, both commonly available for national or other political units. It can, however, be misleading and may obscure more of reality than it reveals. It is an average, and a nation may contain extensive regions that are only sparsely populated or largely undevelopable along with intensively settled and developed districts. A national average density figure reveals nothing about either class of territory. In general, the larger the political unit for which crude or arithmetic population density is calculated, the less useful is the figure.

Various modifications may be made to refine density as a meaningful abstraction of distribution. Its descriptive precision is improved if the area in question can be subdivided into comparable regions or units. Thus, it is more revealing to know that in 1985 New Jersey had a density of 1013 and Wyoming of 5 persons per square mile (390 and 2 per km²) of land area than to know only that the figure for the conterminous United States (48 states) was 80 per square mile (31 per km²) of land. The calculation may also be modified to provide density distinctions between classes of population—rural versus urban, for example. Rural densities in the United States rarely exceed 300 per square mile (115 per km²), while portions of major cities can have tens of thousands of people per square mile.

Another revealing refinement of crude density relates population not simply to total national territory but to that area of a country that is or may be cultivated—that is, to *arable* land. The resulting figure is the **physiological density** and is, in a sense, an expression of

TABLE 5.5

Comparative Densities for Selected Countries (persons per square mile)

Country	Crude density	Physiological density
Australia	5	91
Bangladesh	1926	3058
Canada	7	134
China	287	2606
Egypt	134	6711
India	631	1236
Iran	79	880
Japan	850	6541
Nigeria	304	896
USSR	33	328
United Kingdom	601	2072
United States	67	337

Source: *1987 World Population Data Sheet* (Population Reference Bureau, Inc.), derived from UN Food and Agriculture Organization (FAO) and U.S. Department of State.

population pressure differentially exerted on agricultural land. It is readily apparent from Table 5.5 that there are differences between nations in physiological density and that the contrasts between crude and physiological densities for and between nations point up actual settlement pressures that are not revealed by crude or arithmetic densities alone. But their calculation depends upon uncertain definitions of arable and cultivated land, assume that all arable land is equally productive and comparably used, and includes only one part of the resource base of a nation.

Overpopulation

It is an easy and common step from concepts of population density to assumptions about overpopulation or overcrowding. It is well to remember that **overpopulation** is a value judgment reflecting an observation or conviction that an environment or territory is unable to support its present population. It is not the necessary and inevitable consequence of high density of population. Tiny Monaco, a principality in southern Europe about half the size of New York's Central Park, has a crude density of nearly 40,000 persons per square mile. The sizable (604,000 square miles) Mongolian People's Republic, between China and the Soviet Union, has 3 persons per square mile; Iran, only slightly larger, has 79. Macao, an island possession of Portugal off the coast of China, has

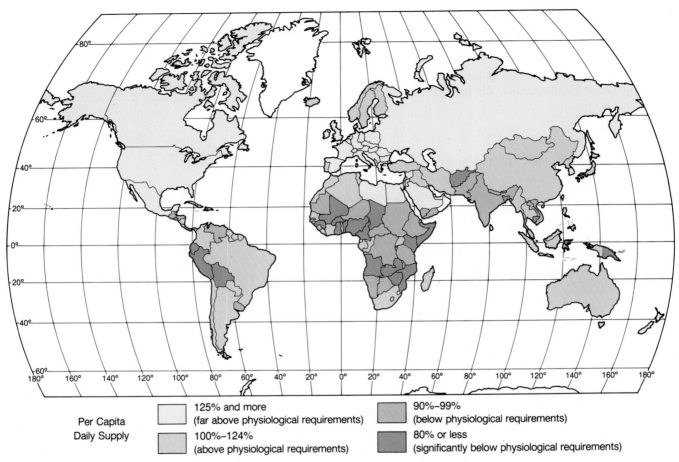

Figure 5.18
Percent of daily required calories.
Source: FAO Production Yearbook.

Per Capita Daily Supply

125% and more (far above physiological requirements)

100%–124% (above physiological requirements)

90%–99% (below physiological requirements)

80% or less (significantly below physiological requirements)

more than 66,000 persons per square mile; the British-owned Falkland Islands off the Atlantic Coast of Argentina count at most one person for every 2.5 square miles of territory. No conclusions about conditions of life, levels of income, adequacy of food, or prospects for prosperity can be drawn from these density comparisons.

Overcrowding is a reflection not of numbers per unit area, but of the **carrying capacity** of land—the number of people an area can support given the prevailing technology. A region devoted to efficient, energy-intensive commercial agriculture that makes heavy use of irrigation, fertilizers, and biocides can support more people at a higher level of living than one engaged in the slash-and-burn agriculture described in Chapter 9. An industrial nation that takes advantage of resources such as coal and iron ore, and has access to imported food, will not feel population pressure at the same density levels as one with rudimentary technology.

Since carrying capacity is related to level of economic development, maps such as Figure 5.16 displaying present patterns of population distribution and density do not suggest a correlation with conditions of life. Many industrialized, urbanized countries have lower densities and higher levels of living than less developed

countries. Densities in the United States—where there is a great deal of unused and unsettled land—are considerably lower than those in Bangladesh, where essentially all land is arable and which, with 1926 persons per square mile, is the most densely populated nonisland nation in the world. At the same time, many African nations have low population densities and low levels of living, whereas Japan combines both high densities and wealth.

Overpopulation can be equated with levels of living or conditions of life that reflect a continuing imbalance between numbers of people and carrying capacity of the land. One measure of that imbalance might be the availability of food supplies sufficient in caloric content to meet individual daily energy requirements and so balanced as to satisfy normal nutritional needs. Although caloric requirements vary according to type of occupation, age, sex, and size of individuals, and even to the temperature conditions they encounter, it generally is accepted that about 2400 calories is the minimum daily consumption need. Figure 5.18 indicates that people in large parts of the world do not achieve that requisite energy intake. Low caloric intake is usually coupled with lack of dietary balance reflecting an inadequate supply of the range and amounts of carbohydrates, proteins, fats, vitamins, and minerals needed for optimum physical and mental development and maintenance of health. As Figure 5.18 indicates, dietary insufficiencies—with inevitable adverse consequences for life expectancy, physical vigor, and intellectual acuity—are most likely to be encountered in the developing nations, which have large proportions of their populations in the young age groups (Figure 5.8).

If those developing nations simultaneously have rapidly increasing population numbers dependent upon domestically produced foodstuffs, the prospects must be for continuing undernourishment and overpopulation. Much of sub-Saharan Africa finds itself in this circumstance. Africa's per capita food production decreased 20% between 1960 and 1985, and a further 30% drop is predicted over the next quarter century as the population–food gap widens (Figure 5.19). Africa is not alone. The international Food and Agricultural Organization projects that by A.D. 2000, no less than 65 separate countries with nearly 30% of the population of the developing world will be unable to feed their inhabitants from their own national territories at the low level of agricultural technology and inputs apt to be employed.

In the modern world, of course, insufficiency of domestic agricultural production to meet national caloric requirements cannot be considered a measure of overcrowding or poverty. Only a few countries are agriculturally self-sufficient. Many of those at or close to that achievement, such as China and India, are not yet among the developed nations of the world. Japan, a leader among the advanced countries, supplies no more than 20% of its food requirements from its own production. Its physiological density is high, as Table 5.5 indicates, but, of course, it does not rely upon an arable land resource for its present development. Largely lacking in either agricultural or industrial resources, it nonetheless ranks well on all indicators of national well-being and prosperity. For nations such as Japan, a sudden cessation of the international trade that permits the exchange of industrial products for imported food and raw materials would be disastrous. Domestic food production could not maintain the dietary levels now enjoyed by their populations and they, more starkly than many underdeveloped countries, would be "overpopulated."

Urbanization

Pressures upon the land resource of nations are increased not just by their growing populations, but by the reduction of arable land that such growth occasions. More and more of world population growth must be accommodated not in traditional rural areas but in cities. During the years 1980 through 1984, while world population grew by over 300 million, the economically active population in agriculture remained constant (Figure 6.5).

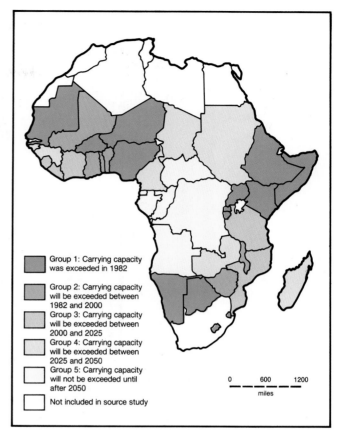

Group 1: Carrying capacity was exceeded in 1982

Group 2: Carrying capacity will be exceeded between 1982 and 2000

Group 3: Carrying capacity will be exceeded between 2000 and 2025

Group 4: Carrying capacity will be exceeded between 2025 and 2050

Group 5: Carrying capacity will not be exceeded until after 2050

Not included in source study

0   600   1200
miles

Figure 5.19

Carrying capacity and potentials in sub-Saharan Africa. The map assumes that (1) all cultivated land is used for growing food; (2) food imports are insignificant; (3) agriculture is conducted by low technology methods.

Source: *Population Growth and Policies in Sub-Saharan Africa.* A World Bank Policy Study. © 1986, The International Bank for Reconstruction and Development/The World Bank, Washington, D.C.

Whether this suggests that the food needs of the world can be met, as they are in such advanced nations as the United States, by an ever smaller proportion of the total work force or, less optimistically, that the world has run out of arable land able to support growing rural populations, the fact remains that in a world where some 44% of employment is agricultural, most new entrants into the labor pool must seek their livelihood off the land.

Increasingly, of course, that means residence and the hope for employment in the city. Largely because of population increases, the number and size of cities everywhere are growing. In 1950, less than 30% of the world's population lived in urban areas; by 1987 more than 43% of a much larger total population were urban dwellers (Figures 5.20 and 11.2). As a result of the attractions of cities in both the promise of jobs and access to health, welfare, and other public services, the *urbanization* (transformation from rural to urban status) of population in developing nations is increasing dramatically. In 1987, more than one-third of their inhabitants were urban, and collectively the less developed nations contained over 60% of the world's urban population.

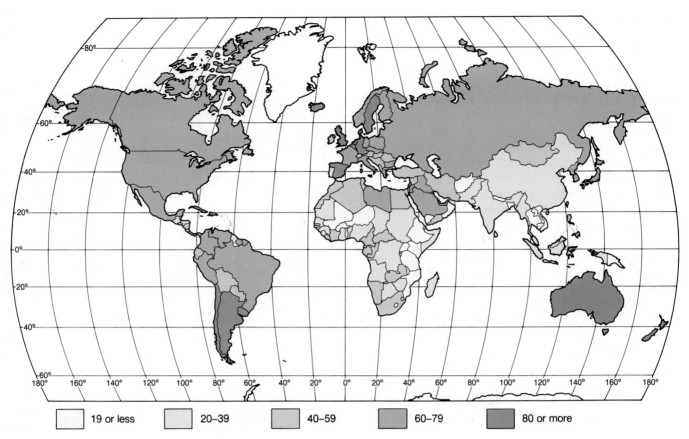

Figure 5.20

The percentage of national population that is classified as urban, 1987. Urbanization has been particularly rapid in the developing continents. In 1950, only 17% of Asians and 15% of Africans were urban; in 1987 the figures were 32% and 30%.

Source: Data from Population Reference Bureau, Inc.

The sheer growth of those cities in people and territory has increased pressures on arable land and adjusted upward both arithmetic and physiological densities. Urbanization consumes millions of acres of cropland each year. In Egypt, for example, urban expansion and development between 1965 and 1985 took out of production as much fertile soil as the Aswan dam made newly available. By themselves, some of these cities, which are surrounded by concentrations of people living in uncontrolled settlements, slums, and shantytowns (Figures 5.21 and 11.25), are among the most densely populated areas in the world. They face massive problems in trying to provide housing, jobs, education, and adequate health and social services for their residents. These and other matters of urban geography will be discussed in Chapter 11.

Population Data and Projections

Population geographers, demographers, planners, governmental officials, and a host of others rely upon detailed population data to make their assessments of present national and world population patterns and to estimate future conditions. Birth rates and death rates, rates of fertility and of natural increase, age and sex composition of the population, and other items all are necessary ingredients for their work.

Population Data

The data that students of population employ come primarily from the United Nations Statistical Office, the World Bank, the Population Reference Bureau, Inc., and, ultimately, from national censuses and sample surveys.

Unfortunately, the data as reported may on occasion be more misleading than informative. For much of the developing world, a national census is a massive undertaking. Isolation and poor transportation, insufficiency of funds, lack of trained personnel to conduct and evaluate the census, high rates of illiteracy limiting the type of questions that can be asked, and populations suspicious of all things governmental serve to restrict the frequency, coverage, and accuracy of population reports.

However derived, detailed data are published by the major reporting agencies for all national units even when those figures are poorly based on fact or are essentially fictitious. For years, data on the total population, birth and death rates, and other vital statistics for Somalia were regularly reported and annually revised. The fact was, however, that Somalia had never had a census and had no system whatsoever for recording births. Seemingly precise data were regularly reported as well for Ethiopia. When that nation had its first-ever census in 1985, at least one data source had to drop its estimate of the country's birth rate by 15% and increase by more than 20% its figure for Ethiopia's total population.

Fortunately, census coverage on a world basis is improving. Almost every modern nation has now had at least one census of its population, and most have been subjected to periodic sample surveys (Figure 5.22).

Figure 5.21

Street life in Bombay. Tens of thousands of people live permanently on the pavements of major Indian cities, most of them supporting themselves by menial work or alms, but unable to find or afford permanent shelter. More arrive each day in cities unable to provide for their most basic needs.

Figure 5.22

Taking the census in rural China. The sign identifies the "Third National Census. Mobile Registration Station."

World Population Projections

While the need for population projections is obvious, demographers face difficult decisions regarding the *assumptions* to be used in preparing them. Assumptions must be made about the future course of birth and death rates and, in some cases, about migration.

Many factors must be considered when projecting a country's population. What is the present level of the birth rate, of literacy, and of education? Does the government have a policy to influence population growth? What is the status of women?

Along with these must be weighed the likelihood of socioeconomic change, for it is generally assumed that as a country "develops," a preference for smaller families will cause fertility to fall to the replacement level of about two children per woman. But when can this be expected to happen in less developed countries? And for the majority of more developed countries with fertility currently below replacement level, can one assume that fertility will rise to avert eventual disappearance of the population and, if so, when?

Predicting the pace of fertility decline is most important, as illustrated by the United Nations long-range projections for Africa. As with many projections, these are issued in a "series" to show the effects of different assumptions. The UN "low" projection for Africa assumes that replacement-level fertility is reached in 2030, which would put the continent's population at 1.4 billion in 2100. If attainment of replacement-level fertility is delayed to 2065, as in the "high" variant, the population is projected to be 4.4 billion in 2100. That difference of 3 billion should serve as a warning that using population projections requires caution and consideration of *all* the possibilities.

However, only about 10% of the developing world's population lives in countries with anything approaching complete systems for registering births and deaths. It has been estimated that 40% or less of live births in Indonesia, Pakistan, India, or the Philippines are officially recorded. It appears that deaths are even less completely reported than births throughout Asia. And whatever the deficiencies of Asian nations, African statistics are still less complete and reliable. It is, of course, on just these basic birth and death data that projections about population growth and composition are founded.

Even the age structure reported for national populations, so essential to many areas of population analysis, must be viewed with suspicion. In many societies, birthdays are not noted, nor are years recorded by the Western calendar. Non-Western ways of counting age also confuse the record. The Chinese, for example, consider a person to be 1 year old at birth, and increase that age by 1 year each (Chinese) New Year's Day. Bias and error arise from the common tendency of people after middle age to report their ages in round numbers ending in zero. Also evident is a bias toward claiming an age ending in the number 5 or as an even number of years. Inaccuracy and noncomparability of reckoning, added to incompleteness of survey and response, conspire to cloud national comparisons in which age or the implications of age are important ingredients.

Projections

For all their inadequacies and imprecisions, however, current data reported for national units form the basis of **population projections.** Projections are not forecasts, and demographers are not the social science equivalent of meteorologists. Weather forecasters work with a myriad of accurate observations applied against a known, tested model of the atmosphere. The demographer, in contrast, works with sparse, imprecise, and missing data applied to human actions that will be unpredictably responsive to stimuli not yet evident.

Population projections, therefore, are based upon assumptions for the future applied to current data that are, themselves, frequently suspect. Since projections are not predictions, they can never be wrong. They are simply the inevitable result of calculations about fertility, mortality, and migration applied to each age cohort of a population now living, and the making of birth rate, survival, and migration assumptions about cohorts yet unborn. Of course, the perfectly valid *projections* of future population size and structure resulting from those calculations may be dead wrong as *predictions.*

Since those projections are invariably treated as scientific expectations by a public that ignores their underlying qualifying assumptions, agencies like the UN that estimate the population of, say, Africa in the year 2025, do so by not one but three projections: high, medium, and low (see "World Population Projections"). Each is completely valid but based upon different assumptions of growth rate. For areas as large as Africa, the medium projection is assumed to benefit by compensating errors and statistically predictable behaviors of very large populations. For individual African countries and smaller populations, the medium projection may be much less satisfying. The usual tendency in projections is to assume that something like current conditions will be applicable in the future; obviously, the more distant the future, the less likely is that assumption to remain true. The resulting observation should be that the further into the future one wishes to project the population structure of small areas, the greater is the implicit and inevitable error.

Population Controls

All population projections include an assumption that at some point in time population growth will cease and approach the replacement level. Without that assumption, future numbers become unthinkably large—for the world at unchecked present growth rates, there would be 1 trillion people three centuries from now, 4 trillion four centuries in the future, and so on. Although there is reasonable debate about whether the world is now overpopulated and about what either its optimum or maximum sustainable population should be, totals in the trillions are beyond any reasonable expectation.

Population pressures do not come from the amount of space humans occupy. It has been calculated, for example, that the entire human race could stand in an area slightly smaller than that of Hong Kong. The problems stem from the food, energy, and other resources necessary to support the population and from the impact on the environment of the increasing demands and the technologies required to meet them. Rates of growth currently prevailing in many countries make it nearly impossible for them to achieve the kind of social and economic development they would like. The demographic equilibrium between high birth rates and high death rates has been upset; at the present time human population is growing at a rate that is impossible to sustain.

Clearly, at some point population will have to stop increasing as fast as it has been. That is, either there will inevitably be imposed the self-induced limitations on expansion implicit in the demographic transition or an equilibrium between population and resources will be established in more dramatic fashion. Recognition of this eventuality is not new. "[The evils of] pestilence, and famine, and wars, and earthquakes have to be regarded as a remedy for nations, as the means of pruning the luxuriance of the human race" was the opinion of the theologian Tertullian during the 2nd century A.D.

Thomas Robert Malthus, an English economist and demographer, put the problem succinctly in a treatise published in 1798: all biological populations have a potential for increase that exceeds the actual rate of increase, and the resources for the support of increase are limited. In later publications, Malthus amplified his thesis by noting the following:

1. **Population is inevitably limited by the means of subsistence.**
2. **Populations invariably increase with increase in the means of subsistence unless prevented by powerful checks.**
3. **The checks that inhibit the reproductive capacity of populations and keep it in balance with means of subsistence are either "private" (moral restraint, celibacy, and chastity) or "destructive" (war, poverty, pestilence, and famine).**

The chilling consequences of Malthus's dictum that unchecked population increases geometrically while food production can increase only arithmetically have been reported throughout human history, as they are today. Starvation, the ultimate expression of resource depletion, is no stranger to the past or present. By conservative estimate, some 50 persons worldwide will starve to death during the 2 minutes it takes you to read this page; half will be children under 5. They will, of course, be more than replaced numerically by new births during the same 2 minutes. Losses always are. All battlefield casualties, perhaps 50 million, in all of humankind's wars over the last 500 years equal only a 6-month replacement period at present rates of natural increase.

Yet, inevitably—following the logic of Malthus, the apparent evidence of history, and our observations of animal populations—equilibrium must be achieved between numbers and support resources. When overpopulation of any species occurs, a population dieback is inevitable. The madly ascending leg of the J-curve is bent by external forces to the horizontal, and the J-curve is converted to an S-curve. It has happened before in human history, as Figure 5.23 summarizes. The S-curve represents a population size consistent with, and supportable by, the exploitable resource base. When the population is equivalent to the carrying capacity of the occupied area, defined earlier, it is said to have reached a **homeostatic plateau.**

In animals, overcrowding and environmental stress apparently release an automatic physiological suppressant of fertility. Although famine and chronic malnutrition may reduce fertility in humans, population limitation

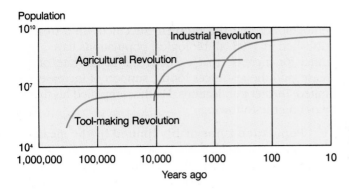

Population

Figure 5.23
The steadily higher homeostatic plateaus (states of equilibrium) that humans have achieved are evidence of their ability to increase the carrying capacity of the land through technological advance. Each new plateau represents the conversion of the J-curve into an S-curve. The diagram is suggestive of the process; it does not represent actual world population figures, which are shown here on a logarithmic scale.

Source: From Deevey, "The Human Population," *Scientific American,* p. 198, September 1960. Copyright © 1960 by Scientific American, Inc. All rights reserved.

Figure 5.24
This poster in a Bangladesh family planning clinic reads: "Don't blame Allah if you have too many children."

usually must be either brutish or self-imposed. The demographic transition to low birth rates matching reduced death rates is cited as evidence that Malthus's first assumption was wrong—human populations do not inevitably grow geometrically. Fertility behavior, it was observed, was conditioned by social determinants, and not solely by biological or resource imperatives.

Although Malthus's ideas were discarded as deficient by the end of the 19th century in light of the European population experience, the concerns he expressed were revived during the 1950s. Observations of population growth in underdeveloped nations and the strain that growth placed upon their resources inspired the viewpoint that improvements in living standards could be achieved only by raising investment per worker. Rapid population growth was seen as a serious diversion of scarce resources away from capital investment and into unending social welfare programs. To lift living standards required that existing national efforts to lower mortality rates be balanced by governmental programs to reduce birth rates. **Neo-Malthusianism,** as this viewpoint became known, has been the underpinning of national and international programs of population limitation primarily through birth control and family planning (Figure 5.24).

Neo-Malthusianism has had a mixed reception. Led by China and India, Asian nations have in general—though with differing successes—adopted family planning programs and policies. In some instances, success has been declared complete. Singapore established its

Population and Family Planning Board in 1965, when its fertility rate was 4.9 lifetime births per woman. By 1986, that rate had declined to 1.7, well below the 2.1 replacement level, and the board was abolished as no longer necessary. Caribbean and South American nations, except the poorest and most agrarian, have also experienced declining fertility rates, though often these reductions have been achieved despite pronatalist views of governments influenced by the Roman Catholic Church. Africa and the Middle East have generally been unresponsive to the Neo-Malthusian arguments because of ingrained cultural convictions among people, if not in all governmental circles, that large families—6 or 7 children—are desirable. Islamic fundamentalism opposed to birth restrictions also is a cultural factor.

Other barriers to fertility control exist. When first proposed by Western nations, Neo-Malthusian arguments that family planning was necessary for development were rejected by many less developed countries. Reflecting both nationalistic and Marxian concepts, they maintained that institutional and structural obstacles rather than population increase hindered development. Some government leaders think that there is a correlation between population size and power, and pursue pronatalist policies, as did Mao's China during the 1950s and early 1960s. And a number of American economists have during the 1980s expressed the view that population growth is a stimulus, not a deterrent, to development. Economic advance, they suggest, may be thwarted by market inadequacies, but will ultimately be achieved through the application of human ingenuity to population–resource problems.

Why Momentum Matters

When the average fertility rate of a population reaches the replacement threshold—about 2.1 births per woman—one would expect births and deaths to be in balance. Eventually they will be, but not immediately. The reason for this is *demographic momentum.*

A population's age structure reflects its past pattern of demographic events—mortality, migration, and especially fertility. A population that has had very high fertility in the years before reaching replacement level will have a much younger age structure than a population with lower fertility before crossing the replacement threshold.

The proportion of young people in a population approaching replacement-level fertility is important because the size of the largest recently born generation as well as the size of the parent generation determine the ultimate size of the total population when births and deaths are finally in balance. Even after replacement-level fertility is reached, births will continue to outstrip deaths so long as the generation that is producing births is disproportionately larger than the older generation where most deaths occur.

China, with a 1985 population of 1.04 billion and 34% under age 15, is among the handful of less developed countries with fertility even close to replacement level. China's total fertility rate was 5 to 6 births per woman as late as the early 1970s but dropped precipitously to 2.1 in 1984. Even with its aggressive fertility control program, China's population is expected to grow to 1.567 billion (a 51% increase over 1985) before stabilizing. Yugoslavia also recently reached replacement-level fertility, but the drop was from a total fertility rate of only 2 to 3 births per woman. Some 24% of its 1985 population of about 23 million was under age 15. Its population is projected to stabilize at about 30 million, 30% over 1985 at the end of the next century.

Thomas W. Merrick, with PRB staff, "World Population in Transition," *Population Bulletin* 41, no. 2 (Population Reference Bureau, Inc., Washington, D.C., 1986). With permission.

Population Prospects

Regardless of population philosophies, theories, or cultural norms, the fact remains that in many parts of the world developing countries are showing significantly declining population growth rates. But reducing fertility levels even to the replacement level of 2.1 births per woman does not mean an immediate end to population growth. Because of the age composition of many societies, numbers of births will continue to grow even as fertility rates per woman decline. The reason is to be found in **demographic** (or **population**) **momentum.**

When a high proportion of the population is young, the product of past high fertility rates, larger and larger numbers enter the child-bearing age each year. One-third or more of the people now on earth are under 15 (Figure 5.8); for Africa as a whole the figure for 1987 was 45%, and in some individual African nations nearly half the population is in that age bracket (as it is in some Asian countries). The consequences of the fertility of these young people are yet to be realized; they will continue to be felt until the now youthful groups mature and work their way through the population pyramid. Inevitably, while this is happening, even the most stringent national policies limiting growth cannot stop it entirely. A nation with a large present population base will experience large numerical increases despite declining birth rates. Indeed, the higher fertility was to begin with, and the sharper its drop to low levels, the greater will be the role played by momentum (see "Why Momentum Matters").

The Narrowing Gap: Population, Health, and Development

World population, we have seen, is on the path to stabilization. Momentum will assure some decades of continuing expansion, and the prospects for fertility reduction to replacement level in sub-Saharan Africa and parts of Asia are not yet clear. But the general pattern seems agreed: continued growth to about 10.2 billion persons worldwide, a level likely to be reached in little over a century (according to the UN long-range medium projection to A.D. 2100). At that new homeostatic plateau, with fertility reduced worldwide to the replacement level, many of the negative population and economic contrasts between today's "developed" and "developing" nations will presumably vanish. Until that new equilibrium between populations and resources is reached, the contrasts will remain—though in diminishing force as more of the world's nations stabilize their growth and pass from "underdevelopment" to "advanced nation" status. Rapid population growth is not necessarily the cause of all problems of poverty, pollution, or resource depletion, but it may greatly intensify the problems that developing nations encounter. World patterns of disease and health illustrate the point.

Taken at their extremes, the stabilized Western nations of advanced economy and the rapidly growing, newly developing countries occupy two distinct worlds

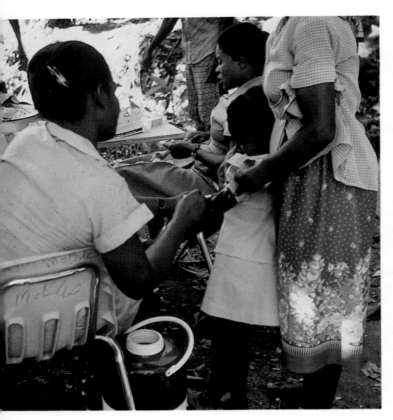

Figure 5.25

The World Health Organization (WHO) is the agency of the United Nations that helps bring modern preventive health care, safe water, and sanitation to the less developed world. WHO workers help to fight certain diseases, improve nutrition and living conditions, and aid developing countries in strengthening their health services. The organization plans to immunize 80% of the world's children (and 100% of newborn babies) against measles, diphtheria, polio, tetanus, whooping cough and tuberculosis by 1990, freeing the world of their threat just as it has been declared free from smallpox. Pictured is preventive health care in a Haitian clinic.

of disease and health. One is an affluent world. Its death rates are low, and the chief killers of its mature populations are cancers, heart attacks, and strokes. The other world is impoverished, crowded, and pathogenic. The deadly dangers of its youthful populations are infectious, respiratory, and parasitic diseases made more serious by malnutrition. In the affluent world, infant mortality rates may range from 8 to 12, and healthy babies can expect to live well into their 70s. In the crowded world of poverty, diarrheal diseases cause more than 5 million children to die each year, and infant mortality rates may register near 200 per 1000. In the poorest sections of the most overcrowded nations, half of all children die before reaching their first birthday.

Poverty, malnutrition, infection, and death existed long before the recent period of explosive population growth in less developed nations. Indeed, part of that growth is attributable to the application of imported medical technologies to save and prolong lives that, in their absence, would have been lost or shortened (Figure

5.25). In the 1960s and 1970s, developing nations with their own and international resources sought to improve health conditions with drugs, pesticides, and centralized hospitals that served mainly their urban populations. Clean water, sanitation, and rural health programs were coupled with family planning efforts to introduce improved living conditions for populations of manageable size and controlled growth. As developmental programs lifted many Asian and Latin American nations out of poverty, life expectancies were increased, infant and childhood deaths were reduced, and general conditions of health and well-being were improved to levels approaching those of the advanced nations.

For the poorest, most explosively growing nations, particularly of Africa, those approaches were unsuccessful. Medical cures and care were inadequate to the needs of explosively expanding populations. Their susceptible cohorts of infants, children, and young women entering child-bearing years were making demands on health services greater than the financial resources of those nations or the international community could support. The cost of drugs was consuming 25%–30% of the health budgets of such low-income countries as Tanzania and Ghana. In 1978, the member nations of the World Health Organization endorsed preventative, rather than curative, health care as a more attainable goal and adopted "health for all by the year 2000" as their official target. It was to be reached through primary health care: low technologies aimed at disease prevention in poorer nations.

The general determinants of health are well known: purchasing power to secure the food, housing, and medical care essential to it; a healthful physical environment that is both sanitary and free from infectious disease; and an educational level sufficient to comprehend the rudiments of nutrition and hygiene. Until recently, the world of the advanced nations that enjoyed these conditions and that of the underdeveloped countries that could only dream of them seemed permanently separated. But family planning, health, and economic developmental programs of national and international scope have begun to reduce the disparities between the two extremes. A bridging world is developing in which at least some of the components of an environment of health have been achieved and the transition to the standards of "health for all" has impressively begun (Figure 5.26).

The extremes of the world pattern of conditions of health are expected. Advanced nations of East and West constitute an elite world of health; the poorest, least developed nations of, particularly, Africa and southern Asia demonstrate the continuing existence of a counter-world of disease and death. What is perhaps surprising is that only 7% of the global population is represented by the 29 nations placed in the lowest quartile of nations ranked according to their conditions of health, while more than 25% of the world's people are in the upper quartile. More than two-thirds of humanity now occupy the transitional

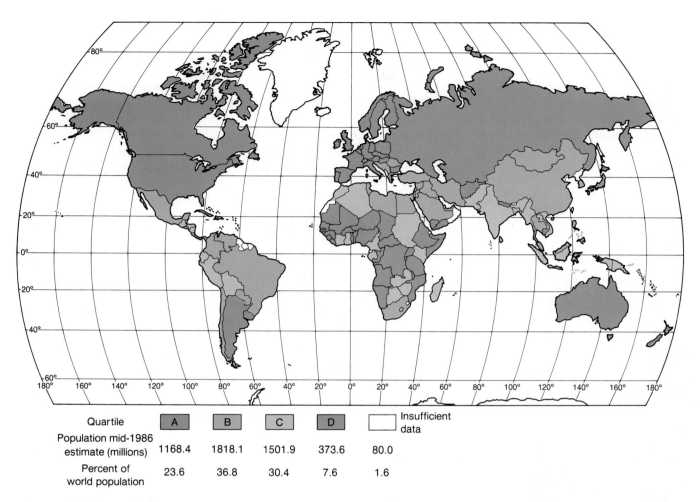

Quartile	A	B	C	D	Insufficient data
Population mid-1986 estimate (millions)	1168.4	1818.1	1501.9	373.6	80.0
Percent of world population	23.6	36.8	30.4	7.6	1.6

Figure 5.26

Composite conditions of health. The indicators of health by which the nations are ranked include: death rate for children aged 1–4; access to safe drinking water; caloric intake as a percentage of FAO standards; and female literacy. A strong correlation has been found between the education of the mother and the health of the family.

Redrawn with permission from Figure 1 of "Spatial Patterns of Disease and Health" by Jerome D. Fellmann, In J. Norwine and A. Gonzalez (eds.), *The Third World*, ©1988 by Allen & Unwin.

ground between the two extremes, and more than half of those have already advanced beyond the midpoint of the composite national rankings. Some two-thirds of the world's population, therefore, are beginning to approach the health standards set by the most advanced nations. The prospect is for a single world of health to accompany the projected new world homeostatic plateau.

Conclusion

Recent "explosive" increases in human numbers and the prospects of continuing population expansion may be traced to precipitous reductions in death rates, increases in longevity, and the impact of demographic momentum on a youthful population largely concentrated in the developing world. Control of population numbers historically was accomplished through a demographic transition first experienced in European societies that adjusted their fertility rates downward as death rates fell and life expectancies increased. The introduction of advanced technologies of preventive and curative medicine, pesticides, and famine relief have reduced mortality rates in developing nations without always a compensating reduction in birth rates. Although population control programs have been differentially introduced and promoted, the present 5 billion human beings will double in number by the year 2100, largely by increases unavoidable because of the size and youth of populations in developing countries.

Eventually, a new balance between population numbers and carrying capacity of the world will be reached, as it has always been following past periods of rapid population increase. When the world eventually returns to replacement-level fertility and stable population totals, it will do so in an environment of reduced differentials between "developed" and "developing" nations in matters of food supply, urbanization rates, employment structure, and conditions of health maintenance and disease avoidance.

Population geography, examining the numbers, composition, distribution, and trends of humans in their spatial context, is the essential human component of the human–environmental relationship of geographic concern and the starting point of all cultural geographic study. But human populations are not merely collections of numerical units nor are they to be understood solely through statistical analysis. Societies are distinguished not just by the abstract data of their numbers, rates, and trends, but by experiences, beliefs, understandings, and aspirations which collectively constitute that human spatial and behavioral variable called *culture*. It is to a discussion of that fundamental human diversity that we next turn our attention.

Key Words

arithmetic density *181*	demographic momentum *189*	mortality rate *167*	population pyramid *169*
carrying capacity *182*		Neo-Malthusianism *188*	rate of natural increase *170*
cohort measures *163*	demographic transition *173*	nonecumene *180*	
crude birth rate (CBR) *164*	dependency ratio *170*	overpopulation *181*	rates *163*
crude death rate (CDR) *167*	doubling time *172*	physiological density *181*	total fertility rate (TFR) *165*
crude density *181*	ecumene *180*	population density *181*	zero population growth *176*
demographic cycle *174*	homeostatic plateau *187*	population projection *186*	
	infant mortality rate *168*		

For Review

1. How do the *crude birth rate* and the *fertility rate* differ? Which measure is the more accurate statement of the amount of reproduction occurring in a population?

2. How is the *crude death rate* calculated? What factors account for the worldwide decline in death rates since 1945?

3. How is a *population pyramid* constructed? What shape of "pyramid" reflects the structure of a rapidly growing country? Of a population with a slow rate of growth? What can we tell about future population numbers from those shapes?

4. What variations do we discern in the spatial pattern of the *rate of natural increase* and, consequently, of population growth? What rate of natural increase would double population in 35 years?

5. How are population numbers "projected" from present conditions? Are projections the same as predictions? If not, in what ways do they differ?

6. Describe the stages in the *demographic cycle.* Where has the final stage of the cycle been achieved? Why do some analysts doubt the applicability of the demographic transition to all parts of the world?

7. Contrast *crude population density* and *physiological density.* For what differing purposes might each usefully be used? How is *carrying capacity* related to the concept of density?

8. What was Malthus's underlying assumption concerning the relationship between population growth and food supply? In what ways do the arguments of *Neo-Malthusians* differ from the original doctrine? What governmental policies are implicit in *Neo-Malthusianism?*

9. Why is *demographic momentum* a matter of interest in population projections? In which world areas are the implications of demographic momentum most serious in calculating population growth, stability, or decline?

Suggested Readings

Bennett, D. Gordon. *World Population Problems.* Champaign, Ill.: Park Press, 1984.

Brown, Lester R., and Jodi L. Jacobson. *Our Demographically Divided World,* Worldwatch Paper 74. Washington, D.C.: Worldwatch Institute, 1986.

Grant, James P., *The State of the World's Children 1987,* Published for United Nations Children's Fund (UNICEF) by Oxford University Press, New York, 1987.

Merrick, Thomas W., with PRB staff. "World Population in Transition," *Population Bulletin* 41, no. 2. Washington, D.C.: Population Reference Bureau, Inc., 1986.

National Academy of Sciences. *Population Growth and Economic Development: Policy Questions.* Washington, D.C.: National Academy Press, 1986.

Newman, James L., and Gordon E. Matzke. *Population: Patterns, Dynamics, and Prospects.* Englewood Cliffs, N.J.: Prentice-Hall, 1984.

Peters, Gary L., and Robert P. Larkin. *Population Geography.* Dubuque, Iowa: Kendall/Hunt, 1983.

Population Reference Bureau, Inc., *Population in Perspective: Regional Views.* Washington, D.C., 1986.

Repetto, Robert. "Population, Resources, Environment: An Uncertain Future," Population Bulletin 42, no. 2. Washington, D.C.: Population Reference Bureau, Inc., 1987.

6 Cultural Diversity

*T*he Gauda's[1] son is eighteen months old. Every morning, a boy employed by the Gauda carries the Gauda's son through the streets of Gopalpur. The Gauda's son is clean; his clothing is elegant. When he is carried along the street, the old women stop their ceaseless grinding and pounding of grain and gather around. If the child wants something to play with, it is given to him. If he cries, there is consternation. If he plays with another boy, watchful adults make sure that the other boy does nothing to annoy the Gauda's son.

Shielded by servants, protected and comforted by virtually everyone in the village, the Gauda's son soon learns that tears and rage will produce anything he wants. At the same time, he begins to learn that the same superiority which gives him license to direct others and to demand their services places him in a state of danger. The green mangoes eaten by all of the other children in the village will give him a fever; coarse and chewy substances are likely to give him a stomachache. While other children clothe themselves in mud and dirt, he finds himself constantly being washed. As a Brahmin (a religious leader), he is taught to avoid all forms of pollution and to carry out complicated daily rituals of bathing, eating, sleeping, and all other normal processes of life.

In time, the Gauda's son will enter school. He will sit motionless for hours, memorizing long passages from Sanskrit holy books and long poems in English and Urdu. He will learn to perform the rituals that are the duty of every Brahmin. He will bathe daily in the cold water of the private family well, reciting prayers and following a strict procedure. The gods in his house are major deities who must be worshipped every day, at length and with great care.[2]

The Gauda and his family are not Americans, as references and allusions in the preceding paragraphs make plain. The careful reader would infer, correctly, that they are Indians—and were one to read more of the book from which this excerpt is taken, it would become evident that the Gauda lives in a village in southern India. The class structure, the religion, the language, the food, and other strands of the fabric of life mentioned in the passage place the Gauda, his family, and his village in a specific time and place. They bind the people of the region together as sharers of a common culture and set them off from those of other areas with different cultural heritages. The 5 billion people who were the subject of Chapter 5 are of a single human family, but it is a family differentiated into many branches each characterized by a distinctive culture.

To writers in newspapers and the popular press, "culture" means the arts (literature, painting, music, and the like). To a social scientist, **culture** is the specialized behavioral patterns, understandings, and adaptations that summarize the way of life of a group of people. In this broader sense, culture is as much a part of the regional differentiation of the earth as are the topography, the climate, and other aspects of the physical environment. The visible and invisible evidences of culture—buildings and farming patterns, language and political organization—are elements of spatial diversity that invite and are subject to geographic inquiry. Cultural differences result in human landscapes with variations as subtle as the differing "feel" of urban Paris, Moscow, and New York or as obvious as the sharp contrasts of rural Sri Lanka and the Prairie Provinces of Canada (Figure 6.1).

Since such differences exist, cultural geography exists, and one branch of cultural geography addresses the question "Why?" Why, since humankind constitutes a single species, are cultures so diverse? What were the origins of the different culture regions we now observe? How, from whatever limited areas in which single culture traits and amalgams developed, were they diffused over a wider portion of the globe? How did people who had roughly similar origins come to display significant areal differences in technology, social structure, and ideology? Finally, why do cultural contrasts between recognizably distinct groups persist even in such presumed "melting-pot" societies as that of the United States or in the outwardly homogeneous, long-established nations of Europe? These questions are the concern of the present chapter.

Components of Culture

Culture is transmitted within a society to succeeding generations by imitation, instruction, and example. As members of a social group, individuals acquire integrated sets of behavioral patterns, environmental and social perceptions, and knowledge of existing technologies. Although one of necessity learns the culture in which one is born and reared, one need not—indeed, probably will not—learn its totality. Age, sex, status, or occupation may dictate the aspects of the cultural whole in which one becomes fully indoctrinated.

1. Gauda = village headman
2. From Alan J. Beals, *Gopalpur: A South Indian Village* (New York: Holt, Rinehart and Winston, 1962). Reprinted by permission of Holt, Rinehart and Winston, CBS College Publishing.

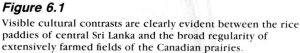

Figure 6.1
Visible cultural contrasts are clearly evident between the rice
paddies of central Sri Lanka and the broad regularity of
extensively farmed fields of the Canadian prairies.

Culture is an interlocked web of such complexity
and pervasiveness that it cannot be grasped and, in fact,
may be misunderstood, by being glimpsed or generalized from limited, obvious traits. Distinctive eating utensils, the use of gestures, or the ritual of religious ceremony
may summarize and characterize a culture for the casual
observer. These are, however, only individually insignificant parts of a totality that can be appreciated only when
the whole is experienced.

The richness and complexity of human life compel
us to seek those more fundamental cultural variables that
give order to societies and to summarize them in meaningful conceptual and spatial terms. We begin with **culture traits,** the smallest distinctive item of culture.
Culture traits are units of learned behavior ranging from
the language spoken, to the tools used, to the games
played. They are the most elementary expression of culture, the building blocks of the complex behavioral patterns of distinctive groups of peoples.

Individual culture traits that are functionally interrelated comprise a **culture complex.** The existence of
such complexes is universal. Keeping cattle is a *culture
trait* of the Masai of Kenya (Figure 6.2). The association
of related traits includes the measurement of personal
wealth by the number of cattle owned; a diet containing
milk and the blood of cattle; and disdain for labor that is
unrelated to herding. The assemblage of these and related traits yields a *culture complex* that describes one

aspect of Masai society. In exactly analogous ways, religious complexes, business behavior complexes, sports
complexes, and others can easily be recognized in American or any other society.

Culture traits and complexes have spatial extent.
When components of culture are plotted on maps, their
regional character is revealed. A **culture region** is an
area that is distinct from surrounding or adjacent areas
in the characteristic specified and mapped. For cultural
geographers, the regions of interest are in some way related to the attributes or activities of people. Examples
include the political organizations that societies devise,
the religions they espouse, the form of economy they
pursue, and even the type of clothing they wear, the
eating utensils they use, or the kind of housing they occupy. There are as many such regions as there are recognized culture traits and complexes for population
groups.

Finally, as we shall see at the conclusion of this
chapter, a set of culture regions showing related culture
complexes and landscapes may be grouped to form a
culture realm. The term recognizes a large segment of
the earth's surface having fundamental uniformity in its
cultural characteristics and showing a significant difference in them from adjacent realms.

Figure 6.2
The Masai of eastern Africa move their herds to seasonally changing pastures. Cattle form the basis of Masai culture, the source of milk and blood in the people's diet and their evidence of wealth and social status.

Interaction of People and Environment

Culture develops in a physical environment that, in its way, contributes to differences among people. In primitive societies, the acquisition of food, shelter, and clothing—all parts of culture—depends on the use of the natural resources at hand. The reaction of people to the environment of a given area, their use of it, and their impact on it are interwoven themes within the discipline of geography. However, geographers have long dismissed as intellectually limiting and demonstrably invalid the ideas of **environmental determinism**—the belief that the physical environment exclusively shapes humans, their actions, and their thoughts. Environmental factors alone cannot account for the cultural variations that occur around the world. Levels of technology, systems of organization, and ideas about what is true and right have no obvious relationship to environmental circumstances.

The environment does place certain limitations on the human use of territory. However, such limitations must be seen not as absolute, enduring restrictions but as relative to technologies, cost considerations, national aspirations, and linkages with the larger world. Human choices in the use of landscapes are affected by group perception of the feasibility and desirability of their occupance and exploitation. These are not circumstances inherent in the land. Mines, factories, and cities were (and are being) created in the formerly almost unpopulated tundra and forests of Siberia, not in response to recent environmental improvement but as a reflection of Soviet developmental programs.

Map evidence does suggest the existence of some environmental limitations on the use of area. The vast majority of the world's population is differentially concentrated on less than one-half of the earth's land surface (Figure 5.16). Areas with relatively mild climates that offer a supply of fresh water, fertile soil, and abundant mineral resources are densely settled, reflecting in part the different carrying capacities of the land under earlier technologies. Even today, the polar regions, high and rugged

Figure 6.3
The physical and cultural landscapes in juxtaposition. Advanced societies are capable of so altering the circumstances of nature that the cultural landscapes they create become controlling environments.

mountains, deserts, and some hot and humid areas contain very few people. If resources for feeding, clothing, or housing ourselves in a region are lacking, or if we do not recognize them there, there is no inducement for people to occupy the territory.

Environments that do contain such resources provide the framework within which a culture operates. Geographers believe that any environments so endowed offer a range of possibilities. The needs, traditions, and level of technology of a culture affect how these possibilities are perceived and what choices are made regarding them. Each society uses natural resources in accordance with its culture. Changes in a group's technical abilities or objectives bring about changes in its perceptions of the usefulness of the land.

Coal, oil, and natural gas have been in their present locations throughout human history, but they were of no use to preindustrial cultures, and there seemed to be no advantage to the places where they were to be found. Not until the Industrial Revolution did these deposits gain importance and come to influence the location of such great industrial complexes as the Midlands in England, the Ruhr in Germany, and the formerly important steel-making districts in parts of northeastern United States. American Indians made one use of the environment around Pittsburgh, while 19th century industrialists made quite another.

People are also able to modify their environment, and this is the other half of the human–environment relationship of geographic concern. Geography, including cultural geography, examines both the reactions of people to the physical environment and their impact on that environment. We modify our environment through the material objects we place on the landscape: cities, farms, roads, and so on (Figure 6.3). The form that these take is the product of the kind of culture group in which we live. The **cultural landscape,** the earth's surface as modified by human action, is the tangible, physical record of a given culture. House types, kinds of public buildings, architectural styles, and the size and distribution of settlements are among the indicators of the use that humans have made of the land.

The more technologically advanced and complex the culture, the greater its impact on the environment. In sprawling urban-industrial societies, the cultural landscape has come to outweigh the natural physical environment in its impact on people's daily lives. It interposes itself between "nature" and humans, and residents of the cities of such societies—living and working in climate-controlled buildings, driving to enclosed shopping malls—can go through life with very little contact with or concern about the physical environment.

The Subsystems of Culture

Understanding a culture fully is perhaps impossible for one who is not part of it. For analytical purposes, however, the traits and complexes of culture—its building blocks and expressions— may be grouped and examined as subsets of the whole. Following the anthropologist Leslie White, we may divide culture into three major subsystems: the technological, the sociological, and the ideological. Together, they comprise the system of culture as a whole. But they are integrated; each reacts on the others and is affected by them in turn.

The **technological subsystem** is composed of the material objects, together with the techniques of their use, by means of which people are able to live. The **sociological subsystem** of a culture is the sum of those expected and accepted patterns of interpersonal relations that find their outlet in economic, political, military, religious, kinship, and other associations. The **ideological subsystem** consists of the ideas, beliefs, and knowledge of a culture and of the ways in which these things are expressed in speech or other forms of communication.

The Technological Subsystem

Examination of variations in culture and in the states and kinds of human existence from place to place centers on a series of commonplace questions: How do the people in an area make a living? What resources and what tools do they use to feed, clothe, and house themselves? Is a larger percentage of the population engaged in agriculture than in manufacturing? Do people travel to work in cars, on bicycles, or on foot? Do they shop for food or grow their own?

These questions concern the adaptive strategies used by different cultures in making a living. In a broad sense they address the technological subsystems at the disposal of those cultures—the tools and other instruments that enable people to feed, clothe, house, defend, and amuse themselves. For most of human history, people lived by hunting and gathering, taking the bounty of nature with only minimal dependence on weaponry, implements, and the controlled use of fire. Their adaptive skills were great, but their technological level was low. They had few specialized tools, could exploit only a limited range of potential resources, and had little or no control of nonhuman sources of energy. Their impact upon the environment was small, but at the same time the homeostatic plateau, the "carrying capacity" of the land discussed in Chapter 5, was everywhere low, for technologies and tools were essentially the same among all groups.

The retreat of the last glaciers some 11,000 years ago marked the start of a period of unprecedented cultural development. From primitive hunting and gathering economies at the outset, humans progressed through the evolution of agriculture and animal husbandry to, ultimately, urbanization, industrialization, and the intricate complexity of the modern technological subsystem.

Cultural diversity among ancient societies reflected the proliferation of technologies that followed a more assured food supply and made possible a more intensive and extensive use of resources. Different groups in differing environmental circumstances developed specialized tools and behaviors to exploit resources recognizably available to them. Beginning with the Industrial Revolution of the 18th century, however, a reverse trend—toward commonality of technology—began. Today, advanced societies are nearly indistinguishable in the tools and techniques at their command. Those differences in technological traditions that still exist between developed and underdeveloped societies reflect in part national and personal wealth, stage of economic advancement and complexity, and, importantly, the level and type of energy used (Figure 6.4).

Many aspects of a society are related to its level of technological development. The terms *standard of living* and *level of living* bring to mind some of these, such as personal income, level of education, food consumption, life expectancy, and availability of health care. The complexity of the occupational structure, the degree of specialization in jobs, the ways natural resources are used, and the degree of industrialization reflect the technology available to a given society.

Not all members of a culture group, of course, are equal beneficiaries of the existing tools and rewards of production. The plaintive comment that "poor people have poor ways" or the ready distinction that we make between the "Gold Coast" and the "slum" indicate that different groups have differential access to the wealth, tools, and resources of the society of which they are a part. On an international scale, we distinguish between "advanced" or "rich" nations, such as Switzerland, and "underdeveloped" or "poor" countries, like Bangladesh, though neither class of states may wish those adjectives to be applied to its circumstances.

In technologically advanced countries, many people are employed in manufacturing or allied service trades. Per capita incomes tend to be high, as do levels of education. These countries collectively wield great economic and political power. In contrast, technologically less advanced countries have a high percentage of people engaged in agricultural production, with much of the agriculture at a subsistence level (Figure 6.5). The

(a)

(b)

Figure 6.4

(a) This Balinese farmer working with draft animals employs tools typical of the low technological levels of subsistence economies. (b) Cultures with advanced technological subsystems use complex machinery to harness inanimate energy for productive use.

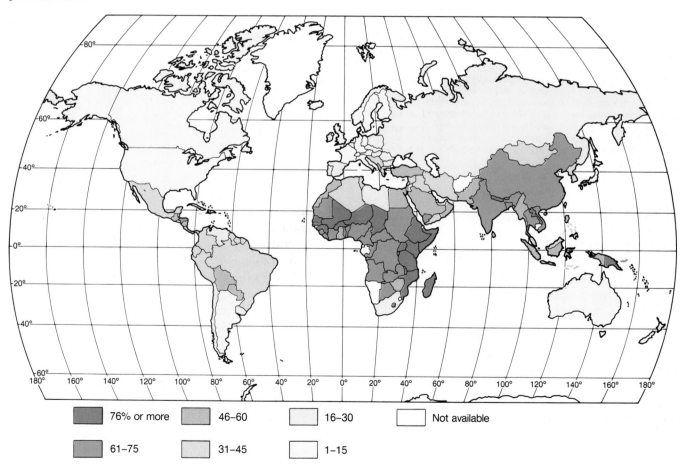

76% or more	46–60	16–30	Not available
61–75	31–45	1–15	

Figure 6.5

Percent of workforce engaged in agriculture. Highly developed economies are characterized by relatively low proportions of their labor forces in the agricultural sector.

Source: Data from World Bank.

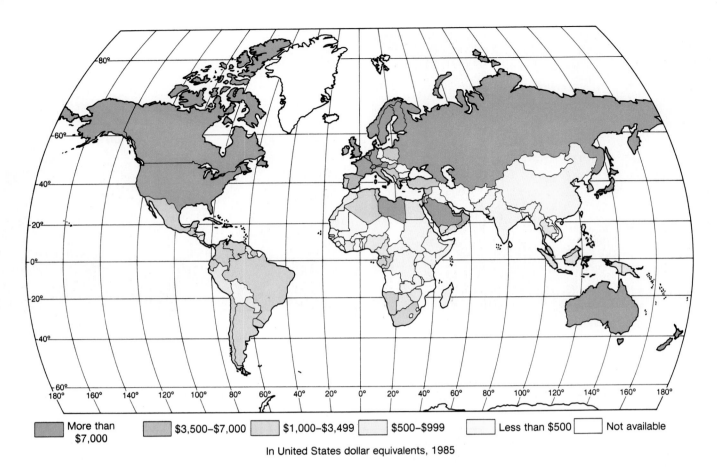

| More than $7,000 | $3,500–$7,000 | $1,000–$3,499 | $500–$999 | Less than $500 | Not available |

In United States dollar equivalents, 1985

Figure 6.6

Gross national product per capita. GNP is a frequently employed summary of degree of technological development, though high incomes in sparsely populated, oil-rich nations may not have the same meaning in subsystem terms as do comparable per capita values in industrially advanced nations. A comparison of this map and Figure 6.5 presents an interesting study in regional contrasts. Figures are for 1985 and are in United States dollar equivalents.

Data from World Bank.

gross national products or GNP (the GNP equals the total value of all goods and services produced by a country during a year) of these nations are much lower than those of industrialized countries. Per capita incomes tend to be low (Figure 6.6), and illiteracy rates high.

Labels such as *advanced–less advanced, developed–underdeveloped, or industrial–nonindustrial* can mislead us into thinking in terms of "either–or" (see Word Games, Page 315). They may also lead the unthinking to the belief that they mean, in general cultural terms, "superior–inferior." This belief is totally improper, since the terms are related solely to economic and technological circumstances and bear no qualitative relationship to such vital aspects of culture as music, art, or religion. Improperly used, these technological comparative terms imply a static rather than a dynamic situation. Many countries are in the process of altering their level of technological development.

Properly understood, however, terms and measures of economic development can reveal important national and world regional contrasts in the implied technological subsystems of different cultures and societies. Apparent in Figure 6.7 is a spatial grouping and division of the world's nations that is persistent and important enough to constitute the basis for organized negotiating blocs at international economic conferences and in the United Nations. Based on a composite of developmental indexes, the map indicates that technological status is high in nearly all European countries, including the USSR, and in Japan, the United States, and Canada—the "North." Most of the less developed countries are in Latin America, Africa, and southern Asia—the "South."

It is important to recognize that national averages conceal internal contrasts. All countries include areas that are at different levels of development. In a highly advanced, industrialized country such as the United States,

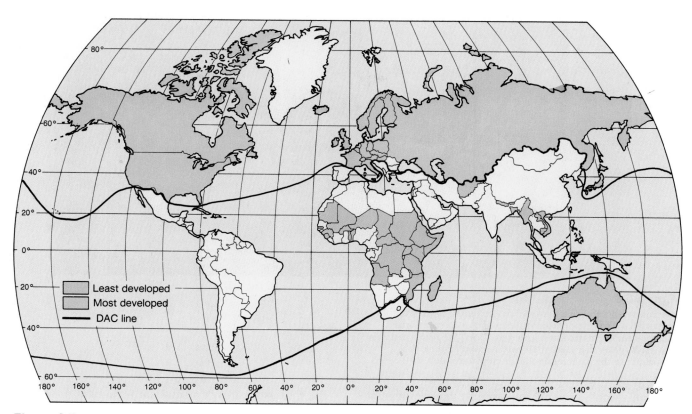

Figure 6.7

A multifactor recognition of "most developed" and "least developed" nations based on United Nations and World Bank data of the mid-1980s. The DAC line is the division between the "developing" and "developed" countries (the "North" and the "South") recognized by the Development Assistance Committee of the Organization for Economic Cooperation and Development (OECD).

there are districts that do not fully share national levels of income or technology. To take the reverse case, many countries that are characterized as technologically less advanced contain regions that are highly developed, with important industries, well-developed transport networks, and people earning relatively high incomes. Technological development is a dynamic concept, and it is most useful and accurate to think of individuals within countries and nations of the world as arrayed along an ever-changing continuum of technological subsystems.

The Sociological Subsystem

Continuum and change also characterize the religious, political, formal and informal educational, and other institutions that comprise the sociological subsystem of culture. Together these define the social organizations of a culture. They specify how the individual functions relative to the group, whether it be family, church, or state.

There are no "givens" in patterns of interaction in any of these associations, except that most cultures possess all these ways of structuring behavior. The importance to the society of the differing behavior sets varies among, and constitutes obvious differences between, cultures. Differing patterns of behavior are learned expressions of culture and are transmitted from one generation to the next by formal instruction or by example and expectation. The story of the Gauda's son that opened this chapter illustrates the point.

Rejection of the sociological patterns that are appropriate to the culture constitutes, beyond narrow limits, unacceptable, deviant behavior subject to formal or informal rebuke, rejection, or punishment. Such censure, like the education of the young, may be seen as the socialization of the untrained or of the rebellious, imparting or reinforcing the culture's sociological system.

Social institutions are closely related to the technological system of a culture group. The description of the cultural environment of the Gauda's son suggested

Figure 6.8

Age and sex differences, not caste or economic status, were the primary basis for the division of labor and of interpersonal relations among hunters and gatherers. Here, San males carry the bows and spears that mark their role as hunters; they also contribute to the gathered food that constitutes 80% of the San diet.

through a single example the certainty that innumerable specific mixes of social structures and ways of behavior are encountered worldwide. Thus, hunter-gatherers have one set of institutions and industrial societies quite different ones. Preagricultural societies tended to be composed of small bands based on kinship ties, with little social differentiation or specialization of function in the band; the San (Bushmen) of the African Kalahari Desert and isolated rain forest groups in Amazonia might serve as modern examples (Figure 6.8).

The revolution in food production occasioned by plant and animal domestication beginning some 10,000 years ago touched off a social transformation that included increases in population, urbanization, job specialization, and structural differentiation within the society. Politically, the rules and institutions by which people were governed changed with the formation of sedentary, agricultural societies. Loyalty was transferred from the kinship group to the state; resources became possessions rather than the common property of all. Equally far-reaching changes occurred after the 18th century Industrial Revolution, leading to the complex of human social organizations that we experience and are controlled by today in "developed" nations and that increasingly affect all cultures everywhere.

Culture is a complexly intertwined whole. Each organizational form or institution affects, and is affected by, related culture traits and complexes in intricate and variable ways. Systems of land and property ownership and control, for example, are institutional expressions of the sociological subsystem. They are, simultaneously, explicitly central to the classification of economies and to

the understanding of spatial and structural patterns of economic development, as Chapter 9 will examine. Again, for each nation the adopted system of laws and justice (Figure 6.9) is a cultural variable identified with the sociological subsystem but extending its influence to all aspects of economic and social organization, including the political geographic systems discussed in Chapter 7.

The Ideological Subsystem

The third class of elements defining and identifying a culture comprises the ideological subsystem. That subsystem does not have the concrete reality of the tools, processes, and developmental levels of the technological subsystem, nor does it have the formal organizational structures making up the sociological subsystem. Instead, it consists of ideas, beliefs, and knowledge and of the ways we express these things in our speech or other forms of communication. Mythologies and theologies, legend, literature, philosophy, folk wisdom, and commonsense knowledge make up this category. Passed on from generation to generation, these statements tell us what we ought to believe, what we should value, and how we ought to act. Beliefs form the basis of the socialization process.

Often we know—or think we know—what the beliefs of a group are from written sources. Sometimes, however, we depend on the actions or objectives of a group to tell us what its true ideas and values are. "Actions speak louder than words" is a recognition of the fact that actions and words do not always coincide. The values of a group cannot be deduced from the written record alone.

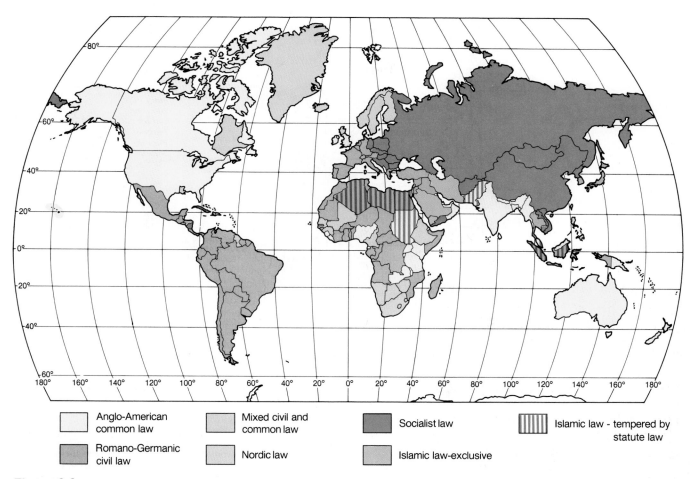

Figure 6.9

Major legal systems of the world. The map reflects 1976
conclusions, modified by changes apparent or declared to the
mid-1980s. *Common law* systems are based on judicial
interpretations of customary practice and equity. In *Romano-
Germanic civil law,* laws are enacted to conform to
preestablished legal principles. In *socialist law* those principles
are rooted in Marxist ideology; the legal system is more
concerned with creation of a collectivist society than in resolving
disputes between individuals. *Islamic law* is based on
fundamentalist interpretations of injunctions contained in the
Koran, the holy book of Islam. In some Muslim nations, it is the
sole recognized legal authority; in others it is usually applied to
interpersonal disputes and grafted onto civil or common law
systems based on European models.

Changes in the ideas that a society holds may affect
the sociological and technological systems just as, for example, changes in technology force changes in the social system. The abrupt alteration of the ideological
structure of Russia from a monarchical, agrarian, capitalistic system to an industrialized, communistic society involved sudden, interrelated alteration in all facets of that
nation's former culture system. The interlocking nature
of all aspects of a culture is termed **cultural integration.**

Culture Change

The recurring theme of cultural geography is change. No
culture has or ever has had a permanently fixed set of
material objects, systems of organization, or even ideologies, although all may be long-enduring in a society at
equilibrium with its resource base and totally isolated
from a need or an influence for change. Such isolation
and stagnation have always been rare. On the whole,
while cultures are essentially conservative, they are simultaneously constantly changing; they are always in a
state of flux. Some changes are major and pervasive. The
transition from hunter-gatherer to sedentary farmer, as we
have seen, markedly affected every facet of the cultures
experiencing that change. Profound, too, has been the
impact of the Industrial Revolution and its associated urbanization on all societies it has touched. Other changes
are so slight individually that they may be, at their in-
ception, almost unnoticed, though cumulatively they may
substantially alter the affected culture. Think of how the
culture of the United States differs today from what you
know it to have been in 1940—not in essentials, perhaps,

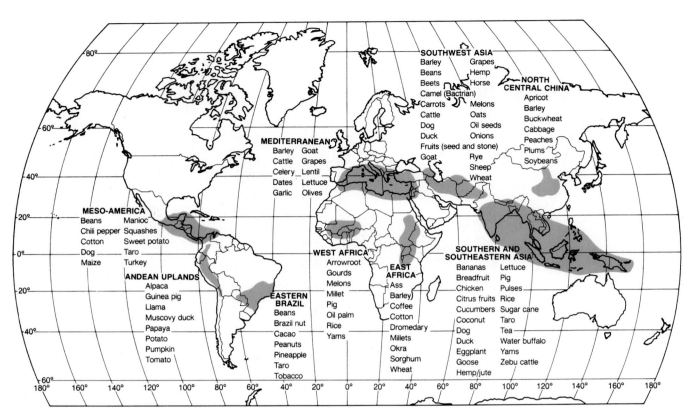

Figure 6.10

Chief centers of plant and animal domestication. The southern and southeastern Asian center was characterized by the domestication of plants such as taro that are propagated by the division and replanting of existing plants (vegetative reproduction). Reproduction by the planting of seeds (for example, maize and wheat) was more characteristic of Meso-America and the Near East. The African and Andean areas developed crops reproduced by both methods. The lists of crops and livestock associated with the separate origin areas are selective, not exhaustive.

but in the innumerable electrical, electronic, and transportation devices that have been introduced and in the social, behavioral, and recreational changes they have wrought. Such cumulative changes occur because the cultural traits of any group are not independent; they are clustered in a coherent and integrated pattern. Change in an apparently minor and limited fashion will have wide repercussions as associated traits arrive at accommodation with the adopted adjustment. Change, both major and minor, within cultures is induced by *innovation* and *diffusion*.

Innovation

Innovation implies changes to a culture that result from ideas created within the social group itself. The novelty may be an invented improvement in material technology, like the bow and arrow or the jet engine. It may involve the development of nonmaterial forms of social structure and interaction: feudalism, for example, or Christianity.

Primitive and traditional societies characteristically are not innovative. In societies at equilibrium with their environment and with no unmet needs, change has no adaptive value and has no reason to occur. Indeed, all societies have an innate resistance to change. Complaints about youthful fads or the glorification of past times are familiar cases in point. However, when a social group is inappropriately unresponsive—mentally, psychologically, or economically—to changing circumstances and to innovation, it is said to undergo **cultural lag.**

Innovation, frequently under stress, has characterized the history of humankind. Growing populations at the end of the Ice Age necessitated an expanded food base. In response, domestication of plants and animals appears to have occurred independently in more than one world area. Indeed, a most striking fact about early agriculture is the universality of its development or adoption within a very short span of human history. Some 10,000 years ago, virtually all of humankind was supported by hunting and gathering. By no later than 2000 years ago, the majority lived by farming. Nonetheless, recognizable areas of "invention" of agriculture have been identified (Figure 6.10). From them, presumably,

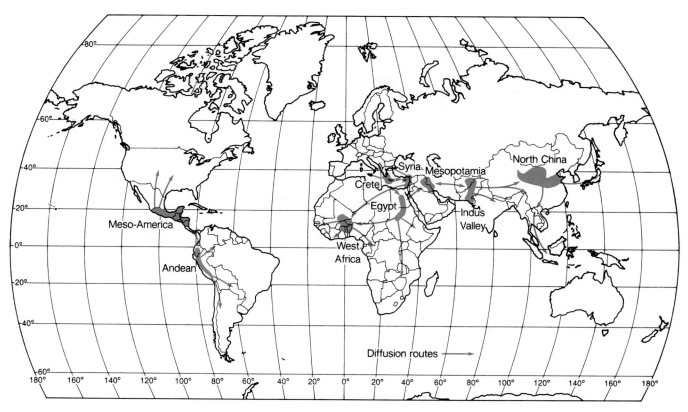

Figure 6.11
Early culture hearths of the Old World and the Americas.

there was a rapid diffusion of food types, production techniques, and new modes of economic and social organization. All innovation has a radiating impact upon the web of culture; the more basic the innovation, the more pervasive are its consequences.

Few innovations in human history have been more basic than the Agricultural Revolution. It affected every aspect of society. Culture altered at an accelerating pace, and change itself became a way of life. Humans learned the arts of spinning and weaving plant and animal fibers. They learned to use the potter's wheel and to fire clay and make utensils; they developed techniques of brick making, mortaring, and construction. They discovered the skills of mining, smelting, and casting metals. Special local advantages in resources or products promoted the development of long-distance trading connections. On the foundation of such technical advancements a more complex exploitative culture appeared, including a stratified society to replace the rough equality of hunting and gathering economies.

The source regions of such social and technical revolutions were initially spatially confined. The term **culture hearth** describes those restricted areas of innovation from which key culture elements diffused to exert an influence on surrounding regions. All hearth areas developed the trappings of *civilization,* including writing (or other form of record keeping), metallurgy, astronomy and mathematics, social stratification, labor specialization, formal governmental systems, and a structured urban society. Several major culture hearths emerged, some as early as 5500 years ago, following the initial revolution in food production. Prominent centers of creativity were located in Egypt, Mesopotamia, the Indus Valley of the Indian subcontinent, and northern China. Other hearths developed in the Americas, west Africa, Crete, and Syria (Figure 6.11).

Documenting Diffusion

The places of origin of many ideas, items, and technologies important in contemporary cultures are only dimly known or supposed, and their routes of diffusion are at best speculative. Gunpowder, printing, and spaghetti are presumed to be the products of Chinese inventiveness; the lateen sail has been traced to the Near Eastern culture world. The moldboard plow is ascribed to 6th century Slavs of northeastern Europe. The sequence and routes of the diffusion of these innovations have not been documented.

In other cases, such documentation exists, and the process of diffusion is open to analysis. For example, hybrid corn was originally adopted by imaginative farmers of northern Illinois and eastern Iowa in the mid-1930s; by the late 1930s and early 1940s the new seeds were being planted as far east as Ohio and north to Minnesota, Wisconsin, and northern Michigan. By the late 1940s, all commercial corn-growing districts of the United States and southern Canada were cultivating hybrid varieties.

Clearly marked as well is the diffusion path of the custom of smoking tobacco, a practice that originated among Amerindians. Sir Walter Raleigh's Virginia colonists, returning home in 1586, introduced smoking in English court circles, and the habit very quickly spread among the general populace. England became the source region of the new custom for northern Europe; smoking was introduced to Holland by English medical students in 1590. Dutch and English together spread the habit by sea to the Baltic and Scandinavian areas and overland through Germany to Russia where, by 1634, laws were being passed to curb the practice of smoking. The innovation continued its eastward diffusion, and within a hundred years, tobacco had spread across Siberia and was, in the 1740s, reintroduced to the American continent at Alaska as both a habit and an item of trade carried by Russian fur traders. A second route of diffusion for tobacco smoking can be traced from Spain, where the custom was introduced in 1558, and from which it spread more slowly through the Mediterranean area into Africa, the Near East, and Southeast Asia.

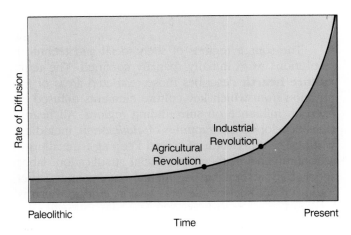

Figure 6.12

The rate of innovation through human history. Hunter-gatherers, living in easy equilibrium with their environment and their resource base, had little need for innovation and no necessity for cultural change. Increased population pressures led to the Agricultural Revolution and the diffusion of the ideas and techniques of domestication, urbanization, and trade. With the Industrial Revolution, dramatic increases in innovation began to alter cultures throughout the world.

In most modern societies, innovative change has become common, expected, and inevitable. The rate of invention, at least as measured by the number of patents granted, has steadily increased, and the period between idea conception and product availability has been decreasing. A general axiom is that the more ideas available and the more minds able to exploit and combine them, the greater the rate of innovation. The spatial implication is that larger urban centers of advanced technologies tend to be centers of innovation, not just because of their size but because of the number of ideas exchanged. Indeed, ideas not only stimulate new ideas, but also create circumstances in which new solutions must be developed to maintain the forward momentum of the society (Figure 6.12).

Diffusion

Diffusion is the process by which an idea or innovation is transmitted from one individual or group to another across space, an expression of the *spatial interaction* discussed more broadly in Chapter 8. Diffusion may assume a variety of forms, each different in its impact on social groups, but basically two processes are involved. Either

Figure 6.13

A street scene in Guatemala City. In modern society, advertising is a potent force for diffusion. Advertisements over radio and television, in newspapers and magazines, and on billboards and signs communicate information about many different products and innovations.

people move, for any of a number of reasons, to a new area and take their culture with them (as the immigrants to the American colonies did), or information about an innovation (like barbed wire or hybrid corn) may spread throughout a culture (Figure 6.13). The former is known as *relocation diffusion,* the latter as *expansion diffusion.*

When expansion diffusion affects almost uniformly all individuals and areas outward from the source region, it is termed *contagious* (or *spatial*) diffusion. The term is reminiscent of the course of infectious diseases (Figure 6.14). *Hierarchical diffusion* results from the transmission and adoption of new ideas or technologies through an established structure of urban centers, levels of political or ecclesiastical authority, or some comparable hierarchy. These and other forms of diffusion are further explored in Chapter 8.

It is not always possible to determine whether the existence of a cultural trait in two different areas is the result of diffusion or of independent (or *parallel*) invention. Cultural similarities do not necessarily prove that

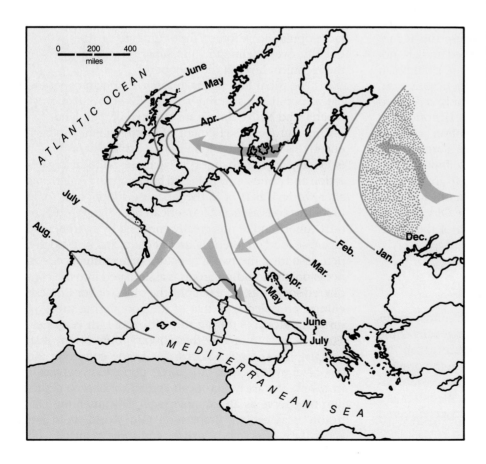

Figure 6.14

An example of "contagious diffusion:" the diffusion pathways of the European influenza pandemic of 1781. The pattern is a wavelike radiation from a nodal origin area.

Figure 6.15

Outward appearance is not the only similarity between Egyptian (left) and Mayan (right) pyramids. In both societies, the pyramids were royal burial sites, goods were placed inside the tombs to accompany the soul on its journey to eternity, and pains were taken to hide the entrance to the tomb.

diffusion has occurred. The pyramids of Egypt and of the Central American Maya civilization, pictured in Figure 6.15, most likely were separately conceived and are not evidence, as some have proposed, of pre-Columbian voyages from the Mediterranean to the Americas. A monument-building culture, after all, has only a limited number of shapes from which to choose. Historical examples of independent, parallel invention are numerous: logarithms by Napier (1614) and Burgi (1620) and the calculus by Newton (1672) and Leibnitz (1675) are but two commonly cited examples. It appears beyond doubt that agriculture was independently invented not only both in the New World and in the Old, but also in more than one culture hearth in each of the hemispheres.

Acculturation

Acculturation is the process by which one culture group undergoes a major modification by adopting many of the characteristics of another, dominant culture group. Indeed, acculturation may involve changes in the original cultural patterns of either or both of the two groups involved in prolonged firsthand contact. Such contact and subsequent cultural alteration may occur in a conquered or colonized region. Very often the subordinate or subject population is forced to acculturate or does so voluntarily, overwhelmed by the superiority in numbers or the technical level of the conqueror.

The tribal Europeans in areas of Roman conquest, native populations in the wake of Slavic occupation of Siberia, and Native Americans stripped of their lands following European settlement of North America experienced this kind of acculturation. In a different fashion, it is evident in the changes in Japanese political organization and philosophy imposed by occupying Americans after World War II or in the Japanese adoption of some more frivolous aspects of American life (Figure 6.16). In turn, American life was greatly enriched by awareness of Japanese food preferences, architecture, and philosophy, demonstrating the two-way nature of acculturation. On occasion, the invading group is assimilated into the conquered society as, for example, the older, richer Chinese culture prevailed over that of the conquering tribes of invading Mongols during the 13th and 14th centuries. The relationship of a mother country to its colony may also result in permanent changes in the culture of the colonizer even though little direct population contact is involved. The early European spread of tobacco addiction may serve as an example (see "Documenting Diffusion"), as can the impact on Old World diets and agriculture of potatoes, maize, and turkeys introduced from America.

Figure 6.16
Baseball, an import from America, is one of the most popular
professional and amateur sports in Japan.

Contact Between Regions

All cultures are amalgams of innumerable innovations
spread spatially from their points of origin and integrated
into the structure of the receiving societies. It has been
estimated that no more than 10% of the cultural items of
any society are traceable to innovations created by its
members, and that the other 90% come to the society
through diffusion. Since, as we have seen, the pace of
innovation is affected strongly by the mixing of ideas
among alert, responsive people and is increased by exposure to a variety of cultures, the most active and innovative historical hearths of culture were those at
crossroad locations and those deeply involved in distant
trade and colonization. Ancient Mesopotamia or classical
Greece and Rome had such locations and involvements,
as did the West African culture hearth after the 5th century and, much later, England during the Industrial Revolution (and the spread of its empire).

Recent changes in technology permit us to travel
farther than ever before, with greater safety and speed,
and to communicate without physical contact more easily
and completely than previously possible. This intensification of contact has resulted in an acceleration of innovation and in the rapid spread of goods and ideas.
Several millennia ago, innovations such as smelting of
metals took hundreds of years to diffuse. Today, worldwide diffusion may be almost instantaneous.

Barriers to diffusion do exist, of course. Such
barriers, called *absorbing* or *interrupting,* are any conditions that hinder either the flow of information or the
movement of people, and thus retard or prevent the acceptance of an innovation. Because of the *friction of distance,* generally the farther two areas are from one
another, the less likely is interaction. The concept of *distance decay* tells us that, all else being equal, the
amount of interaction decreases as the distance between
two areas increases. Distance may be an absorbing barrier. Interregional contact can also be hindered by the
physical environment and by a lack of receptivity by a
contacted culture. Although oceans and rugged terrain
can act as physical barriers slowing or preventing diffusion (and have done so), cultural barriers may be equally
impenetrable. For example, for at least 1500 years, most
California Indians were in contact with cultures using
both maize and pottery, yet they failed to accept either
innovation. Should such reluctant adopters intervene between hearths and receptive cultures, the spread of an
innovation can be delayed. It can also be delayed when
cultural contact is overtly impeded by governments that
interfere with radio reception, control the flow of foreign
literature, and discourage contact between their citizens
and foreign nationals.

Figure 6.17
Motivated by religious conviction that the "good life" must be reduced to its simplest forms, the Amish, including this community in east central Illinois, shun all modern luxuries of the majority secular society around them. Children use horse and buggy, not school bus or automobile, on the daily trip to their rural school.

The more similar two cultural areas are to one another, the greater is the likelihood of the adoption of an innovation, for diffusion is a selective process. The receiving culture may adopt some goods or ideas from the donor society and may reject others. The decision to adopt is governed by the receiving group's own culture. Political restrictions, religious taboos, and other social customs are cultural barriers to diffusion. The French-Canadians, although close geographically to many centers of diffusion, such as Toronto, New York, and Boston, are only minimally influenced by them. Both their language and culture complex govern their selective acceptance of Anglo influences. Traditional groups, perhaps controlled by firm religious conviction, may largely reject culture traits and technologies of the larger society in whose midst they live (Figure 6.17).

Adopting cultures do not usually accept intact items originating outside the receiving society. Diffused ideas and items commonly undergo some alteration of meaning or form that makes them acceptable to a borrowing group. The process of the fusion of the old and new is called **syncretism** and is a major feature of culture change. It can be seen in alterations to religious ritual and dogma made by convert societies seeking acceptable conformity between old and new beliefs. On a more familiar level, syncretism is reflected in subtle or blatant alterations of imported cuisines to make them conform to the demands of America's fast-food franchises.

Cultural Diversity

We began our discussion of culture by recognizing its generalizing subsystems of technological, sociological, and ideological content. We have learned that the distinctive makeup of those subsystems—the combinations and interactions of traits and complexes characteristic of particular cultures—is subjected to, and the product of, change through innovation, diffusion, and integration. Those processes of cultural development and alteration have not, however, led to a homogenized world culture even after thousands of years of cultural contact and exchange since the origins of agriculture. It is true, as we observed earlier, that in an increasingly integrated world, access to the material trappings and technologies of modern life and economy is widely available to all peoples and societies. As a result, important cultural commonalities have developed.

Nevertheless, all of our experience and observation indicate a world still divided, not unified, in culture. Our concern as geographers is to identify those traits of a culture that both have spatial expression and indicate significant differences from other culture complexes. We may reject as superficial and meaningless generalizations derived from trivialities: the foods people eat for breakfast, for example, or the kinds of eating implements they use.

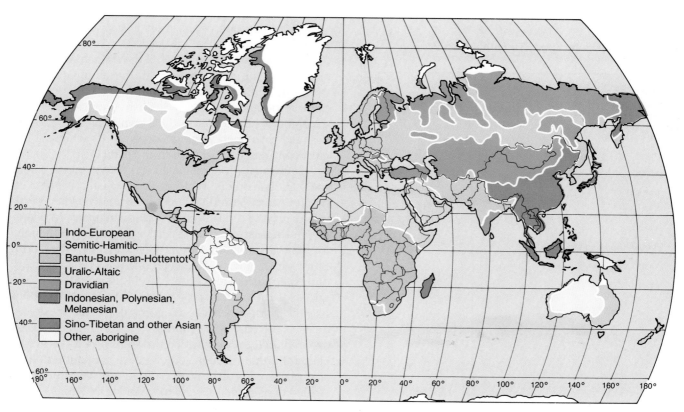

Figure 6.18
The world pattern of major language families. Language families are assumed to group individual tongues that had a common but remote origin.

This rejection is a reflection of the kinds of understanding we seek and the level of generalization we desire. There is no single most appropriate way to designate or recognize a culture or to delimit a culture region. As geographers concerned with world systems, we are interested in those aspects of culture that vary over extensive regions of the world and differentiate societies in a broad, summary fashion. Within culture regions thus recognized, significant elements of cultural diversity may exist. These also invite our attention.

Language

Among all the culture traits, the two of which we are perhaps most immediately made aware are nationality and language. Both are fundamental elements of group recognition and segregation; both are spatially rooted. In all societies, the foreigner is to a greater or lesser degree a person apart, a member of another culture and a citizen of a different country. The foreigner is presumed to embody at least some of the attitudes, customs, and political stances we associate with his or her country of citizenship or origin. We will discuss aspects of nationality and other political affiliations as a special part of cultural differentiation—political geography—in Chapter 7. The other inescapable trait that cannot usually or easily be disguised by adoption of the dress, customs, or citizenship of another country is language, dialect, or accent.

Language, written and oral, makes possible the understandings and shared behavior patterns called culture. Together with the culture of which it is a part and an essential expression, language itself is transmitted to successive generations by learning and imitation. Some anthropologists argue that the language of a society structures the perceptions of its speakers. By the words that it contains and the concepts that it can formulate, language is said to determine the attitudes, the understandings, and the responses of the society to which it belongs. Language, therefore, may be both a cause and a symbol of cultural differentiation.

If that conclusion is true, one aspect of cultural heterogeneity may be easily understood. The roughly 5 billion people on earth speak several thousand languages. From our limited experience, we may find it difficult to name more than 25 or 30 of the languages currently in use. Just by being informed that as many as 1000 languages are spoken in Africa, we immediately have a clearer understanding of the political and social problems in that continent. Europe alone has more than 100 languages and dialects, and language differences were a basis, at least in part, for national delimitation in the political restructuring of Europe after World War I. Language, then, is a hallmark of cultural diversity (Figure 6.18).

TABLE 6.1

Languages Spoken by More Than 30 Million People
(1986)

Language	Millions of speakers
Mandarin (China)	775
English	420
Hindi[a]	290
Spanish	287
Russian	215
Arabic	173
Bengali (Bangladesh, India)	168
Portuguese	163
Malay-Indonesian	128
Japanese	123
German	117
French	113
Urdu[a] (Pakistan, India)	82
Punjabi (India, Pakistan)	71
Korean	66
Italian	64
Tamil (India, Sri Lanka)	63
Telugu (India)	62
Marathi (India)	61
Cantonese (China)	60
Wu (China)	57
Javanese	52
Turkish	51
Vietnamese	50
Min (China)	46
Thai	44
Ukrainian	42
Polish	41
Swahili (East Africa)	38
Kannada (India)	38
Gujarati (India)	37
Malayalam (India)	33
Tagalog (Philippines)	32
Persian (Iran, Afghanistan)	32
Burmese	30

[a]Hindi and Urdu are basically the same language, Hindustani. Written in the Devangari script, it is called *Hindi*, the official language of India; in the Arabic script it is called *Urdu*, the official language of Pakistan.

It is also one of the most fiercely defended elements of a culture. The official Committee for the Defense of the French Language of the French Academy is charged by the government with the elimination from the French language of foreign words and phrases, even those in popular use. Many Spanish-Americans demand the right of instruction in their own language, and the Basques have been waging a civil war to achieve a linguistically based separation from Spain.

Obviously, some languages have many more speakers than others. Table 6.1 lists those that are spoken by more than 30 million people. Even this level of diversity may be reduced if we group languages into families. A **linguistic family** is a group of languages thought to have a common origin. For example, the Indo-European family of languages includes, among many others, English, Greek, Hindi, and Russian. All told, languages in the Indo-European family are spoken by about half the world's peoples. Linguists think that these languages derived from a common ancestor language called *proto–Indo-European,* which was spoken by tribes in east central Europe (though some believe the southern Russian plains eastward to the Caspian Sea were the more likely site of origin) several thousand years ago. As these groups of people spread across Europe and Asia and settled isolated from one another, their languages evolved differently.

Within a linguistic family, we can distinguish *subfamilies.* The Romance languages (including French, Spanish, and Italian) and the Germanic languages (such as English, German, and Dutch) are subfamilies of Indo-European. The languages in a subfamily often show similarities in sounds, grammatical structure, and vocabulary even though they are mutually unintelligible. English *daughter,* German *Tochter,* and Swedish *dotter* may serve as examples.

Linguists also make a distinction between the language as it is spoken by perhaps millions of people and the standard or official version, that is, the form carrying official governmental or educational sanction. In Arab countries, for example, classical Arabic is the language of the mosque, of education, and of the newspapers, and colloquial Arabic is used at home, in the street, and at the market. The United States, English-speaking Canada, Australia, and the United Kingdom all have different forms of standard English. In England, the English spoken by the announcers of the British Broadcasting Corporation is regarded as the linguistic standard.

Dialects are the regional variations of a language. Vocabulary, pronunciation, rhythm, and the speed at which the language is spoken may set groups of speakers apart from one another and, to a trained observer, clearly mark the origin of the speaker. In George Bernard Shaw's play *Pygmalion,* on which the musical *My Fair Lady* was based, Professor Henry Higgins can identify the London neighborhood of origin of a flower girl by listening to her vocabulary and accent. An American specialist in linguistics could probably do the same with a native speaker of American English. Figure 6.19 indicates the variation in usage associated with just one phrase. Southern English and New England speech are among the dialects spoken in the United States that are the most easily recognized by their distinctive accents. There may be so much variation that some dialects are almost unintelligible to other speakers of the same language. Effort is required for Americans to understand Australian English or that spoken in Liverpool, England, or in Glasgow, Scotland. An interesting United States example is discussed in the regional study, "Gullah as Language," Page 415.

Like all elements of culture, languages follow processes of diffusion and evolution. Linguistic diffusion is usually a result of migration or conquest; trade can also be a contributing factor. English owes its form to the Celts, the original inhabitants of the British Isles, and to successive waves of invaders, including the Latin-speaking Romans; the Teutonic Angles, Saxons, and Danes; and the French-speaking Normans. The English language continued to evolve in the centuries following the Norman conquest, just as did the Gaelic of Britons who had been pushed into the outlying, relatively isolated areas of Ireland, Wales, and Scotland. The spread of English worldwide reflects the establishment of English overseas colonies and the long-term dominance of English in world trade and, more recently, in science. From some 7 million speakers 400 years ago on a west European island, English is now the native tongue of over 400 million persons and the official language of more than 40 countries.

Language embodies the culture complex of a people, reflecting both environment and technology. Arabic has 80 words related to camels, an animal on which a regional culture relied for food, transport, and labor; Japanese contains over 20 words for various types of rice. Russian is rich in terms for ice and snow, indicative of the prevailing climate of its linguistic cradle, and the 15,000 tributaries and subtributaries of the Amazon River have obliged the Brazilians to enrich Portuguese with words that go beyond "river." Among them are *paraná* (a stream that leaves and reenters the same river), *igarapé* (an offshoot that runs until it dries up), and *furo* (a waterway that connects two rivers).

Language is the most important medium for transmitting culture. It enables parents to teach their children what the world they live in is like and what they must do to become functioning members of society. A common language fosters unity among people. It promotes a feeling for a region; if it is spoken throughout a country,

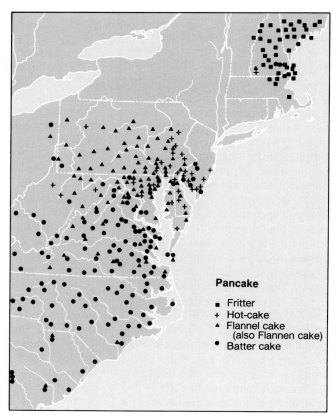

Figure 6.19

Maps such as this are used to record variations over space and among social classes in word usage, accent, and pronunciation. The differences are due not only to initial settlement patterns but also to more recent large-scale movements of people, for example, from rural to urban areas and from the South to the North. Despite the presumed influence of national radio and television programs in promoting a "general" or "standard" American accent and usage, regional and ethnic language variations persist.

it fosters nationalism. For this reason, languages often gain political significance and serve as a focus of opposition to what is perceived as foreign domination. Although nearly all people in Wales speak English, many also want to preserve Welsh because they consider it an important aspect of their culture. They think that if the language is forgotten, their entire culture may also be threatened. French-Canadians have received government recognition of their language and now have established it as the official language of Quebec Province; Canada itself is officially bilingual. In India, with 15 constitutional languages and 1652 other tongues, serious riots broke out in 1965 among people expressing opposition to the imposition of Hindi as the single official national language. In the Soviet Union protest marches and demonstrations in the Georgian Republic in the Caucasus marked the exclusion of reference to Georgian as the official language in the republic's revised constitution.

Pidgins and Creoles

Language is rarely a barrier in communication between peoples, even those whose native tongues are mutually incomprehensible. Bilingualism or multilingualism may permit skilled linguists to communicate in a jointly understood third language, but long-term contact between less able populations may require the creation of new language—a *pidgin*—learned by both parties. A pidgin is an amalgamation of languages, usually a simplified form of one, such as English or French, with borrowings from another, perhaps non-European, native language. In its original form, a pidgin is not the native tongue of any of its speakers; it is a second language for everyone who uses it.

Generally restricted to such specific functions as commerce, administration, or work supervision, most pidgins developed as a result of European trade contacts and colonization. Pidgins may have originated in the 17th century in the seaports of southern China, where British traders developed a compromise language between their own English and the Chinese of their trading partners and employees. Indeed, *pidgin* itself is thought to be a Cantonese corruption of the English word *business*. Pidginized forms of French, Portuguese, and English were used in the slave trade as well as among American slaves of different linguistic backgrounds. Pidgins are not restricted to the past, however. Such is the variety of languages spoken among the some 270 ethnic groups of Zaire that a special tongue called Lingala, a hybrid of Congolese dialects and French, has been created to permit, among other things, issuance of orders to army recruits drawn from all parts of the country.

Pidgins are characterized by a highly simplified grammatical structure and a sharply reduced vocabulary. They tend to eliminate the difference between masculine and feminine forms of nouns and adjectives, use only a single form of the verb (such as the infinitive or the past participle), and indicate person and number by independent pronouns rather than by verb inflections. The limited vocabulary is adequate to express basic ideas but not complex concepts. "Mefella e-go golong salwater, lookawtime fish blong kai-kai. Em e-good here, e-goodfella fish," a Melanesian might say, reporting on a successful fishing trip.

Once a pidgin has become the first language of a group of speakers—who may have lost their former native tongue through disuse—creolization has occurred. A *creole* is a language that has evolved from a pidgin to become a distinctive language that is the first languge of a society. In their development, creoles acquire a complex grammatical structure and enhanced vocabulary.

Creole languages have proved useful integrative tools in linguistically diverse areas. Several have become symbols of nationhood. Swahili, a pidgin formed from a number of Bantu dialects, originated in the coastal areas of East Africa and spread by trade during the period of English and German colonial rules. When Kenya and Tanzania gained independence, they made Swahili the national language of administration and education. Other examples of creolization are Afrikaans (a pidginized form of 17th century Dutch used in the Republic of South Africa); Haitian Creole (the language of Haiti, derived from the pidginized French used in the slave trade); and Bazaar Malay (a pidginized form of the Malay language, a version of which is the official national language of Indonesia).

Bilingualism or multilingualism complicates national linguistic structure. Areas are considered bilingual if more than one language is spoken by a significant proportion of the population. In some countries—Belgium, for example, or Switzerland—there is more than one official language. In many others, such as the United States, only one language may have implicit or official government sanction, although several others are spoken. Speakers of one of these may be concentrated in restricted areas (for example, most speakers of French in Canada live in Quebec Province). Less often, they may be distributed fairly evenly throughout the country. There may be roughly equal numbers of speakers of each language, or one group may be a significant majority. When this is the case, the minority group may have to learn the language of the majority to receive an education, to obtain certain types of employment, or to fill a position in the government. In some countries, the language in which instruction, commercial transactions, and government business take place is neither a native language nor the majority language, but yet another tongue. This is the position of English in many former colonies, such as Bangladesh, Nigeria, and Ghana.

Figure 6.20

This solitary worshiper in a Penang, Malaysia, mosque demonstrates the depth of faith shared with his 750 million coreligionists worldwide. Many rules concerning daily life are given in the Koran, the holy book of the Muslims. All Muslims are expected to observe the five pillars of the faith: (1) repeated saying of the basic creed; (2) prayers five times daily facing Mecca; (3) a month of daytime fasting (Ramadan); (4) almsgiving; and, if possible; (5) a pilgrimage to Mecca.

Religion

Language may be an ardently defended symbol of cultural identity, but it yields to religion as a cultural rallying point. French Catholics and French Huguenots (Protestants) freely slaughtered each other in the name of religion in the 16th century. English Roman Catholics were hounded from the country after the establishment of the Anglican church. Religious enmity between Muslims and Hindus forced the partition of the Indian subcontinent after the departure of the British in 1947. And the 1980s witnessed continuing religious confrontations between, among many others, Catholic and Protestant Christian groups in Northern Ireland; Muslim sects in Lebanon, Iran, and Iraq; Muslims and Jews in Palestine and Lebanon; Christians and Muslims in the Philippines; and Buddhists and Hindus in Sri Lanka.

No element of the sociological subsystem has been more pervasive in its influence upon every aspect of culture than formalized religious belief. Even highly structured religious convictions and principles are undeniably part of the ideological subsystem of culture. Further, they strongly influence attitudes toward the tools and rewards of the technological subsystem. Finally, the organizational expressions of religion in church edifice and religious hierarchy, in public ceremony and symbol, in formal group and individual observances, and even in political accommodations or responses are clearly institutional. Religion bridges the full range of cultural identification.

As bridging institutions, religions affect all facets of a culture, openly or indirectly. Most of the various Hindu sects have strict dietary rules; Muslims do not eat pork; Roman Catholics until recently ate no meat on Fridays. Christians regard Sunday as the day of rest; for Jews the day is Saturday and for Muslims, Friday. These are small examples of the control that religious beliefs exert over people's daily lives. The list could be expanded many times.

There are other important ways in which religions shape believers' lives and thoughts. As formalized views about the relation of the individual to the world and to the hereafter, each religion carries a distinct conception of the meaning and value of this life and most contain strictures about what must be done to achieve salvation (Figure 6.20). These rules become interwoven with the traditions of a culture. One cannot understand India without a knowledge of Hinduism or Israel without an appreciation of Judaism.

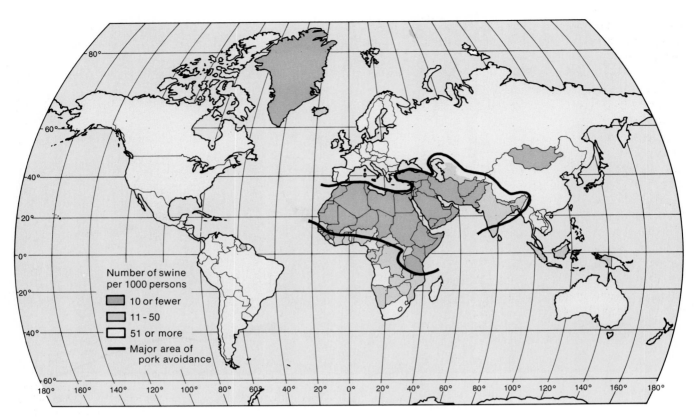

Figure 6.21

Religious prohibition against the consumption of pork, particularly among those of the Jewish and Muslim faiths, finds spatial expression in the incidence of swine production. Because, except for the Soviet Union, production figures are national summaries, the map does not faithfully report small-area distributions of either religious affiliation or animal raising.

Sources: F. J. Simoons, *Eat Not This Flesh*, University of Wisconsin Press, 1961; United Nations, Food and Agricultural Organization; *Narodnoye khozyaystvo SSR.*

Even *economic patterns* may be intertwined with religion. Religious restrictions on food and drink affect the economic structure of an area, very strongly in some places, less so in others. The kinds of animals that are raised or avoided (Figure 6.21), the crops that are grown, and the importance of those crops in the daily diet are often determined by religious beliefs. To take another example, the frequently used term *Protestant ethic* connotes a presumed attitude toward economic progress that is quite different from that of, say, Hinduism. Individualism, hard work, and the acquisition of material wealth are said to characterize the former; in Hinduism, there is considerably less concern about the accumulation of material wealth in this life.

In some countries religion and *government* are connected through a state religion. Buddhism, for example, is the official religion in Burma and Thailand. Fundamentalist Muslim nations are increasing in number. In a few countries, the government leaders are church officials. In others, the political parties have religious affiliations that are reflected in the voting patterns of the

electorate, as in Northern Ireland. Laws, too, may reflect religious beliefs. The penalty attached to inadvertently killing a cow, for example, is very high in Nepal, where the cow is a sacred animal. Many aspects of Israeli life are controlled by religiously based and interpreted law, while in the United States (and other nations) regulations controlling or prohibiting Sunday business hours or the sale of alcoholic beverages are rooted in or defended on grounds of religion and morality. The adoption of Islamic law by nation-states is partially recorded in Figure 6.9.

In a few countries, *political ideologies* have a quasi-religious role. They have many of the elements of a religion, including a set of beliefs, ethical standards, revered leaders, an institutional organization, and a body of literature akin to scripture that is interpreted by special people. Adherents may display an almost religious fervor in their desire to proselytize (convert nonbelievers) and to root out heretical beliefs and practices. In this fashion, Marxism-Leninism may be viewed as the state religion of the USSR, China, and a number of countries in eastern Europe, Africa, and Southeast Asia. It combines a belief in the redeeming role of the proletariat with an ethos of work and sacrifice to obtain future benefits for the state. Rituals (parades and ceremonies) and shrines reinforce the beliefs (Figure 6.22). Communism has even had its own schisms, most notably between Moscow and Beijing (Peking, China), and Moscow and Tirana (Albania), and distinctive sects (for example, the "Titoism" of Yugoslavia).

Settlement patterns are also linked to religion. The members of a religion tend to be areally concentrated, a fact readily apparent from Figure 6.23. Note the predominance of Hinduism in India and of Buddhism in most of Southeast Asia. Although no map distinction is made between the subsets of Christians, Protestantism has predominated in the United States and Australia, and South America has been a Roman Catholic area. Within a country, the members of a particular religion may also display areal concentration. Thus, in urban Northern Ireland, Protestants and Catholics reside in separate areas whose boundaries are clearly understood and respected.

Figure 6.22
People lined up to visit Lenin's tomb in Red Square, Moscow. V. I. Lenin was a leader of the Communist revolution of 1917, which marked the end of the rule of the czars and the establishment of the Union of Soviet Socialist Republics. Lenin's body is preserved in a glass-covered coffin inside the tomb, which is visited by thousands of people each day. Millions also make a pilgrimage to Lenin's birthplace on the Volga River. The tomb and Lenin's early home may be said to be sacred shrines for Soviet Communists.

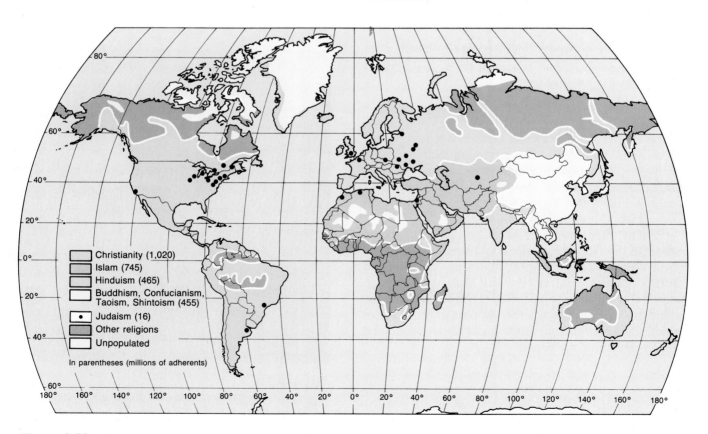

Figure 6.23
The distribution of the major religions. Although there are hundreds of religions, those named in the map legend include about half the people of the world.

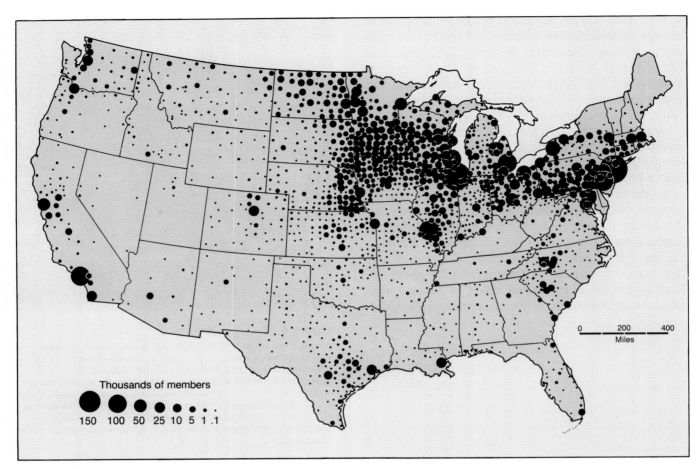

Figure 6.24

The distribution of reported Lutherans in the United States. Although Lutheran church members may be found in every state, their great preponderance is in the upper Midwest and the urbanized Northeast.

In the United States, Jews and Unitarians tend to live in urban areas, and Mennonites are found mainly in rural settings. More Baptists live in the South than elsewhere, and Mormons are found chiefly in Utah and portions of the adjoining states. Figure 6.24 suggests the regional concentration of American Lutherans.

Many other instances of religion as an important strand in the fabric of culture could be cited. Religious ideas affect the role of women in a society, marriage and divorce customs and taboos, and the importance attached to education (Figure 6.25). Our ideas about the significance of this life, whether we have only one life or many, and what we should strive for in life all are molded in part by religious precepts.

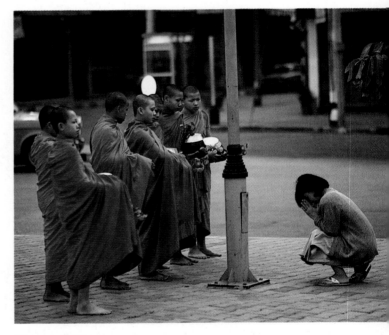

Figure 6.25

In some countries, such as Burma and Thailand, all males are expected to spend at least a year of their lives as Buddhist monks. These men, with shaved heads and dressed in their distinctive saffron-colored robes, are a familiar sight on the streets of Bangkok. Emphasis on monastic training has led to a high rate of literacy among males, an example of the effect of religion on other facets of culture.

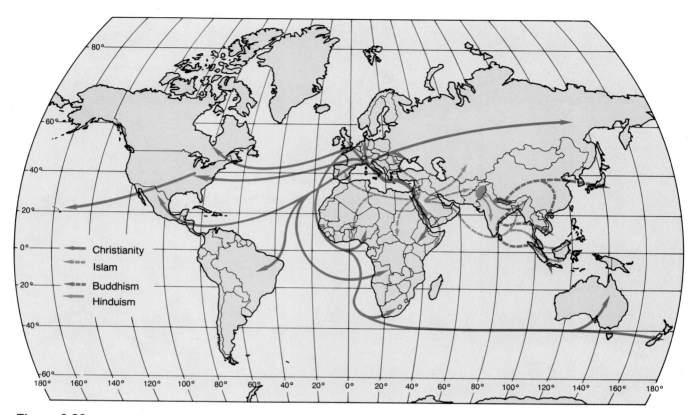

Figure 6.26
The origin areas and diffusion patterns of the four major religions. Christianity achieved its worldwide dispersion through colonization and vigorous missionary activity; Islam spread through conquest and trade. Trade and missionaries slowly diffused Buddhism through eastern and southeastern Asia, while the limited spread of Hinduism reflects the restricted migration pattern of its adherents.

Religion is a strong force for promoting *cultural stability.* Through rituals and scripture, time-honored ways of doing things are fostered. Religions in general act to retard cultural changes. Much of the diversity within a given religion has occurred as its believers have been faced with changing times or conditions. All religions contain such diversity. Most Americans are familiar with at least the names of the various denominations of Protestantism: Lutheran, Unitarian, Baptist, Presbyterian, and Episcopalian, to name only a few. Each is characterized by beliefs or practices that distinguish it from the rest. Similarly, Islam contains several, frequently antagonistic, sects including the Sunnites, the Shiites, and the Wahhabites. Judaism has spawned Orthodox, Conservative, and Reform branches. Hinduism contains a variety of creeds and practices. Often this diversity has been caused by different interpretations of religious doctrine, sometimes by differences that evolved when a major religion came into conflict with native cultures.

The study of religion aids us in understanding the diverse cultural patterns we see on the face of the earth. Religions both promote and hinder contact between groups. Although religious differences may be the focus of animosities and conflicts, religions have also promoted contact between regions by inspiring believers to convert others, through either conquest or peaceful missionary efforts. Many of the major religions have thus acted as unifying forces, bringing large numbers of people formerly more diverse to a belief in a common faith (Figure 6.26).

Ethnicity

Any discussion of cultural diversity would be incomplete without mention of **ethnicity.** Based on the root word *ethnos*—meaning "people" or "nation"—the term usually refers to the ancestry of a particular people who have in common distinguishing characteristics associated with their heritage. No single trait denotes ethnicity. Recognition of ethnic communities may be based on language, religion, national origin, unique customs, or an ill-defined concept of "race" (see "The Question of Race,"

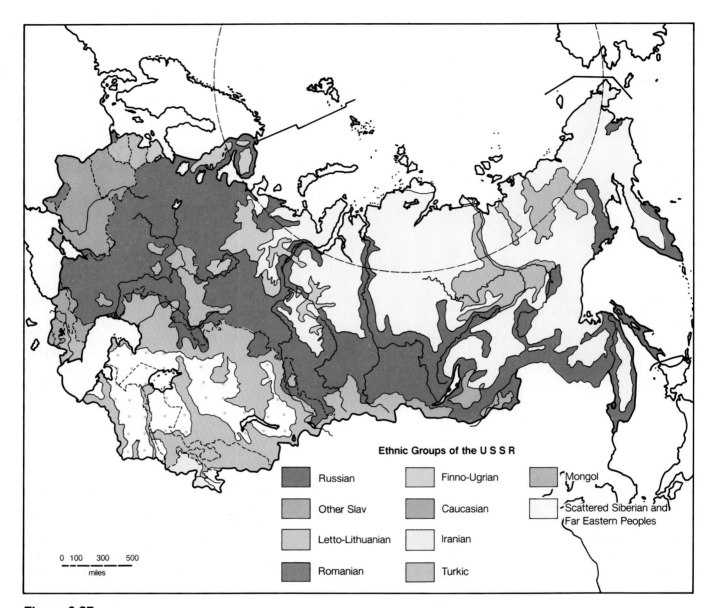

Ethnic Groups of the U S S R

- Russian
- Other Slav
- Letto-Lithuanian
- Romanian
- Finno-Ugrian
- Caucasian
- Iranian
- Turkic
- Mongol
- Scattered Siberian and Far Eastern Peoples

0 100   300   500
miles

Figure 6.27

More than 100 non-Slavic ''nationalities'' and ethnic groups have homelands within the boundaries of the USSR. Made part of the czarist empire, only a fraction of them have separate political recognition under the Soviet constitution. Their growing populations pose increasing challenges to established Slavic, particularly Russian, domination.

Page 225). Whatever the unifying thread, ethnic groups may strive to preserve their special shared ancestry and cultural heritage through the collective retention of language, religion, festivals, cuisines, traditions, and in-group work relationships, friendships, and marriages. Those preserved relationships and associations are fostered by, and support, *ethnocentrism*—the feeling that one's own ethnic group is superior.

Normally, reference to ethnic communities is recognition of their minority status within a nation or region dominated by a different majority culture group. We do not identify Koreans living in Korea as an ethnic group because theirs is the dominant culture in their own land. Koreans living in Japan, however, constitute a discerned

and segregated group in that foreign nation. Ethnicity, therefore, is evidence of areal cultural diversity and a reminder that culture regions are rarely homogeneous in the characteristics displayed by all of their occupants.

Territorial segregation is a strong and sustaining trait of ethnic identity, and one that assists groups to retain their distinction. On the world scene, indigenous ethnic groups have developed over time in specific locations and have established themselves in their own and others' eyes as distinctive peoples with defined homeland areas. The boundaries of most countries of the world encompass a number of racial or ethnic minorities (Figure 6.27). Their demands for special territorial recognition have sometimes increased with advances in economic development and self-awareness, as Chapter 7

Religion as Politics

The seizure of the American embassy in Tehran, Iran, in November 1979 led to 444 days of captivity for a group of United States diplomats, civilian employees, and military personnel. It was also a signal event in a revolution that overthrew the ruling shah, Mohammad Reza Pahlavi, and established an 80-year-old Islamic religious fundamentalist, Ayatollah Ruholla Khomeini, as national leader.

Most Iranians are Shiite Muslims, adherents of a sect of Islam that emphasizes duty, severity, and martyrdom. Supporters of the revolution believed that the Islamic regime was the only way to free Iran from the kind of foreign influence—political, economic, and cultural—that

it had experienced for over a century at the hands of Britain, then Russia, and finally the United States. A sense of persecution and humiliation resulted from foreign control. More important, such control had led to secularization, evidenced by the Western life-style that the elite had adopted and the Western legal systems by which the nation was guided. The economic modernization and land-reform programs undertaken by the shah conflicted with the fundamentalist tenets of the religious leaders and threatened their wealth and power. One of the first acts of the revolutionary government was to dismiss thousands of female employees and to order all women to wear either *chadors* (a head-to-toe covering) or headscarves when they left their houses. The women were to return to their traditional roles as wives and mothers.

The war between Iran and Iraq was another consequence of the revolution. Begun in 1980, it became for Iranians a war of faith against an enemy nation dominated by Sunni Muslims, a rival sect. (The Shiites' patron, Hossein, died 1300 years ago in a battle with Sunni Muslims.) The Iranians hoped to spread their version of Islamic fundamentalism by establishing Iranian-style Islamic republics in Iraq and also in neighboring Persian Gulf states.

What can we learn from the Iranian upheavals? Certainly that Islam is not a monolithic faith; that religion has great power as a cultural rallying point; and that serious cultural conflict can occur when technology, social systems, and ideologies are not in accord or when change based on foreign values appears to threaten cherished local beliefs.

---

points out. In 1820 more than half the population of Europe belonged to ethnic groups without recognized cultural autonomy or political independence. By the late 20th century, this figure had dropped to only 3%.

Increasingly in a world of movement, ethnicity is less a matter of indigenous populations and more one of outsiders in an alien culture. Immigrants, legal and illegal, and refugees from war, famine, or persecution are a growing presence in nations throughout the world. Immigrants to a country typically have one of two choices. They may hope for **assimilation** by giving up many of their past cultural traits, losing their distinguishing characteristics, and merging into the mainstream of the dominant culture. Or they may try to retain their distinctive cultural heritage. In either case, they usually settle initially in an area where other members of their ethnic group live, as a place of refuge and learning (Figure 6.28). With the passage of time they may leave their protected community and move out among the general population. Or, retaining their ethnic identity and resisting full assimilation, they may collectively relocate or expand their area of occupance (Figure 6.29). The "Chinatowns" and "Little Italys" of North American cities have provided the support systems essential to new immigrants in an alien culture region. Japanese, Italians, Germans, and other ethnics have formed agricultural colonies

in Brazil in much the same spirit. Such ethnic enclaves may provide an entry station, allowing both individuals and the groups to which they belong to undergo sufficient cultural and social modifications to enable them to operate effectively in the new, majority society. Sometimes, of course, settlers have no desire to assimilate or are not allowed to, so that they and their descendants form a more or less permanent subculture in the larger society. The Chinese in Malaysia belong to this category. Ethnicity in the context of nationality is discussed more fully in Chapter 7.

Other Aspects of Diversity

Culture is the sum total of the way of life of a society. It is misleading to isolate, as we have done, only a few elements of the technological, sociological, and ideological subsystems and imply by that isolation that they are identifying characteristics differentiating culture groups. Economic developmental levels, language, religion, and ethnicity all are important common distinguishing cultural traits, but they tell only a partial story. Other suggestive, though perhaps less pervasive, basic elements exist.

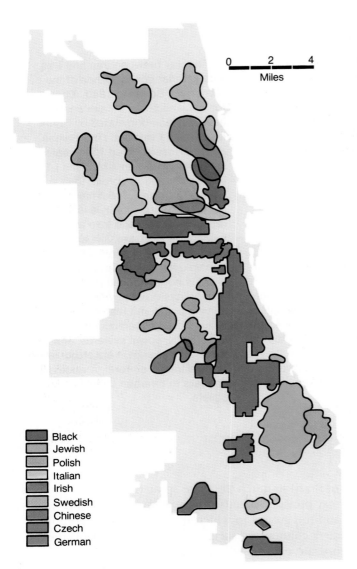

Figure 6.28
The distribution of ethnic groups in Chicago, 1957. Most large cities of the United States are ethnically heterogeneous, particularly in the industrial North, to which came successive waves of immigrants. The heterogeneity is usually expressed by a well-defined series of homogeneous neighborhoods rather than by a smooth intermixture of peoples. On this map, black areas are defined as neighborhoods with 25% or more black inhabitants. They have expanded significantly in size since 1957.

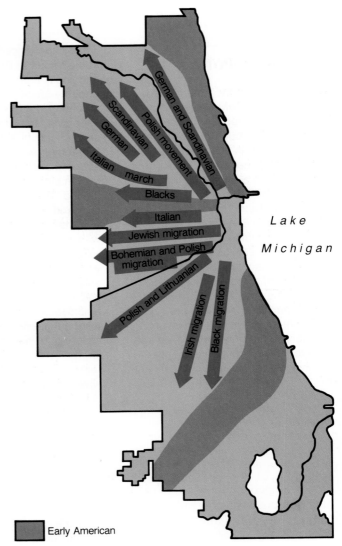

Figure 6.29
The outward expansion of blacks and ethnic groups in Chicago. "Often," Samuel Kincheloe observed a half century ago, "[minority] groups first settle in a deteriorated area of a city somewhere near its center, then push outward along the main streets."

Architectural styles in public and private buildings are evocative of region of origin even when they are encountered in indiscriminate juxtaposition in American cities. The Gothic and New England churches, the neoclassical bank, and the skyscraper office building suggest not only the functions they house but the culturally and regionally variant design solutions that gave them form. The Spanish, Tudor, French Provincial, or ranch-style residence may not reveal the ethnic background of its American occupant, but it does constitute a culture statement of the area and the society from which it diffused.

Music, food, games, and other evidences of the joys of life, too, are cultural indicators associated with particular world or national areas. Music is an emotional form of communication found in all societies, but, being culturally patterned, it varies among them. Instruments, scales, and types of compositions are technical forms of variants; the emotions aroused and the responses evoked among peoples to musical cues are learned behaviors. The Christian hymn means nothing emotionally to a pagan New Guinea clan. The music of a Chinese opera may be simply noise to the European ear. Where there is sufficient similarity between musical styles and instrumentation, blending (syncretism) and transferal may

The Question of Race

Human populations may be differentiated on the basis of acquired cultural or inherent physical attributes. Culture is learned behavior transmitted to the successive generations of a social group through imitation and through that distinctively human capability, speech. Culture summarizes the way of life of a group of people and may be adopted by members of the group irrespective of their individual genetic heritage, or race.

The spread of human beings over the earth and their occupation of different environments were accompanied by the development of those physical variations in the basic human stock that are commonly called *racial.* In very general terms, a **race** may be defined as a population subset whose members have in common some hereditary biological characteristics that set them apart physically from other human groups. Since the reference is solely to genes, the term *race* cannot be applied meaningfully to any human

attribute that is acquired. Race, therefore, has no significance in reference to national, linguistic, religious, or other culturally based classifications of populations.

Although it has been argued that race is a misleading concept, there does exist a common understanding that populations are recognizably grouped on the basis of differences in pigmentation, hair characteristics, facial features, and other traits largely related to variations in soft tissue. Some subtle skeletal differences among peoples also exist. Such differences form the basis for the division by some anthropologists of humanity into racial groups. Caucasoid (or "white"), Negroid, Mongoloid, Amerindian, and Australian races (and others) have been recognized in a continuing process of classification modification and refinement. Racial differentiation as commonly understood is old and can reasonably be dated at least to the Paleolithic (100,000 to about 11,000 years ago) spread and isolation of population groups.

If all humankind belongs to a single species that can freely interbreed and produce fertile offspring, how did any apparent differentiation occur? Among several causative factors commonly discussed, two appear to be most important: *genetic drift* and *adaptation.* Genetic drift refers to an inheritable trait (such as a distinctive shape of the nose) that appears by chance in one group and becomes accentuated through inbreeding. If two populations are spatially too far apart for much interaction to occur (*isolation*), a trait may develop in one but not in the other. Adaptation to particular environments is thought to account for the rest of racial differentiation, although it should be recognized that we are only guessing here. One cannot say with certainty that any racial characteristic is the result of such evolutionary changes. Genetic studies, however, have suggested a relationship between solar radiation and blood type, and between temperature and body size, for example.

occur. American jazz represents a blend; calypso and flamenco music have been transferred to the North American scene. Foods identified with other culture regions have similarly been transferred to become part of the culinary environment of the American "melting pot."

These are but a few additional minor statements of the variety and the intricate interrelationships of that human mosaic called culture. Individually and collectively they are, in their areal expressions and variations, the subject matter of the cultural geographer.

Culture Realms

Our discussion of culture has had one consistent and recurrent message: culture has spatial expression in all of its details and composites. The individual culture traits we examined and mapped show the subdivision of earth space into special-purpose regions. Of course, the same trait—the Christian religion, perhaps, or socialist law—may be part of more than one culture, but each separate

culture will be marked by a distinctive complex of such individual traits, setting it off spatially from adjacent cultures with their own identifying composites of traits.

If two or more culture complexes have a number of such traits in common, a **culture system** may be recognized as a larger spatial reality and generalization. Multiethnic societies, perhaps further subdivided by linguistic differences, varied food perferences, and a host of other internal differentiations, may nonetheless share enough common characteristics of the subsystems of culture to be recognizably—to themselves and others—distinctive cultural entities. Certainly, citizens of "melting-pot" United States would identify themselves as *Americans,* together comprising a unique culture system on the world scene.

Culture regions and complexes are elements in the spatial hierarchy of cultural geography. They may, at a still higher level of generalization, be combined into composite world regions, into *culture realms.* At that level of summary, cultural specifics become obscured and

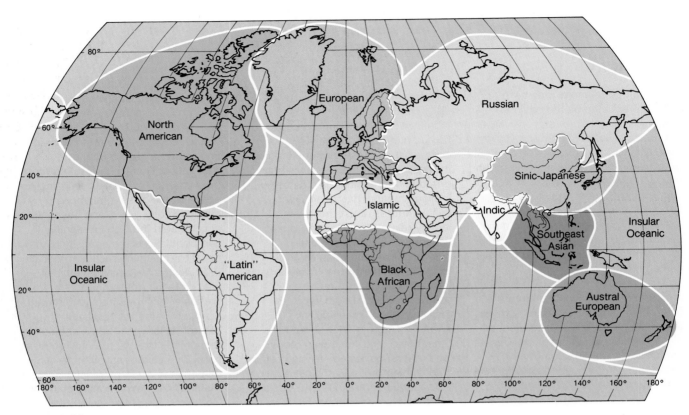

Figure 6.30

Culture realms of the modern world. This is just one of many possible subdivisions of the world into multifactor cultural regions.

perceptions of world regional differentiation come into play. Culture realms attempt to document those perceptions of world-scale cultural contrasts by identifying groups of culture systems with enough distinctive characteristics in common to set them apart from other realms with differing sets of identifying generalizations.

Clearly, our present data base is inadequate to the task of definitive world cultural regionalization. Political structure, economic orientation, patterns of behavior, levels of urbanization—all aspects of contemporary culture— are yet to be considered. A preliminary recognition of composite culture realms may be attempted, however, on the basis of the fundamental differentiating characteristics already discussed. Language and religion are ingrained and persistent attributes of populations that color their acceptance, and regulate their adoption, of modern economies and technologies. "Developing" realms are modernizing in ways distinctively their own, taking on the trappings of an integrated world community, but remaining clearly separated in underlying culture traits.

Weakening the validity of the claim that uniform and distinct culture realms may be distinguished is the inevitable *cultural heterogeneity* resulting from the past

and present migrations of peoples. Earlier European colonization and emigration and the more recent massive refugee relocations of the 20th century are striking departures from the slow migrations and assimilations of earlier human history. All culture regions are impermanent, at least judged against the standards of regional permanency of the physical geographer dealing with landforms or climates. The culture regions of Amerindian North America, so slowly developed over thousands of years, were nearly totally obliterated and replaced within the space of 300 years. In its turn, the "melting pot" of contemporary North America has produced not a culture uniform in all respects, but a collection of peoples whose ethnic and cultural differences are increasingly a matter of pride and emphasis. Similar confused and indistinct mixtures cloud the picture at the national and international scale in the Soviet Union, in South America, in Southeast Asia, and in Africa.

Nevertheless, overriding internal religious, linguistic, and national-origin differences within broad world regions are expressions of cultural commonality recognizable in map analysis and explicit in popular understanding. Figure 6.30 is offered as a world subdivision into culture realms, regionally discrete areas that are more alike internally than they are like other realms.

Conclusion

The world's cultural diversity is suggested by spatial variations in the technological, sociological, and ideological subsystems that societies have developed or adopted, and by the diversity of cultural traits and complexes that they display. As we have seen, the mosaic of culture elements and outlooks may be summarized into culture realms, internally diverse but containing elements of pervasive uniformity to distinguish them from other major earth areas. Each of the principal traits of culture discussed in this chapter may serve as evidence of its own and related aspects of the spatial variation in the human condition which is the special province of the cultural geographer.

Cultural differentiation and regionalization are not limited to the subsystems, traits, and complexes discussed in this chapter. Of nearly comparable importance as a determinant of societal identity at all levels of generalization are the political affiliations entered into or imposed upon individuals and groups. Formal political organization, the allocation and exercise of power it implies, and the role of nationality as a prime element of cultural variation are topics dealt with by political geography, to which we next turn our attention.

Key Words

acculturation *210*	culture complex *197*	environmental	race *225*
assimilation *223*	culture hearth *207*	determinism *198*	sociological
barriers to diffusion *211*	culture realm *197*	ethnicity *221*	subsystem *200*
cultural integration *205*	culture region *197*	ideological	syncretism *212*
cultural lag *206*	culture system *225*	subsystem *200*	technological
cultural landscape *199*	culture traits *197*	innovation *206*	subsystem *200*
culture *196*	diffusion *208*	linguistic family *214*	

For Review

1. What is included in the concept of *culture?* How is culture transmitted? What personal characteristics affect the aspects of culture that any single individual acquires or fully masters?

2. What is included in the *technological subsystem* of culture? In the *sociological?* In the *ideological?* Are these subsystems fully distinct and separable? Explain.

3. In what ways is the technological level that has been achieved by a society an indicator of culture, as the term is employed in this chapter? Briefly discuss why technology is unsatisfactory as a prime determinant of cultural differentiation.

4. What are the common characteristics of *culture hearths?* What, in the cultural geographic sense, is meant by *innovation?* By *diffusion?* What forms of diffusion can you name and describe? Give some examples of how distance, the physical environment, and people themselves act as barriers to diffusion.

5. In what ways are *culture traits* part of and distinct from *culture complexes?* What kind of region does a *culture realm* represent? What, in your opinion, is the validity (if any) of the culture realm concept in the modern world?

6. Why might language be considered the dominant differentiating element of culture separating different societies?

7. In what ways may religion affect other cultural traits of a society? In what cultures or societies does religion appear to be a growing influence? What might be the economic developmental consequences of that growth?

8. How does *acculturation* occur? Is *ethnocentrism* likely to be an obstacle in the acculturation process? How do acculturation and *assimilation* differ?

Suggested Readings

Broek, Jan O.M., and John W. Webb. *A Geography of Mankind,* 3d ed. New York: McGraw-Hill, 1982.

Brown, Lawrence A. *Innovation Diffusion: A New Perspective.* New York: Methuen, 1982.

Clarke, Colin; David Ley; and Ceri Peach (eds.). *Geography and Ethnic Pluralism.* London: George Allen and Unwin, 1984.

DeBlij, Harm J., and Peter O. Muller. *Human Geography: Culture, Society, and Space,* 3d ed. New York: John Wiley, 1986.

International Bank for Reconstruction and Development/ The World Bank. *World Development Indicators,* Annex to *World Development Report.* New York: Oxford University Press, annual.

Jackson, W. A. Douglas. *The Shaping of Our World.* New York: John Wiley, 1985.

Jordan, Terry G., and Lester Rowntree. *The Human Mosaic: A Thematic Introduction to Cultural Geography,* 4th ed. New York: Harper & Row, 1986.

Rooney, John F., Jr., Wilbur Zelinsky, and Dean R. Louder (eds.). *This Remarkable Continent: An Atlas of United States and Canadian Society and Cultures.* College Station, Tex.: Texas A&M University Press, 1982.

7 Political Geography

They met together in the cabin of the little ship on the day of the landfall. The journey from England had been long and stormy. Provisions ran out, a man had died, a boy had been born. Although they were grateful to have been delivered to the calm waters off Cape Cod that November day of 1620, their gathering in the cramped cabin was not to offer prayers of thanksgiving but to create a political structure to govern the settlement they had come to establish. The Mayflower Compact was an agreement among themselves to "covenant and combine our selves togeather into a civill body politick . . . to enacte, constitute, and frame such just and equall lawes, ordinances, acts, constitutions, and offices . . . convenient for ye generall good of ye Colonie. . . ." They elected one of their company governor, and only after those political acts did they launch a boat and put a party ashore.

The land they sought to colonize had for more than 100 years been claimed by the England they had left. The New World voyage of John Cabot in 1497 had invested their sovereign with title to all of the land of North America and a recognized legal right to govern his subjects dwelling there. That right was delegated by royal patent to colonizers and their sponsors, conferring upon them title to a defined tract and the right to govern it. Although the Mayflower *settlers were originally without a charter or patent, they recognized themselves as part of an established political system. They chose their governor and his executive department annually by vote of the General Court, a legislature composed of all freemen of the settlement.*

As the population grew, new towns were established too distant for their voters to attend the General Court. By 1636 the larger towns were sending representatives to cooperate with the executive branch in making laws. Each town became a legal entity, with election of local officials and enactment of local ordinances the prime purpose of the town meetings that are still common in New England today.

The Mayflower Compact, signed by 41 freemen as their first act in a New World, was the first step in a continuing journey of political development for the settlement and for the larger territory of which it became a part. From company patent to crown colony to rebellious commonwealth under the Continental Congress to state in a new nation, Massachusetts (and Plymouth Plantation) were part of a continuing process of governance of people in area—part, that is, of a dynamic political geography.

Political geographers analyze the organization and distribution of political phenomena in their areal expression. They seek to relate those concerns to other spatial evidences of culture, such as religion, language, and ethnicity. Nationality is a basic element in cultural variation among people, and political geography traditionally has had a primary interest in states: in the spatial patterns reflecting the exercise of central governmental control, in questions of boundary delimitation and effect, and in the economic and organizational viability of states.

Some definitions are in order. On the international scene, **state** refers to an independent political unit occupying a defined, permanently populated territory and having full sovereign control over its internal and foreign affairs. Not all recognized territorial entities are states. Antarctica, for example, has neither established government nor permanent population; it is, therefore, not a state. Nor are *colonies* or *protectorates* recognized as states. Although they have defined extent, permanent inhabitants, and—usually—some degree of separate governmental structure, they lack full control over all of their internal and external affairs. *Country* is sometimes used as a synonym for the territorial and political concept of "state." **Nation** refers to people, not to political structure. Nation is a cultural concept and implies a group of people occupying a particular region or area, bound together by a strong sense of unity arising from shared beliefs and customs. Language and religion may be unifying elements, but even more important are an emotional conviction of cultural distinctiveness and a sense of ethnocentrism.

The state is only one geographic expression of the legally bounded spaces within which humans organize themselves—or have organization imposed on them—to advance their individual and collective interests. As these interests characteristically have spatial roots reflecting different levels of administrative concern, from the local school district or voting precinct to supranational economic unions or military alliances, political-geographic inquiry takes place in a framework defined by scale. In this chapter, we consider some of the problems of defining political jurisdictions, asking why political power is fragmented and seeking the elements that lend cohesion to a political entity. We begin with local administrative organization and end with international political systems.

Political Systems

The ways in which we live our lives, conduct our daily affairs, and make plans for our future are affected by decisions made by public agencies. Where children go to school, how old one must be to vote or purchase alcoholic beverages, and how much we pay in taxes are examples of such decisions. In these American instances, the controlling decisions are made by many different organizations, ranging from the local school board through state and federal agencies to the Congress of the United States. Each system or organization operates within a carefully defined geographic area. Congress can make laws for the entire country, but a school board has the power to operate only within its school district.

Even these limited examples suggest that populations are associated with—or are effectively controlled by—two different classes of organizational entities. On the one hand are those, such as Congress, state legislatures, county governments, or city councils, that establish general laws or objectives that apply to the legally bounded territories under their control. In turn, legislatively enacted general bodies of law are converted into specific programs and detailed actions by the second class of control: elected or appointed administrative or regulatory agencies, such as the Interstate Commerce Commission, state departments of conservation, or local sanitary districts.

Particularly at the lower political-administrative levels, jurisdictional boundaries are largely determined by the stated objectives of the organizational units, and at all levels, they reflect spatial expressions of differing common interests. Political organizations are given the responsibility of identifying people's goals, which may be economic, social, educational, or military, and are endowed with the power to pursue them. If the goal is attained, such as the building of a dam, the organization may cease to exist. More often, however, the goals are long-run and may even be unattainable—for example, "to insure domestic tranquillity"—in which case the controlling political mechanism constitutes a continuing effort.

Political systems operate within specific geographic areas. The limits of those areas usually are not evident on the landscape. The boundaries of the wards in a city, the municipal limits, and even most state boundaries are not physical entities that can be seen or touched, and yet people have emotional feelings about many territorial units. In the United States, most people identify strongly with their town and their state. Most people have a feeling that they are in a "foreign," or at least a different, area when they cross from their own state into another.

Political entities purposely try to instill feelings of allegiance in their constituents, for such feelings give the system strength. People who have such allegiance are likely to accept the rules governing behavior in the area and to contribute to the decision-making process. The forces that promote cohesiveness in a region are examined later in this chapter. Let us look now at the problems encountered in delimiting boundaries on the local level.

Local Political Organization: The Districting Problem

There are more than 85,000 local governmental units in the United States. Slightly more than half of these are municipalities, townships, and counties. The remainder are school districts, water-control districts, airport authorities, sanitary districts, and other special-purpose bodies. Around each of these districts, boundaries have been drawn. Although the number of districts does not change greatly from year to year, many boundary lines are redrawn in any single year.

For example, rulings of the U.S. Supreme Court have led to the redrawing of thousands of attendance boundaries of school districts. The court's decision in 1954 in *Brown v. Board of Education of Topeka, Kansas,* held that the doctrine of "separate but equal" school systems for blacks and whites was unconstitutional because separate educational institutions were inherently unequal. Subsequent rulings broadened the decision to mean that no racially segregated school system is allowed, whether it occurs by design or inadvertently as the result of de facto segregation in housing. The boundaries of school attendance zones have had to be adjusted to ensure that schools will be racially integrated.

The Supreme Court's "one person, one vote" ruling in *Baker v. Carr* (1962) signified the end of overrepresentation of sparsely populated rural districts in state legislatures and the U.S. House of Representatives. It has led to the frequent adjustment of electoral districts within states and cities, so that they contain roughly equal numbers of voters. Reapportionment is occasioned by shifts in population, as areas gain or lose people. **Gerrymandering** describes the division of an area into voting districts in such a way as to give one political party an unfair advantage in elections, to fragment voting blocs, or to achieve other nondemocratic objectives. However, the drawing of boundaries is tremendously complicated, and it is not always clear what is "unfair."

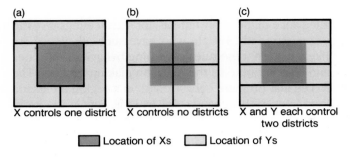

Figure 7.1

The problem of drawing boundaries. Xs and Ys might represent Republicans and Democrats, urban and rural voters, blacks and whites, or any other groups.

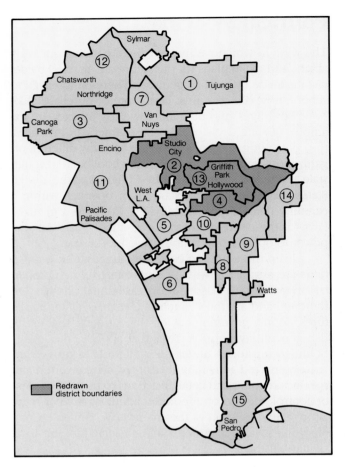

Figure 7.2

City Council districts in Los Angeles created by the 1982 redistricting plan had to be redrawn when it was ruled that they violated the Voting Rights Act by splitting the city's Hispanic population. Although Hispanics constituted more than 30% of the city's population, they were a majority, and elected a council member, in only one district, the 14th. In 1986 the City Council agreed to redraw the district boundary lines to create a second predominantly Hispanic council district and a third district in which Hispanics are likely to be a majority in a few years.

How boundaries are delimited depends on the desired ends. The power of a group of people may be maximized, minimized, or effectively nullified by the way in which the lines are drawn. Assume that X and Y represent two groups with different policy preferences. In Figure 7.1a, the Xs are concentrated in one district and have one representative of four; in Figure 7.1b, the Xs are dispersed, are not a majority in any district, and stand the chance of not electing any representative at all. The power of the Xs is maximized in Figure 7.1c, where they control two of the four districts.

Figure 7.1 depicts an idealized district, square in shape with a uniform population distribution and only two groups competing for representation. In actuality, municipal voting districts are oddly shaped to consider such factors as the city limits, current population distribution, historic settlement patterns, and transportation routes. In any large city, many groups vie for power. A U.S. Justice Department lawsuit filed successfully against the Los Angeles City Council claimed that the council's 1982 district boundary lines, shown in Figure 7.2, violated the Voting Rights Act because they diluted the voting strength of the city's Hispanic residents by dispersing them among several council districts. The council agreed to draft a redistricting plan, to take effect in 1987, although another ethnic political conflict is likely when the council redistricts itself after the 1990 census. As we noted, the way boundaries are delimited depends on the ends that are desired. Minorities seek representation in proportion to their numbers, so that they will be able to elect representatives who are concerned about and responsive to ethnic community issues. But ethnically homogeneous districts are impossible when—as in the case of Los Angeles—there are only 15 council members representing 3 million people and over 80 ethnic groups.

Boundary drawing is never easy, particularly when political groups want to maximize their representation. Furthermore, the boundaries that we may want for one set of districts may *not* be those that we want for another. For example, sewage districts must take natural drainage features into account, whereas police districts may be based on the distribution of the population or the number of miles of street to be patrolled, and school attendance zones must take into account the numbers of school-aged children and the capacities of individual schools.

Too Many Governments

If you are a property owner in Wheeling Township in the city of Arlington Heights in Cook County, Illinois, here's who divvies up your taxes: the city, the county, the township, an elementary school district, a high school district, a junior college district, a fire protection district, a park district, a sanitary district, a forest preserve district, a library district, a tuberculosis sanitarium district, and a mosquito abatement district.

Lest you attribute this to population density or the Byzantine ways of Cook County politics, it's not that much different elsewhere in Illinois—home to more governmental units than any other state in the United States. According to the latest figures from the U.S. Bureau of Census, there are 6,467 local government units in Illinois. Second-place Pennsylvania has 5,198 governments—1,269 fewer than Illinois.

Illinois has 102 counties, 1280 municipalities, 1434 townships, 1049 school districts, and 2602 single-function special districts ranging from Chicago's Metropolitan Sanitary District to the Caseyville Township Street Lighting District. Most of these governments have property-taxing power. Some also impose sales or utility taxes.

This proliferation is in part a historical by-product of good intentions. The framers of the state's 1870 constitution, wanting to prevent overtaxation, limited the borrowing and taxing power of local governments to 5% of the assessed value of properties in their jurisdictions. When this limit was reached and the need for government services continued to grow with population, voters and officials circumvented the constitutional proscription by creating new taxing bodies—special districts. Illinois' special districts grew because they could get around municipality debt limitations and because they could be fitted to service users without regard to city or county boundaries.

Critics say all these governments result in duplication of effort, inefficiencies, higher costs, and higher taxes. Supporters of special districts, townships, and small school districts argue that such units fulfill the ideal of a government close and responsive to its constituents.

Adapted from J. M. Winski and J. S. Hill, "Illinois: A Case of Too Much Government," *Illinois Business,* Winter 1985, pp. 8–12.

Ideally, political boundary making involves the practical application of geographic principles of regionalization, and geographers have been professionally involved in resolving delimitation problems. They were important contributing members of President Woodrow Wilson's staff after World War I, giving the American representatives at the Paris Peace Conference advice on boundary making in eastern Europe, where new nations were created from the defeated Austro-Hungarian Empire and from the western reaches of the former Russian Empire. A geographer was appointed special master by the federal district court in Seattle, Washington, in the early 1970s to reapportion congressional and legislative districts in the state of Washington when the two political parties in the state legislature were unable to agree on new district boundaries. These are just two examples of the real-life application of the skills of political and regional geographers.

Regional Organization: Political Fragmentation in the United States

The United States is subdivided into great numbers of political administrative units whose areas of control and influence are spatially limited. The 50 states are partitioned into more than 3000 counties ("towns" in some New England states and "parishes" in Louisiana), most of which are further subdivided into townships, each of which has a still lower level of spatial influence and governing power. This political fragmentation is further increased by the existence, within both urban and rural areas, of nearly innumerable special-purpose districts whose boundaries rarely coincide with the standard major and minor civil divisions of the nation, or even with each other. Each district represents a form of political allocation of territory to achieve a specific aim of local need or legislative intent.

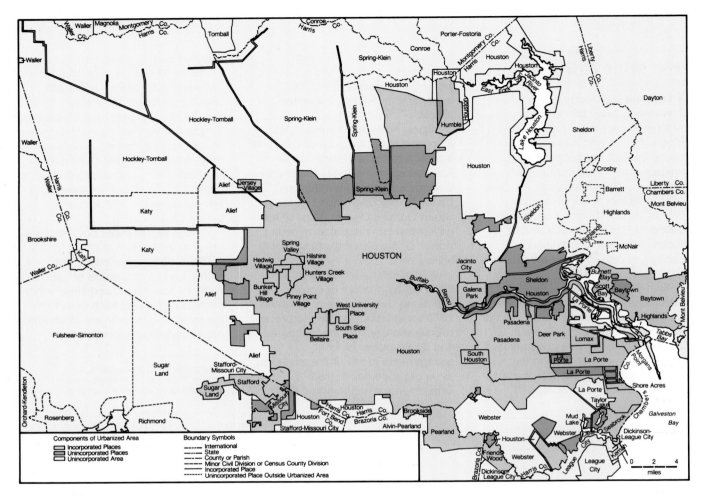

Figure 7.3

The complex intermixture of incorporated and unincorporated areas in the Houston metropolitan region. Larger than the state of Delaware, the city of Houston has jurisdiction over 2080 square miles (5400 km²) of territory.

Most Americans live in large and small cities. These, too, are subdivided, not only into wards or precincts for voting purposes but also into special districts for such functions as fire and police protection, water and electricity supply, education, recreation, and sanitation. These districts almost never coincide with one another, and the larger the urban area, the greater the proliferation of small, special-purpose governing and taxing units.

Our largest recognized urban areas are the 358 metropolitan statistical units of the Bureau of the Census in the United States and Puerto Rico, which may extend over several municipalities (the "central city" and its suburbs) and, often, several counties. (A review of the different categories of U.S. metropolitan areas will be found in Chapter 11.) Figure 7.3 shows one such extended "city." Some, such as the New York consolidated metropolitan statistical area, even straddle state lines.

Counting all of the general- and special-purpose legal units, the Chicago metropolitan area alone has over 1000 governments.

The proliferation of governmental units in small areas results in a complex mosaic of controls that often hinders achievement of the very aim behind their creation. The theory is that if control is kept at the local level, the citizens are assured a voice in government. Yet even the best-informed citizens cannot follow the workings of all of the many boards and agencies that are supposed to represent them. They may be able to keep up with the local transit and planning commissions, but may have to neglect the city council, the board of health, and the school board, to name only a few. Thus, true understanding and extensive participation are impossible, one reason for the low turnout of voters in most local elections.

The existence of such a great number of districts in metropolitan areas causes inefficiency in public services and hinders the orderly use of space. Zoning ordinances, for example, are determined by each municipality. They are intended to allow the citizens to decide how land is to be used and, thus, are a clear example of the effect of political decisions on the division and development of space. Zoning policies dictate the areas where industries may and may not be located, and whether they will be light or heavy industries; the sites of parks and other recreational areas; the location of business districts; and what kind of housing is allowed and where it can be built. Unfortunately, in large urban areas, the efforts of one community may be hindered by the practices of neighboring communities. Thus land zoned for an industrial park or perhaps a slaughterhouse in one city may abut land zoned for single-family residences in an adjoining municipality. Each district pursues its own interests, which may not coincide with those of the larger region.

Inefficiency and duplication of effort characterize not just zoning but many of the services provided by local governments. The efforts of one community to avert air and water pollution may be, and often are, counteracted by the rules and practices of other towns in the region, although state and national environmental protection standards are now reducing such potential conflicts. Social as well as physical problems spread beyond city boundaries. Thus nearby suburban communities are affected when a central city lacks the resources to maintain high-quality schools or to attack social ills. The provision of health care facilities, electricity and water, transportation, and recreational space affects the whole region and, many professionals think, should be under the control of a single unified metropolitan government ("Unigov" or metro-government).

Such consolidation of control has long been proposed, but its introduction is prohibited by law in some states and in all instances involves the surrender of jealously protected local authority and autonomy to a larger unit. In addition, the removal of decision making farther from the citizens can have negative side effects, increasing feelings of isolation from government. A variety of approaches to regionwide administration and planning has been tried in the United States. Nashville—Davidson County, Tennessee, and Indianapolis—Marion County, Indiana, are among the areas where city-county consolidation has occurred. A different system operates in Metropolitan Dade County, Florida, which includes Miami and 26 other municipalities. There, local governments have retained some powers (such as the provision

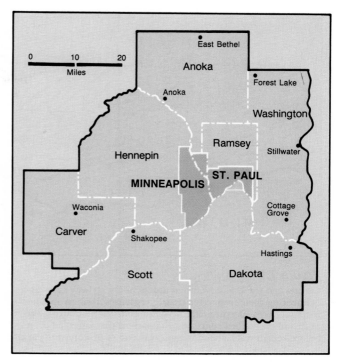

Figure 7.4

The Metropolitan Council of the Twin Cities Area, created by the Minnesota state legislature in 1967, is responsible for the orderly development of a seven-county area that contains over 300 local units of government, including Minneapolis and St. Paul. The council coordinates such services as sewers and solid-waste disposal and the development of highways and parks. It reviews all land use plans of individual governments and can delay their implementation for 60 days while it attempts to coordinate them with regional goals.

of police and fire protection), but others (like land use planning and tax collection) have been transferred to the county. Yet another approach to regionalism is represented by the Metropolitan Council of the Twin Cities Area. It is charged with coordinating the planning and development of a 2800-square-mile (7300–km²) region (Figure 7.4), but it does not consolidate or federate the local governments, and its members are appointed rather than elected.

Problems of local governmental control are not unique to the United States. In every metropolitan area, decisions must be made about what services are to be provided, how and where they will be provided, and who is to do the providing. In most countries, however, these decisions are not left to the smallest of local units. Central authorities have the power to decide when a reorganization of the political system is necessary and to redraw the boundaries of regions.

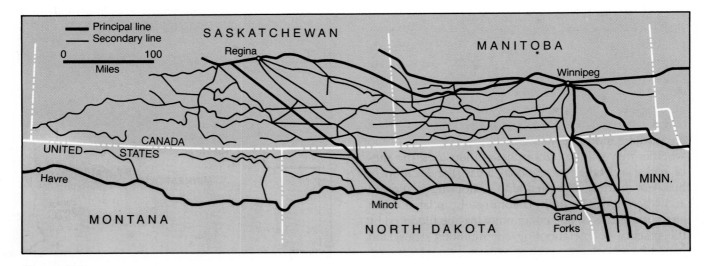

Figure 7.5

Canada and the U.S. developed independent railway systems connecting their respective prairie regions with their separate national cores. Despite extensive rail construction during the 19th and early 20th centuries, the pattern that emerged even before recent track abandonment was one of discontinuity at the border. Note how the political boundary reduced spatial interaction between adjacent territories. Many branch lines approached the border, but only 8 crossed it. In fact, for over 300 miles (480 km), no railway bridged the boundary line.

In Great Britain, a royal commission was appointed in 1974 to study the problem of the relationship of the over 1000 local districts that existed. Each district was directly responsible to the central government. The commission redrew county lines and gave control over certain functions—such as education, planning, and transportation—to the counties. Within the counties, district councils were established that were responsible for providing other services, including housing and garbage collection. Believing that certain functions demanded a higher level of organization, the commission created a limited number of regional authorities with control over health services and water supply. The commission's plan attempted to strike a balance between centralization and decentralization of responsibility and control, based on the recognition that some functions should be controlled and provided locally, and others areawide.

Regional arrangements are one approach to solving some of today's governmental problems by attacking areawide concerns, equalizing both the tax burden and the delivery of services. A second type of solution is to decentralize political authority to allow the citizens greater control over, or input into, governmental decisions that affect them and to bring services and government closer to the people. New York City, Chicago, and Los Angeles are among the cities that have experimented with decentralization of schools and with community involvement in decision making. A number of large cities have established "little" or neighborhood city halls to improve communication with the citizens, to increase the effectiveness of services, and to reduce neighborhood tensions. A third type of decentralization is the citizens' advisory commission or review board. Usually appointed rather than elected, the members of such boards may represent various segments of the community or specific groups (minorities, young people, environmentalists, and so on); by reflecting the community's wishes, an advisory board may effectively influence policy.

The Nation-State

Although the local control of government in the United States creates areal differences in the behavior of the citizens, the most profound contrasts in culture tend to occur between, rather than within, countries. The border between the United States and Canada has more impact both on people's lives and on the cultural landscape than do state boundaries in the United States (Figure 7.5). Similarly, people separated by the Mexico—United States border lead vastly different lives. San Diego resembles Los Angeles more than it does Tijuana, Mexico, although Tijuana is also a large city and is only 15 miles (25 km) from San Diego.

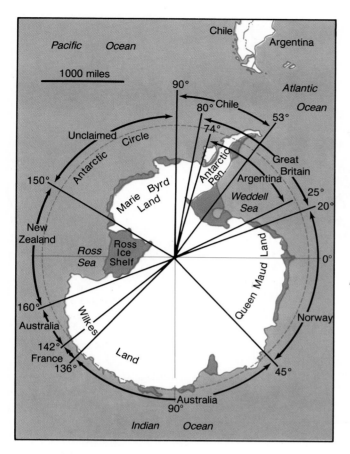

Figure 7.6
Territorial claims in Antarctica. Although 7 countries claim
sovereignty over portions of Antarctica, and 3 of the claims
overlap, the continent has no permanent inhabitants. In 1961, 12
governments, including the 7 claimants, signed the Antarctic
Treaty, which freezes the territorial disputes, bans military use of
the continent, and designates it as a theater of scientific research.
The discovery of mineral deposits and offshore food and fuel
resources may test the strength of the treaty, which by 1986 had
accumulated a total of 18 signatories.

The Idea of the Nation-State

The word "state," we have seen, denotes territory, and
the word "nation" implies a separate people. It follows,
therefore, that the composite term **nation-state** refers
to a state whose territory is identical to that occupied
by a particular nation or people. For the nation-state,
nationalism—the concept of allegiance to a single
country—is the cohesive emotion unifying its inhabitants and providing a sense of identity akin to the ethnocentrism of cultural groups.

One of the most significant elements in cultural geography is the nearly complete division of the earth's land
surface into separate national units, as shown on the world
political map inside the front cover. Even Antarctica is
subject to the rival territorial claims of seven countries,
although these claims, with their implication of sovereign control, have not been pressed because of the Antarctic Treaty of 1961 (Figure 7.6). A second element is
that this division into country units is relatively recent.
Although nation-states have existed since the days of early
Egypt and Mesopotamia, only in the last century has the
world been almost completely divided into independent
governing entities. Now people everywhere accept the
idea of the state, and its claim to sovereignty within its
borders, as normal.

The idea of the modern state was developed by political philosophers in the 18th century. Their views gave
rise to the concept that people owe allegiance to a state
and the people it represents rather than to its leader, such
as a king or a feudal lord. The new concept coincided in
France with the French Revolution and spread over
western Europe, to England, Spain, and Germany. The
term *nation* implies a feeling of loyalty and attachment
to a country and a willingness to abide by its rules. It also
implies a commonality of culture, history, purpose, and
territory. Many of the countries that were created in Europe after World War I resulted from an attempt to redraw boundaries that coincided with such feelings.

Many states are the result of European expansion
during the 17th, 18th, and 19th centuries, when much of
Africa, Asia, and the Americas was divided into colonies.
Usually these colonial claims were given fixed and described boundaries where none had earlier been formally defined. Of course, precolonial native populations
had relatively fixed home areas of control claimed by
themselves (and recognized by neighboring groups),
within which there was recognized dominance and
border defense and from which there were, perhaps, raids
of plunder or conquest of neighboring "foreign" territories. Beyond understood tribal territories, great empires arose, again with recognized outer limits of
influence or control: Mogul and Chinese; Benin and Zulu;
Incan and Aztec. Upon them, where they still existed,
and upon the less formally organized spatial patterns of
effective tribal control, European colonizers imposed
their arbitrary new administrative divisions of the land.

Figure 7.7

The discrepancies between tribal and national boundaries in Africa. Tribal boundaries were ignored by European colonial powers; the result has been significant ethnic diversity in nearly all African countries.

In fact, tribes that had little in common were often joined in the same colony (Figure 7.7). The new divisions, therefore, were not usually based on meaningful cultural or physical lines. Instead, the boundaries simply represented the limits of the colonizing empire's power.

As these former colonies have gained political independence, they have retained the idea of the state. They have generally accepted—in the case of Africa, by a conscious decision to avoid precolonial territorial or ethnic claims that could lead to war—the borders established by their former European rulers. The problem that many of the new countries face is to develop feelings of nationalism and loyalty to the state among their arbitrarily associated citizens. Zaire, the former Belgian Congo, contains more than 250 frequently antagonistic tribes. Only if past tribal animosities can be converted into an overriding spirit of national cohesion will countries like Zaire truly be nation-states.

State Cohesiveness

What are those forces that can make a state cohesive, that enable it to function and give it strength? We have already mentioned one of them: a feeling of collective distinction from all other peoples and lands, that is, an identification with the state and an acceptance of national goals that is called nationalism. A sense of unity binding the people of a state together is necessary to overcome the divisive forces present in most societies.

National Symbols

Iconography is the study of the symbols that bind a people together. National anthems and other patriotic songs, flags, national flowers and animals, colors, and rituals are all developed as symbols of a state in order to attract allegiance (Figure 7.8). They ensure that all citizens, no matter how diverse the population may be, will have at least these symbols in common. They impart a

Figure 7.8
The ritual of the pledge of allegiance is just one way in which schools in the United States seek to instill a sense of nationalism in students.

sense of belonging to a political entity called, for example, Japan or Canada. In some countries, certain documents, such as the Magna Charta in England or the Declaration of Independence in the United States, serve the same purpose. Symbols such as these are significant insofar as ideologies and beliefs are an important aspect of culture. When a culture is very heterogeneous, composed of people with different customs, religions, and languages, belief in the national unit can help weld them together.

Institutions That Promote Nationalism

Schools, particularly elementary schools, are among the most important of the institutions that promote national feelings. Children learn the history of their own country and relatively little about other countries. Schools are expected to inculcate the society's goals, values, and traditions; allegiance to the state is accepted as the norm. As a rule, schools teach youngsters to identify with their country rather than with the world or with humanity as a whole.

Other institutions that promote nationalism are the armed forces and, sometimes, a state church. The military organization fulfills a primary state goal: the provision of security, both internal and external. A high percentage of most states' budgets is spent to secure such protection. The armed forces are of necessity taught to identify with the state. They see themselves as protecting the state's welfare from what are perceived to be its enemies.

In some countries, the religion of the majority of the people may be designated a state church. In such cases the church sometimes becomes a force for cohesion, helping to unify the population. This is true of the Roman Catholic church in the Republic of Ireland, Islam in Pakistan, and Judaism in Israel. In countries like these, the religion and the church are so identified with the state that belief in one is transferred to allegiance to the other.

The schools, the armed forces, and the church are just three of the institutions that teach people what it is like to be members of a state. These institutions operate primarily on the level of the ideological system. By themselves, they are not enough to give cohesion, and thus strength, to a state.

Organization

A second necessity for cohesion, in addition to the feeling of nationalism, is effective organization of the state. Can it provide security from external aggression and internal conflict? Are its resources distributed and allocated in such a way as to promote the economic welfare of its citizens? Are there institutions that encourage consultation and the peaceful settlement of disputes? How firmly established are the rule of law and the power of the courts? Is the system of decision making responsive to the people's needs?

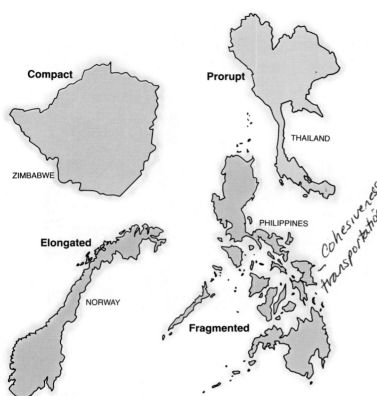

Figure 7.9
Shapes of states. The national territories are not drawn to scale.

Let us look at just one set of these questions, that dealing with the state and the economy. Today as never before, the promotion of economic well-being is seen as a task of government. People expect the government to act in such a way as to enhance their welfare. Although the degree to which the state is involved in the economy varies from country to country, from the public ownership of the means of production to private ownership, in every country there is some governmental role. Tariff barriers restrict trade, protecting home industries from foreign competition; direct aid is given to certain industries in many countries; some nations may try to attract foreign investment, whereas others strictly control or discourage it. Governments extend subsidies to those engaged in certain kinds of agriculture. They may sponsor programs aimed at increasing agricultural yields or at supporting basic industry.

The best efforts of government, however, are conditioned by certain elements over which it has little control. One of these elements is the physical makeup of the territory of the nation. The area that the state occupies

may be large, as is true of China, or small, as is true of Liechtenstein. The larger in area the state is, the greater is the chance that it will have resources such as fertile soil and minerals from which it can benefit. On the other hand, a very large country may have vast areas that are inaccessible, sparsely populated, and hard to govern. Thus *size* alone is not critical in determining a country's stability and strength.

Potentially important is a nation's *relative location*. If it is located on major trade routes, it will benefit materially. It will also be exposed to the diffusion of new ideas and new ways of doing things.

A country's *shape* may affect its patterns of organization. The most efficient form would be a circle, assuming no major topographic barriers, with the capital located in the center. In such a country, all places could be reached from the center in a minimal amount of time and with the least expenditure for roads, railway lines, and so on. It would also have the shortest possible borders to defend. Uruguay and Poland have roughly circular, **compact** shapes (Figure 7.9). **Prorupt** states are nearly compact but possess one or sometimes two narrow extensions of territory. Burma and Thailand have such shapes.

The least efficient shape is represented by countries like Norway and Chile, which are long and narrow. In such **elongated** states, the parts of the country far from the capital are likely to be isolated because great expenditures are required to link them to the core. These countries are also likely to encompass more diversity of climate, resources, and peoples than compact states, perhaps to the detriment of national cohesion or, perhaps, to the promotion of economic strength.

A fourth class of shapes is called **fragmented.** This class includes countries composed entirely of islands (e.g., the Philippines and Indonesia), countries that are partly on islands and partly on the mainland (Italy and Malaysia), and those that are chiefly on the mainland but whose territory is separated by another state (the United States). Pakistan was a fragmented nation until 1971, when the eastern part of the country broke away from the west and declared itself the independent state of Bangladesh.

The *location of the capital* city, the seat of central authority, sometimes becomes a matter of concern. All other things being equal, a capital located in the center of the nation provides equal access to the government, facilitates communication to and from the capital, and enables the government to exert its authority easily. Many capital cities, like Washington, D.C., were centrally located when they were designated as capitals but lost their centrality as the state expanded.

The Ministates

Totally or partially autonomous political units that are small in area and population pose some intriguing questions. Should size be a criterion for statehood? What is the potential of ministates to cause friction among the major powers? Under what conditions are they entitled to representation in international assemblies like the United Nations?

About half the world's independent countries contain fewer than 5 million people; of these, 37 have under 1 million. Nauru has about 8000 inhabitants on its 8.2 square miles. Many are insular territories located in the West Indies and the Pacific Ocean (such as Grenada and Tonga Islands), but Europe (Vatican City and Andorra), Asia (Bhutan and Singapore), and Africa (Gambia and Equatorial Guinea) have their share.

Many statelets are vestiges of colonial systems that no longer exist. Some of the small states in West Africa and on the Arabian peninsula fall into this category. Others, such as Mauritius, served primarily as refueling stops on transoceanic voyages. However, some occupy strategic locations (such as Bahrain, Malta, and Singapore), and others contain valuable minerals (Kuwait, Nauru, and Trinidad). The possibility

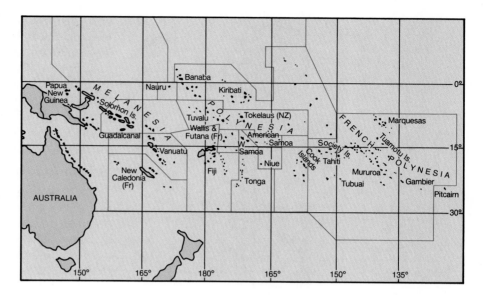

of claiming 200-mile-wide zones of sea adds to the attraction of yet others.

The smallness of the regions, combined with their strategic or economic value, can invite unwanted attention from larger neighbors. The 1982 war over the Falkland Islands demonstrated the ability of such areas to bring major powers into conflict and to receive world attention out of all proportion to their size and population. Timor and Macao are among the areas whose control may be disputed in the future.

The proliferation of tiny countries raises the question of their representation and their voting weight in international assemblies.

Should there be a minimum size necessary for participation in such bodies? Should nations receive a vote proportional to their population? The influence of the United States and other major powers in the United Nations has already been eroded by the small nations. Although the United States pays 25% of the UN budget and has about one and one-half times the population of all the small nations combined, its vote can be balanced by that of any of them. The fact that as many as 50 additional territories may gain independence in the foreseeable future underscores the seriousness of this concern.

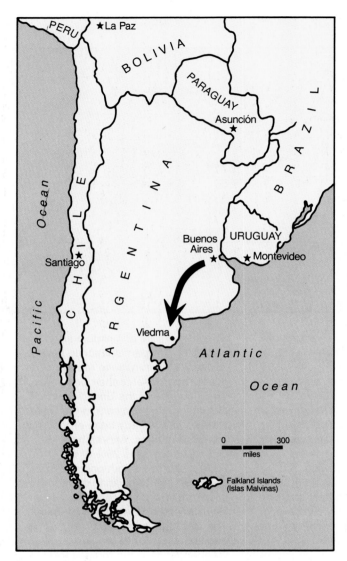

Figure 7.10
The relocation of Argentina's capital city from Buenos Aires to Viedma was proposed in 1986 as a way of creating a more even national distribution of government resources and attention.

Some capital cities have been relocated, at least in part to achieve the presumed advantages of centrality. A **forward-thrust capital** city is one that has been deliberately sited in a state's frontier zone, signaling the country's interest in extending its influence. In Turkey, the capital was moved to Ankara from Istanbul; in Brazil, to Brasília from Rio de Janeiro. In 1986 Argentina's president proposed moving the capital from Buenos Aires to a more central location (Figure 7.10; see also "Transferring the National Capital," Page 245).

It should be noted that the shape of a governing unit and the location of its capital affect political units other than nation-states. Thus states or even municipalities that have elongated or prorupt shapes experience the administrative difficulties associated with such shapes. And capital cities in compromise locations exist all over the United States. In Pennsylvania, Ohio, Michigan, and Illinois, the capitals were moved from previous sites to more central locations. For over 20 years, Alaskans have debated proposals to move the capital from Juneau closer to Anchorage, the center of population and commerce.

Transportation and Communication

A state's transportation network fosters political integration. When transportation brings about interaction between areas, the people may be tied economically and socially to each other. The role of a transportation network in uniting a country has been recognized since ancient times. The saying that all roads lead to Rome had its origin in the impressive system of roads that linked Rome to the rest of the empire. Centuries later, a similar network was built in France, linking Paris to the various departments of the country. Often the capital city is better connected to other cities than the outlying cities are to one another. In France, for example, it can take less time to travel from one city to another by way of Paris than by direct route.

Roads and railroads have played a historically significant role in promoting political integration. In the United States and Canada, they not only opened up new areas for settlement but increased interaction between rural and urban areas. The Soviet Union anticipates comparable benefits of economic, social, and political interaction and territorial development with the completion through Siberia of the Baikal–Amur Mainline Railroad (BAM). The cost—some $38 billion—of the 2000-mile (3200-km) project (Figure 7.11) is expected to yield great returns by tapping a development area of 600,000 square miles (1.5 million km²) with some of the world's richest deposits of minerals and timber, by providing access to a planned 13 major new industrial centers, and by establishing a militarily secure link between the western and the extreme far-eastern sections of the USSR.

Because transportation systems play a major role in a state's economic development, it follows that the more economically advanced a country is, the more extensive its transport network is likely to be. By fostering interdependence between regions, a transport network keeps regions from having to be self-sufficient. Each can specialize in the production of the goods and services for

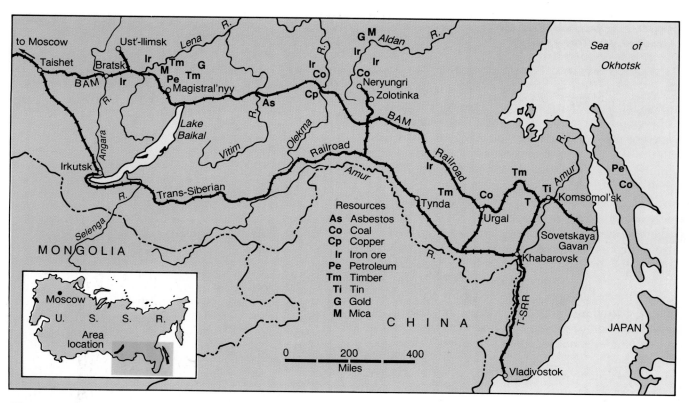

Figure 7.11

The Baikal-Amur Mainline Railroad will be a vital link in the long-term development of Siberian resources undertaken by the Soviet Union. Although a ceremonial golden spike marking official completion of the project was driven on October 1, 1984, years of tunnel work still will be required before the route is fully operational.

which it is best suited. Specialization results in an overall increase in productivity. The concept of *comparative advantage,* to be discussed in Chapter 9, helps to explain how this system works. At the same time, the higher the level of development, the more money there is to be spent on building transport routes. In other words, the two feed on one another.

Transportation and communication are encouraged within a state and are curtailed or at least controlled between states. The mechanisms of control include restrictions on trade through tariffs or embargoes, legal barriers to immigration and emigration, and limitations on travel through passports and visa requirements. A pointed illustration is the Australian railroad system (Figure 7.12). Until their federation in 1901, the states of Australia were independent of one another. As competitors for British markets, they had no desire to foster mutual interaction. Different rail gauges, rather than a single, standard gauge, were used, so that people and goods traveling between states had to change trains at the border.

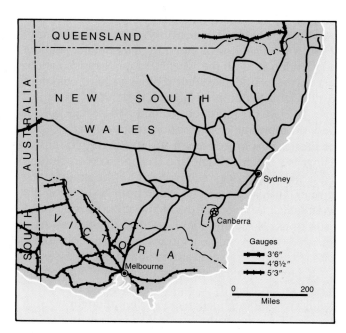

Figure 7.12

Railway gauges in eastern Australia. Gauges of different widths were established to hinder interaction between the states of Australia, which until federation in 1901 owed primary allegiance to Great Britain rather than to one another.

Ethnic Separatism and Intrastate Conflict

The concept of the state and the sense of nationalism that overrides small-group consciousness have suffered erosion and have been subject to compromise in all parts of the world in recent years. Since 1945 ethnic-group awareness and identification with a region rather than with the state have created currents of unrest and division, threatening established national patterns of cohesion. In Western Europe alone, five countries (the United Kingdom, France, Belgium, Italy and Spain) are affected by **ethnic separatism**—that is, each has separatist political movements demanding regional autonomy (Figure 7.13). Of course, ethnoregional loyalties long antedate 1945, having, in fact, presented problems for almost all European states at one time or another. Czechoslovakia, Greece, Norway, Finland, Estonia, and many of the Balkan countries (such as Bulgaria, Romania, and Yugoslavia) housed autonomist movements that were successful in achieving independence, etiher permanently or temporarily.

The separatist movements now active in Western Europe reject the existing sovereign state in which they have been included, often involuntarily, and claim to be the core of a separate national group. Their basic demand is for **regional autonomy**—usually in the form of self-government or "home rule" rather than complete independence. Both the Welsh and the Scottish nationalist parties would like their regions to have commonwealth status within the United Kingdom, for example.

The two preconditions common to all regional autonomist movements are *territory* and *nationality*. First, the group must be concentrated in a core region that it claims as a national homeland. Second, certain cultural characteristics must provide a basis for the group's perception of separateness and cultural unity. These might be language, religion, or distinctive group customs and institutions. Normally, these cultural differences have persisted over several generations and have survived despite strong pressures toward assimilation. As discussed in Chapter 6, both language and religion promote feelings of group identity at the same time that they foster exclusivity. Although language is often thought of as an essential basis for nationhood, it is not necessarily so; language is of little consequence to Scottish nationalists, for example. The regional language may be spoken by a majority, as in Catalonia, or by a minority, as in Wales (21%) and Brittany (10%). In the latter two cases, saving

Figure 7.13

Regions in Western Europe demanding autonomy. Each was once an independent entity.

the language is a priority of the separatists. In addition to language and religion, separate institutions or the memory of them may form a cultural rallying point for the group. Scotland's distinctive legal system, Catalonia's parliament, and the Basques' ancient privileges of tax collecting are examples of such systems.

Other characteristics common to many separatist movements are a *peripheral location* and *social and economic inequality*. As Figure 7.13 indicates, the troubled regions in Western Europe tend to be peripheral, and their location away from the seat of central government engenders feelings of alienation and neglect. Second, the dominant culture group is often seen as an exploiting class that has suppressed the local language, controlled access to the civil service, and taken more than its share of wealth and power. Poorer regions like Brittany, Wales, and Corsica complain that they have lower incomes and greater unemployment than prevail in the sovereign state, and that "outsiders" control key resources and industry.

Transferring the National Capital

President Raúl Alfonsín of Argentina recently announced a plan to move the nation's capital from Buenos Aires to the small city of Viedma, about 500 miles to the south. . . . The proposed transfer highlights the recurrent puzzle of why governments throughout history have felt compelled to uproot themselves and move elsewhere.

Argentina's national government is presently entrenched in Buenos Aires, an urban agglomeration of over 10 million people. Viedma, in contrast, has a population of about 34,000 (less than the number of government employees in just one ministry). . . . The move to Viedma, according to Alfonsín, will create a more even national distribution of resources and government attention.

Alfonsín's proposal has a number of historical precedents, including capital transfers in Brazil, Australia, Pakistan, and the United States. Each of the precedents sheds some light on Alfonsín's decision. . . .

Thirty years ago President Kubitschek of Brazil acted upon a constitutional provision for relocating the capital to the country's uninhabited interior and away from the urban sprawl of Rio de Janeiro and São Paulo. Brasília . . . is now the center of national government activity and . . . has established itself as an urban center for Brazil's vast interior.

Australia's capital was established at its present site as a result of rivalry between the country's two largest cities, Sydney and Melbourne. Canberra, located roughly halfway between these two cities, was intended as a demonstration of the national government's commitment to unify Australia's independent-minded states.

Pakistan shifted its capital from Karachi to Islamabad in 1965. . . . Islamabad . . . was to be the capital of a new nation formed from contiguous Muslim-majority districts. . . . The capital transfer away from the hot, humid climate of the coast and to the country's cultural heartland also served to emphasize Pakistan's interest in the disputed northern region.

In the early years of the United States, the founding fathers decided to move their capital from New York City to a more centralized location to serve better all of the 13 states and to separate the capital from the administration of individual state governments. . . .

A general look at motives for capital transfers reveals several common themes. First, a capital transfer represents a "fresh start" for a national government concerned with establishing its own mark on a country's future rather than simply continuing previous institutions. For some developing countries, a new, interior capital represents a physical break from a colonial heritage which was centered around the port cities (where local resources were exported to the "Mother Country"). The capital is often moved to a more centralized hinterland location to demonstrate the government's commitment to develop all parts of the country equally.

Abridged from William B. Wood, "Transferring the National Capital," *Focus*, Summer 1986, pp. 31–32.

Not all the areas shown in Figure 7.13 are poor; the Basque country, Catalonia, and Scotland are relatively rich regions where the separatists believe that they could exploit their resources for themselves and do better without the constraints imposed by the central state.

Analysts have suggested a number of factors to account for the resurgence of separatism in Western Europe. First, because the impact of industrialization and modernization tends to be uneven, areas that are not favored use nationalist demands to protest their position. Second, decolonization has reduced the prestige and the power of many European states since 1945, inadvertently providing regions with successful examples of anticolonial nationalism. Finally, recent years have seen a general widespread reaction against the growth of centralized government, which is seen as intruding into too many aspects of daily life.

Ethnic separatist movements do not affect only Western Europe. Scarcely a week goes by without minority agitation in some part of the world. The Basques of Spain and the Bretons of France have their counterparts in the Arabs in Israel, the Sikhs in India, the Chinese in Malaysia, and many others. For over a decade, for example, the island nation of Sri Lanka has been wracked by a bitter and bloody ethnic conflict between a minority Tamil community and a government dominated by the majority Sinhalese population. With a population of about 12 million, the Sinhalese are predominantly Buddhist, while the approximately 3 million Tamils are mostly Hindu and Christian. The goal of the Tamil guerrillas is to create an independent state in the northeast section of the country; more moderate Tamils desire autonomy for a Tamil province in the north and some sort of power

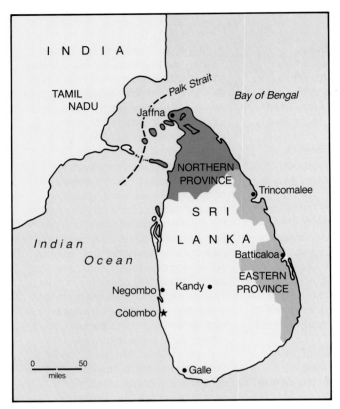

Figure 7.14
Traditional Tamil areas in Sri Lanka. Although Tamils are not a majority in the eastern province, militant Tamils demand a "homeland" embracing the northern and eastern provinces.

sharing in the ethnically mixed eastern areas (Figure 7.14). The situation is complicated by the fact that the south Indian state of Tamil Nadu, the home of over 50 million Tamils, has received some 100,000 Tamil refugees from Sri Lanka and is the base of Tamil guerrillas. Ending the insurgency in Sri Lanka is thus likely to require the cooperation of India.

Boundaries: The Limits of the State

We noted at the beginning of this discussion of the nation-state that the earth's land surface has now been divided among the various nation-states. They are separated from one another by international *boundaries,* or lines that establish the limit of each state's jurisdiction and authority. Boundaries indicate where the sovereignty of one state ends and that of another begins. Within its own bounded territory, a state administers laws, collects taxes, provides for defense, and performs other such governmental functions. Thus, the location of the boundary determines the kind of money people in a

given area use, the legal code to which they are subject, the army they may be called upon to serve in, and the language and perhaps the religion children are taught in school. These examples suggest how boundaries serve as powerful reinforcers of cultural variation over the earth's surface.

Before boundaries were delimited, nations or empires were likely to be separated by *frontier zones.* Although ill defined and fluctuating, these areas marked the effective end of a state's authority. They often were uninhabited or only sparsely populated and were liable to change with shifting settlement patterns. In North America, the frontier moved westward as pioneers continued to push back the limits of settlement. Many present-day boundaries lie in former frontier zones, and in that sense the boundary line has replaced the broader frontier as a marker of a state's authority.

Classification of Boundaries

Geographers have traditionally distinguished between "natural" and "artificial" boundaries. **Natural (or physical) boundaries** are those based on recognizable physiographic features such as mountains, rivers, and lakes. Although they might seem to be attractive as boundaries because they actually exist in the landscape and are visible dividing elements, many natural boundaries have proved to be unsatisfactory. That is, they do not effectively separate states.

Many international boundaries lie along mountain ranges, for example in the Alps, Himalayas, and Andes, but while some have proved to be stable, others have not. Mountains are rarely total barriers to interaction. They are crossed by passes, roads, and tunnels. High pastures may be used for seasonal grazing, and the mountain region may be the source of water for hydroelectric power. Nor is the definition of a boundary along a mountain range a simple matter. Should it follow the crests of the mountains or the *water divide* (a ridge dividing two drainage areas)? The two do not always coincide.

Rivers can be even less satisfactory as boundaries. In contrast to mountains, rivers foster interaction. River valleys are likely to be agriculturally or industrially productive, and to be densely populated. Nor is it clear precisely where the boundary should lie: along the right or left bank, along the center of the river, or perhaps along the middle of the navigable channel? Any decision tends to complicate the use of the river by people of the states involved. Furthermore, what happens to the boundary when the river changes its course, floods, or dries up?

The alternative to natural boundaries are **artificial or geometric boundaries.** Frequently delimited as sections of parallels of latitude or meridians of longitude, they are found chiefly in Africa, Asia, and the Americas.

Figure 7.15
Like its historical predecessors, Hadrian's
Wall and the Great Wall of China, the
Berlin Wall is a demarcated boundary.
Unlike them, it cuts across a large city.
The Berlin Wall is a *subsequent* and
superimposed boundary.

The western portion of the United States–Canada boundary line, which follows the 49th parallel, is an example of a geometric boundary. Many were established when the areas in question were colonies, the land was only sparsely settled, and detailed geographic knowledge of the border region was lacking.

Boundaries can also be classified according to when they were laid out with respect to the development of the cultural landscape. An **antecedent boundary** is one drawn across an area before it is well populated, that is, before most of the cultural landscape features developed. To continue our earlier example, the western portion of the United States–Canada boundary is such an antecedent line. Boundaries drawn after development of the cultural landscape are termed **subsequent.** They may be either **consequent**—also called **ethnographic**—coinciding with some cultural divide, such as religion or language, or **superimposed** on an existing cultural pattern.

When Great Britain prepared to leave the Indian subcontinent after World War II, it was decided that two independent states would be established in the region: India and Pakistan. The boundary between the two countries, defined in the partition settlement of 1947, was thus both a subsequent and a superimposed line. As millions of Hindus migrated from the northwestern portion of the subcontinent to seek homes in India, millions of Muslims left what would become India for Pakistan. In a sense, they were attempting to insure that the boundary would be consequent, that is, that it would coincide with a division based on religion. This boundary example is more fully discussed in Political Regions in the Indian Subcontinent, Page 416.

Stages in the Development of Boundaries

There are three distinct stages in boundary line development, but any individual boundary need not proceed through all three stages, and decades or even centuries can elapse between the stages. **Boundary definition** is a general agreement between two states about the allocation of territory, a verbal description of the boundary and the area through which it passes. If the area is only sparsely settled or otherwise lacks value, there is usually little pressure to proceed to the second stage, **delimitation** of the boundary. This involves actually plotting, as precisely as possible, the boundary line on maps or aerial photographs. In Africa and Asia, delimitation became important when former colonies became independent nation-states. Finally, the boundary may actually be marked on the ground in a process of **demarcation.** The markers may be intermittent, like poles or pillars, or continuous fences or walls. Most international boundaries are not demarcated—for example, that between the United States and Canada for most of its length. The Iron Curtain, in contrast, is marked along much of its length by fences, watch towers, and mine fields (Figure 7.15).

If a former boundary line that no longer functions as such is still marked by some landscape features or differences on the two sides, it is termed a **relict** boundary. The abandoned castles dotting the former frontier zone between Wales and England are examples of a relict boundary. They are also evidence of the disputes that sometimes attend the process of boundary making.

Figure 7.16

Areas of international dispute in Latin America. Among the countries disputing the precise location of their boundaries are Argentina and Chile, Argentina and Uruguay, Venezuela and Guyana, and Honduras and El Salvador.

Boundary Disputes

Boundaries create many opportunities for conflict. In recent years, almost half of the world's sovereign states have been involved in border disputes with neighboring countries. States are far more likely to have disputes with neighboring countries than with those that are more distant. It follows that the more neighbors a state has, the greater the likelihood of conflict. A large number of adjoining states increases the opportunity for disputes, increases a state's feelings of vulnerability and the probability that it will be attacked.

Although the causes of border disputes are many and varied, we can identify four distinct types of border conflict. *Positional* disputes occur when states disagree about the interpretation of documents that define a boundary and/or the way the boundary was delimited. Such disputes typically arise when the boundary is antecedent, preceding effective human settlement in the border region. Once the area becomes populated and gains value, the exact location of the boundary becomes important.

The boundary betwen Argentina and Chile, originally defined during Spanish colonial rule, was to follow the highest peaks of the southern Andes and the watershed divides between east- and west-flowing rivers. Because the terrain had not been adequately explored, it wasn't apparent that the two do not always coincide. In some places, the water divide is many miles east of the highest peaks, leaving a long, narrow area of several hundred square miles in dispute (Figure 7.16). During the late 1970s, Argentina and Chile nearly went to war over the disputed territory, whose significance had been increased by the discovery of oil and natural gas deposits. Argentina has also contested its border with Uruguay in the Rio de la Plata estuary. Argentina claims that the international boundary is in the deepest part of the river, while Uruguay contends that it lies in the middle of the estuary.

Territorial disputes over the ownership of a region commonly arise when a boundary that has been superimposed on the landscape divides an ethnically homogeneous population. Each of the two states then has some justification for claiming the territory inhabited by the ethnic group in question. The Balkan countries of eastern

Figure 7.17
Ethnic minority areas in the Balkans, where separatist influences are a destabilizing force.

Europe offer numerous examples of such territorial disputes. They provided the sparks that helped ignite both World Wars, and although the region is less in the news today, it is far from stable. Ethnic minority problems fueled by historic enmities affect all the Balkan countries (Figure 7.17).

Yugoslavia's 6 federated republics and 2 autonomous provinces contain 5 major ethnic groups in addition to smaller ethnic minorities, 3 major religions, and 3 official national languages. Nearly 2 million ethnic Albanians live in Yugoslavia. They form a majority in the Kosovo region, and the minority Serbs and Montenegrins have charged that Albania is fomenting agitation in the province in the hope of detaching it from Yugoslavia. To the north lies Vojvodina, populated largely by former Hungarians, and in the south of the country is Macedonia. From 1941 to 1944 it was occupied by Bulgaria, which has not renounced all claims to the area.

Bulgaria itself has disputes with Turkey, which charges that 1 million Turks in southeastern Bulgaria are being brutally assimilated, and with Romania, which

claims the Romanians in the Dobruja region of Bulgaria—as well as those in the Moldavian S.S.R.—have been separated from their homeland. Romania's relations with Hungary have long been troubled by Romania's annexation in 1920 of Transylvania. As Figure 7.17 indicates, there are other ethnic minority areas in Austria, Albania, and islands in the Aegean Sea.

Closely related to territorial conflicts are *resource disputes*. Neighboring states are likely to covet the resources—whether they be valuable mineral deposits, fertile farmland, or rich fishing grounds—lying in border areas and to disagree over their use. In recent years, the United States has been involved in disputes with both its immediate neighbors, Mexico and Canada, over the shared resources of the Colorado River and the Georges Bank fishing grounds (see "The Georges Bank" on Page 252).

Functional conflicts arise when neighboring states disagree over policies to be applied along a boundary. Such policies may concern immigration, the movement of traditionally nomadic groups, customs regulations, or land use. In Central America, relations between Honduras and El Salvador, two countries that had long disputed their common boundary, worsened in the late 1970s when Honduras expelled Salvadoran farmers who had illegally occupied parts of western Honduras. More recently, U.S. relations with Mexico have been affected by the increasing number of illegal aliens (estimated to be over 1 million per year) entering the United States from Mexico.

The Seas: Open or Closed?

We have seen that boundaries play an important role in defining those areas in which certain activities may or may not be allowed to take place. However, we have considered only boundaries on land. As water covers about two-thirds of the earth's surface and people have used the seas since ancient times, it is not surprising that boundaries across water are also important.

A basic question of increasing international concern involves the right of nations to use water and the resources that it contains. The waters in a country, such as rivers and lakes, have traditionally been regarded as being within the sovereignty of that country. Oceans, however, are not within a single nation's borders. Should they, then, be open to all states to use, or may a single country claim sovereignty and limit access and use by other countries?

National Claims

For most of human history, the two-thirds of the earth's surface covered by the oceans remained effectively outside individual national control or international jurisdiction. The seas were a common highway for those daring enough to venture on them, an inexhaustible larder for fishermen, and a vast refuse pit for the muck of civilization. By the end of the 19th century, however, most coastal countries claimed sovereignty over a continuous belt 3 or 4 nautical miles wide (a nautical mile equals 1.15 statute miles or 1.85 km). Such sovereignty, though recognizing the rights of others to innocent passage, permitted national protection of coastal fisheries, allowed the enforcement of quarantine and customs regulations, and made claims of neutrality effective during other people's wars. The primary concern was with security and unrestricted commerce. No separately codified law of the sea existed, however, and none seemed to be needed until after World War I.

A League of Nations Conference for the Codification of International Law, convened in 1930, inconclusively discussed maritime legal matters and served to identify areas of concern that were to become increasingly pressing after World War II. Important among these was an emerging shift from interest in commerce and national security to a preoccupation with the resources of the seas, an interest fanned by the Truman Proclamation of 1945. Motivated by a desire to exploit offshore oil deposits and to prevent individual states such as Texas from claiming and taxing such deposits, the federal government under this doctrine laid claim to all resources on the continental shelf contiguous to its coasts.

In many respects the Truman Proclamation set off the great sea rush of the 20th century as other nations, many claiming even broader areas of control, hurried to annex marine resources. By 1947 Chile, Peru, and Mexico had proclaimed their control over 200 nautical miles (370 km) of offshore waters, a region containing the rich fisheries of the Humboldt (Peru) Current. Some African countries claimed absolute sovereignty over both mineral and fishing rights for 100 nautical miles (185 km) or more offshore. To protect its offshore sources of such staples as haddock, herring, and cod from being depleted through indiscriminate fishing by large, mechanized foreign fishing fleets, and to show other countries what such claims would mean, the United States in 1976 claimed sovereignty over fishing rights for 200 miles offshore. Although foreign fleets are not prohibited from fishing in that zone, they must obtain permits covering the specific varieties and amounts to be taken.

By the mid-1980s, nearly a third of all the oceans had been arbitrarily appropriated by some 60 coastal states, and seabed claims were being extended even farther. These extended claims have aggravated old disputes and created new ones, particularly among important

Figure 7.18

A U.S. Coast Guard cutter escorts a foreign fishing vessel to port after seizing it for violation of the Fisheries Conservation and Management Act. In 1976 the United States extended its claim of jurisdiction over fishery resources from 12 to 200 nautical miles (22 to 370 km) from its shoreline. Congress was impelled by a need to conserve seafood stocks before they were depleted to the point of exhaustion.

fishing nations (Figure 7.18). Britain first challenged Iceland's claim to exclusive fishing rights in a 50-mile (80-km) zone by sending warships to protect British fishing boats and then retaliated by establishing a 200-mile fishing limit around the British Isles. Unlicensed United States fishermen have been arrested for fishing within the 200-mile limit claimed by Peru. An international boundary dispute resulted from overlapping United States and Canadian claims in the Gulf of Maine, one of the richest fishing grounds off the East Coast. Argentina and Chile have contested the control of three tiny islands in the Beagle Channel with claims to a large area of strategic and potential economic importance at stake.

Arguments over sovereignty are becoming increasingly frequent as countries use the oceans more heavily as a source of food, minerals, and offshore oil and gas. Because competition for those resources has become keener, there is a growing need for countries to agree on how the resources of the ocean will be used.

Modern fishing techniques and rapid processing capabilities in large "factory" ships enable nations to harvest vast quantities of fish at great distances from their home ports. They also make it possible to *overfish,* that is, to take in fish faster than they can reproduce and consequently to decrease the number available in future years. Between 1950 and 1970, the world's annual fish catch quadrupled, and the regenerative abilities of many food species were endangered. Several species have already become extinct, and it is more than likely that others, as well as some types of whales, will become extinct in the very near future.

International agreements to limit the amount of fish taken in a specified period of time help alleviate the danger of overfishing, but the political mechanisms for making and enforcing such agreements are inadequate.

The Georges Bank

One of the richest fishing grounds in the world for scallops, haddock, cod, and flounder is the Georges Bank, a 9000-square-mile (23,300 km²) expanse east of Cape Cod in the Gulf of Maine. For centuries, Canadians and New Englanders shared the grounds with little conflict. Then the United States in 1976, and Canada a year later, extended their claims over fishing rights in waters up to 200 miles offshore, creating the basis for a bitter boundary dispute. The claims overlapped in the eastern portion of the Georges Bank around a section called the Northeast Peak, generally agreed to be the most productive grounds. Approximately half the Georges Bank haddock and pollock, a third of the flounder and scallops, a quarter of the cod, and the best swordfish and lobster were harvested in the Northeast Peak, where deep waters provide rich harvests even in winter months.

Canada and the United States tried without success to settle the dispute with a treaty providing for joint fishing and management. They eventually agreed to refer the conflict to the International Court of Justice at The Hague, Netherlands. Canada claimed slightly less than half of the disputed area; the United States claimed the whole bank.

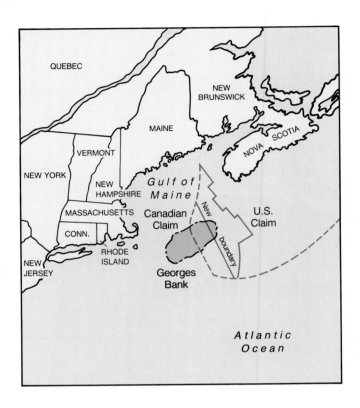

Perhaps predictably, the decision issued by the World Court in October 1984 satisfied neither nation entirely. It awarded the United States about 80% of the Georges Bank. Canada received the remainder, but that portion includes the Northeast Peak. Fishermen from both countries are confined to their side of the new line. U.S. fishermen contend they are now crowded into more restricted, less productive waters, while Canadians deplore their loss of access to most of the bank.

The Georges Bank issue reflects many political-geographic concerns: the significance of boundaries to nation-states and their potential to cause disputes; the growing significance of maritime boundaries as states seek both to protect and to exploit the resources of the sea; and the use of international authorities to resolve disputes peacefully.

The Japanese have agreed, by treaty, to limit their taking of North Pacific salmon to waters west of 175° W longitude, but new techniques for open-sea capture of that species continue to endanger North American stocks. The International Whaling Commission, given the power to regulate that industry in the southern hemisphere since 1946, has regularly set the quotas on desirable species so high as to fail to guarantee their survival; indeed, the blue whale has been declared officially extinct. In 1978 a 13-nation draft convention to conserve "Antarctic marine living resources," notably krill (a tiny shrimp-like crustacean), was signed, in this case before the serious depletion of a potentially invaluable marine source of protein. As with many other such agreements, however, the signatory group was a self-selecting society whose arrangements do not have the force of international law.

Countries do not have to sign agreements if they do not want to, and nonsignatory countries can make useless the treaties that others observe. Some ship-owners transfer their vessels to foreign registry to avoid agreements binding on their own country. The quotas

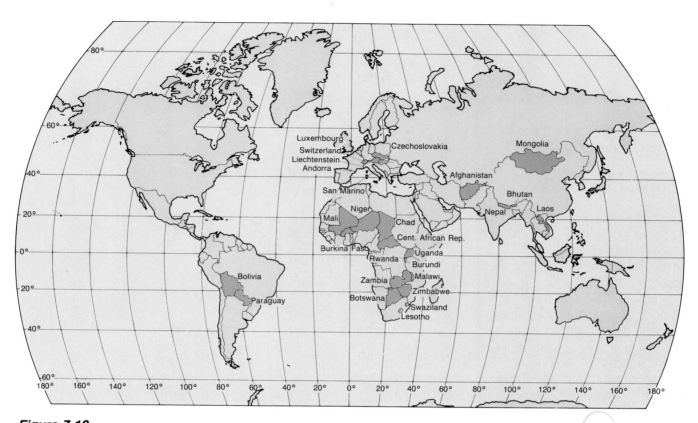

Figure 7.19

These landlocked countries are among those that advocate international control of deep-seabed natural resources. The formation of 200-nautical-mile exclusive economic zones as proposed in the Law of the Sea Convention will do little to enable landlocked states to participate in the exploitation of the seas.

set in the treaties may be so high as to permit overfishing anyway. In addition, such agreements are hard to police; individual fishermen may think that it is not in their own best interests to observe quotas that they have not agreed to. Finally, there is no international body, not even the International Court of Justice, that has the power to enforce the recommendations that it makes. Until such a body exists to make decisions to which member countries must adhere, and until suitable policing mechanisms are developed, there is little hope of avoiding overfishing.

In the years to come, competing claims to the mineral resources on the ocean floor may occasion more disputes than fishing does. Petroleum and natural gas are in increasing demand as energy supplies. Many countries claim sovereignty over the continental shelf where such resources are known to exist and are relatively accessible. However, who should have the right to exploit the resources of the ocean floor: internationally licensed companies or countries contiguous to those waters? When natural gas was discovered under the North Sea in waters not claimed by any country, the contiguous nations

agreed to a partitioning of the sea. Such a peaceful solution may prove to be the exception rather than the rule, and in any case, it begs the question whether such resources are available for use by all or only by those favored by an accident of location (Figure 7.19).

An International Law of the Sea

Unrestricted extensions of jurisdiction and territorial disputes over proliferating claims to maritime space and resources led to a realization that the established principles of international law needed to be buttressed by specific agreement on a law of the sea. A United Nations conference, convened in 1958, resolved some issues but left others, such as the width of territorial waters and the creation of exclusive fishing zones, unsettled. In 1967 an explosive new issue was injected into international debate on the control of maritime resources: the concept, advocated by the underdeveloped and the landlocked nations, that the seabed and the resources it contains should be recognized as the "common heritage of mankind."

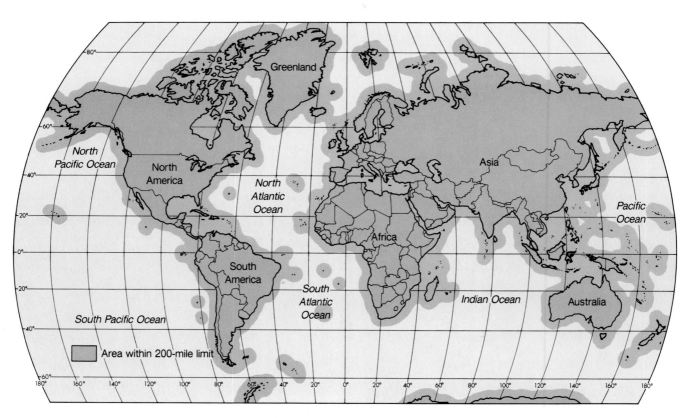

Figure 7.20
The 200-nautical-mile exclusive economic zones (EEZs) claimed by coastal states. Provided for by the 1982 United Nations Convention on the Law of the Sea, the EEZs include about one-third of the sea in the world. Overlapping claims may take decades to negotiate.

From 1973 to 1981 the United Nations Conference on the Law of the Sea met about twice a year. Delegates from over 150 countries attempted to achieve consensus on a treaty that would govern the use of the world's oceans and seabed. The meetings culminated in a draft treaty in 1982. The **Law of The Sea Convention** will not take effect until one year after 60 countries have ratified it; by early 1987 fewer than 30 had done so.

The treaty delimits territorial boundaries and rights, setting a 12-mile (19-km) maximum for territorial waters. Ships of all nations will enjoy the right of innocent passage through the territorial sea. The new code also affirms the rights of free passage through more than 100 international straits that would otherwise fall within the 12-mile limit over which coastal nations could exercise sovereignty. Passages 24 miles wide or less do not become territorial waters; all ships will have a right of passage through, under, and over them.

The treaty gives a legal basis to the 200-mile zones of exploitation already claimed, allowing countries to fish and to extract minerals in **exclusive economic zones** (EEZs) extending 200 miles from their coasts (Figure 7.20). Furthermore, such countries will have exclusive rights to the resources lying within the continental shelf when this extends farther, up to 350 miles (560 km) beyond their coasts. In 1983 the United States, although refusing to sign the International Law of the Sea Treaty, did proclaim a 200-mile-wide coastal economic zone in which it will exercise exclusive rights over all resources, including minerals as well as fisheries. This claimed EEZ is a reaffirmation of the 1945 claim, extending control where applicable beyond the continental shelf to the full 200-mile limit.

About 36% of the world's ocean is within 200 miles of coastlines, and it contains the great majority of the sea's known resources, including all the offshore oil and natural gas, the readily accessible minerals, and about 90% of the world's current fish catch. Ten countries (the United States, Australia, New Zealand, Indonesia, Canada,

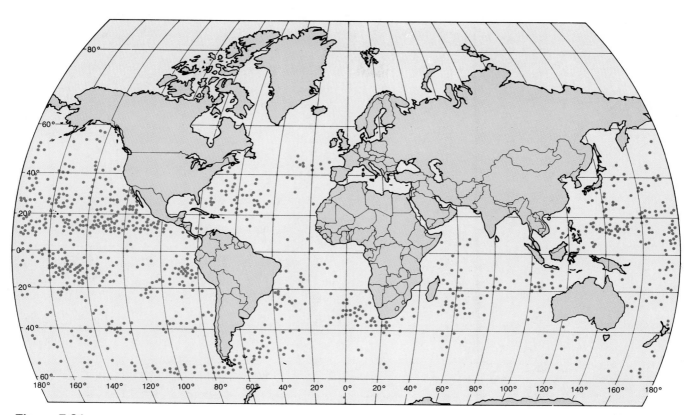

Figure 7.21

Known manganese nodule deposits. The largest field yet discovered stretches from the coast of California to Hawaii. Most deposits lie in international waters outside the proposed exclusive economic zones of coastal countries. A new international seabed authority would manage their exploitation.

the USSR, Japan, Brazil, Mexico, and Chile) will together acquire more than half the entire economic zone. The treaty gives landlocked and other "geographically disadvantaged" nations (for example, those with short coastlines or narrow shelves) limited fishing rights in the exclusive economic zones of the coastal states but no access to the nonliving resources of the zone.

The single most controversial provision of the treaty is that involving *seabed mining* beyond the limits of national jurisdiction. Indeed, for several years, controversy over this issue blocked agreement on the whole new sea-law code, and it may yet result in the failure of industrial nations to ratify the treaty. At stake is control of a vast mineral treasure in the form of *manganese nodules* scattered over thousands of square miles of ocean bottom. Potato-sized, crumbly accretions of blackish ore containing up to 27 separate elements, the nodules have 4 minerals of particular interest: manganese (a crucial ingredient in steel making), copper, cobalt, and nickel. Their potential value has been conservatively estimated at more than $3 trillion. Most deposits lie in international waters outside the jurisdiction of individual countries (Figure 7.21).

Under the treaty, the deep seabed and its resources are beyond national jurisdiction and are viewed as the common heritage of humankind, to be managed for the benefit of all the peoples of the earth. Specific clauses, championed by about 110 developing countries, provide that exploitation of these resources be managed by an international agency that will pick the sites to be mined, with industrial countries providing the technology and guaranteeing the profitability of the mining operations. A large portion of the benefits would be used to aid the economic development of the poorer nations. The United States and a few other industrial nations that possess the technology and the capital necessary to mine the deep seabeds want to be able to do so through private investment. They fear that the developing countries would, by their numbers, dominate the mining agency.

Should countries such as the United States, the USSR, and the United Kingdom fail to ratify the treaty, it is unclear how effective the new sea law can be. The result may well be an intensification of national rivalries and mounting political turmoil.

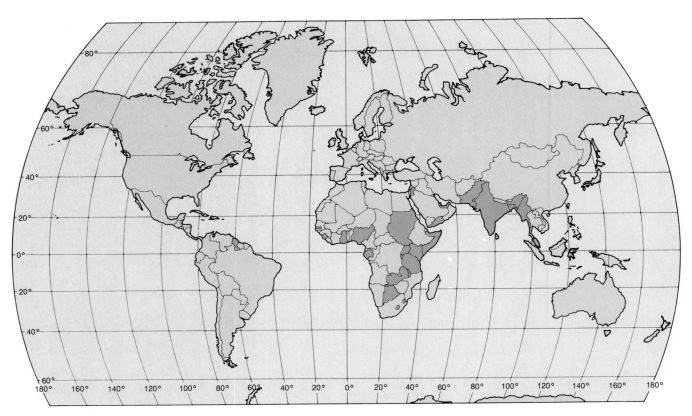

Figure 7.22

Some 43 former possessions, colonies, dominions, and protectorates of Great Britain have achieved independent statehood since 1945. This map shows how large and far-flung was the territory controlled, well into the 20th century, by a single small European nation-state. At its greatest extent, the British Empire covered almost a quarter of the globe and contained nearly a quarter of the world's population.

Aspects of International Political Systems

One of the most significant trends in recent years has been the rapid increase in the number of sovereign states. At the beginning of World War II, in 1939, there were about 70 independent countries. By 1970, the number had more than doubled, and by mid-1987, 159 independent states had membership in the United Nations.

The Dissolution of Empires

Most of the newly formed states were once colonies of one or another of the European empires. Expanding political consciousness in Asia, Africa, and South America brought a demand for change and, subsequently, an end to the colonial era. Faced with strong nationalisitc feelings on the part of the inhabitants, and sometimes recognizing the legitimacy of those feelings, convinced at other times that the colonies were no longer assets either economically or politically, or responding to the pressures of world opinion, the European countries relinquished their control.

From the former British Empire and Commonwealth, there have come, just in the period since World War II, the independent countries of India, Pakistan, Bangladesh, Malaysia, and Singapore in Asia, and Ghana, Nigeria, Kenya, Uganda, Tanzania, Malawi, Botswana, Zimbabwe, and Zambia in Africa. Even this extensive list is not complete. These and other now-independent, former British colonies are shown in Figure 7.22. A similar process has occurred in most of the former overseas possessions of France, the Netherlands, Spain, and Portugal.

Empires have been created—and have fallen—throughout history: the Roman, the Ottoman, and the Austro-Hungarian, to name only three. Each was, for a while, very strong; each eventually dissolved. What is there, then, that is significant about the dissolution of empires in this century?

The answer lies in the speed of the process, in the areal extent over which it has occurred, and in what has happened to the former colonies. The modern European empires, comprising vast parts of the globe, dissolved almost simultaneously in a few years' span after World War II. The colonies of those empires have now become independent states; this was not a consequence when previous empires broke apart before the concept of

Figure 7.23
Under the auspices of the United Nations, soldiers from many different countries have staffed peacekeeping forces and military observer groups in an effort to halt or mitigate conflicts. Among the areas where they have served are Indonesia, Korea, Cyprus, and the Dominican Republic. Shown here are soldiers on duty in Lebanon.

nationalism became dominant. Nationalism is now a truly global phenomenon; there will soon be very few people left who are not citizens of independent states.

New International Systems

In many ways, countries are now weaker than ever before. Many are economically weak, others are politically unstable, and some are both. Strategically, no country is safe from military attack, for technology now enables us to shoot weapons halfway around the world. Some people believe that no national security is possible in the atomic age. The recognition that a country cannot by itself guarantee either its prosperity or its own security has led to increased cooperation among states. In a sense, these cooperative ventures are replacing the empires of yesterday. They are proliferating quickly, and they involve countries everywhere.

Worldwide Organizations

The United Nations (UN) is the only organization that tries to be universal, and even it is not all-inclusive. The United Nations is the most ambitious attempt ever undertaken to bring together the world's nations in international assembly and to promote world peace. It is stronger and more representative than its predecessor, the League of Nations. It provides a forum where countries may meet to discuss international problems and a mechanism, admittedly weak but still significant, for forestalling disputes or, when necessary, for ending wars (Figure 7.23). The United Nations also sponsors 40 programs and agencies aimed at fostering international cooperation with respect to specific goals. Among these are the World Health Organization (WHO), the Food and Agriculture Organization (FAO), and the United Nations Educational, Scientific, and Cultural Organization (UNESCO).

The United Nations would be stronger if countries were willing to surrender enough of their sovereignty to make it a true world government—specifically, their sovereignty with respect to the ability to wage war. The weakness of the United Nations stems partly from its legal inability to make and enforce a world law. There is little world law, and there is no permanent world police force. Although there is recognized international law adjudicated by the International Court of Justice, rulings by this body are sought only by nations agreeing beforehand to abide by its arbitration. The United Nations has no authority over the military forces of individual countries, it cannot take prompt and effective action to counter aggression, and thus there is no guarantee of peace.

Countries have shown themselves to be more willing to surrender some of their sovereignty to participate in smaller multinational systems. These systems may be economic, military, or political, and many have been formed since 1945.

Regional Alliances

Cooperation in the economic sphere seems to come more easily to states than does political or military cooperation. Among the most powerful and far-reaching of the economic alliances are those that have evolved in Europe, particularly the Common Market. It had several forerunners.

Shortly after the end of World War II, the Benelux countries (Belgium, the Netherlands, and Luxembourg) formed an economic union to create a common set of tariffs and to eliminate import licenses and quotas. At about the same time were formed the Organization for European Economic Cooperation, which coordinated the distribution and use of Marshall Plan funds, and the European Coal and Steel Community, which integrated the development of that industry in the member countries. A few years later, in 1957, the **European Economic Community,** or Common Market, was created, composed at first of only six countries: France, Italy, West Germany, and the Benelux countries.

To counteract these Inner Six, as they were called, other countries joined in the European Free Trade Association (EFTA). Known as the Outer Seven, they were the United Kingdom, Norway, Denmark, Sweden, Switzerland, Austria, and Portugal (Figure 7.24). In 1973 the United Kingdom, Denmark, and Ireland applied for and were granted membership in the Common Market. They were joined by Greece in 1981; Spain and Portugal entered in 1986 (Figure 7.25). The European Economic Community directly influences the lives of 325 million Europeans and, as the world's largest trading bloc, affects the lives of countless millions of others. A free-trade zone has been established over much of Western Europe. Not only have tariff barriers among the member states been eliminated, but there is much freer movement of labor and services as well. In addition, there is some coordination of planning with respect to transportation and agriculture. The Common Market is developing shared policies in fields as varied as energy, fisheries, the environment, and scientific research. The member states are also moving toward common foreign policies.

We have traced this developmental process not because it is important to remember all the forerunners of the Common Market, but to illustrate the rapidity with which regional alliances are made. Countries come together in an alliance, some drop out, and others join. New treaties are made, and new alliances emerge. It seems

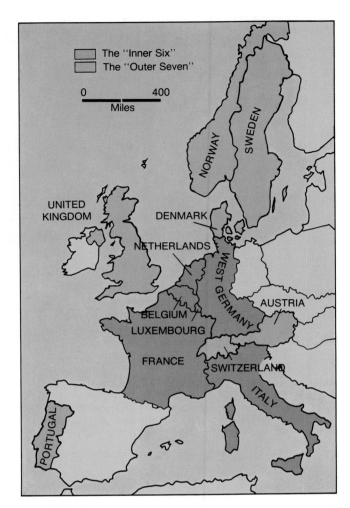

Figure 7.24
The original Inner Six and Outer Seven of Europe.

safe to predict that although the alliances themselves will change, the idea of economic associations has been permanently added to that of political and military alliances, which is as old as the concept of nations.

Three further points about economic unions are worth noting. The first—and it applies to military and political alliances as well—is that the formation of an alliance in one area often stimulates the creation of another alliance by countries left out of the first. Thus the union of the Inner Six gave rise to the treaty among the Outer Seven. Similarly, a counterpart to the Common Market is the Council of Mutual Economic Assistance, also known as Comecon, which links the Communist countries of Eastern Europe and the USSR through trade agreements.

Second, the new economic unions tend to be composed of contiguous states (Figure 7.26). This was not the case with the recently dissolved empires, which included far-flung territories. Contiguity facilitates the movement of people and goods. Communication and transportation are simpler and more effective among adjoining countries than among those far removed from one another.

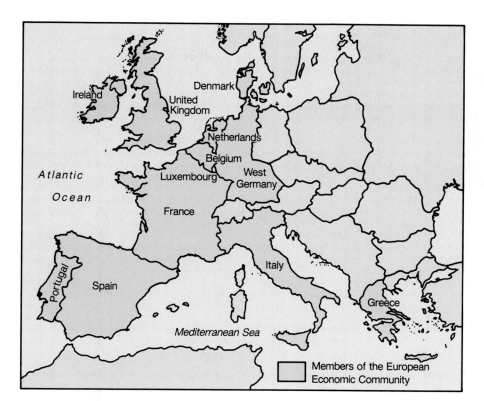

Figure 7.25

The members of the European Economic Community have taken many steps to integrate their economies and coordinate their policies in such areas as transportation and agriculture. The 7 members of EFTA have signed free-trade agreements with the EEC, and some 60 states in Africa, the Caribbean, and the Pacific are affiliated with the EEC by the Lomé Convention, which provides for development aid and access to EEC markets.

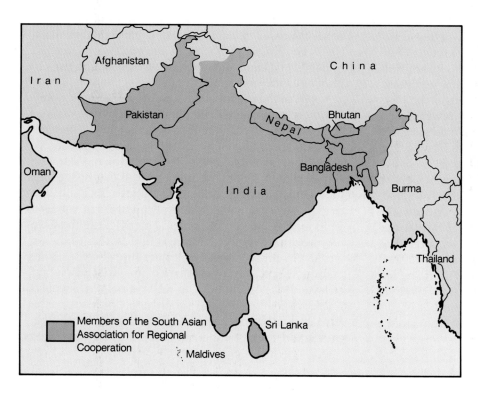

Figure 7.26

Members of the South Asian Association for Regional Cooperation, formed in December 1985 to promote cooperation in such areas as agriculture and rural development, transportation and telecommunications. The 7 countries contain about one-fifth of the world's population. Until the formation of the association, South Asia was the only major part of the world with no organization for regional cooperation.

Figure 7.27

Members of the Union of Banana Exporting Countries. Founded in 1974, the union accounts for about half the world's banana exports. The cartel has established a multinational trading corporation in an attempt to challenge the dominance of 3 large U.S. multinational corporations, which together control some 70% of world trade in bananas.

Finally, it does not seem to matter whether countries are homogeneous or heterogeneous in their economies, as far as joining economic unions is concerned. There are examples of both. If the countries are dissimilar, they may complement each other. This is the basis for the European Common Market. Dairy products and furniture from Denmark are sold in France, freeing that country to specialize in the production of machinery and clothing, although the Common Market is faced by expensive gluts of, for example, butter and wine, which are produced in more than one country and are basic products of the politically potent national farm groups. On the other hand, countries that produce the same raw materials find that by joining together in an economic alliance, they are able to enhance their control of the markets and the prices for their products (Figure 7.27). The Organization of Petroleum Exporting Countries (OPEC), discussed in Chapter 10, is a case in point. Less successful attempts to form commodity cartels and price agreements between producing and consuming nations are represented by the International Tin Agreement, the International Coffee Agreement, and others.

Countries form alliances for other than economic reasons. Strategic, political, and cultural considerations may also foster cooperation. Military alliances are based on the principle that in unity there is strength. Such pacts usually provide for mutual assistance in the case of aggression. Once again, action breeds reaction when such an association is created. The formation of the North Atlantic Treaty Organization (NATO), a defensive alliance of many European countries and the United States, was countered by the establishment of the Warsaw Treaty Organization, which joins the USSR and its satellite countries of Eastern Europe. Both pacts allow the member states to base armed forces in one another's territories— a relinquishment of a certain degree of sovereignty unique to this century.

Military alliances depend on the perceived common interests and political goodwill of the countries involved. As political realities change, so, too, do the strategic alliances. NATO was altered in the 1960s as a result of policy disagreements between France and the United States. The conflict over Cyprus between Greece and Turkey, both NATO members, and their response to the American reaction to that dispute altered the original cohesiveness of that alliance. The organization similar to NATO in Asia, SEATO (Southeast Asia Treaty Organization), was disbanded in response to changes in the policies of its member countries.

All international alliances recognize communities of interest. In economic and military associations, common objectives are clearly seen and described, and joint actions are agreed on with respect to the achievement of those objectives. More generalized common concerns or appeals to historical interests may be the basis for alliances that are primarily political. Such alliances tend to be rather loose, not requiring their members to yield much power to the union. Examples are the Commonwealth of Nations (formerly the British Commonwealth), composed of many former British colonies and dominions, and the Organization of American States, both of which offer economic as well as political benefits. There are many examples of abortive political unions that have foundered precisely because the individual countries could not agree on questions of policy and were unwilling to subordinate individual interests to make the union succeed. The United Arab Republic, the Central African Federation, the Federation of Malaysia and Singapore, and the Federation of the West Indies fall within this category.

Observers of the world scene have wondered about the prospect that "superstates" will emerge from the many international alliances that now exist. Will a United States of Europe, for example, evolve out of the many pacts that now bind the countries of Western Europe economically, militarily, and culturally? Will unification in one sector trigger unification in the others, eventually creating ties of mutual interest so strong that a common government is the logical next step? No one really knows, but as long as the individual state is regarded as the highest form of political and social organization (as it is now) and as the body in which sovereignty rests, such unification is unlikely.

However, Western Europe is the area to watch in this regard. If a successful international government is going to be formed, it will probably be there. It is there that unification, including partial monetary unification, has proceeded so far and in so many different sectors that eventual political union is a possibility. In fact, a quasi-parliament, the Council of Europe, has existed since 1949, but as now constituted, it lacks the power to make and enforce laws.

Conclusion

The political subdivision of the world into nation-states and lower orders of territorial jurisdiction constitutes an expression of cultural separation and identity as pervasive as that inherent in language, religion, or ethnicity. Indeed, political concerns impinge more forcefully on the affairs of peoples and societies than do the more subtle expressions of cultural differentiation, as the reading of any newspaper will show. It is also apparent that control over our lives is tending to shift to increasingly higher levels of political organization. The culmination of this process is seen in the many international systems that are evolving. At the same time, as nations become interlocked in their economic systems and are tied together by a global communications network, there is increasing though sometimes reluctant recognition of the fact that developments in one part of the world affect the rest of it and that certain elements of sovereignty must, in the national and the international interest, be surrendered.

That changing pattern of political control necessarily affects the behavior not only of the nation-states directly involved but also of the individual citizens of those countries. Behavioral geography, the pattern of individual and group action against the background of their culture, is our next topic of concern.

Key Words

antecedent boundary 247
artificial boundary 246
boundary definition 247
boundary delimitation 247
boundary demarcation 247
compact state 240
consequent boundary 247
elongated state 240
ethnic separatism 244
ethnographic boundary 247
European Economic Community 258
exclusive economic zone (EEZ) 254
forward-thrust capital 242
fragmented state 240
geometric boundary 246
gerrymandering 231
iconography 238
Law of the Sea Convention 254
nation 230
nationalism 237
nation-state 237
natural boundary 246
physical boundary 246
prorupt state 240
regional autonomy 244
relict boundary 247
state 230
subsequent boundary 247
superimposed boundary 247

For Review

1. What reasons can you suggest for the great political fragmentation of the United States? As few other nations have such a multiplicity of local governing and administrative units, is such subdivision—in your opinion—an evidence of internal weakness? Why?

2. What is meant by *metro-government?* What would appear to be its advantages? Its disadvantages?

3. Why, in your opinion, were the cultural realms suggested on Figure 6.30 generally drawn so that they are composed of whole, rather than parts of, states? Comment on how political jurisdiction reinforces other cultural contrasts between the United States and Mexico.

4. What are some of the devices by which national cohesion and identity are achieved? What internal cultural characteristics might serve to disrupt or destroy the cohesiveness that nation-states attempt to foster?

5. What factors account for the wide variety of regionalist movements seeking some or more autonomy within their established state? What characteristics are common to all or most regionalist autonomist movements? Where are some of these tension areas in Western Europe? Why do they tend to be on the periphery rather than at the national core in terms of socioeconomic development and political power?

6. How may boundaries be classified? How do they create opportunities for conflict? Describe and give examples of three types of border disputes.

7. What are some of the pressing issues facing the states attempting to codify an international law of the sea? What are some of the provisions of the United Nations Law of the Sea Convention? Which provision is the most controversial, and why? Do you feel that the interests of your country would be best served by tight international control of maritime resources? Why or why not?

8. What made the post-1945 dissolution of empires unique in the political history of the world?

9. What types of international organizations and alliances can you name? What were the purposes of their establishment? What generalizations can you make regarding economic alliances?

Suggested Readings

Bennett, D. Gordon. *Tension Areas of the World.* Delray Beach, Fla.: Park Press, 1982.

Boyd, Andrew. *An Atlas of World Affairs.* 7th ed. New York: Methuen, 1985.

Burnett, Alan D., and Peter J. Taylor, eds. *Political Studies from Spatial Perspectives.* Chichester, England: John Wiley, 1981.

Clarke, C.; D. Ley; and C. Peach, eds. *Geography and Ethnic Pluralism.* London: George Allen & Unwin, 1984.

Glassner, M. I., and Harm J. de Blij, *Systematic Political Geography.* New York: John Wiley, 1980.

Johnston, R. J., *Political, Electoral and Spatial Systems: An Essay in Political Geography.* Oxford, England: Clarendon Press, 1979.

Johnston, R. J., and P. J. Taylor, eds. *A World in Crisis: Geographical Perspectives.* Oxford, England: Basil Blackwell, 1986.

Taylor, Peter J., *Political Geography: World Economy, Nation-States and Localities.* London: Longman, 1985.

Williams, C. H., ed. *National Separatism.* Cardiff: University of Wales Press, 1982.

8 Behavioral Geography

Early in January of 1849 we first thought of migrating to California. It was a period of National hard times . . . and we longed to go to the new El Dorado and "pick up" gold enough with which to return and pay off our debts. Our discontent and restlessness were enhanced by the fact that my health was not good. . . . The physician advised an entire change of climate thus to avoid the intense cold of Iowa, and recommended a sea voyage, but finally approved of our contemplated trip across the plains in a "prairie schooner." Full of the energy and enthusiasm of youth, the prospects of so hazardous an undertaking had no terror for us, indeed, as we had been married but a few months, it appealed to us as a romantic wedding tour. *

So begins Catherine Haun's account of her journey from Clinton, Iowa, to California in 1849, a trip that was to last 9 months and cover 2400 miles (3900 km). The Hauns were just two of the 250,000 people who traveled across the continent on the Overland Trail in one of the world's great migrations. The dangers inherent in such a trip were numerous; thousands were to die en route, and many others stopped or turned back short of their goal. The migrants faced at least 6 months, and often more, of grueling travel over badly marked routes that crossed swollen rivers, deserts, and mountains. The weather was often foul, with hailstorms, drenching rains, and summer temperatures that could exceed 110° F (43° C) inside the covered wagons. Wagon breakdowns were frequent. Graves along the route were a silent testimony to the lives claimed by buffalo stampedes, Indian skirmishes, cholera epidemics, and other disasters.

What inducements were so great as to make emigrants leave behind all that was familiar and risk their lives on an uncertain venture? Catherine Haun's account is unusual in that it gives the reasons for their trip. She alludes to economic hard times; the depression that swept the United States in 1837 inaugurated a

prolonged period of bank closures, depressed prices for agricultural goods, and high rates of unemployment. The Hauns hoped to strike it rich by mining gold. Other migrants were attracted by reports of free land and rich soil in the Oregon and California territories, of productive fishing grounds, and of ample furs for trapping. Like other migrants, the Hauns were also attracted by the climate in the West, which was said to be always sunny and free of disease. Finally, like most who undertook the trip along the Overland Trail, the Hauns were young, moved by restlessness and a sense of adventure.

Catherine Haun's story is unique only in its particulars. As had her predecessors since the beginnings of humankind, she and her family acted in space and over space on the basis of acquired information and awareness of opportunity. Her story summarizes the content of this chapter, a survey of one of the most exciting thrusts of geographic inquiry: how individuals make spatial behavioral decisions and how those separate decisions may be summarized by models and generalizations to explain collective actions.

In this chapter we pose a question basic to the themes that we have been exploring in preceding chapters: What considerations influence how individual human beings use space and act within it? Implicit in this question are subsidiary analytical concerns: How do individuals view (perceive) their environment? How is information transmitted through space and acted on? How might all the separate decisions of many individuals be summarized, so that we may understand the order that underlies the apparent chaos of individual action?

Two aspects of that action concern us. The first is the daily or temporary use of space—the journeys to stores, work, or school. The second is the long-term commitment related to decisions to travel, to migrate, or to settle away from the home territory. This latter aspect suggests a concern about time. Spatial actions of humans are not instantaneous; they operate over time as an expression of process or system. Elements of both aspects are embodied in how individuals perceive space and act within it. Those perceptions and actions are the concern of human geography in general, and interest in the specific topics of behavioral geography is expressed—and in some instances has been discussed in related contexts—in other chapters of this book.

*From Catherine Haun, "A Woman's Trip Across the Plains in 1849," as quoted in Lillian Schlissel, *Women's Diaries of the Westward Journey,* Schocken Books, New York, 1982. By permission of the Huntington Library, San Marino, Calif.

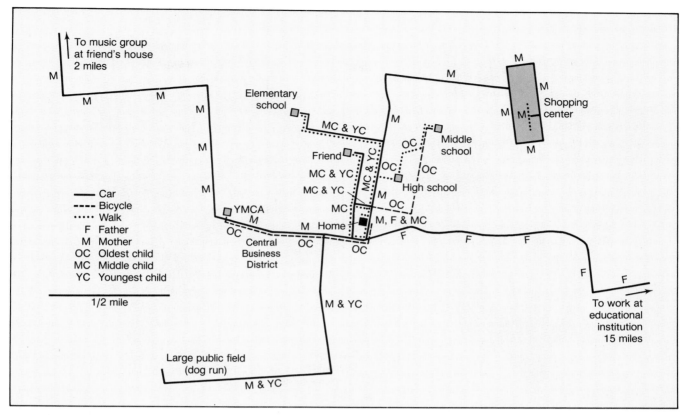

Figure 8.1

Activity space for each member of a family of five for a typical weekday. Routes of regular movement and areas repeatedly visited help to foster a sense of territoriality and to color one's perceptions of space.

Individual Activity Space

We saw in Chapter 7 that groups and nations draw boundaries around themselves and divide space into territories that are, if necessary, defended. The concept of **territoriality**—the emotional attachment to, and the defense of, home ground—has been seen by some as a root explanation of much human action and response. It is true that some collective activity appears to be governed by territorial defense responses: the conflict between street groups in claiming and protecting their "turf" (and their fear for their lives when venturing beyond it) and the sometimes violent rejection by ethnic urban neighborhoods of an encroaching black, Hispanic, or other population group. But for most people, our personal sense of territoriality is a tempered one; homes and property are regarded as defensible private domains but are opened to innocent known or unknown visitors, or to those on private or official business. Nor do we confine our activities so exclusively within controlled home territories as street-gang members do within theirs. Rather, we have a more or less extended home range, an **activity space** within which we move freely on our rounds of regular activity, sharing that space with others

who are also about their daily affairs. Figure 8.1 suggests a probable activity space for a suburban family of five for a day. Note that the activity space for each individual for one day is rather limited, even though two members of the family use automobiles. If one week's activity were shown, more paths would have to be added to the map, and in a year's time, several long trips would probably have to be noted. Because long trips are taken irregularly, we will confine our idea of activity space to often-visited places.

The kinds of activities in which individuals engage can be classified according to type of trip: journeys to work, to school, to shops, for recreation, and so on. People in nearly all parts of the world make these same types of journeys, though the spatially variable requirements of culture and economy dictate their frequency, duration, and significance in the time budget of an individual. Figures 8.2, 8.3, and 8.4 illustrate this point. Figure 8.2 depicts variations in travel patterns in two different culture groups in rural Ontario, Canada. It suggests that "modern" rural Canadians, who want to take advantage of the variety of goods offered in the regional capital, are willing to travel longer distances than are people of a

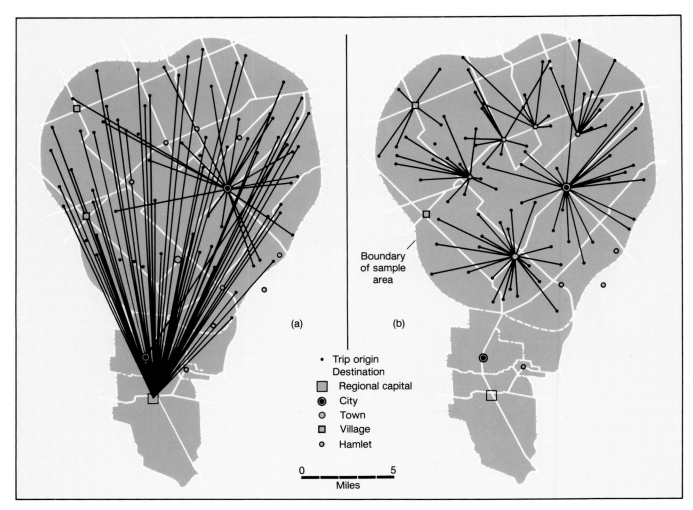

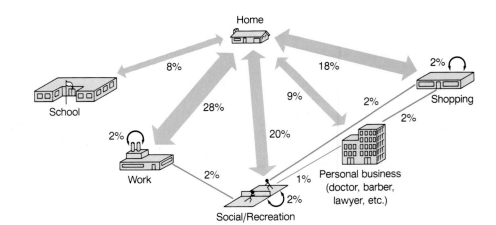

Figure 8.2

Travel patterns for purchase of clothing and yard goods of (a) rural cash-economy Canadians, and (b) Canadians of the Old Order Mennonites sect. These strikingly different travel behaviors demonstrate the great variation that may exist in the action spaces of different culture groups occupying the same territory.

Figure 8.3

Chicago travel patterns. The numbers are the percentages of all urban trips taken in Chicago. The greatest single movement is the journey to and from work. Over 96% of all trips are represented on the diagram.

Data from Chicago Area Transportation Study, *1970 Travel Characteristics.*

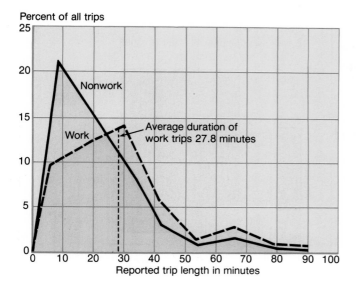

Figure 8.4

The frequency distribution of work and nonwork trip lengths in minutes in Toronto. Work trips are usually longer than other internal city journeys. For North America, 20 minutes seems to be the upper limit of commuting time before people begin to consider time seriously in their choice of home site. Most nonwork trips are very short.

traditionalist culture who have different tastes in clothing and consumer goods and whose demands are satisfied in local settlements. In addition, one must be aware that, in this case, the traditionalists do not own cars, thus limiting their spatial range. Figures 8.3 and 8.4 suggest the importance of the journey to work among urban populations in two different cities.

The types of trips that individuals take and thus the extent of their activity space depend on at least three variables: their stage in the life cycle; the means of mobility at their command; and the demands or opportunities implicit in their daily activities. The first, *stage in the life cycle,* refers to membership in specific age groups. Preschoolers stay close to home unless they accompany their parents. School-age children usually travel short distances to primary schools and longer distances to upper-level schools. After-school activities tend to be limited to walking or bicycle trips to nearby locations. High school students are usually more mobile and take part in more activities than do younger children. Adults responsible for household duties make shopping trips and trips related to child care as well as journeys away from home for social, cultural, or recreational purposes. Wage-earning adults usually travel farther from home than other family members. Elderly people normally do not find it feasible or desirable to have extended activity spaces.

The second variable that affects the extent of activity space is *mobility,* or the ability to travel. An informal consideration of the cost and effort required to overcome the friction of distance is implicit. Where incomes are high, automobiles are available, and the cost of fuel is reckoned to be a minor item in the family budget, mobility may be great and individual action space large. In societies where cars are not a standard means of personal conveyance, the daily nonemergency action space may be limited to the shorter range afforded by the bicycle or by walking. Obviously, both intensity of purpose and the condition of the roadway affect the execution of movement decisions.

The mobility of individuals in countries or in sections of countries with high incomes is relatively great; their activity space horizons are broad. These horizons, however, are not limitless. There are a fixed number of hours in a day, most of them consumed in performing work, preparing and eating food, and sleeping. In addition, there are a fixed number of road, rail, and air routes, so that even the most mobile individuals are constrained in the amount of activity space that they can use. No one can easily claim the world as his or her activity space.

A third factor limiting activity space is the individual assessment of the availability of possible activities or *opportunities.* In the subsistence economies discussed in Chapter 9, the needs of daily life are satisfied at home; the impetus for journeys away from the residence is minimal. If there are no stores, schools, factories, or roads, one's expectations and opportunities are limited, and one's activity space is therefore reduced. In impoverished nations or neighborhoods, low incomes limit the inducements, opportunities, destinations, and necessity of travel.

Distance and Human Interactions

People make many more short-distance trips than long ones. If we drew a boundary line around our activity space, it would be evident that trips to the boundary are taken much less often than short-distance trips around home. Think of activity space as more intensively used near one's home place or base and as declining in use with increasing distance from the base. This is the principle of **distance decay**—the exponential decline of an activity or function with increasing distance from its point

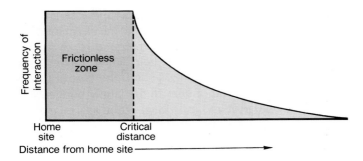

Figure 8.5

This general diagram indicates how distance is observed by most people. For each activity, there is a distance beyond which the intensity of contact declines. This is called the *critical distance*, if distance alone is being considered, or the *critical isochrone*, if time is the measuring rod. For the distance up to the critical distance, a frictionless zone is identified in which time or distance considerations do not effectively figure in the trip decision.

of origin. The tendency is for the frequency of trips to fall off very rapidly beyond an individual's critical distance. Figure 8.5 illustrates this principle with regard to journeys from the home site. The **critical distance** is the distance beyond which cost, effort, and means play an overriding role in our willingness to travel.

For example, a small child will make many trips up and down his or her block but is inhibited by parental admonitions from crossing the street. Different but equally effective constraints control adult behavior. Daily or weekly shopping may be within the critical distance of an individual, and little thought may be given to the cost or the effort involved, but shopping for special goods is relegated to infrequent trips, and cost and effort are considered.

Spatial Interaction and the Accumulation of Information

The critical distance is different for each person. The variables of life-cycle stage, mobility, and opportunity, together with an individual's interests and demands, help to define how much and how far a person will travel. On the basis of these variables, we can make inferences about the likelihood that a person will gain more or less information about his or her activity space and the space beyond.

We gain information about the world from many sources (Figure 8.6). Although information obtained from radio, television, and newspapers is important to us, face-to-face contact is assumed to be the most effective means

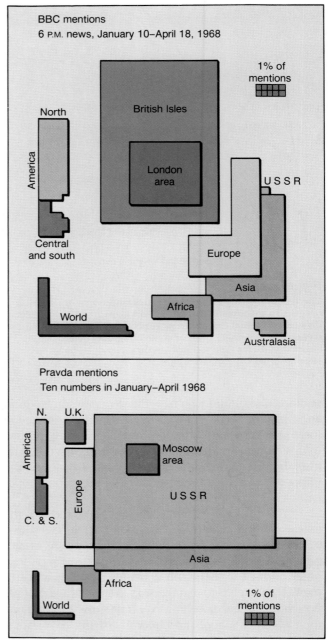

Figure 8.6

View of the world from the British Broadcasting Company and *Pravda*. Pick up a newspaper and see how much of it is devoted to local, national, and international news. You will discover that in general, as distance increases, the information about places decreases. Perhaps the reason why there is more London news on the BBC than Moscow news in *Pravda* is that London contains a greater proportion of the population of the British Isles than does Moscow of the USSR.

of communication. If we combine the ideas of activity space and distance decay, we see that as the distance away from the home place increases, the number of possible face-to-face contacts usually decreases. We expect more human interactions at short distances than at long distances. Where population densities are high, as in cities (particularly central business districts during business hours), the spatial interaction between individuals can be at a very high level, which is one reason why these centers of commerce and entertainment are often also centers for the development of new ideas.

Barriers to Interaction

The fact that the possible number of interactions is high, however, does not necessarily mean that the effective occurrence of interactions will be high. In crowded areas, people commonly set psychological barriers around themselves so that only a limited number of interactions take place. The barriers are raised in defense against information overload and for psychological well-being. We must have a sense of privacy in order to filter out the information that does not directly concern us. As a result, individuals tend to reduce their interests to a narrow range when they find themselves in crowded situations, allowing their wider interests to be satisfied by use of the communications media.

Within any population, there are those who will not *innovate,* that is, who will not accept new ideas. In some societies—particularly traditional, dispersed agricultural societies—the number of nonadopters for any innovation may be very high indeed. In other societies where innovation and change are given high status, and where the innovations are not risky, ideas spread much more rapidly.

Cost represents another barrier to interaction. Relatives, friends, and associates living long distances apart may find it difficult to afford to interact. The frequency and time allocated to telephone communication, a relatively inexpensive form of interaction, is very much a function of the rate structure, which, of course, favors short-distance interaction.

Finally, there are the obvious barriers to interaction such as mountains, oceans, rivers, and differing religions, languages, ideologies, and political systems.

Spatial Interaction and Innovation

The probability that new ideas will be generated out of old ideas is a function of the number of available old ideas in contact with one another. People who specialize in a particular field of interest seek out others with whom they wish to interact. The gathering place has a higher density than previously. Crowded central cities are characteristically composed of specialists in very narrow fields of interest. Consequently, under short-distance, high-density circumstances, the old ideas are given a hearing and new ideas are generated by the interaction. New inventions and new social movements usually arise in circumstances of high spatial interaction. An exception, of course, is the case of intensely traditional societies—Japan in the 17th and 18th centuries, for example—where the culture rejects innovation and clings steadfastly to customary ideas and methods.

The culture hearths—introduced in Chapter 6—of an earlier time were the most densely settled, high-interaction centers of the world in their day. Today, the great national and regional capital cities attract people who want or need to interact with others in special-interest fields. The association of population concentrations and the expression of human ingenuity has long been noted. The home addresses recorded for patent applicants by the U.S. Patent Office over the last century indicate that the inventors were typically residents of major urban centers, presumably closely in contact and exchanging ideas with those in shared fields of interest. It still appears that the metropolitan centers of the world attract those who are young and ambitious, and that face-to-face or word-of-mouth contact is important in the creation of new ideas and products. The recent revolution in communications that now allows for inexpensive interaction by a variety of telephone services and computer equipment have suggested to some that the traditional importance of cities as collectors of creative talent may decline in the future.

Figure 8.7

The major paths of the early migration of Germans to America constitute a diffusion process. The first colonists landed and many settled in Boston, New York, Philadelphia, Baltimore, and Charleston. The arrows show main routes of inland migration (diffusion) until the latter part of the 19th century. The migrants carried with them such aspects of their culture as religion, language, and food preferences.

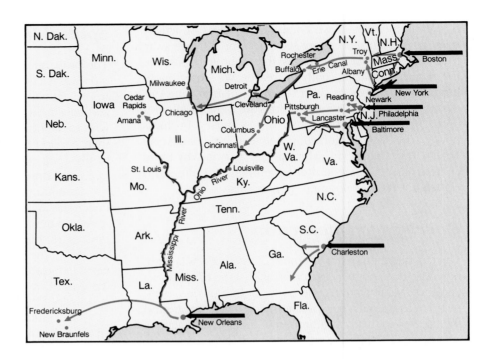

Spatial Diffusion

When a product or an idea is dispersed outward from a center of origin, the process is known as **spatial diffusion.** Ideas generated in a center of activity will remain there unless some process is available for their spread. Innovations—ideas that change people's behavior—spread in various ways. Some new concepts are so obviously advantageous that they are quickly put to use by those who can profit from their employment. A new development in petroleum extraction may promise such material reward as to lead to its quick adoption by all of the major petroleum companies, irrespective of their distance from the point of introduction. The new strains of wheat and rice that were part of the "Green Revolution" were quickly made known to agronomists in all cereal-producing countries, but they were more slowly taken up in poor countries, which could benefit most from them, partly because the farmers had difficulty paying for them.

Many innovations are of little consequence by themselves, but sometimes the widespread adoption of seemingly inconsequential innovations brings about large changes when viewed over a period of time. A new tune, "adopted" by a few people, may lead many individuals to fancy that tune plus others of a similar sound, which in turn may have a bearing on dance routines, which in turn may bear on clothing selection, which in turn may affect retailers' advertising campaigns and consumers'

spending patterns. Eventually a new cultural form will be identified that may have an important impact on the thinking processes of the adopters and on those who come into contact with the adopters. Notice that a broad definition of innovation is used, but notice also that what is important is whether or not innovations are adopted.

In spatial terms, we may identify several different diffusion processes. These are distance-affected diffusion, population-density-affected diffusion, and hierarchical diffusion.

Distance-Affected Diffusion

Let us suppose that an automobile enthusiast develops a gasoline additive that noticeably improves the performance of his or her car. Suppose further that the enthusiast shows friends and associates the invention and that they in turn tell others. This process is similar to the spread of ripples in calm water. The innovation will continue to diffuse until barriers are met (that is, people not interested in adopting the new idea) or until the area is saturated (that is, no more people can be contacted because the boundaries of the island or the country or the activity spaces are reached). This **distance-affected diffusion** process follows the rules of distance decay at each step: Short-distance contacts are more likely than long-distance contacts, but the idea may have spread far from the original site over time. In like manner, humans, by migration, carry with them their cultural characteristics that may diffuse along the route of movement or at the destination (see Figure 8.7 and Chapter 6).

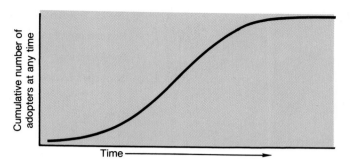

Figure 8.8

The diffusion of innovations over time. The number of adopters of an innovation rises at an increasing rate until the point at which about one-half of the total who ultimately decide to adopt the innovation have done so. At this point, the number of adopters increases at a decreasing rate.

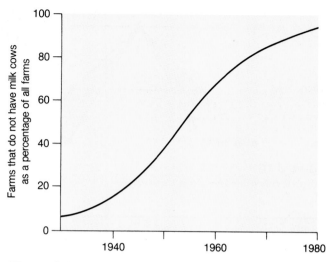

Figure 8.9

Farmers in Illinois found it disadvantageous to herd dairy cattle during the period 1930 to 1978. The idea of discontinuing having milk cows was adopted slowly, but increased rapidly from 1940 to 1960. Compare this real example with the theoretical diagram shown in Figure 8.8.

A number of characteristics of this kind of diffusion are worth noting. If the idea has merit in the eyes of potential adopters, the number of contacts of adopters with potential adopters will compound. Consequently, the innovation will spread slowly at first and then more and more rapidly, until there is saturation or a barrier is reached. The incidence of adoption is represented by the S-shaped curve in Figure 8.8 and shown for an actual case in Figure 8.9. In a similar manner, the area in which the adopters are located will at first be small, and then the area will enlarge at a faster and faster rate. The spreading process will slow as the available areas and/or people decrease. The process is precisely the same as the spread of a contagious disease (Figure 6.14).

Population-Density-Affected Diffusion

If an inventor's idea fell into the hands of a commercial distributor, the diffusion process would follow a somewhat different course from that discussed previously. The distributor might "force" the idea into the minds of individuals by using the mass media. If the media were local in impact, such as newspapers, then the pattern of adoptions would be similar to that described above. If, however, a nationwide television, newspaper, or magazine advertising campaign were undertaken, the innovation would become known in numbers roughly corresponding to the population density. Where more people live, there would, of course, be more potential adopters. Economic or other barriers may also affect the diffusion. One immediately sees, however, why large TV markets are so valuable and why national advertising is so expensive.

This type of **population-density-affected diffusion** process may act together with the distance-decay process. Many of those who accept the innovation after learning of it in the mass media will tell others, so that the distance-decay effect will begin to take over soon after

the original contact is made. Each type of medium has its own level of effectiveness. Advertisers have found that they must repeat messages time and again before the messages are accepted as important information. This fact says something about the effectiveness of the mass media as opposed to, say, face-to-face contact. It may also be a comment on the type of innovations advertised in the mass media.

Hierarchical Diffusion

The third way in which innovations are spread combines some aspects of the first two plus the inclusion of a new element: the hierarchy. A *hierarchy* is a classification of objects into categories so that each category is increasingly complex or has increasingly higher status. Hierarchies are found in many systems of organization, such as government offices (the organizational chart); universities (instructors, professors, deans, and so on); and urban centers (villages, towns, regional centers, metropolises).

As an example, let us suppose that a new way of processing passengers is adopted at a major airport. Information on the innovation is spread among managers at comparably sized airports. Eventually, the innovation may be adopted at the airports of smaller cities, and so on. The innovation was spread among those in the

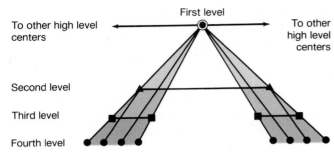

Figure 8.10

A four-level communication hierarchy.

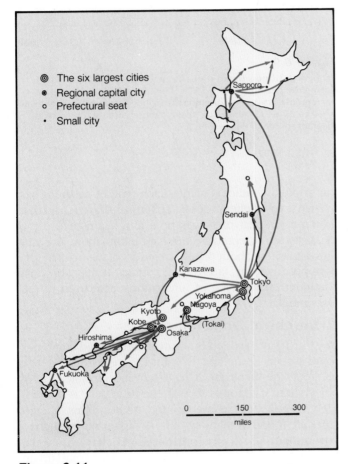

Figure 8.11

Rotary Clubs, an international service association, were established in the large cities of Japan during the 1920s. New clubs were established under the sponsorship of the original ones. This map shows the pattern of diffusion. Note the regional and spatial effects of the diffusion process.

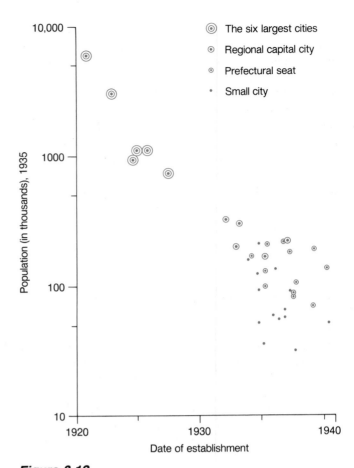

Figure 8.12

This diagram shows the hierarchical diffusion of Rotary Clubs in Japan. The largest cities were the first centers of Rotary Clubs, followed by cities at lower and lower levels of urban population and city functions.

highest level of the airport hierarchy, and then further diffusion took place when it was possible to spread to lower levels in the hierarchy. A hypothetical scheme showing how a four-level hierarchy might be connected in the flow of information is shown in Figure 8.10. Note that the lowest-level centers are connected to higher-level centers but not to each other. Note, too, that connections might bypass intermediate levels and link only with the highest-level center.

Hierarchical diffusion often takes place simultaneously with the other two kinds of diffusion. One might expect variations when the density of high-level centers is great and when the distances between centers are short. A quick and inexpensive way to spread an idea is to communicate information about it at high-order hierarchical levels. Then the three types of diffusion processes may be used most effectively; even while an idea is diffusing through a high level in the hierarchy, it is also spreading outward from the high-level centers. Consequently, low-level centers that are a short distance from high-level centers may be apprised of the innovation before more distant medium-level centers. People living in suburbs and small towns near a large city are privy to much that is new in the large city, as are individuals in other large cities half a continent away. Figures 8.11 and 8.12 show these patterns for a case taken from Japan.

All these forms of diffusion operate in the spread of culture. They add their impact to the enlargements of activity space, and the consequences are the spatial interaction and innovation discussed earlier in this chapter. We should also recall from Chapter 6 that migration, invasions, selective cultural adoptions, and cultural transference aid the diffusion of innovation. These broader movements and exchanges represent interactions of people beyond their usual activity spaces. For a further understanding of these interactions, movements, and migrations, it is useful to consider how individuals think of areas that are essentially unfamiliar to them.

Perception of the Environment

The term **environmental perception** refers to our awareness, as individuals, of home and distant places, and the beliefs we have about them. It involves our feelings, reasoned or irrational, about the complex of natural and cultural characteristics of an area. Whether our view accords with that of others or truly reflects the "real" world seen in abstract descriptive terms is not the major concern. Our perceptions are the important thing, for the decisions people make about the use of their lives are not necessarily based on reality but on their *perceptions of reality*.

Psychologists and geographers are interested in determining how we arrive at our environmental perceptions. We shall focus attention on perceptions of places beyond our normal activity space, although the question of how we view areas with which we are quite familiar is also of interest. In technologically advanced societies, television and radio, magazines and newspapers, books and lectures, travel brochures, and hearsay all combine to help us develop a mental picture of unfamiliar places. Again, however, the most effectively transmitted information seems to come from word-of-mouth reports. These may be in the form of letters or visits from relatives, friends, and associates. Probably the strongest lines of attachment to relatively unknown areas develop through the information supplied by family members and friends.

As we know, our knowledge of close places is greater than our knowledge of far-away places. But barriers to information flow give rise to *directional biases*. Not having friends or relatives in one part of a country may represent a barrier to individuals, so that interest in and knowledge of the area beyond the "unknown" region are sketchy. In the United States, both northerners and southerners tend to be less well informed about each other's areas than about the western part of the country.

Traditional communication lines in the United States follow an east–west rather than a north–south direction, the result of early migration patterns, business connections, and the pattern of the development of principal cities.

Mental Maps

When information about a place is sketchy, blurred pictures develop. These images influence the impression we have of places and cannot be discounted. Many important decisions are made on the basis of incomplete information or biased reports. We might say that each individual has a **mental map** of the world. No single person, of course, has a true and complete image of the world; therefore, there can be no completely accurate mental map. In fact, the best mental map that most individuals have is that of their own activity space.

No one can reproduce on paper an exact replica of the mental image that he or she might have of an area. The study of mental maps must by necessity be indirect. If we want to know how particular people envisage their town, we must either ask them questions about the town or ask them to draw sketch maps. Although the result will not be a completely accurate picture of what they have in mind, it will be suggestive of the mental map.

Whenever individuals think about a place or how to get to a place, they produce a mental map. What they believe to be unnecessary details are left out, and only the important elements are incorporated. Those elements usually include awareness that the object or the destination does indeed exist, some conception of the distances separating the starting point and the named object(s), and a feeling for the directional relationships between points. A mental route map might also include reference points to be encountered on the chosen path of connection or on alternate lines of travel. Although mental maps are highly personalized, people with similar experiences tend to give similar answers to questions about the environment and to produce roughly comparable sketch maps.

Awareness of places is usually accompanied by opinions about them, but there is no necessary relationship between the depth of knowledge and the perceptions held. In general, the more familiar we are with a locale, the more sound will be the factual basis of our mental image of it. But individuals form firm impressions of places with which they have no personal experience at all, and these ideas may color travel or migrational decisions.

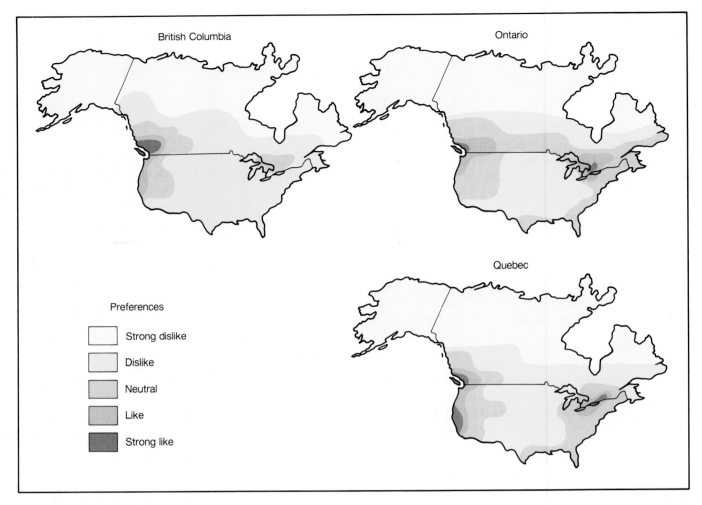

Figure 8.13

Each of these maps shows the residential preference of a sampled group of Canadians from the provinces of British Columbia, Ontario, and Quebec, respectively. Note that each group of respondents prefers its own area, but that all like the Canadian and United States west coasts.

One way to ascertain how individuals see the environment is to ask them what they think of various places. For instance, they may be asked to rate places according to desirability—perhaps residential desirability—or to make a list of the 10 best and the 10 worst cities in a region such as the United States. With the exception of a few glamour spots, such as Aspen, Colorado, certain regularities appear in such studies. Figure 8.13 presents some Canadian college students' ideas about desirable places to live. These and comparable mental maps suggest that near places are preferred to far places unless a lot of information is available about the far places. Places with similar cultural forms are preferred, as are places with high standards of living. Individuals tend to be indifferent to unfamiliar places, and to dislike unfamiliar areas that have competing cultural interests (such as disliked political and military activities) or a physical environment known to be unpleasant.

Mentally, people tend to place their own location in a central position, to increase the size of things nearby, and to decrease the size of all else. The regional study, The Yurok World View (Page 418) demonstrates these conclusions in the collective mental map of a primitive society. Figure 8.14 gives two examples of the way people are affected by the familiar. Also, as we grow older our perspectives change (Figure 8.15).

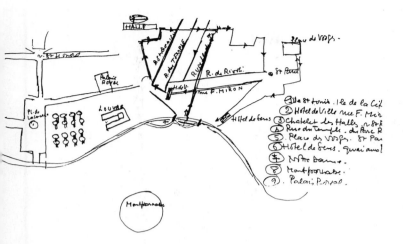

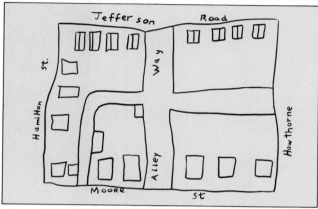

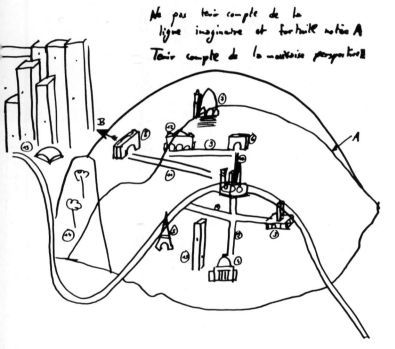

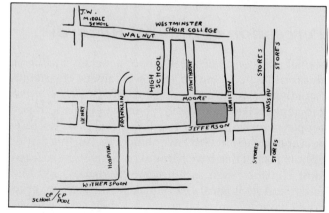

Figure 8.14

Two views of Paris. Several residents of Paris were asked to draw a map of the city. The first response is from a 50-year-old woman who ignored most of Paris and focused on the neighborhood in which she lived. The second sketch was done by a 25-year-old commercial artist who had recently left the university. Note how the skyscrapers at the city's edge are highlights—features absent in the maps of the older Parisians.

Figure 8.15

Three children, aged 6, 10, and 13, who lived in the same house, were asked to draw maps of their neighborhood. No further instructions were given. Notice how perspectives broaden and neighborhoods expand with age. For the 6-year-old, the "neighborhood" consisted of the houses on either side of her own. The square block on which she lived was the neighborhood for the 10-year-old. The wider activity space of the 13-year-old is also evident. The square block that the 10-year-old drew is colored red in the 13-year-old's sketch.

Figure 8.16

High seas have undermined this house in Aptos, California.

Perception of Natural Hazards

Mental maps of home areas do not generally include as an overriding concern an acknowledgment of potential natural dangers. An intriguing area of research to which geographers have addressed themselves deals with how people perceive *natural hazards,* defined as processes or events in the physical environment that are not caused by humans but that have harmful consequences to them. Most climatic (such as hurricanes, tornados, and blizzards) and geological (earthquakes and volcanic eruptions) hazards cannot be prevented; their consequences may be disastrous. The hurricanes that struck the delta area of Bangladesh in 1970 and again in 1985 left at least 500,000 people dead; the 1976 earthquake in the Tangshan area of China devasted a major urban-industrial complex with 700,000 to 1 million casualties. These were major and exceptional natural hazards, but more common occurrences are experienced and are apparently discounted by those in their areas of effect. Johnstown, Pennsylvania, has suffered recurrent floods, and yet its residents rebuild; violent storms frequently strike the Gulf and East coasts of the United States, and people remain or return.

Why do people choose to settle in high-hazard areas, in spite of the potential threat to their lives and property? Why do hundreds of thousands of people live along the San Andreas fault in California, build houses in Pacific coastal areas with a known risk of severe erosion during storms (Figure 8.16), or farm in flood-prone areas along the Mississippi River? What makes the risk worth taking? People perceive hazardous areas differently and there are many reasons that some people choose to settle in them. Of major importance is the fact that specific hazards are relatively rare. Many people think that the likelihood of an earthquake, flood, or other natural calamity is sufficiently remote that it is not economically feasible to protect themselves against it. They are also influenced by the fact that the scientists who study such hazards may themselves differ on the probability of an event or on the damage that it may inflict. And, in fact, the prediction of hazards is not an exact science; it is based on the calculation of the probability of occurrences of uncommon events.

People are also influenced by their past experiences in high-hazard areas. If they have not suffered much damage in the past, they may be optimistic about the future. If past damage has been great, they may think that

TABLE 8.1

Common Responses to the Uncertainty of Natural Hazards

Eliminate the hazard	
Deny or denigrate its existence	*Deny or denigrate its recurrence*
"We have no floods here, only high water."	"Lightning never strikes twice in the same place."
"It can't happen here."	"It's a freak of nature."

Eliminate the uncertainty	
Make it determinate and knowable	*Transfer uncertainty to a higher power*
"Seven years of great plenty. . . . After them seven years of famine."	"It's in the hands of God."
"Floods come every five years."	"The government is taking care of it."

Ian Burton and Robert Kates, "The Perception of Natural Hazards in Resource Management." Reprinted with permission from *3 Nat. Res. J.* 435 (1964), published by the University of New Mexico School of Law, Albuquerque, N.M.

the probability of similar occurrences in the future is low (Table 8.1). People's memories can be short. In the year following an earthquake, for example, a sense of security grows, building codes or their interpretation are relaxed, and population in the area increases.

High-hazard areas are often sought out because they possess desirable topography or scenic views, as do the Atlantic and Pacific coasts. Once people have purchased property in a known hazard area, they may be unable to sell it for a reasonable price even if they want to. They think that they have no choice but to remain and protect their investment. The cultural hazard—loss of livelihood and investment—appears to be more serious than whatever natural hazards there may be.

Migration

When the continental glaciers began their retreat some 11,000 years ago, the action space and awareness space of Stone Age humans were limited. Due to the pressures of numbers, availability of food, changes in climate, and migration of game, those spaces were collectively enlarged to encompass the world. An important aspect of human history has been the migration of peoples, the evolution of their separate cultures, and the diffusions of those cultures and their components by interchange and communication. We have touched on portions of that story in Chapter 6. These past collective movements have,

of course, their modern counterparts. The settlement of North America, Australia, and New Zealand involved great long-distance movements of peoples and the impact of their cultures on largely natural landscapes. The flight of refugees from past and recent wars, the settlement of Jews in Israel, the movement south in Africa from the drought-stricken Sahel, the recent migration of workers to the United States from overpopulated Mexico, and many other examples of mass movement come quickly to mind. In all cases, societies intermixed, ideas diffused, and history was altered. The vast numbers who suffered the hardships of movement to the American West, like Catherine Haun, were the predecessors of the regionally relocating job seekers and retirees of today.

In the study of those individual actions that result in the collective responses summarized in behavioral geography, two types of long-distance movement are recognized. The first is not considered migration; it is simply the *planned two-way trip*. It includes the exciting, perhaps historically decisive, daring journey of exploration about which we read in history, or the more common contemporary vacation, business, or social trip in which modern societies indulge so freely. The latter sort of trip, of course, enhances the mental maps and enlarges the awareness space of the participants and may be the prelude to migration. It may also contribute to the diffusion of information through cultural contact, but its individual impact is small because it is so transitory.

Much more important is **migration:** a relocation of both residential environment and activity space. Naturally, the length of the move and its degree of disruption of normal household activities raise distinctions important in the study of migration. A change of residence from the central city to the suburbs certainly changes the activity space of schoolchildren and of adults in many of their nonworking activities, but the workers may still retain the city—indeed, the same place of employment there—as an action space. On the other hand, immigration from Europe to the United States and the massive farm-to-city movements of rural Americans late in the last and early in the present centuries meant a total change of all aspects of behavioral patterns.

The Decision to Migrate

The decision to move is a cultural and temporal variable. Nomads fleeing the famine and spreading deserts in the Sahel obviously are motivated by different considerations than those of the executive receiving a job transfer to Chicago, the resident of Appalachia seeking factory employment in the city, or the retired couple searching

Figure 8.17
Some of the approximately 350,000 Vietnamese "boat people" who fled their country after 1975 to seek a new life in a new location. The Office of the United Nations High Commissioner for Refugees in 1987 estimated that worldwide there were over 12 million refugees, people who had left their countries because of the well-founded fear of being persecuted for their political views, economic activities, or membership in certain social or religious groups.

for sun and sand. In general, people who decide to migrate are seeking better economic, political, or cultural conditions or certain amenities. Of course, as in the case of the Hauns, the reasons for migration are frequently a combination of several of these.

Economic causes have impelled more migrations than any other single category. If migrants face unsatisfactory conditions at home (such as unemployment or famine) and believe that the economic opportunities are better elsewhere, they will consider a move. To a nomadic herdsman, better opportunities might be abundant grass and water; to a Western emigrant, rich soil; to a modern worker, the promise of a high-paying job in a Los Angeles suburb. Attractions like these are called **pull factors,** and conditions that contribute to dissatisfaction at home are called **push factors.** Very often migration is a result of both push and pull factors.

The desire to escape war and persecution at home and to pursue the promise of freedom in a new location is a political incentive for migration (Figure 8.17). Americans are familiar with the history of settlers who immigrated to North America seeking religious and political freedom. In more recent times, the United States has received hundreds of thousands of refugees from countries such as Hungary (following the uprising of 1956), Cuba (after its takeover by Fidel Castro), and Vietnam (after the fall of South Vietnam in 1975). The massive movements of Hindus and Muslims across the Indian subcontinent in 1947, when Pakistan and India were established

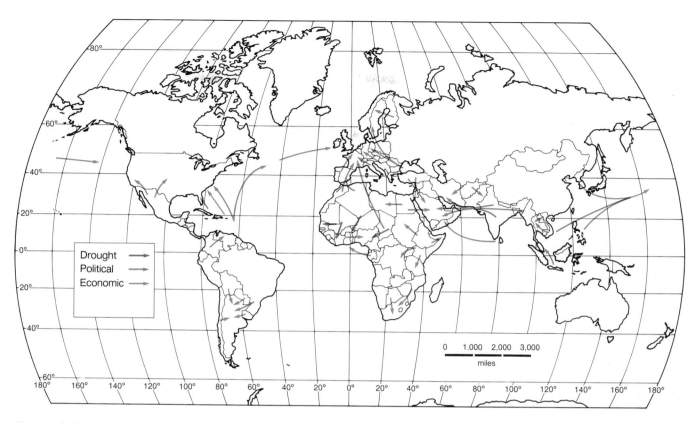

Figure 8.18

Major recent international migration streams. Most of those shown are economic, that is, people moving to the location of available jobs. In the case of the Vietnamese, Afghan, and Sahel migrations, however, politics or drought is the fundamental cause.

as governing entities, and the exodus of Jews fleeing persecution in Nazi Germany in the 1930s are other examples of politically inspired moves. Figure 8.18 identifies recent major migrations.

Migration normally involves a hierarchy of decisions. Once people have decided to move and have selected a general destination (such as America or the West), they must still choose a particular site at which to settle. At this scale, cultural variables can be important pull factors. Migrants tend to be attracted to areas where the language, the religion, and the racial or ethnic background of the inhabitants is similar to their own. This similarity can help the migrants to feel at home when they arrive at their destination and may make it easier to find a job and to become assimilated into the new culture. The Chinatowns and little Italys of large cities attest to the drawing power of cultural factors, as we saw in Chapter 6.

Another set of inducements comes under the general heading *amenities,* the particularly attractive or agreeable features of a place. Amenities may be natural (mountains, oceans, climate, and the like) or cultural (the arts and music opportunities available in large cities).

They are particularly important to relatively affluent people seeking "the good life." They help account for the attractiveness for retirees of the so-called Sunbelt states in this country; a similar movement to the south coast has also been observed in countries such as the United Kingdom and France.

The significance of the various factors also varies according to the age, sex, education, and economic status of the migrants. For the modern American, reasons to migrate have been summarized into a limited number of categories that are not mutually exclusive. They include (1) changes in life cycle (getting married, having children, getting a divorce, or needing less dwelling space when the children leave home); (2) changes in the career cycle (getting a first job or a promotion, receiving a job transfer, or retiring); (3) forced migrations associated with urban development, construction projects, and the like; (4) neighborhood changes from which there is flight (pressures from new and unwelcome ethnic groups, building deterioration, street gangs, and so on); and (5) changes of residence associated with individual personality. Some people simply seem to move often for no

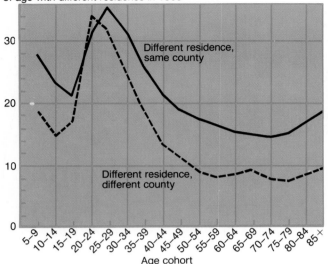

Percent of 1970 population more than 5 years
of age with different residence in 1965

Different residence,
same county

Different residence,
different county

Age cohort

Figure 8.19

Percent of 1970 population more than 5 years of age with different residence in 1965. Young adults figured most prominently in both short- and long-distance moves between 1965 and 1970. Mobility is distinctly age-related.

easily discernible reason, whereas others, *stayers,* settle into a community permanently. Of course, a country like the Soviet Union, with its limitations on emigration, a tightly controlled system of internal passports, restrictions on job changes, and severe housing shortages, would see a totally different set of summary migration factors.

The conditions that contribute to mobility tend to change over time. The group that in most societies has always been the most mobile is young adults (Figure 8.19). They are the members of society who are launching careers and making their first decisions about occupation and location. They have the fewest responsibilities of all adults and are not as strongly tied to family and institutions as older people. Most of the major voluntary migrations have been composed primarily of young people who suffered from lack of opportunities in the home area and who were easily able to take advantage of opportunities elsewhere.

The concept of **place utility** helps us to understand the decision-making process that potential migrants undergo. It is the value that an individual places on each known, potential migration site. The decision to migrate is a reflection of the appraisal by a potential migrant of the current home site as opposed to other sites of which something is known. The individual may also adjust to conditions at the home site and thus decide not to migrate.

In the evaluation of the place utility of each potential site, the decision maker considers not only present relative place utility but also expected place utility. The

evaluations are matched with the individual's aspiration level, that is, the level of accomplishment or ambition that the person sees for himself or herself. Aspirations tend to be adjusted to what an individual considers attainable. If one is satisfied with present circumstances, search behavior is not initiated. If, on the other hand, there is dissatisfaction, a utility is assigned to each of the possible new sites. The utility is based on past or expected future rewards at the various sites. Because the new places are unfamiliar, the information received about them acts as a substitute for the personal experience of the home site. The decision maker can do no more than sample information about the new sites, and, of course, there may be sampling errors.

One goal of the potential migrant is to minimize uncertainty. Most decision makers either elect not to migrate or postpone the decision unless uncertainty can be sufficiently lowered. Most migrants reduce uncertainty by imitating the successful procedures that others have followed. We see that the decision to migrate is not a perfunctory, spur-of-the-moment reaction to information. It is usually a long, drawn-out process based on a great deal of sifting and evaluation of data.

An example of some of these observations and generalizations in action can be seen in the case of the large numbers of young Mexicans who have migrated both legally and illegally to the United States over the last 20 years. Faced with rural poverty and overpopulation at home, they regard the place utility there as minimal. Their space-searching ability is, however, limited by lack of money and by lack of alternatives in their native land. With a willingness to work and with aspirations for success—perhaps wealth—in the United States, they are eager listeners to friends and relatives who tell of numerous job opportunities in the United States. Hundreds of thousands, presented with the glittering prospect of a job, low-paying though it might be, quickly place high utility on perhaps a temporary relocation (maybe 5 or 10 years) to the United States. Many know that dangerous risks are involved if they attempt to enter the country illegally, but the rewards are worth the risk. Their arrival indicates their assignment of higher utility to the new site than to the old.

In the 20th century, nearly all countries have experienced a movement of population to cities from rural areas. The migration has presumably paralleled the number of perceived opportunities within the cities and convictions of absence of place utility in the rural districts. Perceptions, of course, do not necessarily accord with reality. The enormous influx of rural folk to major urban areas in the developing countries and the economic destitution of many of those in-migrants suggest recurrent gross misperceptions, faulty information, and sampling error. In Chapter 11, we discuss the reasons for this phenomenon.

Rational Decision Making: An Example

An example of a structure developed for making a rational decision whether or not to migrate can be found in the case of the selection by a hypothetical unmarried male who is a recent college graduate. This person partitioned his consideration into four major aspects: monetary compensation, geographic location, travel requirements, and the nature of the work. Each of these aspects was subdivided, and some were subdivided again, as illustrated in the table.

Each category was given a value on a scale from 0 to 1.0, according to the importance that the potential migrant placed on it. The restriction that the total of the values must add up to 1.0 enabled the graduate to multiply across any path to find the true relative worth of the 15 categories. In the table, we see that the graduate placed the greatest emphasis on starting salary and the kind of work to be done at the outset. Retirement and insurance benefits were least important.

Once the graduate had a clearer idea of his values, he could take each of his job opportunities and evaluate them according to the best information available about each subcategory. For each possible job, he assigned a value to every category. According to the system 90–100 is excellent, 80–90 is good, 70–80 is fair, and so on. By multiplying each of these by the relative worth of each category and summing each column, the graduate could make an objective decision. For example, for job opportunity No. 1, 95 × .040 = 3.80; 80 × .059 = 4.72; and so on. For job opportunity No. 1, the score was 71.8; for No. 2, the value was 71.2; and for No. 3, it was 77.3.

This decision was based on the subject's own value system. Each person's categories and weighting systems would be different. In the case illustrated, the graduate finally ranked the local opportunity ahead of the national and regional ones. The ranking led to the decision not to migrate at the present time.

				Relative worth placed on each category	Rating on job opportunities		
					1 Large national company	2 Regional office	3 Local job
Nature of work	Current and future work features		Management training program	.040	× 95 = 3.80	80 = 3.20	50 = 2.00
			Variety of work	.059	× 80 = 4.72	75 = 4.43	70 = 4.13
			Technical challenge	.049	× 80 = 3.92	80 = 3.92	95 = 4.66
	Immediate work features			.132	× 70 = 9.24	80 = 10.56	90 = 11.88
Travel requirements	Long trips away from office		Trip lengths	.082	× 90 = 7.38	70 = 5.74	60 = 4.92
			Proportion time away	.054	× 80 = 4.32	70 = 3.78	60 = 3.24
	Daily commuting characteristics			.034	× 50 = 1.70	70 = 2.38	100 = 3.40
Geographic location			Climate	.034	× 80 = 2.72	70 = 2.38	60 = 2.04
			Proximity to leisure-time activities	.068	× 90 = 6.12	80 = 5.44	60 = 4.08
			Proximity to relatives	.068	× 60 = 4.08	70 = 4.76	100 = 6.80
Monetary compensation	Future salary prospects		10-year increase	.035	× 90 = 3.15	80 = 2.80	80 = 2.80
			3-year increase	.064	× 60 = 3.84	70 = 4.48	80 = 5.12
	Immediate prospects	Fringe benefits	Retirement	.009	× 95 = 0.86	70 = 0.63	60 = 0.54
			Insurance	.014	× 95 = 1.33	70 = 0.98	60 = 0.84
		Starting salary		.209	× 70 = 14.63	75 = 15.68	100 = 20.90
				1.000	71.81	71.16	77.35

Problems in the Sunbelt

Since 1970 there has been a net in-migration of over 12 million people to the region known as the Sunbelt, the tier of southern states stretching from the Atlantic to the Pacific Coast. This region captures the bulk of the nation's migrants and immigrants from other countries. Arizona's population grew to 2.7 million in 1980 from 1.8 million a decade before. Florida's growth rate for the decade was almost 50%. Ironically, the rapid growth of the Sunbelt causes problems that threaten some of the very qualities that make the area attractive.

Low energy costs, low wages, and low taxes attracted businesses and industries from the North to the Sunbelt. Both blue- and white-collar workers from the North followed the companies, drawn by the promise of jobs and the particular attractions of the region: warmth and sunshine, unspoiled land, recreational opportunities, and other amenities.

Although the Sunbelt continues to attract retirees, companies, and workers, its advantages are eroding rapidly, and growth itself is the reason.

With economic growth have come higher wages. In some parts of the region, taxes have risen already, and they will rise more as the states and cities try to cope with the services that population and economic growth demand. Across the Sunbelt, the rapid influx of people has strained the capacity of sewage treatment plants. Depending on which city is considered, poor roads, inadequate fire and police protection, and limited mass transit are problems. Florida and the Southwest face water shortages as well.

Many Sunbelt cities, although prosperous, exhibit the kinds of problems that we associate with northern and eastern cities: crowded freeways and traffic jams, noise, urban sprawl, rising crime rates, unemployment, poor inner cities, and air pollution. The clean air that attracted people with allergies,

asthma, and other respiratory problems to Arizona is being polluted not only by cars and mining operations but also by the foliage that the migrants planted to remind them of home. For example, the mulberry, a fast-growing tree that provides good shade, produces pollen levels four times as great as those of other plants. Over all, Arizona has experienced a tenfold increase in airborne pollen since 1960. The problems caused by growth tend to inhibit further growth. The high crime rate in Florida, for example, hurts the huge tourist industry, and higher wages in the Southeast deter new companies from moving there.

A city does not create urban problems independently of the people who live there. When a substantial number of those people move to a new city, the potential for the reappearance of the problems that they hoped to leave behind is considerable.

Barriers to Migration

Paralleling the incentives to migration is a set of disincentives, or barriers. They explain the fact that many people do not choose to move even when conditions are bad at home and are known to be better elsewhere. Migration depends on a knowledge of the opportunities in other areas. People with a limited knowledge of these opportunities are less likely to migrate than are better-informed individuals. Other barriers include physical features, the costs of moving, ties to individuals and institutions in the original activity space, and government regulations.

Physical barriers to travel include seas, mountains, swamps, deserts, and other natural features. In prehistoric times, they played an especially significant role in limiting movement. Thus the spread of the ice sheets across most of Europe in Pleistocene times was a barrier to both migration and human habitation. Physical barriers to movement have probably assumed less importance only within the last 400 years. The developments that made possible the great age of exploration, beginning about A.D. 1500, and the technological developments associated with industrialization have enabled people to conquer space more easily. With industrialization came improved forms of transportation that made

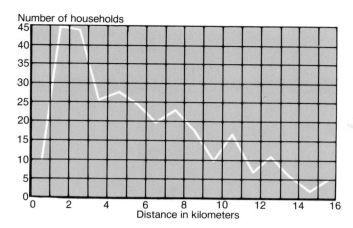

Figure 8.20

Distance between old and new residences in the Asby area of Sweden. Notice how the number of movers decreased with increasing distance.

travel faster, easier, and cheaper. Still, as the account of conditions on the Overland Trail illustrated, only a century and a half ago travel could be arduous, and in some parts of the world it remains so today.

Economic barriers to migration include the cost of travel and of establishing a residence elsewhere. Frequently, the additional expense of maintaining contact with those left behind is also a cost factor that must be considered. Normally all of these costs increase with the distance traveled and are a more significant barrier to travel for the poor than for the rich. Many immigrants to this country were married men who came alone; when they had acquired enough money, they sent for their families to join them. This phenomenon is still evident among recent immigrants from the Caribbean area to the United States, and among Turks, Yugoslavs, and West Indians who have settled in Northern and Western European countries.

The *cost* factor limits long-distance movement, but the larger the differential between present circumstances and perceived opportunities, the more individuals are willing to spend on moving. Figure 8.20 is an example of the effect of distance on relocation. For many people, especially older people, the differentials must be extraordinarily high for movement to take place.

Cultural factors also contribute to decisions not to migrate. Family, religious, ethnic, and community relationships defy the principle of differential opportunities. Many people will not migrate under any but the most pressing circumstances. The fear of change and human inertia—the fact that it is easier not to move than to do so—may be so great that people consider but reject a move. Ties to one's own country, cultural group, neighborhood, or family may be so strong as to compensate for the disadvantages of the home location. Returning migrants may convince potential leavers that the opportunities are not in fact better elsewhere—or, even if they are better, that they are not worth entering an alien culture or sacrificing home and family.

Restrictions on immigration and emigration consitute *political* barriers to migration. Many governments frown on movements into or outside their own borders. Recognizing that emigration might be economically or politically disadvantageous, many nations restrict outmigration. These restrictions may make it impossible for potential migrants to leave, and they certainly limit the number who can do so.

On the other hand, countries suffering from an excess of workers often encourage emigration. The huge migration of people to the Americas in the late 19th and early 20th centuries is a good example of perceived opportunities for economic gain far greater than in the home country. Many European countries were overpopulated, and their political and economic systems stifled domestic economic opportunity at a time when people were needed by American entrepreneurs hoping to increase their wealth in untapped resource-rich areas.

Countries where per capita incomes are high or are perceived as high are generally the most desired international destinations. In order to protect themselves against overwhelming migration streams, such countries have restrictive policies on immigration. In addition to absolute quotas on the number of immigrants (usually classified by country of origin), a country may impose other requirements, such as the possession of a labor permit or sponsorship by a recognized association.

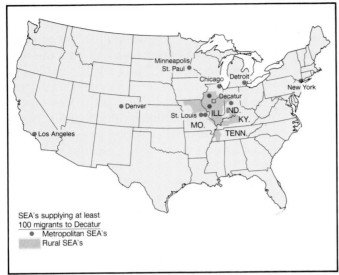

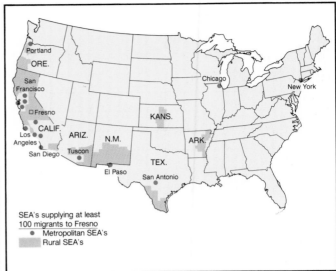

Figure 8.21

The in-migration fields of Decatur, Illinois, and Fresno, California. For Decatur, as expected, most of the in-migrants come from nearby rural areas and cities. For Fresno, the pattern is similar with an important addition—rural areas in the Southwest, Kansas, and Arkansas. This secondary pattern represents the constant westward movement of population in the United States.

Patterns of Migration

Several geographic concepts deal with patterns of migration. The first of these is the migration field. For any single place, the origin of its in-migrants and the destination of its out-migrants remain fairly stable spatially over time. Areas that dominate a locale's in- and out-migration patterns constitute the **migration fields** for the place in question (Figure 8.21). As we would expect, areas near the point of origin make up the largest part of the migration field. However, places far away, especially large cities, may also be prominent. These characteristics of migration fields are functions of the hierarchical movement to larger places (as we shall see) and the fact that there are so many people in large metropolitan areas that one might expect some migration into and out of them from most areas within a country.

Migration fields do not conform exactly to the diffusion concepts mentioned earlier. As shown by Figure 8.22, some migration fields reveal a distinctly **channelized** pattern of flow. The channels link areas that are socially and economically tied to one another by past migration patterns, by economic trade considerations, or by some other affinity. As a result, flows of migration along these channels are greater than would otherwise be the case. The movements of blacks from the southern United States to the North, of Scandinavians to Minnesota and Wisconsin, of Mexicans to such border states as California, Texas, and New Mexico, and of retirees to Florida and Arizona are all examples of channelized flows.

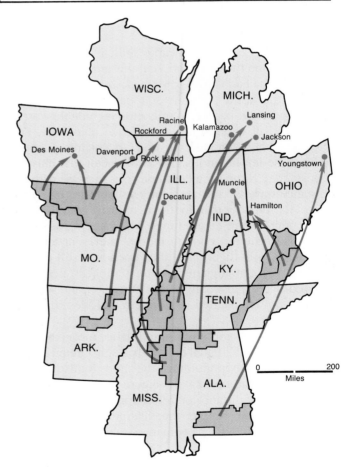

Figure 8.22

Channelized migration flows from the rural South to Midwestern cities of medium size. Distance is not necessarily the main determinant of flow direction. Perhaps through family and friendship links, the rural southern areas are tied to particular Midwestern destinations.

Third World International Migration

The young merchant assessed his options. If he remained at home, there was little chance that his life would improve materially. But if he were willing to take the risk, he could travel surreptiously to the border, slip across it at night, and enter one of the richest countries in the region . . . a country whose high rate of economic growth was marked by modern office buildings, highways, and well-stocked stores. If he weren't deported, he could expect to earn many times the annual wages he received at home.

The story of a Mexican immigrant to the United States? Not at all. Marcel Amao is Malian, and his destination was the Ivory Coast, which along with Nigeria is the richest of the West African countries. In recent years, immigrants have flooded into the Ivory Coast at such a rate that they now constitute about one-third of the country's population of 10 million. By the thousands, they come not only from Mali but also from Benin, Burkina Faso, Ghana, and Senegal.

That migration echoes many of the themes explored in this chapter. Most migrants leave their home for economic reasons; "pushed" by poverty, they are "pulled" by the chance to better their lot in the Ivory

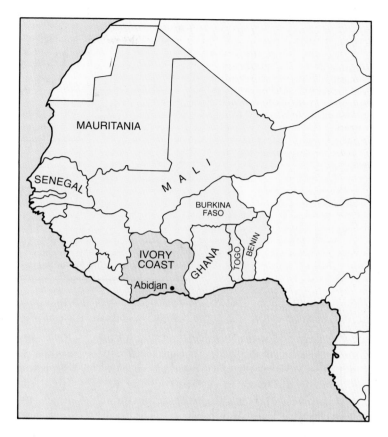

Coast. The migration is not strictly distance-affected; emigrants from Benin and Senegal bypass neighboring countries and cross two or three national borders before entering the Ivory Coast. The movement is somewhat channelized. In the capital city of Abidjan, people from Benin tend to be cooks and cabinet makers, while immigrants from the impoverished Sahelian nations of Burkina Faso and Mali find work as street sweepers and nightwatchmen, laundrymen, and sellers of cloth. Finally, migrants are both "stayers" and "leavers." In the words of its president, the Ivory Coast "is a country without a passport: one comes, one leaves, one stays, but more often one stays."

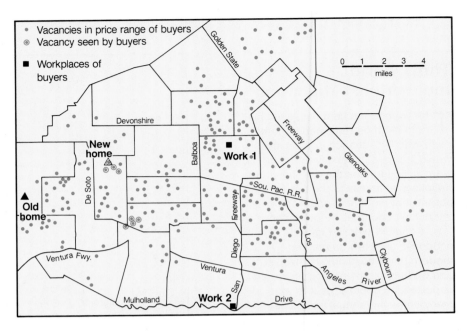

Figure 8.23
An example of spatial search in the San Fernando Valley area of Los Angeles. The dots represent the house vacancies in the price range of a sample family. Note (1) the relationship of the new house location to the workplaces of the married couple, (2) the relationship of the old house location to the new house location, and (3) the limited spatial search space. This example is typical of intra-urban moves.

Redrawn by permission from *Annals of the Association of American Geographers*, vol. 76, 1986, J. O. Huff.

Sometimes the channelized migration is specific to occupational as well as to ethnic groups. For example, nearly all newspaper vendors in New Delhi, in the north of India, come from one small district in Tamil Nadu, in the south of India. Most construction workers in New Delhi come either from Orissa, in eastern India, or Rajasthan, in the northwest, and taxicab drivers originate in the Punjab. The diamond trade of Bombay, India, is dominated by a network of about 250 related families who come from a small town several hundred miles to the north.

Return migration refers to those who, soon after migrating, decide to return to their point of origin. If freedom of movement is not restricted, it is not unusual for as many as 25% of all migrants to return to their place of origin. Unsuccessful migration is sometimes a result of inability to adjust to the new environment; more often, it is a result of false expectations based on distorted mental images of the destination at the time of the move. Myths, secondhand and false information, and people's own exaggerations contribute to what turn out to be mistaken decisions to move. Although return migration often represents the unsuccessful adjustment of individuals to a new environment, it does not necessarily mean that negative information about a place returns with the migrant. It usually means a reinforcement of the channel, as communication lines between the unsuccessful migrant and would-be migrants take on added meaning and understanding.

In addition to channelization, the influence of large cities causes migration fields to deviate from the distance-decay pattern. The concept of **hierarchical migration** helps in understanding the nature of migration fields. Earlier we noted that sometimes information diffuses according to a hierarchical rule, that is, from city to city at the highest level in the hierarchy and then to lower levels. Hierarchical migration, in a sense, is a response to that flow. The tendency is for individuals in domestic relocations to move up in the hierarchy, from small places to larger ones. Very often, migrants skip levels on the way up; only in periods of general economic decline is there considerable movement down the hierarchy. The suburbs of large cities are considered part of the metropolitan areas, so that movement from a town to a small suburb would be considered moving up the hierarchy. From this pattern, we can envisage information flowing to a place via a hierarchical routing. Once the information is digested, some respond by migrating to the area from which the information came.

Spatial Search

Many of the ideas presented in this chapter are related to people's decision making in a geographic environment; that is, after evaluating alternatives, people come to a conclusion about where to satisfy their desires. Geographers call the process by which the alternatives are evaluated **spatial search.** The seeker may or may not actually travel to the various destinations; one may pursue the search by evaluating the information that comes to one's residence. For example, searching for a new residence may entail reading newspaper advertisements.

The quality of the decision is very much a function of the quality of the available information. Some decisions, like the decision to buy a new home, are so important that individuals take months or even years to gather information. And, of course, in such circumstances, one is usually obliged to visit each of the possible home sites (Figure 8.23).

Decision making is based on a relationship between information availability and the preferences and motivations of the individual decision maker. Unfortunately, many decisions are made with inadequate or distorted information. Gathering information is a time-consuming activity. Many people are unable or unwilling to use their time for information gathering; many disappointments are based on insufficient information. Of course, even if the time were available and people were willing to gather information, the needed information often might just be unavailable. Many return migrants are responding to an inadequate understanding of the migration location.

Decision makers evaluate information by assessing the utility of each of the possible outcomes of the various courses of action. They depend on their image—perhaps a mental map—of the thing or place being considered, and they use some type of scaling or preference criteria, keeping in mind the uncertainties that exist. Their motivations are based on their belief and value systems. For example, does the decision maker expect instant gratification or is she or he more inclined to expect gratification in several weeks, months, or years? Does one think that family satisfaction is more important than individual satisfaction? And so on.

In a study of the search for housing and residential choice in Toronto, many of the variables and factors involved in spatial search were evaluated. Among other things, the study found that those looking for apartments spent much less time looking than did potential home owners. The difference was ascribed to the importance people place on buying as opposed to renting. It was found that apartment dwellers, even though they may complain about noise, inadequate space, and other problems, are not much less satisfied with their decision than home owners. Most young apartment dwellers think of their quarters as temporary and therefore set lower standards of comfort in order to take advantage of lower cost housing, more convenience, and less responsibility.

In terms of the search itself, people spend longer deciding to look for new housing than they do in actually inspecting and choosing it. Most home buyers look at seven or more housing units, compared to only about three for those who eventually choose apartments. A Los Angeles study concluded that those home buyers not leaving the metropolitan area altogether search for homes within a short distance of both their present home and their workplace. The limiting factor, it was determined, is the income level of the decision maker.

Conclusion

In this discussion of behavioral geography, we have emphasized the factors that influence how individuals use space. At the beginning of the chapter, we posed a number of questions, used as our theme. In considering how individuals view their environment, we spoke of the nature of the information available and of the age of people, their past experiences, and their values. The question of differences in the extent of space used led us to the concept of activity space. The age of an individual, his or her degree of mobility, and the availability of opportunities all play a role in defining the limits of individual activity space. We saw, however, that as distance increases, familiarity with the environment decreases. The concept of critical distance identifies the distance beyond which the decrease in familiarity with places begins to be significant.

How space is used is a function of all of the factors mentioned, but certain factors having to do with the diffusion of innovations indicate what opportunities exist for individuals living in various places. Distance, density, and hierarchical diffusion influence the geographic direction that cultural change will take. Finally, we examined how the ability of many to make well-reasoned, meaningful decisions is a function of the utility that they assign to places and the opportunities at those places.

Behavioral geographers view individual action in a special way. There is strong emphasis on information flow and the frictional effects of distance. In many respects, it is an interdisciplinary field in that the combined ideas of geographers, psychologists, economists, and other social scientists are crucial to an understanding of spatial behavior. Such behavior has an economic component, and it is that to which we next turn our attention.

Key Words

activity space *265*
channelized migration *284*
critical distance *268*
distance-affected diffusion *270*
distance decay *267*
environmental perception *273*
hierarchical diffusion *272*
hierarchical migration *286*
mental map *273*
migration *277*
migration field *284*
place utility *280*
population-density-affected diffusion *271*
push and pull factors *278*
return migration *286*
spatial diffusion *270*
spatial search *286*
territoriality *265*

For Review

1. What factors affect the areal extent of a person's activity space?

2. Recall the places that you have visited in the past week. In your movements, were the distance-decay and critical-distance rules operative? What variables affect an individual's critical distance?

3. Briefly distinguish between distance-affected diffusion, population-density-affected diffusion, and hierarchical diffusion. In what ways, if any, were these forms of diffusion in operation in the culture hearths discussed in Chapter 6?

4. On a blank piece of paper, and without any maps to guide you, draw a map of the United States, putting in state boundaries wherever possible; this is your mental map of the country. Compare it with a standard atlas map. What conclusions can you reach?

5. What considerations affect a decision to migrate? What is *place utility,* and how does its perception induce or inhibit migration?

6. What common barriers to migration exist? Why do most people migrate within their own country?

7. Some migration fields show a channelized flow of people. Select a particular channelized migration flow (such as the movement of Scandinavians to the United States, or people from the Great Plains to California, or southern blacks to the North) and explain why a channelized flow developed.

Suggested Readings

Brown, Lawrence A. *Innovation Diffusion: A New Perspective.* New York: Methuen, 1981.

Cox, Kevin R., and Reginald G. Golledge. *Behavioral Problems in Geography Revisited.* New York: Methuen, 1982.

Downs, Roger M., and David Stea. *Maps in Mind: Reflections on Cognitive Mapping.* New York: Harper & Row, 1977.

Fligstein, Neil. *Going North: Migration of Whites and Blacks from the South, 1900–1950.* New York: Academic Press, 1981.

Gold, John R. *An Introduction to Behavioural Geography.* New York: Oxford University Press, 1980.

Gould, Peter R. and Rodney White. *Mental Maps.* 2d ed. Boston: Allen and Unwin, 1986.

Jakle, John A., Stanley Brunn, and Curtis C. Roseman. *Human Spatial Behavior.* Prospect Heights, Ill.: Waveland Press, 1987.

Lewis, G. J. *Human Migration: A Geographical Perspective.* London: Croom Helm, 1982.

Saarinen, Thomas F. *Environmental Planning: Perception and Behavior.* Boston: Houghton Mifflin, 1976.

Part Three

The Locational Tradition

Given, then, our population-map, what has it to show us? Starting from the most generally known before proceeding towards the less familiar, observe first the mapping of London—here plainly shown, as it is properly known, as Greater London—with its vast population streaming out in all directions— east, west, north, south—flooding all the levels, flowing up the main Thames valley and all the minor ones, filling them up, crowded and dark, and leaving only the intervening patches of high ground pale. . . . This octopus of London, polypus rather, is something curious exceedingly, a vast irregular growth without previous parallel in the world of life—perhaps likest to the spreadings of a great coral reef. Like this, it has a stony skeleton, and living polypes—call it, then, a "man-reef" if you will. Onward it grows, thinly at first, the pale tints spreading further and faster than the others, but the deeper tints of thicker population at every point steadily following on. Within lies a dark crowded area; of which, however, the daily pulsating centre calls on us to seek some fresh comparison to higher than corraline life.[1]

[1]From *Patrick Geddes: Spokesman for Man and the Environment*, ed. by Marshall Stalley. New Brunswick, N.J.: Rutgers University Press, 1972, p. 123.

Thus, in the lavish prose of the early 20th century, did Patrick Geddes, "spokesman for man and the environment," capsulize the spread of people in the London region. The physical environment, the vibrating human life of cities, and change in human settlement patterns are all implicit in his remarks, as they are in the locational tradition of geography.

A theme that has run like a thread through the first eight chapters of this book may aptly be termed the *theme of location*. Distributions of climates, landforms, culture hearths, and religions were points of interest in our examinations of physical and cultural landscapes. In Part III of our study, the analysis of location becomes our central concern rather than just one strand among many, and the locational tradition of geography is brought to the fore.

In the study of economic, resource, or urban geography, a central question has to do with *where* certain types of human activities take place, not just in absolute terms but also in relation to one another. In studying the distribution of a given activity, such as making iron and steel, the concern is with identifying the locational pattern, analyzing it to see why it is arranged the way it is, and searching for underlying principles of location. Because the elements of culture are integrated, such a study often must take into consideration the way in which the location of one element affects the locations of others. For example, certain characteristics of the labor pool have helped to alter the distribution of the textile industry in America, and this alteration has had economic consequences for both New England and the South, and has affected the location of other economic activities as well.

In Chapter 9, "Economic Geography," our attention is directed to the location of economic activities as we seek to answer the question of why they are distributed as they are. What forces operate to make some regions extremely productive and others less so? Broad types of economic systems, such as commercial and planned, are analyzed, and we take a close look at the locational characteristics of some important manufacturing and agricultural industries.

The resources upon which humankind draws and depends for sustenance and development are not distributed uniformly in quantity or quality over the earth. Chapter 10, "Resource Geography," centers on the relationship between the demands people place upon the spatially varying environment—demands that are culture-bound—and the ability of the environment to sustain those

demands. In an increasingly integrated economic and social world of growing consumption demands, both access to resources, particularly energy and mineral resources, and their wise and productive use loom ever more important in human affairs. The understanding of resources in both their distributive and their exploitative characterstics is a theme of growing interest in the locational tradition of geography.

The increasing urbanization of the world's population represents another stage of the processes of cultural development, economic change, and environmental control, aspects of which have been the topics of separate chapters. In Chapter 11, urban areas are viewed in two ways. First, we focus upon cities as points in space with patterns of distribution and specializations of functions that invite analysis. Second, we consider cities as landscape entities with specialized arrangements of land uses resulting from recognizable processes of urban growth and development. Within their confines, and by means of the interconnected functional systems that they form, cities summarize a present stage of economic patterns and resource use of humankind. The consideration of urban geography thus logically concludes our examination of the locational tradition of geography.

9 Economic Geography

*T*he crop bloomed luxuriantly that summer of 1846. The disaster of the preceding year seemed over, and the potato, the sole sustenance of some 89 million Irish peasants, would again yield in the needed bounty. Yet, within a week, wrote Father Mathew, "I beheld one wide waste of putrefying vegetation. The wretched people were seated on the fences of their decaying gardens . . . bewailing bitterly the destruction that had left them foodless." Colonel Gore found that "every field was black," and an estate steward noted that "the fields . . . look as if fire has passed over them." The potato was irretrievably gone for a second year; famine and pestilence were inevitable.

Within five years, the settlement geography of the most densely populated country in Europe was forever altered. The United States received a million immigrants, who provided the cheap labor needed for the canals, railroads, and mines that it was creating in its rush to economic development. New patterns of commodity flows began as American maize (corn) for the first time found an Anglo-Irish market—as part of Poor Relief—and then entered a wider European market that had also suffered general crop failure in that bitter year.

Within days, a microscopic organism, the cause of the potato blight, had altered the economic and human geography of two continents. It had done so by a chain of interlocking causes and consequences dramatically demonstrating that all of those physical and cultural patterns that make up that world view called geography are really parts of one whole. Central among those patterns are those that the economic geographer isolates for special study.

Simply stated, economic geography is the study of how people earn their living, how livelihood systems vary by area, and how economic activities are interrelated and linked. It is, of course, vastly complex, as many lengthy volumes attest. In fact, we cannot really comprehend the totality of the economic pursuits of 5 billion human beings. We cannot examine the infinite variety of productive activities found on each unit area of the earth's surface; nor can we trace all their innumerable interrelationships, linkages, and flows. Even if that level of understanding were possible, it would be valid for only a fleeting instant of time.

The economic geographer seeks consistencies. The goal is to discover generalizations that will aid in understanding the maze of economic variations that characterize human existence. The search may be regional or topical—contrasts between rain forest and grassland or between agricultural and manufacturing industries. The search may be, as it is here, directed to recognizing the types of economic systems identified with generalized patterns of production and consumption, and an examination of the location of economic activities within those systems. Ultimately, as geographers, economic geographers address the how and the why of variations in the spatial patterns of economic activity.

In the compass of a single chapter, we can only hope to suggest the breadth of economic geographic interests and their centrality to all human geographic inquiry. More will be left out than included of the range of concerns, contributions, and conclusions of economic geographers. Some matters, particularly those dealing with the ways people earn their living in cities, are deferred to a later chapter. The intent is introduction; the hope is for a new awareness of the dynamic, interlocking diversity of human enterprise and of the increasing interdependence of differing national and regional economic systems. The potato blight, apparently of concern only to one small island, ultimately affected the economies of continents. The depletion of America's natural resources and the "de-industrialization" of its economy are altering the relative wealth of nations, flows of international trade, domestic employment and income patterns, and more. Economic geographic awareness is the essential first step to understanding otherwise inexplicable and unsettling changes taking place in established national, regional, and local economic structures and expectations (Figure 9.1).

Figure 9.1
These Japanese cars are part of a continuing flow of imported goods capturing a share of the market traditionally held by American manufacturers. Free trade in commercial economies may lead to short-term national dislocations; in an interdependent world, "stand-alone" economies are not possible.

Types of Economic Systems

Broadly speaking, there are three major types of economic systems: subsistence, commercial, and planned. None of them is "pure"; none exists in isolation in an increasingly interdependent world. Each, however, has certain underlying characteristics that mark it as a distinctive form of resource management and economic control.

In a **subsistence economy** system, goods and services are created for the use of the producers and their kinship groups. There is, therefore, little exchange of goods and only limited need for markets. In a **commercial economy,** as the term is used here, the producers or their agents market goods and services, and the laws of supply and demand determine their price and quantity. Market competition is the primary force shaping the location patterns involved. Trading, bargaining, and bidding are characteristic of the system. A **planned economy** is one in which the producers or their agents market the goods and services, usually to a government agency that controls their supply and price. The quantities produced and the locational patterns of production are carefully programmed by central planning departments.

In actuality, few people are members of only one of these systems. A farmer in India may produce rice and vegetables privately for the family's consumption, but very often the family will save some of the produce to sell. Members of the family may make cloth or other handicrafts that are marketed either by the farmer or by a trader in a local marketplace. With the money derived from the sale of the goods the Indian peasant can buy, among other things, clothes for the family, tools, or fuel. Thus, that Indian farmer is a member of at least two systems: subsistence and commercial. Similarly, in Eastern Europe a Polish farmer may produce a specified amount of grain for a government food agency and another amount for the family; any surplus may be sold in a nearby marketplace. That Polish farmer is a member of all three types of systems.

In the United States, government controls on the production of various types of goods and services (such as growing wheat or tobacco, producing alcohol, constructing and operating nuclear power plants, or engaging in licensed personal and professional services) mean that the United States does not have a purely commercial economy. To a limited extent, it is a planned one. Example after example would show that there are very few people in the world who are members of only one type of economic system.

Figure 9.2

Independent food sellers in modern China. By 1986 the country's nearly 11 million registered private businesses exceeded the total number of private enterprises operating in 1949 when the Communists took power. Since government price controls on most food items were removed in May 1985, free markets have multiplied. In Beijing, free markets retail one-third of the city's sales of vegetables, eggs, and beef, while in more rural districts up to two-thirds of fruits, vegetables, and meat are sold privately. The new private enterprises are labor intensive and help relieve unemployment in a nation where services account for less than 20% of gross national product (for India, the figure is nearly 40%).

Nonetheless, in a given country, one of the three economic systems tends to dominate. It is relatively easy to classify countries subjectively by that dominance even while recognizing that some elements of the other two systems may exist within the controlling scheme. It is generally agreed that Western Europe, Japan, and the United States are among the areas characterized by commercial or market economies. Although in varying degrees governmental intervention may direct some aspects of their operations, their resources have entered the marketplace of supply and demand, and their factories have traditionally been privately owned.

In the planned economies of Eastern Europe and the Soviet Union, the means of production—land, minerals, farms, and mills—are property of the state, although some variant ownership forms (private and cooperative) are permitted in a few nations such as Poland and Yugoslavia. China has a governmentally controlled economic system, but with so many of the Chinese producing agricultural goods for their immediate working groups, one must recognize the large degree of subsistence economy in the nation. Furthermore, in recent years a relaxation of central controls in China has added a growing market-economy component. China now permits privately owned small businesses (Figure 9.2). It allows some seller competition and market control of prices; a growing private, rather than communal, control of agriculture; and a partial "open door" to foreign investment and trade.

Many of the countries of Latin America, Africa, Asia, and the Middle East have until the present been dominated by subsistence economies. These areas, however, are undergoing rapid change. In fact, a most striking aspect of recent history is the extraordinary change in economic systems and development characteristic of many of these countries. It is convenient to separate the nations of these areas into two groups: those with abundant resources per capita—for example, the oil-rich Middle Eastern countries (the "have" countries) and the nations with poor resource bases (the "have-not" countries). Where subsistence economies still dominate, there is a uniformity in their nature, but the forces that are promoting change within them differ.

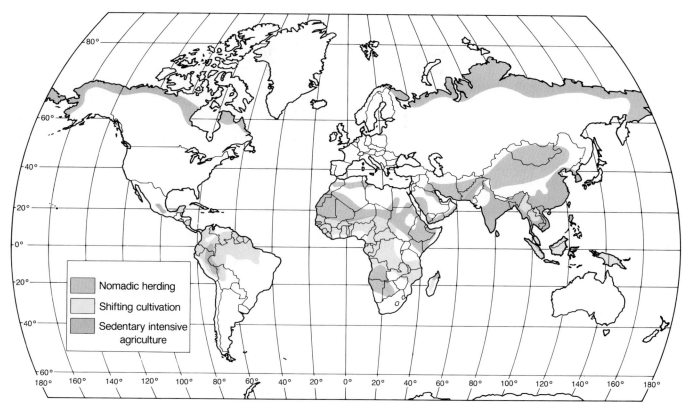

Figure 9.3

Subsistence agricultural areas of the world. Nomadic herding, supporting relatively few people, was the ages-old way of life in large parts of the dry and cold world. Shifting or swidden agriculture maintains soil fertility by tested native practices in tropical wet and wet-and-dry climates. Large parts of Asia support millions of people engaged in sedentary intensive cultivation, with rice and wheat the chief crops.

Much more difficult is equating these separate systems with differing national levels of technological advancement. Although it is true that subsistence-dominated countries are usually less developed technologically than either commercial or planned economies (see Figure 6.7), in recent years there has been a strong movement among them to increase the level of administrative control over their economies. More and more they are adopting advanced technologies and simultaneously becoming either more commercial, more planned, or, as is usually the case, a combination of both.

Subsistence Economic Systems

By definition, a *subsistence* economic system involves nearly total self-sufficiency on the part of its members. Production for exchange is minimal, and each family or close-knit social group relies upon itself for its food and most other essential requirements. Farming for the immediate food needs of the family is, even today, the predominant occupation of humankind. In most of Africa, much of Latin America, and most of Asia outside of the Soviet Union, the majority of people are primarily concerned with feeding themselves from their own land and livestock.

Subsistence systems are, then, overwhelmingly agricultural—more precisely, rural—and technologically "underdeveloped." Within that generalized description, two chief types of subsistence economic systems may be recognized: extensive and intensive. The variants within each type are several; the essential contrast between the two types, however, is realizable yield per acre used and, therefore, the potential population that can be supported. **Extensive subsistence agriculture,** including livestock specializations, involves large areas of land and minimal labor input. Both product per acre and population densities are low. **Intensive subsistence agriculture,** in which livestock rearing is less significant, involves the cultivation of small landholdings through the expenditure of great amounts of labor. Yields per unit area and population densities are both high (Figure 9.3).

Extensive Subsistence Systems

Of the several types of *extensive subsistence* systems—varying one from another in their intensities of land use—two are of particular interest.

Nomadic herding—the wandering but controlled movement of livestock solely dependent upon natural forage—where it still remains is the most extensive type of land use system (Figure 9.3). Over large portions of the Asian semidesert and desert areas and on the fringes of and within the Sahara, a relatively small number of people graze animals not for market sale but for consumption by the herder group. Sheep, goats, and camels are most common, while cattle, horses, and yaks are locally important. The reindeer of Lapland are part of the same system (Figure 9.4). Whatever the animals involved, their common characteristics are hardiness, mobility, and an ability to subsist on sparse forage. The animals provide a variety of products: milk, cheese, and meat for food; fibers and skins for clothing; skins for shelter; and excrement for fuel. For the herder, they represent primary subsistence. Nomadic movement is tied to sparse and seasonal rainfall or to cold temperature regimes and to the areally varying appearance and exhaustion of forage. Extended stays in any given location are neither desirable nor possible.

As a type of economic system, nomadic herding is declining. Many economic, social, and cultural changes are causing nomadic groups to alter their way of life or to disappear entirely. On the Arctic fringe of the Soviet Union, herders are now members of state or collective herding enterprises. In the Scandinavian North, Lapps (or Saami) are engaged in commercial more than in subsistence livestock farming. In the Sahel (from the Arabic word meaning "borderland") region of Africa (Figure 4.20), oases formerly controlled by nomads have been taken over by farmers. Some herdsmen have themselves become settled farmers along southern Sahelian floodplains; others have sought employment in newly discovered or exploited mineral deposits, especially petroleum. But probably of most significance, the great droughts of recent decades have forever altered the formerly nomadic way of life of thousands. In Mauritania, in 1965, 58% of the population were classified as nomads; by 1975, that figure had diminished to 25%, and it further declined after the great drought of 1977. Similar reductions are recorded for Mali, Chad, and the Sudan. The droughts and destructive human pressures upon the environment

Figure 9.4

The Roman historian Tacitus first mentioned the nomadic Lapps (or Saami) of northern Scandinavia in A.D. 98. Less than 10% of the some 70,000 Saami still are herders, and they are no longer truly nomadic. The half-million reindeer they tend browse on lichens that were contaminated in 1986 by fallout from the Chernobyl nuclear accident; radiation above acceptable levels will affect the reindeer meat into the 1990s.

extending the desert margin brought to conclusion evolutionary changes already in progress to reduce or eliminate a traditional way of life and form of land use. Total disappearance of nomads would, of course, leave large expanses of arid land unproductive and, at best, sparsely populated.

A much differently based and distributed form of extensive subsistence agriculture is found in all warm, moist, low-latitude areas of the world (Figure 9.3). There, largely in the tropical rain forest and tropical wet-and-dry environments discussed in Chapter 3, many people engage in a kind of nomadic farming. Through clearing and use, the soils of those areas come to be leached of many of their nutrients (as the nutrients are dissolved and removed by percolating water), and farmers cultivating them need to move on after harvesting several crops. In a sense, they rotate fields rather than crops to maintain productivity. This type of **shifting cultivation** has a number of names, the most common of which are *swidden* (an English localism for "burned clearing") and *slash-and-burn*.

Characteristically, the farmers hack down the natural vegetation, burn the cuttings, and then plant such crops as maize, millet (a cereal grain), rice, starchy

Figure 9.5

A swidden plot in the Ivory Coast of West Africa. The pile of second-growth slash has just been set afire; stumps and trees left in the clearing will remain after the burn.

manioc or cassava, yams, and sugar (Figure 9.5). Increasingly included in many of the crop combinations are such high-value, labor-intensive commercial crops as coffee, providing the cash income that is evidence of the growing integration of all peoples into exchange economies. Initial yields—the first and second crops—may be very high in *biomass* (the quantity of living organic matter in a given area) in comparison with other world agricultural forms, but the yields quickly become lower with each successive planting on the same plot. As that occurs, cropping ceases, native vegetation is allowed to reclaim the clearing, and gardening shifts to another newly prepared site. The first clearing will ideally not be used again for crops until many years later, after natural fallowing has replenished its fertility (see "Swidden Agriculture").

Nearly 5% of the world's people are still predominantly engaged in tropical shifting cultivation on more than one-fifth of the world's land area (Figure 9.3). Since the essential characteristic of the system is the intermittent cultivation of the land, each family requires a total occupance area equivalent to the garden plot in current use plus all land left fallow for regeneration. Population densities are traditionally low, for much land is needed to support few people, but here as elsewhere, population density must be considered a relative term. In actuality, although overall density is low, persons per square mile of *cultivated* land may be high.

Shifting cultivation is one of the great agricultural systems of the world. It is found on the islands of Kalimantan (Borneo), New Guinea, and Sumatera (Sumatra) in Indonesia. It is part of the economy of the uplands of South Asia in Vietnam, Thailand, Burma, and the Philippines. Nearly the whole of Central and West Africa away from the coasts, Brazil's Amazon Basin, and large portions of Central America are all noted for this type of extensive subsistence agriculture. Many of these areas are among the most isolated in the world, though some are found near major cities. Roads, railroads, and telephones are largely absent (Figure 9.9). As modern technology is introduced, these cultivators will inevitably be reduced in number. It may be argued that shifting cultivation is a highly efficient cultural adaptation where land is abundant in relation to population, and where levels of technology and capital availability are low. As those conditions change, the system becomes less viable. The basic change, as Chapter 5 showed, is that land is no longer abundant in relation to population in many of the less developed wet, tropical nations. Their growing populations have cleared and settled the forest lands formerly used only intermittently in swidden cultivation.

Swidden Agriculture

When, after months of discussion, a garden site has been selected, the swidden farmer begins to remove unwanted vegetation. The first phase of this process consists of slashing and cutting the undergrowth and smaller trees with bush knives. The principal aim is to cover the entire site with highly inflammable dead vegetation, so that the later stage of burning will be most effective. Because of the threat of soil erosion the ground must not be exposed directly to the elements at any time during the cutting stage. During the first months of the agricultural year, activities connected with cutting take priority over all others; it is estimated that the time required ranges from 50 to 200 labor-hours per hectare, with most effort expended in reclearing woody second-growth vegetation and least time needed to clear virgin forest. The average-sized swidden site thus requires 20 to 80 work-hours for clearing.

Once most of the undergrowth has been slashed, chopped to hasten drying, and spread to protect the soil and assure an even burn, the larger trees must be felled with steel axes or otherwise handled so that unwanted shade will be removed. The time and labor needed for this work vary with the type of vegetation. In primary growth, as much as 300 to 400 labor-hours per hectare may be required; in bamboo stands, less than 50 and in mixed woody second growth, about 150. The successful felling of a real forest giant is a dangerous activity and requires great skill. Felling in second growth is usually less dangerous and less arduous. Some trees are merely trimmed but not felled, both to reduce the prodigious amount of labor and to leave trees to reseed the swidden during the subsequent fallow period.

The crucial and most important single event in the agricultural cycle is swidden burning. The main firing of a swidden is the culmination of many weeks of preparation in spreading and leveling chopped vegetation, preparing firebreaks to keep flames from escaping into the jungle, and allowing time for the drying process. An ideal burn rapidly consumes every bit of litter. The swidden firer tries to encircle the swidden completely, thrusting a torch into the litter every 3 or 4 meters. He must always look in toward the center of the field, the direction of planned convergence of the flames. Quickly, clouds of dense smoke billow up over the swidden; in no more than an hour or an hour and a half, only smoldering remains are left.

The Hanunóo note the following as the benefits of a good burn:

(1) removal of unwanted vegetation, resulting in a cleared swidden; (2) extermination of many animal and some weed pests; (3) preparation of the soil for *dibble* (any small hand tool or stick to make a hole) planting by making it softer and more friable; (4) provision of an evenly distributed cover of wood ashes, good for young crop plants and protective of newly planted grain seed. Within the first year of the swidden cycle, an average of between 40 and 50 different crops have been planted and harvested.

Between two-thirds and three-fourths of the total time covered by an average complete swidden cycle of, perhaps, twenty years consists of controlled fallowing. The most critical feature of swidden agriculture is the maintenance of soil fertility and structure. The solution is to pursue a system of rotation of one to three years in crop and ten to twenty in woody (or bush fallow) regeneration. When population pressures mandate a reduction in the length of fallow period, productivity of the region tends to drop as soil fertility is lowered, marginal land is used, and environmental degradation occurs. The balance is delicate.

The above depiction of shifting cultivation in the Philippines is drawn from *Hanunóo Agriculture*, by Harold C. Conklin (FAO Forestry Development Paper No. 12).

Intensive Subsistence Systems

Over one-half of the people of the world are engaged in intensive subsistence agriculture, which predominates in the areas shown on Figure 9.3. As a descriptive term, *intensive subsistence* is no longer fully applicable to a changing way of life and economy in which the distinction between subsistence and commercial is less and less valid. While families may still be fed primarily with the produce of their individual plots, the exchange of farm commodities within the system is considerable. Production of foodstuffs for sale in rapidly growing urban markets is increasingly vital for the rural economies of "subsistence farming" areas and for the sustenance of the growing proportion of national and regional populations no longer themselves engaged in farming. Within this increasingly mixed—but traditionally and still importantly subsistence system—hundreds of millions of Indians, Chinese, Pakistanis, Bangladeshis, and Indonesians plus further millions in other Asian, African, and Latin American nations are small-plot producers of rice, wheat,

Figure 9.6
Transplanting rice seedlings requires arduous hand labor by all members of the family. The newly flooded diked fields, previously plowed and fertilized, will have their water level maintained until the grain is ripe. This photograph was taken in India; the scene is repeated wherever subsistence wet-rice agriculture is practiced.

maize, millet, or pulses (peas, beans, and other legumes). Most live in monsoon Asia, and we will devote our attention to that area.

Intensive subsistence farmers are concentrated in such major river valleys and deltas as the Ganges and the Chang Jiang (Yangtze), and in smaller valleys close to coasts—level areas with fertile alluvial soils. These warm, moist districts are well suited to the production of rice, a crop that under ideal conditions can provide large amounts of food per unit of land. Rice also requires a great deal of time and attention, for planting rice shoots by hand in standing fresh water is a tedious art (Figure 9.6). In the cooler and drier portions of Asia, wheat is grown intensively, along with millet and, less commonly, upland rice.

Rice is known to have been cultivated in parts of China and India for more than 7000 years. Today, wet, or lowland, rice is the mainstay of subsistence agriculture and diets of populations from Sri Lanka and India to Taiwan, Japan, and Korea. Almost exclusively used as a human food, it constitutes half the diet of over 1.6 billion

people, with another 400 million relying on it for between one-quarter and one-half of their food intake. Its successful cultivation depends upon the controlled management of water, relatively easy in humid tropical river valleys with heavy, impermeable, water-retaining soils, though more difficult in upland and seasonally dry districts. The necessary water management systems have throughout Asia left their regionally distinctive marks on the landscape. Permanently diked fields to contain and control water, levees against unwanted water, and reservoirs, canals, and drainage channels to control its availability and flow are common sights. Terraces to extend level land to valley slopes are occasionally encountered as well (Figure 5.17).

Intensive subsistence farming is characterized by large inputs of labor per unit of land, by the promise of high yields in good years, by small plots, and by the intensive use of fertilizers, mostly animal manure (see "The Economy of a Chinese Village"). For food security and dietary custom, some other products are also grown.

Figure 9.7
Water buffalo are a vital part of the agricultural system in much
of Southeast Asia, where they serve as draft animals as well as
sources of milk, meat, and fertilizer. Here they are used to work a
rice paddy in Burma.

Vegetables and some livestock are part of the agricultural
system, and fish may be reared in rice paddies and ponds.
Cattle are a source of labor and of food (Figure 9.7). Food
animals include swine, ducks, and chickens, but since
Muslims eat no pork, hogs are absent in their areas of
settlement. Hindus generally eat no meat, but the large
number of cattle in India are vital for labor, as a source
of milk, and as producers of fertilizer and fuel.

The introduction of improved health care in this
century has greatly influenced population growth rates
in these countries of intensive subsistence agriculture.
Lowered infant and crude death rates have helped to increase population enormously. Historically high birth
rates have only recently shown decrease. Rising population, of course, puts increasing pressure on the land;
the response has been to increase the intensity of production. Lands formerly considered unsuitable for agriculture by reason of low fertility, inadequate moisture,
difficulty of clearing and preparation, isolation from settlement areas, and other factors have been brought into
cultivation.

To till those additional lands, a price must be paid.
Any economic activity incurs an additional (marginal)
cost in labor, capital, or other unit of expenditure to bring
into existence an added unit of production. When the
value of the added (marginal) production at least equals

The Economy of a Chinese Village

The village of Nanching is in subtropical southern China on the Zhu River delta near Guangzhou (Canton). Its pre-Communist subsistence agricultural system is described by a field investigator, whose account is condensed here. The system is found in its essentials in other rice-oriented societies.

In this double-crop region, rice was planted in March and August and harvested in late June or July and again in November. March to November was the major farming season. Early in March the earth was turned with an iron-tipped wooden plow pulled by a water buffalo. The very poor who could not afford a buffalo used a large iron-tipped wooden hoe for the same purpose.

The plowed soil was raked smooth, fertilizer was applied, and water was let into the field, which was then ready for the transplanting of rice seedlings. Seedlings were raised in a seedbed, a tiny patch fenced off on the side or corner of the field. Beginning from the middle of March, the transplanting of seedlings took place. The whole family was on the scene. Each took the seedlings by the bunch, ten to fifteen plants, and pushed them into the soft inundated soil. For the first 30 or 40 days the emerald green crop demanded little attention except keeping the water at a proper level. But after this period came the first weeding; the second weeding followed a month later. This was done by hand, and everyone old enough for such work participated. With the second weeding went the job of adding fertilizer. The grain was now allowed to stand to "draw starch" to fill the hull of the kernels. When the kernels had "drawn enough starch," water was let out of the field, and both the soil and the stalks were allowed to dry under the hot sun.

Then came the harvest, when all the rice plants were cut off a few inches above the ground with a sickle. Threshing was done on a threshing board. Then the grain and the stalks and leaves were taken home with a carrying pole on the peasant's shoulder. The plant was used as fuel at home.

As soon as the exhausting harvest work was done, no time could be lost before starting the chores of plowing, fertilizing, pumping water into the fields, and transplanting seedlings for the second crop. The slack season of the rice crop was taken up by chores required for the vegetables, which demanded continuous attention, since every peasant family devoted a part of the farm to vegetable gardening. In the hot and damp period of late spring and summer, eggplant and several varieties of squash and beans were grown. The green-leafed vegetables thrived in the cooler and drier period of fall, winter, and early spring. Leeks grew the year round.

When one crop of vegetables was harvested, the soil was turned and the clods broken up by a digging hoe and leveled with an iron rake. Fertilizer was applied, and seeds or seedlings of a new crop were planted. Hand weeding was a constant job; watering with the long-handled wooden dipper had to be done an average of three times a day, and in the very hot season when evaporation was rapid, as frequently as six times a day. The soil had to be cultivated with the hoe frequently as the heavy tropical rains packed the earth continuously. Instead of the two applications of fertilizer common with the rice crop, fertilizing was much more frequent for vegetables. Besides the heavy fertilizing of the soil at the beginning of a crop, usually with city garbage, additional fertilizer, usually diluted urine or a mixture of diluted urine and excreta, was given every 10 days or so to most vegetables.

Adapted from C. K. Yang, *A Chinese Village in Early Communist Transition.* The MIT Press, Cambridge, Mass. © Copyright 1959 by the Massachusetts Institute of Technology.

the added cost, the effort may be undertaken. In past periods of lower population pressure, there was no incentive to extend cultivation to less productive or more expensive, unneeded lands. Now, circumstances are different. Possibilities for land conversion to farming are, however, limited in many intensive subsistence agricultural economies. More than 60% of the population of the Third World lives in countries in which some three-quarters of potential arable land is already under cultivation and where undeveloped cultivable land has low potential for settlement and use. When population pressures dictate land conversion, serious environmental deterioration may result. Clearing wet tropical forests in the Philippines and Indonesia has converted dense woodland to barren desolation within a very few years as soil erosion and nutrient loss have followed forest destruction. In Southeast Asia, some 25 million acres (10 million hectares) of former forest land is now wasteland, covered by useless sawgrasses that supply neither forage nor food nor fuel.

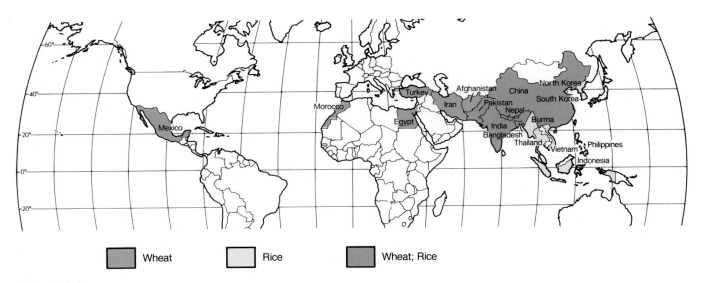

Wheat   **Rice**   **Wheat; Rice**

Figure 9.8

Chief beneficiaries of the Green Revolution. Among the
eleven nations adopting new rice varieties and cropping
technologies, average yields increased by 52% between 1965 and
1983; for the rest of the world, they actually dropped by 4%
during the same period. Wheat yields increased 66% for the 9
reporting countries (excluding Mexico); for the rest of the world
they grew only 29%.

The Green Revolution

Increased productivity of existing cropland, rather than
expansion of cultivated area, has accounted for most of
the growth of agricultural production over the past quarter
century. **Green Revolution** is the shorthand reference
to a complex of seed and management improvements
adapted to the needs of intensive agriculture that have
brought larger harvests from a given area of farmland. In
the 1970s alone, the world's average annual grain production increased by 2.6% per year; increases in per acre
yields accounted for some 80% of that growth. These yield
increases and the improved food supplies they represent
have been particularly important in densely populated,
subsistence farming areas heavily dependent upon rice
and wheat cultivation (Figure 9.8). Genetic improvements in these two grains have formed the basis of the
Green Revolution. Dwarfed varieties have been developed that respond dramatically to heavy applications of
fertilizer, that resist plant diseases, and that can tolerate
much shorter growing seasons than traditional native varieties. Adopting the new varieties and applying the irrigation, mechanization, fertilization, and pesticide
practices demanded in the new "high-input" agriculture

increased rice yields between 1965 and 1984 by more
than 50% in India, 80% in Indonesia and Pakistan, and
75% in the Philippines. Similar gains were reported for
wheat-producing areas: 80% yield increases in Bangladesh, 50% in India and Pakistan. In South and Southeast
Asia, the new strains of wheat have been planted on some
80% of the wheat area; between 40 and 50% of rice area
is planted to the new varieties.

The increases in food availability made possible
through the Green Revolution have averted catastrophic
shortages and famines predicted during the 1960s and
1970s for the rapidly expanding populations of subsistence agricultural countries. But a price has been paid.
Traditional agriculture is being displaced. With it is lost
the food security that distinctive locally adapted native
crop varieties (*land races*) provided and the nutritonal
diversity and balance that multiple-crop intensive gardening assured. Traditional subsistence farming, wherever practiced, was oriented toward risk minimization.
Many differentially hardy varieties of a single crop guaranteed some yield whatever adverse weather, disease, or
pest problems might occur. Commercial agriculture aims
at profit maximization, not minimal food security. Poor
farmers who could not afford the capital investment that
the Green Revolution demands have been displaced by

a commercial monoculture, traditional rural society has been disrupted, and landless peasants have been added to the urbanizing populations of affected nations. To the extent that land races are lost to monoculture, the genetic diversity of food crops is reduced. "Seed banks" rather than native cultivation are increasingly needed to preserve genetic diversity for future plant breeding and as insurance against catastrophic pest or disease susceptibility of inbred varieties.

Summary There are variations on the systems we have described. For example, there are still some groups of hunters and gatherers in the Arctic north, in the deserts, and in the rain forest. Subsistence fishing villages abound along the coasts of South Pacific islands and in southeast Asia. Wherever commercial economies have had an impact, these subsistence systems have been modified.

Commercial Economic Systems

Modifications of subsistence systems have inevitably made them more complex by imparting to them at least some of the diversity and linkages of activity that mark the commercial system. The *commercial* system—that of advanced economies—possesses the complex specializations and interconnections that the subsistence economies lack. The results of specialized labor inputs are hierarchically and spatially connected in interdependent patterns of production, exchange of goods and services, and ultimate consumption. By its very complexity, the commercial economic system has invited the most serious attention of economists and economic geographers. The body of theory and the models designed to explain the system's formal and informal mechanisms are vast and far beyond the scope of this introduction. The purpose here is merely to outline the system's central elements.

Three important characteristics of commercial systems are *specialization, profit,* and *interdependence.* Individuals, industries, and even areas specialize, allowing producers to be efficient, so that their unit costs of making and handling goods are as low as possible. Low prices theoretically bring higher sales and greater profits. Because producers must cooperate with each other to ensure efficient marketing procedures and increased profits, they are interdependent. Business people recognize, of course, that profits allow them to invest more capital in production and thereby reap even greater rewards. The profit motivation can result in savings, investment, and increased production. From the geographer's point of view, specialization, interdependence, and profit imply that those engaged in business select efficient locations for their activities, locations that are the generating points of trade.

Clearly some sites are much more advantageous than others for commercial activities. Location at a favorable site allows for low production costs and perhaps large-scale production. Some locations have both raw material resources and markets nearby. Water and a food supply may be present, too, as well as a productive labor force. Transportation is a key variable; in fact, no advanced economy can flourish without a well-connected transport system (Figure 9.9). Outmoded vehicles or silted harbors, potholed roads or expensive toll roads, all put restraints on trade by raising transportation costs. The advantages of a site may be so great that an industry or a specialized form of agriculture can overcome these restraints, but all site costs are considered very carefully by profit-motivated entrepreneurs.

Not only is there specialization at sites, but countries themselves specialize. If one country can gain an advantage in a particular enterprise because of low costs, it is motivated to increase its degree of specialization. The concept of **comparative advantage** is helpful to understanding specialization and trade. Assume that two countries both have a need for, and can successfully produce domestically, two commodities. Further assume that there is no transport cost consideration. No matter what its cost of production of either commodity, Country A will choose to specialize in only one of them if, by that specialization and through exchange with Country B for the other, A stands to gain more than it loses. The key to comparative advantage is the use of resources in such a fashion as to gain, by specialization, a volume of production and a selling price that permit exchange for a needed commodity at a cost level that is below that of the domestic production of both.

At first glance, the concept of comparative advantage may at times seem to defy logic. For example, Japan may be able to produce airplanes and home appliances more cheaply than the United States, thereby giving it an apparent advantage in both. But it benefits both countries if they specialize in the product in which they have a comparative advantage. In this instance, Japan's lower labor costs make it more profitable for Japan to specialize in the volume production of appliances and to buy airplanes from the United States. Resource-poor countries tend to develop specializations that other, better endowed countries are less inclined to pursue. In this way, poorer nations enter international trade.

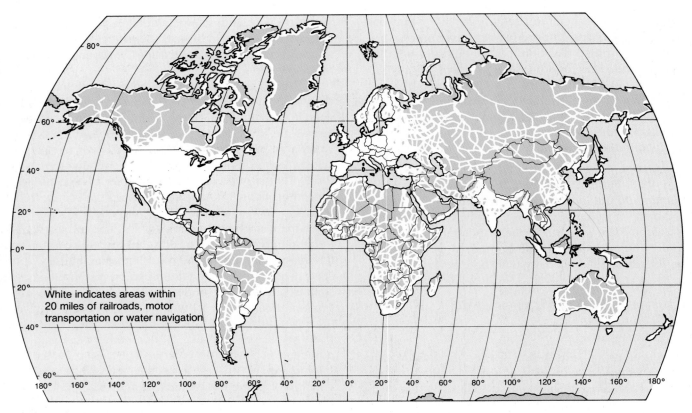

White indicates areas within
20 miles of railroads, motor
transportation or water navigation

Figure 9.9
Accessibility is a key measure of economic development and of
the degree to which a world region can participate in
interconnected market activities. Isolated areas of nations with
advanced economies suffer a price disadvantage because of high
transportation costs. Lack of accessibility in subsistence
economic areas slows their modernization and hinders their
participation in the world market.

Often the thought of a particular country brings to
mind products associated with it. Usually these are the
"specialized" goods that enter international trade. French
wines, Scotch whisky, teakwood products from Thailand,
shoes from Italy, coffee from Brazil, and television sets
from Japan are all examples of products that have become specializations in the nations named. When those
specializations are simultaneously directly competitive
with similar commodities produced within importing
nations, developed domestic market economies may
suffer. The impact of imports of steel, automobiles, textiles, shoes, and other foreign-produced goods upon
American industry and the nation's trade deficit is widely
reported.

When other nations' comparative advantages reflect lower labor, land, raw material, and capital costs,
manufacturing activities may voluntarily relocate from
higher-cost market locations to lower-cost foreign production sites. Such voluntary *out-sourcing* (producing
parts or products abroad for domestic sale) by American
manufacturers has employment and areal economic consequences like those resulting from successful competition by foreign companies or from ongoing industrial
development decisions favoring one section of the
country over others. There will be growth and development in favored world or national regions and job loss
and depression in areas in decline. A North American case
in point is along the northern border of Mexico. In the
1960s Mexico enacted legislation permitting foreign
(specifically American) companies to establish "sister"

Figure 9.10

American manufacturers, seeking lower labor costs, have
frequently established manufacturing and component assembly
operations along the international border in Mexico. The finished
or semifinished products are brought duty-free into the United
States. An employee in an automated General Motors subsidiary
assembly plant in Cuauhtemoc, Mexico, is shown here.

plants within 12 miles (19 km) of the U.S. border for the
duty-free assembly of products destined for reexport. By
the end of 1985 some 1100 such assembly plants had
been established to produce a diversity of goods including electronic products, textiles, furniture, leather
goods, toys, and automotive parts. The plants generated
direct and indirect employment for some half-million
Mexican workers, many of them replacing United States
employees (Figure 9.10).

Within both national and international economies,
transportation costs are central to the spatial patterns of
production, explaining the location of a large variety of
economic activities. Water-borne transportation is nearly
always cheaper than any other, and the enormous amount
of commercial activity that takes place on coasts or on
navigable rivers leading to coasts is an indication of that
cost advantage. When railroads were developed and the
commercial exploitation of inland areas could begin,
coastal sites continued to be important as more and more
goods were transferred there between low-cost water and
land transport. Even today, resources at many potential
sites have not been developed simply because the costs
of moving the goods to markets would be too high. Every
increase or decrease in transportation charges has an effect on commercial location.

In the rare instance when transportation costs become a negligible factor in production and marketing,
an economic activity is said to be *footloose.* Some manufacturing activities are located without reference to raw
materials; for example, the raw materials for electronic
products such as computers are so valuable, light, and
compact that transportation costs have little bearing on
where production takes place.

Agriculture

Agriculture fully shares with industry the basic characteristics of commercial economic systems. These are
specialization—by enterprise (farm), by area, and even
by nation; *profit*—rather than self-sufficiency and subsistence; and *interdependence*—with suppliers and buyers
through the market mechanism. Like industrialists, commercial farmers attempt to reap the highest profits possible from their labors. To a greater extent than in
industry, perhaps, elements of seeming irrationality may
affect their decisions and thwart the maximization of net
return that they seek. Among those elements are a less
certain knowledge of the economic factors affecting production and marketing decisions, less complete control
over the conditions of production (such as weather and
field conditions), and—for the smaller if not the corporate farmer—a greater influence of tradition or personal
preference on production decisions.

The major difficulty that farmers face is that when
they prepare their fields for planting, they do not know
what the market price will be for their produce when the
crops are ready for harvesting months or even years later.
That price reflects conditions over which they individually have no control, which are not related to their production costs, and which, indeed, may reflect not just
national but international circumstances. In selling, the
farmer inevitably stands at a disadvantage. The producers

are many and small; the buyers are few, large, and better informed. At the market, the latter obviously have an advantage. The price is set with no room for bargaining if the producer chooses to sell immediately.

Nonetheless, as part of the commercial system, farmers strive for as much information as they can get in order to make sound economic judgments. Thus, the crop or the mix of crops and livestock that individual farmers produce is a result of a careful appraisal of profit possibilities. Markets must be assessed and prices predicted, the physical nature of farmland and the possible weather conditions must be evaluated. The costs of production (fuel, fertilizer, capital equipment, labor) must be reckoned. Farmers do not necessarily grow what they want to grow. Of the hundreds of possible crops that may be grown on a given farm, only a handful can bring profit to the producer.

The market for any good in a commercial society is basically a function of supply and demand, though this idealization is only partially realized in modern economies, with their strong infusion of governmentally administered constraints. If there is a glut of wheat on the market, for example, the price per ton will come down and the area sown to it should diminish. It will also diminish if governments, responding to economic or political considerations, impose acreage controls. If, on the other hand, an extended dry spell reduces the amount and the quality of the wheat produced, the market price for wheat will increase, and more farmers may be tempted to grow it. Again, this expected response to market conditions may be subverted by such government-imposed constraints as price controls and crop agreements. To a limited extent, farmers may withhold their product from the market in the hope of improved prices. However, this option lasts for only a short time for livestock farmers, whose ongoing feed costs might prove ruinous; it does not exist at all for the producer of perishable fruits and vegetables.

Agriculture in even the least controlled of commercial economies is strongly affected by governmental policies and in many is as much diverted from theoretical production and marketing patterns as it is in the most tightly planned economies. The political power of farmers in the European Economic Community (EEC) has secured for them generous product subsidies and for the Community immense unsold stores of butter, wine, and grains. The EEC sets artifically high prices for domestic wheat and provides to farmers subsidies for the wheat they export at less than market price. Similar price supports and export subsidies are in effect in other nations, including the United States. The result is international tension as market competition builds to dispose of unwanted farm surpluses and a distortion in traditional international agricultural commodity flows (Figure 9.16).

Figure 9.11

Open storage of corn. In the world of commercial agriculture, supply and demand are not always in balance. Both the bounty of nature in favorable crop years and the intervention of governmental programs that distort production decisions can create surpluses for which no market is readily available.

In the United States, programs of farm price supports, acreage controls, financial assistance, and other governmental involvements in agriculture are of long standing and have an equally distorting effect (Figure 9.11).

Early in the last century, before such governmental influences were the norm, Johann Heinrich von Thünen observed that lands of apparently identical physical properties were used for different agricultural purposes. Seeking regularities governing farmers' individual land use decisions, von Thünen deduced that the uses to which parcels were put was a function of the differing values (rents) placed upon seemingly identical lands. Those differences, he determined, reflected the cost of overcoming the distance separating a given farm from a central market town. The greater the distance, the higher was the operating cost to the farmer, since transport charges had to be added to other expenses. When production costs plus transport costs just equaled the value of the commodity at the market, a farmer was at the economic margin of cultivation. A simple exchange model emerged: the greater the transportation cost, the lower the rent that could be paid for land to remain competitive in the market. At the economic margin, the *economic rent* earned by a cultivated parcel dropped to zero. Closer to the market, lowered transport costs increased the income of a farmer and he could afford to pay a higher price for land based upon its location (*locational rent*), just as he could afford to pay more for land of superior fertility.

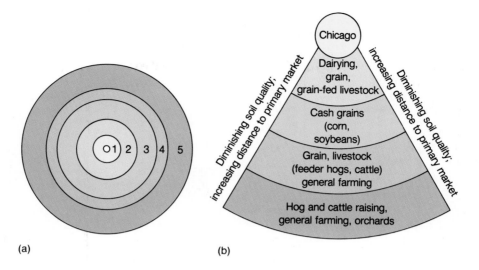

(a)   (b)

Figure 9.12

(a) **Von Thünen's schema.** Recognizing that as distance from the market increases, the value of land decreases, von Thünen developed a descriptive model of intensity of land use that holds up reasonably well in practice. The most intensively produced agricultural crops are found on the land close to the market; the less intensively produced commodities are located at more distant points. In the diagram, circle 1 is an area of dairying and truck gardening (horticulture); circles 2, 3, and 4 represent various crop-rotation farming systems, each of which includes grain. Circle 4 is farmed less intensively than are circles 3 and 2. Circle 5 represents the area of extensive stock raising. As the market at the center increases in size, the agricultural speciality areas are displaced outward, but the relative position of each is retained. Compare this diagram with Figure 9.13. (b) **A schematic view of the von Thünen zones in the sector south of Chicago.** There, farmland quality decreases southward as the boundary of recent glaciation is passed and hill lands are encountered in southern Illinois. On the morainic margins of the city near the market, dairying competes for space with grain and livestock farming and suburbanization; southward into flat, fertile central Illinois, cash grains dominate. In southern Illinois, livestock rearing and fattening, general farming, and some orcharding are the rule.

This understood, a rational pattern of crop decisions was revealed that was based not only on land quality but also on land cost. **Von Thünen's model,** shown in Figure 9.12, helps explain the changing crop patterns and farm sizes evident on the landscape at increasing distances from major cities. Farmland close to markets takes on high value, is used *intensively* for high-value crops, and is subdivided into relatively small units. Land far from markets is used *extensively* and in larger units. The effect of these relationships on a national basis is seen on Figure 9.13.

Intensive Commercial Agriculture

Farmers who apply large amounts of capital (for machinery and fertilizers, for example) and/or labor per unit of land engage in **intensive commercial agriculture.** The crops that justify such costly inputs are characterized by high yields and high market value per unit of land; they include fruits, vegetables, and dairy products, all of which are highly perishable. Near most medium-sized and large cities, there are **truck farms** that produce a wide range of vegetables, and dairy farms. Since the produce is perishable, transport costs increase because of the special handling that is needed, such as refrigerated trucks and custom packaging. This is another reason for locations close to market. Note the distribution of truck farming in Figure 9.14.

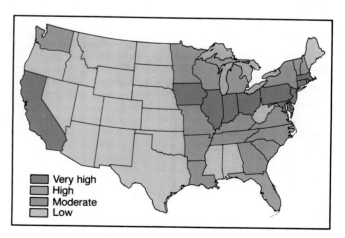

Figure 9.13

Relative value per acre of farmland and buildings. In a generalized way, per acre valuations support von Thünen's model. The major metropolitan markets of the Northeast, the Midwest, and California are in part reflected by high rural property valuations, and fruit and vegetable production along the Gulf Coast increases land values there. National and international agricultural-goods markets, soil productivity, climate, and terrain characteristics are also reflected in the map patterns.

Data from *Statistical Abstract of the United States.*

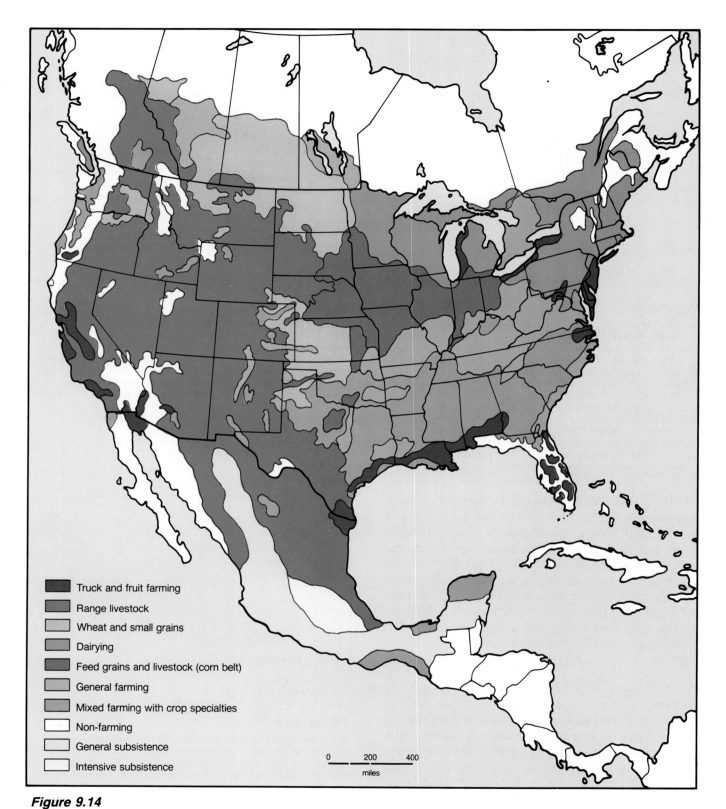

Legend

- Truck and fruit farming
- Range livestock
- Wheat and small grains
- Dairying
- Feed grains and livestock (corn belt)
- General farming
- Mixed farming with crop specialties
- Non-farming
- General subsistence
- Intensive subsistence

0   200   400
miles

Figure 9.14

Generalized agricultural regions of North America.

Sources: Bureau of Agricultural Economics; Canada Department of Agriculture; Mexico. Secretaría de Agricultura y Recursos Hidráulicos.

Livestock-grain farming involves growing grain to be fed on the producing farm to livestock, which constitutes the farm's cash product. In Western Europe three-fourths of cropland is devoted to production for animal consumption; in Denmark, 90% of all grains are fed to livestock for conversion not only into meat but also into butter, cheese, and milk. Livestock-grain farmers also work their land intensively, but the value of their product per unit of land is usually less than that of the truck farm. Consequently, in the United States at least, livestock-grain farms are farther from the main markets than are truck and dairy farms.

Normally the profits for marketing livestock (chiefly hogs and beef cattle in the United States) per pound are greater than those for selling corn or other feed, such as alfalfa and clover. As a result, farmers convert their corn into meat on the farm by feeding it to the livestock, efficiently avoiding the cost of buying grain. They may also convert farm grain at local feed mills to the more balanced feed that modern livestock rearing requires. Where land is too expensive to be used to grow feed, especially near cities, feed must be shipped to the farm. The livestock-grain belts of the world are close to the great coastal and industrial-zone markets. The "corn belt" of the United States and the livestock region of Western Europe are two examples. Lower-cost land traditionally allowed livestock to be shipped to the large European market from as far away as Argentina and New Zealand.

Extensive Commercial Agriculture

Farther from the market, on less expensive land, there is less need to use the land intensively. Cheaper land gives rise to larger farm units. **Extensive commercial agriculture** is typified by large wheat farms and livestock ranching.

There are, of course, limits to the land use explanations attributable to von Thünen's model. While it is evident from Figure 9.13 that farmland values decline westward with increasing distance from the northeastern market of the United States, they show no corresponding increase with increasing proximity to the massive West Coast market region. The western lands are characterized by extensive agriculture, but as a consequence of environmental, not distance, considerations. Climatic regionalization, discussed in Chapter 3, obviously affects the productivity and the potential agricultural use of an area, as do associated soils regions and topography. In

the United States, of course, increasing distance westward from eastern markets is fortuitously associated with increasing aridity and the beginning of mountainous terrain. In general, rough terrain and dry climates, rather than simple distance from market, underlie the widespread occurrence of extensive agriculture under all economic systems.

Livestock ranching differs significantly from livestock-grain farming and, by its commercial orientation and distribution, from the nomadism it superficially resembles. A product of the 19th century growth of urban markets for beef and wool in Western Europe and northeastern United States, ranching has been primarily confined to areas of European settlement. It is found in western United States and adjacent sections of Mexico and Canada (Figure 9.14); the grasslands of Argentina, Brazil, Uruguay, and Venezuela; the interior of Australia; the uplands of South Island, New Zealand; and the Karoo of South Africa. All except New Zealand and the humid pampas of South America have semiarid climates. All, even the most remote from markets, were a product of improvements in transportation by land and sea, refrigeration of carriers, and of meat-canning technology. In all ranching regions, livestock range (and the area of traditional ranching) has been reduced as crop farming has encroached on their more humid margins, as pasture improvement has replaced less nutritious native grasses, and as grain fattening has supplemented traditional grazing.

In livestock ranching areas, young cattle or sheep are allowed to graze over thousands of acres. In the United States, when the cattle have gained enough weight so that weight loss in shipping will not be a problem, they are sent to livestock-grain farms or to feedlots near slaughterhouses for accelerated fattening. Since ranching can be an economic activity only where alternative land uses are nonexistent and land quality is low, ranching regions of the world characteristically have low population densities, low capitalizations per land unit, and relatively low labor requirements.

Large-scale wheat farming requires sizable capital inputs for planting and harvesting machinery, but the inputs per unit of land are low; wheat farms are very large. In North America, the spring wheat (planted in spring, harvested in autumn) region includes the Dakotas,

Figure 9.15

Contract harvesters follow the ripening wheat northward through the plains of the United States and Canada.

eastern Montana, and the southern parts of the Prairie Provinces of Canada; the winter wheat (planted in fall, harvested in mid-summer) belt focuses on Kansas and includes adjacent sections of neighboring states (Figure 9.15). Argentina is the only South American nation to have comparable large-scale wheat farming. In the eastern hemisphere, the system is fully developed only in the Soviet Union on vast state farms east of the Volga River (in northern Kazakhstan and the southern part of Western Siberia) and in southeastern and western Australia.

Because wheat is an important crop in many agricultural systems, large-scale wheat farms face competition from commercial and subsistence supplies produced throughout the world. Wheat may be stored, if care is taken, so that wheat prices reflect not only current crop conditions but also the amount available in grain elevators. World trade in wheat is considerable, but in recent years increasing production and export subsidies have altered traditional trade patterns. Figure 9.16 shows that the United States, Canada, Australia, and Argentina are usually wheat exporters, while the USSR, India, and China are frequent wheat importers. In recent years, Western European countries have, through government subsidization of production, turned from importers to exporters of the grain.

Special Crops

Proximity to the market does not guarantee the intensive production of high-value crops, should terrain or climatic circumstances prevent it. Nor does great distance from the market inevitably determine that extensive agriculture on low-priced land will be the sole agricultural option. Special circumstances, most often climatic, make some places far from markets intensively developed agricultural areas. Two special cases are agriculture in Mediterranean climates and in plantation areas (Figure 9.17).

Most of the arable land in the Mediterranean basin itself is planted to grains, and much of the agricultural land is used for grazing. **Mediterranean agriculture** as a specialized farming economy, however, is known for grapes, olives, oranges, figs, vegetables, and other commodities that cannot be grown easily in and around the great industrial zones, which generally are found in areas of cooler climate. These crops need warm temperatures all year round and a great deal of sunshine in the summer. The distribution of Mediterranean agriculture is indicated in Figure 9.17. These are among the most productive agricultural lands in the world. Farmers can regulate their output in sunny areas such as these because storms and other inclement weather problems are infrequent. Also, the precipitation regime of Mediterranean climate areas, as shown in Chapter 3, lends itself to the controlled use of water. Of course, much capital must be spent for the irrigation systems—another reason for the intensive use of the land for high-value crops.

Climate is also considered the vital element in the production of what are commonly, but imprecisely, known as **plantation crops.** The implication of *plantation* is the introduction of a foreign element—investment, management, and marketing—into an indigenous culture and economy, often employing an alien labor force. Even the crops, although native to the tropics, were frequently foreign to the areas of plantation establishment: African coffee and Asian sugar in the western

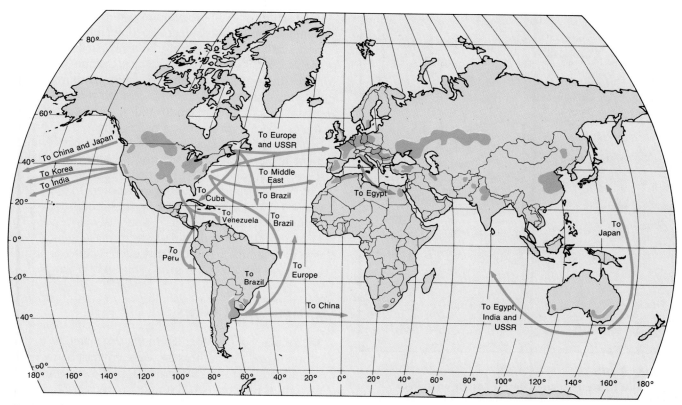

Figure 9.16

Wheat-growing areas and expected patterns of world trade in wheat. The arrows mark characteristic origins and flows of wheat in international trade. The volume of movements as well as the origin and destination nations vary from year to year.

Recently, Third World successes of the Green Revolution and subsidized surpluses of wheat in Europe have altered traditional patterns of demand and trade.

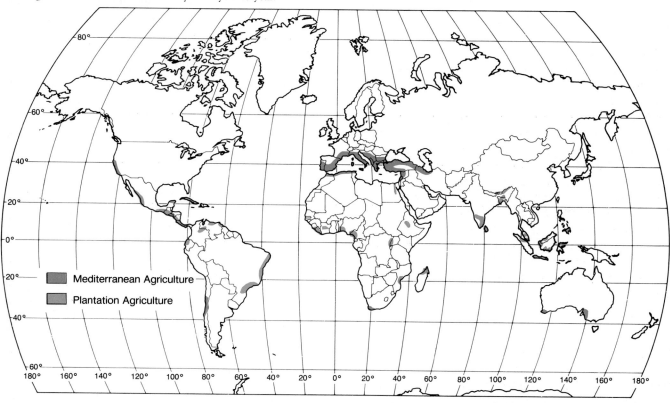

Figure 9.17

Special crop commercial agriculture. Areas of Mediterranean agriculture have roughly similar climates and specialize in similar commodities, such as grapes, oranges, olives, peaches, and vegetables. The specialized crops of plantation agriculture are influenced by both physical geographic conditions and present or, particularly, former (colonial) control of area.

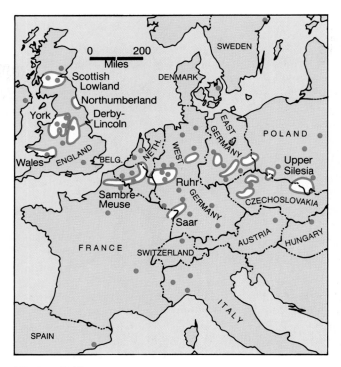

Figure 9.18
Coalfields in Western and Central Europe and major manufacturing cities. Although a number of cities are distant from coalfields, the majority of leading industrial centers are at or near them.

hemisphere, and American cacao, tobacco, and rubber in southeast Asia and Africa are examples. Entrepreneurs in Western countries such as England, France, the Netherlands, and the United States became interested in the tropics partly because they afforded them the opportunity to satisfy a demand in temperate lands for vegetative commodities not producible in the market areas. "Plantation" becomes an inappropriate term where the foreign element is lacking, as in cola nut production in Guinea, spice growing in India or Sri Lanka, or sisal production in the Yucatán, though custom and convenience frequently retain the term even where native producers of local crops dominate.

The major plantation crops and the areas where they are produced include tea (India and Sri Lanka); jute (India and Bangladesh); rubber (Malaysia and Indonesia); cacao (Ghana and Nigeria); cane sugar (Cuba and the Caribbean area, Brazil, Mexico, India, and the Philippines); coffee (Brazil and Colombia); and bananas (Central America). As Figure 9.17 suggests, for ease of access to shipping, most plantation crops are cultivated along or near coasts.

Summary Agriculture shares with all economic-geographic phenomena the characteristic of constant, frequently unpredictable change. As with all such changes, the consequences for other human patterns may be great. A case in point has been the enormous impact of changes

in agricultural technology upon patterns of population. Significant migrations occur as people, released from the land, search for jobs in industrial zones. In the early years of this century in the United States, the introduction of increasingly efficient farm machinery made redundant millions of farm workers, who, in their turn, joined the urban labor force in the industrial Northeast. After World War II, the improved cotton picker released great numbers of southern blacks, who also sought new employment in northern cities, altering the urban ethnic mix there. In Europe, recent mechanization and farm consolidation have induced migrant laborers to move from the south and the east to the northern industrial zones, frequently changing their country of residence. The mechanization and monoculture of the Green Revolution have similarly substantially altered patterns of rural and urban settlement in many Third World nations.

Power and Ores

The primacy of power and mineral raw materials in their economies is a clear measure of the contrast between the developed and the underdeveloped economies (although, as the box "Word Games" reports, such descriptive terms as "developed" and "underdeveloped" are in this and other respects not fully satisfactory contrasts). There is no meaningful distinction between the commercial and planned systems of advanced technology in importance and use of power and minerals, though there are contrasts in their respective reactions to cost considerations. Commercial systems are relatively more sensitive to changes in the market price of these important components of production.

Power

Except for the brief localized importance of waterpower, coal was the form of inanimate energy upon which the Industrial Revolution was founded and developed. The vastness of world coal deposits—some 10 trillion (10^{13}) metric tons, estimated to be sufficient for several thousand years at present rates of extraction—assure coal's future importance, particularly as supplies of petroleum and natural gas approach exhaustion. At the time of the Industrial Revolution, it became clear that it was more costly to transport coal than manufactured goods. Coalfields became efficient locations for production because they offered low total transportation costs. As a result, most of the great older industrial belts of the world are located on or very near coal deposits. The Ruhr and English Midlands fields, shown in Figure 9.18, and the Pittsburgh-Cleveland region in the United States are examples of early major industrial zones. They not only supplied the coastal cities with manufactured products but became major markets themselves because of their attraction for labor. Industries that were not related directly to coal but to the coal-related industries began to

Word Games

Since World War II, the international community has searched for acceptable descriptive terms to classify countries according to their economic circumstances, their degrees of modernization and industrialization, and their sometimes shifting political affiliations. The Western nations led the way in devising adjectives to describe the nonindustrial countries: backward, poor, undeveloped, underdeveloped, less developed, developing, rapidly developing, and a host of other euphemisms have been introduced. They have also been applied to problem areas within developed countries, as the regional study, "Appalachia," (Page 419) suggests. Invariably the terms have been criticized by those they described as imprecise, patronizing, or degrading. Frequently, the terms in vogue have been expected to serve more than one descriptive function. The United Nations, in grouping countries for its statistical series, has tried to make distinctions based both on degee of development and on ideology of government. In their turn, individual countries have sought identity not only as at a particular stage of development or possessing a special political philosophy, but also as having particular resource, economic, or trade interests to promote. The result is a contradictory collection of imprecise descriptors fully satisfactory to no one. Author Gorän Palm, as reported in the UN journal *Development Forum,* reviews some of the current terminology. In our international vocabulary, he tells us:

"The developed world" is contrasted to "the Third World." "Industrialized countries" are contrasted to "underdeveloped countries." "The rich in the North" are contrasted to "the poor in the South."

UN trade statistics commonly divide the world into three areas or "economic classes": industrialized countries or developed (Western) countries; developing countries or underdeveloped countries; socialist countries, Eastern bloc countries, or planned economies.

In addition to North America and Western Europe, the "industrialized countries" include Japan, Australia, New Zealand, South Africa and sometimes Israel. The "developing countries" are mainly the nearly one hundred nations of Asia, Africa, Latin America and Oceania which, within the framework of the UN, form the Group of 77 (sometimes identified with the non-aligned states). ["Nonaligned"] includes several nations with certain ties to the Soviet Union or China, such as Guinea, Yugoslavia, India, Tanzania, [and] also members of the "Eastern bloc." Lastly, the "socialist countries" include all nations of Eastern Europe except Yugoslavia, a handful of countries in Asia and a single country (Cuba) in Latin America.

This may seem like a good enough preliminary classification but on closer inspection we find that the first group of countries is the only one to be given an indisputably positive label. To be labelled "underdeveloped" or to represent a "developing country" is less gratifying. Such labels imply that one is somehow a little inferior or backward, that one, say, lives in tribes, worships holy cows, is afraid of contraceptives, doesn't know one's own best interests.

To be reckoned a member of the "Eastern bloc" or one of the "communist countries" is not much better. It amounts to being excluded or discriminated [against] on political grounds. Such countries are not part

of the "development" and "industrialization" so craved throughout the southern hemisphere.

Labels like "socialist countries" and "planned economies" are more accurate [than "Eastern" or "communist."], but here, too, a different norm is applied to these countries than to the other groups. They are identified not according to their level of development, but according to ideology and the nature of their economic system.

If we take industrialization as the criterion, any differences between the Eastern and Western countries evaporate. A similar [equivalence of terms] might be extended to all the countries outside the industrial world. Most of the countries of Latin America, Africa, and Asia depend on the production of foodstuffs and primary materials for industry for their livelihood. To sidestep the problem of ideology, we may conveniently categorize the world into "industrialized countries" versus "raw materials-producing countries."

It is perfectly acceptable to say "industrialized countries" just as long as they are contrasted not to "developing countries" but to "agrarian" or "raw materials-producing countries." It is fine to say "socialist countries" as long as they are contrasted to "capitalistic countries." Fine to say "underdeveloped" just as long as it is contrasted to "overdeveloped." . . . The problem lies in the misleading pairs of opposites so frequently used.

Source: "Lexical lapses on the line," United Nations, Centre for Economic & Social Information/OPI, *Development Forum,* Vol. 6, No. 10 (November–December 1978), p. 6.

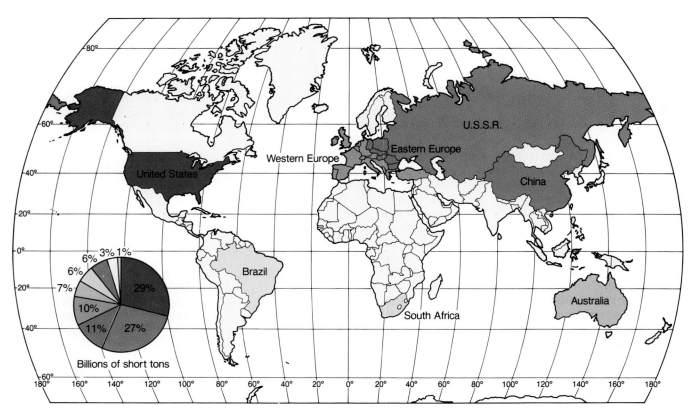

Figure 9.19
World pattern of coal reserves. Each of the nations and regions named contains at least 10 billion short tons, or 1% of the world's total, of recoverable coal—that is, reserves that can be recovered under current economic conditions with current technology. Together, the named areas have more than 97% of the world's recoverable coal deposits.

Source: Data from *International Energy Annual,* U.S. Energy Information Administration.

cluster in these zones, and industrial agglomerations of great size with complex linkages developed. The associated large industrial cities represented an attraction for a further increase in market-oriented manufacturing activity. Coal still plays a role in industrial development, and though no longer dominant, it still accounts for nearly one-half of the world's inanimate energy. As petroleum and natural gas become less plentiful, coal may once again assume a major position in the world's industrial energy budget. Should that occur, the distribution of recoverable coal reserves favors those world areas already industrially advanced (Figure 9.19).

Petroleum, first extracted commercially in the 1860s in both the United States and in present-day USSR, became a major power source only early in this century. The rapidity of its adoption as both a favored energy resource and a raw material, along with the limited size and the speed of depletion of known and probable reserves, suggest that it cannot long retain its present position of importance in the energy budget of nations

(Table 9.1). On a world basis, petroleum accounted for 47% of commercial energy in 1973 but had dropped to 40% by 1984. The decline reflected not only cost increases of oil (consumption rose again to 41% in 1986 as oil prices declined) but also improvements in the efficiency of oil use. With its high energy value in relation to its bulk, petroleum has tended to be transported to, and refined at, consuming areas already established (Figure 9.20). Relatively few industrial complexes of major size have developed with petroleum deposits serving as the prime attraction. Where there has been such a spatial relationship—on or near producing fields or along major pipelines—the orientation has been to petroleum as a raw material, not as a fuel.

There is a strong desire in the developed countries of both commercial and planned economies to develop power supplies alternative to and supplementing coal and petroleum. Nuclear and solar energy are two candidates, though neither has as yet fulfilled the promise initially claimed for them. At present, safety factors, construction and operating costs, and public fears of the former, and

TABLE 9.1

Commercial Energy Consumption, 1986 (million tons of oil equivalent)

Region and country	Oil	Natural gas	Coal	Hydro-electric	Nuclear	Total	% by oil
Anglo-America	813.9	459.5	471.1	157.2	129.7	2031.4	40.1
Latin America	215.4	71.2	22.1	82.2	1.5	392.4	54.9
Western Europe	585.1	192.5	250.0	101.1	138.6	1267.3	46.2
Africa	81.7	28.5	66.2	16.1	1.0	193.5	42.2
Australasia	31.3	17.9	40.7	9.9	—	99.8	31.4
Middle East	108.3	48.0	2.2	2.6	—	161.1	67.2
Southeast Asia	117.3	18.6	39.1	7.3	12.0	194.3	60.4
South Asia	56.3	17.2	111.7	20.0	1.3	206.5	27.3
Japan	204.4	36.4	70.2	19.6	41.4	372.0	54.9
China	99.2	12.1	531.2	28.2	—	670.7	14.8
USSR	445.0	505.3	376.2	52.5	35.2	1414.2	31.5
World total	2881.0	1507.1	2309.1	519.4	372.7	7589.3	38.0

Note: Commercial energy includes only commercially traded fuels; excluded are wood, peat, animal waste, and other biomass important in many less developed countries.

Unreported figures for smaller countries are included in world totals. Source: British Petroleum Company, *BP Statistical Review of World Energy, 1987.*

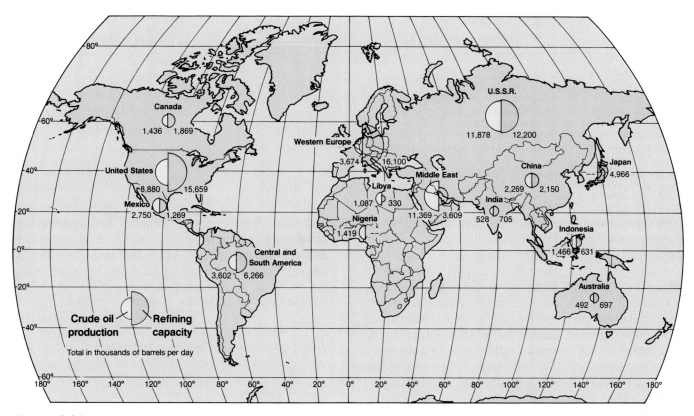

Figure 9.20

Daily petroleum production and refining capacity.
Refineries tend to be market oriented; petroleum is exploited wherever found. The result, of course, is an enormous trade in crude petroleum.

Source: Data from *International Energy Annual,* U.S. Energy Information Administration.

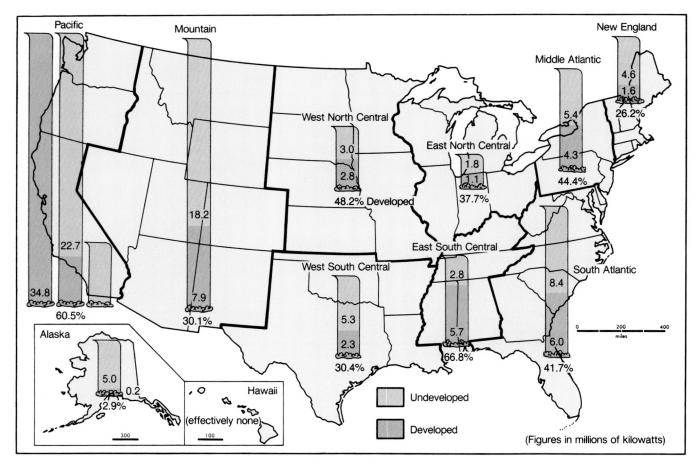

Figure 9.21

Conventional hydroelectric capacity. Waterpower, developed
and undeveloped, in the United States. The greatest potential for
hydroelectric power is in the Mountain and Pacific states.
Industries may take greater advantage of this potential in the
future, although dam construction increasingly meets objections
from environmental protection groups and alternate users of the
water in question.

Source: Data from U.S. Federal Energy Regulatory Commission,
*Hydroelectric Power Resources of the United States, Developed and
Undeveloped,* January 1, 1984.

costs and low efficiencies of the latter, detract from their
widespread use, matters discussed in detail in Chapter
10.

Hydroelectric power is developed where sites for
dams are feasible, though, of course, not all such sites
can or will be used. Although rugged terrain and a steep
natural fall of water are frequently associated with hydropower potential, comparable descents and potential
can be created by damming and impounding streams of
relatively low gradient, though at the cost of flooding vast
areas of land. Hydroelectric power sites are not plentiful,
and as Figure 9.21 shows, developable potential is often
remote from existing markets. As expected, the greatest
degree of use of hydropower potential has been in the

technically advanced world regions: 50% in the United
States and 44% in Europe. The unrealized potential is in
Africa, Latin America, and Asia (Figure 9.22), where rainfall is plentiful and where great rivers run undammed,
but where industrial demands and available capital are
limited (see "Hydropower Negatives," Page 129).

Ores

Transportation costs play a great role in determining
where *low-value minerals* will be mined. Minerals such
as gravel, limestone for cement, and aggregate are found
in such abundance that they have value only when they
are near the site where they are to be used. For example,
gravel for road building has value if it is at the road-
building site but not otherwise. Transporting gravel
hundreds of miles is an unprofitable activity.

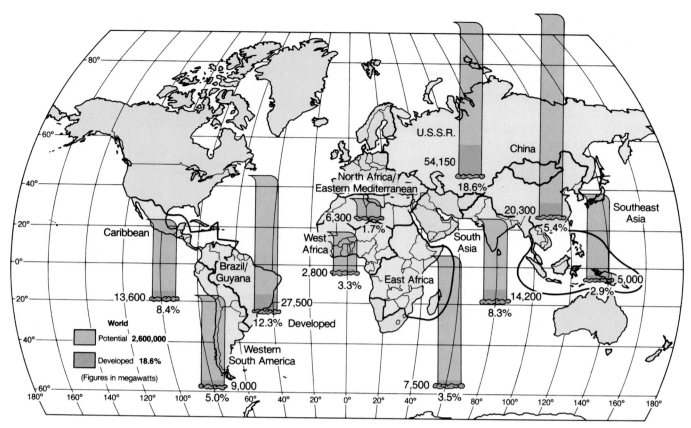

Figure 9.22

Annual hydro potential and use, selected regions. Less than 20% of the world's estimated 2.5 million megawatts of hydroelectric potential had been developed by the early 1980s, with most installed capacity found in Europe and North America. The difference between potential and utilization for countries and regions shown on the map is no guarantee of the exploitation of hydroelectric power in developing countries. Lack of market, high construction and debt servicing costs, and fears of ecological damage limit major hydropower developments in many regions. China, Brazil, and the Soviet Union (eastern) have vast potentials, rapidly growing demand, and a continuing commitment to dam construction. Their programs and those of other nations continually alter the world and regional ratios of potential and development.

The production of other minerals, especially *metallic minerals* such as copper, lead, and iron ore, is affected by a balance of three forces: the quantity available, the richness of the ore, and the distance to markets. Even if these conditions are favorable, mines may not be developed or even remain operating if supplies from competing sources are more cheaply available in the market. Between 1980 and 1985, more than 25 million tons of iron ore-producing capacity was permanently shut down in the U.S. and Canada. Similar declines occurred in North American copper, nickel, zinc, lead, and molybdenum mining as market prices fell below domestic production costs. The developed industrial nations of commercial economies, whatever their former or even present mineral endowment, find themselves at a competitive disadvantage against Third World producers with lower cost labor and state-owned mines with abundant, rich reserves.

When the ore is rich in metallic content, it is profitable to ship it directly to the market for refining. But, of course, the highest-grade ores tend to be mined first.

Consequently, the demand for low-grade ores has been increasing in recent years as richer deposits have been depleted. Low-grade ores are often upgraded by various types of separation treatments at the mine site to avoid the cost of transporting waste materials not wanted at the market. Concentration of copper is nearly always mine oriented; refining takes place near areas of consumption.

The large amount of waste in copper (98%–99% or more of the ore) and in most other industrially significant ores should not be considered the mark of an unattractive deposit. Indeed, the opposite may be true. Many higher-content ores are left unexploited because of the cost of extraction or the smallness of the reserves in favor of using large deposits of even very low-grade ore. The attraction of the latter is a reserve of sufficient size to justify the long-term commitment of development capital and, simultaneously, to assure a long-term source of supply.

At one time, high-grade magnetite iron ore was mined and shipped from the Mesabi area of Minnesota. Those deposits are now exhausted. Yet immense amounts of capital have been invested in mining the virtually un-limited supplies of low-grade iron-bearing rock (tac-onite) still remaining and processing this ore into high-grade iron-ore pellets. Such investments do not assure the profitable exploitation of the resource. The metals market is highly volatile, and rapidly and widely fluc-tuating prices can quickly change profitable mining and refining ventures to losing undertakings. Marginal gold and silver deposits are opened or closed in reaction to trends in precious metals prices. Taconite *beneficiation* (waste material removal) in the Lake Superior region has virtually ceased in response to the decline of the U.S. steel industry. In commercial economies, cost and market controls dominate economic decisions.

Manufacturing

In market economies, entrepreneurs seek to maximize profits by locating manufacturing activities at sites of lowest total input costs (and high revenue yields). In order to assess the advantages of one location over an-other, industrialists must evaluate the most important **variable costs.** They subdivide their total costs into cat-egories and note how each cost will vary from place to place. In different industries, transportation charges, labor rates, power costs, plant construction or operation ex-penses, the interest rate of money, or the price of raw materials may be the major variable cost. The industri-alist must look at each of these and by a process of elim-ination eventually select the lowest-cost site. If the producer then determines that a large enough market can be reached cheaply enough, the location promises to be profitable.

But in the economic world, nothing remains con-stant. Because of a changing mix of input costs, produc-tion techniques, and marketing activities, most initially profitable locations do not remain advantageous. Migra-tions of population, technological advances, and changes in the demand for products affect industrialists and in-dustrial locations greatly. The abandoned mills and fac-tories of New England or of the steel towns of Pennsylvania, even the "de-industrialization" of America itself in the face of foreign competition, are testimony to the impermanence of the "best" locations.

The concern with variable costs as a determinant in industrial location decisions has inspired an extensive theoretical literature. Most of it is based upon and ex-tends ideas presented by the German location economist **Alfred Weber.** Figure 9.23 gives an idea of his analytic technique, which is employed in the following brief dis-cussions of a few of the characteristic industries in ad-vanced commercial economies.

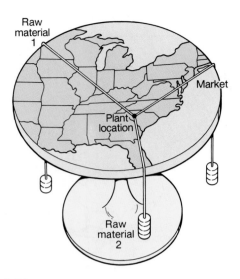

Figure 9.23
Plane table solution to a plant location problem. When a weight is allowed to represent the "pull" of raw material and market locations, an equilibrium point is found on the plane table. That point is the location at which all forces balance each other. It represents the least-cost location—the place where, theory suggests, the plant should be located. Alfred Weber developed the analytic techniques for determining least-cost locations for such simple cases and for instances where costs other than transportation charges—labor and power costs, for example—must be taken into account.

Iron and Steel Industry

The metal-making and metalworking industries were the keystones of the Industrial Revolution and the unmistak-able evidence of industrial development of nations and peoples. The foundation stone of industrial power in the modern world is frequently seen to be the iron and steel industry. Small wonder that grandiose steel-mill projects appeared so frequently and often so irrationally in the development plans of nations newly emerging from sub-sistence economies. The steel mill serves as the sym-bolic mark of national pride and achievement, even if, as is often the case, it requires production and export subsidies to maintain efficient levels of operation.

A primary raw material in steel making is, of course, iron ore, but equally important are coke (a processed form of certain coals); limestone; and ferroalloys, such as nickel, tungsten, molybdenum, or chrome. The iron ore, limestone, and coke are heated to yield iron, which is then treated in open-hearth furnaces or by the basic oxygen process to remove excess carbon. To the re-sulting steel, in its liquid form, one or more ferroalloys are added to give the steel the specific qualities required for its ultimate use. In recent years, scrap metal has often been used in the production process in place of, or in combination with, iron and the ferroalloys.

Plant, labor, and capital are not major variable costs affecting location in the basic iron and steel industry within market economies because they do not vary mark-edly within the bounds of a single country. One might

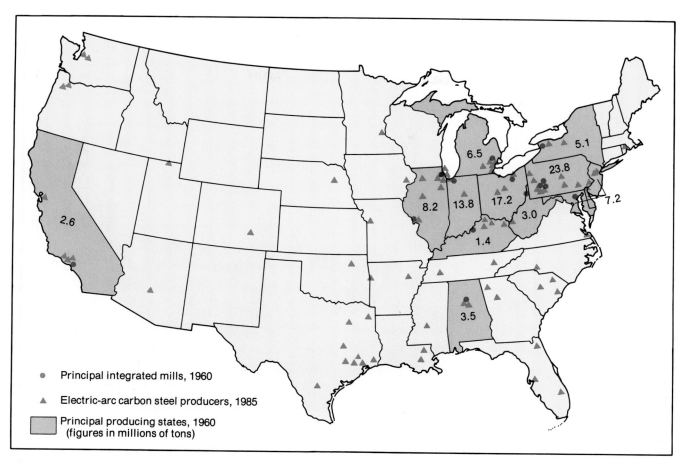

Figure 9.24

Raw steel production points, 1960–1985. The changing pattern of steel production. A quarter-century of fundamental change is recorded on the map. In 1960, 93% of the nation's raw steel was produced in the shaded states, and the electric furnace process accounted for only 8% of the raw steel poured. By 1985, 33 mainland states had one or more electric-arc furnaces (Hawaii and Puerto Rico each had an additional one), bringing steel production closer to regional markets and benefiting from lower production costs. The 84 electric furnace plants in 1985 ranged in size from mini-mills with less than 500,000 tons per year of raw steel capacity to maxi-mills with capacities above 1 million tons. Together, electric furnaces had production capabilities of 41 million tons per year and poured 34% of American steel.

Sources: Data from *Purchasing*, 1985 and 1986; and American Iron and Steel Institute.

assume, therefore, that the industrialist would attempt to locate close to the raw materials or where the transport costs of collecting the necessary ingredients were minimized. Such was the case early in the Industrial Revolution when spatial orientation to, particularly, fuel (initially charcoal, later coking coal) was the rule. Metallurgical industries became concentrated, epitomizing the industrial strength of the regions and nations of their localization: the Midlands of England, the Ruhr district of Germany, and the Donets Basin of the Soviet Union were classic examples. Increasingly in the more advanced economies the growing demand has been, not for great volumes of common grades of steel, but for diversified supplies of specialty grades of metal needed in regional markets. The industry has become more spatially dispersed.

"Mini-, midi-, and maximills" using electric furnaces to convert scrap metal into steel, by-passing iron making altogether, have taken increasing shares of the total steel supply in both market and planned economies. From their first widespread appearance in America about 1960, the more than 80 electric furnaces located throughout the nation by the mid-1980s were supplying more than one-third of the American steel market at the same time that the companies comprising "Big Steel" were closing major integrated mills and reducing their production capacity (Figure 9.24). In the Soviet Union, the world's largest steel-producing nation (and one of the largest importers of specialty steels and steel products) the program for the 1986–1991 Five-Year Plan calls

Figure 9.25

This idled blast furnace in Cleveland, Ohio, typifies the closed mills, lost jobs, and eroded regional economies that mark the wrenching restructuring of the American iron and steel industry.

TABLE 9.2

American Consumption of Steel Mill Products, 1960–1985

Year	Total consumption (000 net tons)	Percent imported
1960	71,511	4.7
1965	100,553	10.3
1970	97,109	13.8
1975	89,016	13.5
1980	95,247	16.3
1985	96,367	25.2

Source: American Iron and Steel Institute, *Annual Statistical Report*.

for increasing the role of electric-arc furnace minimills and gradually phasing out of the open-hearth steel furnaces of the ore-to-product integrated mills.

International competition, too, has altered traditional markets for steel. Stagnant domestic demand and cheaper imports have, in the United States, resulted in plant closures, massive unemployment in the "steel towns" of Pennsylvania, Ohio, West Virginia, and other traditional homes of the industry, and wrenching disruption of local and regional economies (Figure 9.25). Between 1979 and 1985, 250,000 steel industry jobs were lost in the United States, and steel imports rose to supply 25% of the domestic market (Table 9.2). An increasingly integrated world economy inevitably alters established national economic patterns.

Aluminum Industry

The major costs in the aluminum industry are raw materials, power, plant, labor, and capital. The question is "Which of these is a major variable cost?" To answer that question, we need first to sketch how aluminum is made. It is ultimately derived from the mineral *bauxite*. Because the bauxite contains industrial waste as well as aluminum, it is changed—under high temperatures and pressures—to a material called *alumina* (aluminum oxide). To extract the aluminum, massive charges of electricity are sent through the alumina while it is in solution. Electrical power accounts for between 30% and 40% of the cost of producing aluminum and is the major variable cost influencing plant location.

The source of electrical power is irrelevant in the locational decision. Electricity need only be available in great amounts at low cost. The familiar mental association of aluminum plants and waterpower sites, though sometimes accurate, is not necessarily valid. Extreme examples of the influence of power costs to the exclusion of other cost considerations are the Kitimat plant on the west coast of Canada and the Bratsk plant near Lake Baikal in eastern Siberia. Both are far from bauxite supplies and

Figure 9.26

Between 1981 and 1986, seven United States aluminum smelters closed permanently, and the nation changed from being a net exporter to an importer of aluminum materials. Uncompetitively high electric power costs and a declining share of the domestic market reduced production to 68% of mill capacity in 1986. Canadian mills, with lower power costs, produced at 96% of capacity in that year.

Source: Data from *American Metal Marketing* and U.S. Department of Commerce.

from markets, but both are adjacent to huge hydroelectric power sources. Yet well over three-quarters of the aluminum refined in the United States uses power from other than hydroelectric sources (Figure 9.26).

Aluminum, with relatively high value per unit weight, is easily moved and traded on the world market. In a climate of increasing international competition, producers with low energy costs tend to have an unbeatable advantage. In the United States, plants served by utilities using fossil fuels (coal, oil, natural gas) have power costs more than twice the average per kilowatt hour of foreign competitors, many of which are governmentally owned

or enjoy subsidized energy costs. The result has been a closure of reduction plants that relied on fossil-fuel electricity (Figure 9.25), and the threatened closure even of hydro-based plants of the Pacific Northwest as the Bonneville Power Authority raises the formerly low industrial electric rates it charged to the producers.

Transportation costs for alumina are about as important as those for iron ore in the steel industry. Direct-shipping iron ore averages above 50% iron content, while enriched pellets average 65% iron; alumina averages about 50% aluminum content. However, finished aluminum per unit weight is much more valuable than an

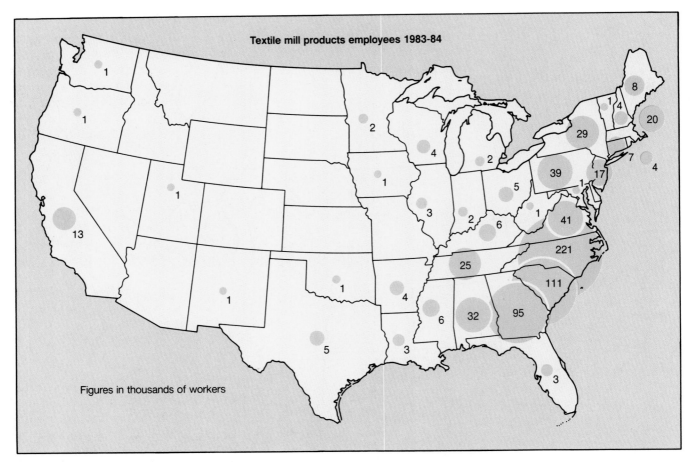

Figure 9.27

Textile mill products employees, 1983–1984. The location of these workers in the United States reflects both the lower wage rates in the nonunion plants of the South and the traditional northeastern mills and markets.

Data from *County Business Patterns*.

Textile Industry

equal weight of steel. Consequently, customer location is not as important for the aluminum as for the steel industry, and American producers are spatially free to invest in new reduction facilities in Canada, particularly Quebec, where lower-cost hydroelectricity is abundantly available.

The major costs in the textile, or cloth-making, industry are plant, labor, capital, and raw materials. The raw material is cotton, wool, or any of a number of artificial fibers. There is little waste, and the transportation costs of raw materials and finished goods, while important, are not a major variable. Among the cost factors, therefore, labor is the locational determinant.

A great deal of labor is necessary to turn the raw materials into a finished product, but it need not be skilled labor. In the United States, nearly all of the cotton and wool textile industries were located in New England through the 19th century. The industry was originally founded along rural New England waterpower sites during the 1820s and 1830s, drawing on a labor pool made up largely of underemployed daughters of the many families working small farms of the area. Later, immigrants provided the surplus unskilled labor supply in the cities and towns of the area. The industry first began to decline in New England when coal (which the area did not have) replaced waterpower. The decline accelerated as plants became obsolete and as wage levels in that area increased. When it became clear that a surplus of cheap labor was available in the Piedmont district of the South, the industry quickly moved in that direction, following a trend begun soon after the Civil War. Today, the United States textile industry that remains in the face of foreign competition is concentrated in the Piedmont (Figure 9.27).

Much of that competition comes from the newly industrializing countries of Asia (including the major suppliers of the American market: Taiwan, South Korea, and Hong Kong) where the textile industry is seen as a near-ideal employer of an abundant labor force. The growth of the industry to significant proportions in such areas is creating unwanted and resisted competition to textile manufacturers in established commercial economies. The advantage now enjoyed by rapidly industrializing nations may decline as it has in former world centers limited-skill, assembly-line operations are requisite for volume production of clothing for a mass, highly price-competitive market. Wage rates in 1984 for apparel production workers in the 20 major exporters of garments of the United States ranged from a low of 2% (Bangladesh) to a high of 25% (Singapore) of the level of their United States counterparts. Those 20 low-wage countries accounted for nearly 90% of imported apparel, which, in turn, captured 60% of the domestic apparel market.

Although the production process may be dispersed to low-wage areas, the design, sales, and administration operations of apparel companies tend to be localized in the major urban centers that constitute the apparel markets. In the United States that location has been importantly, but not exclusively, New York City, where nearly 40% of manufacturers and jobbers in the women's outerwear industry are located and where many additional companies maintain offices, design studios, and showrooms. Traditionally concentrated in the "Garment District" of midtown Manhattan where showrooms and wholesalers still aggregate, the center of production—cutting, sewing, pressing—has shifted to the Lower East Side area of Chinatown (Figure 9.28).

The hundreds of New York factories and contract sewing plants both within the Garment District and elsewhere, as well as those located in the nearby areas of New Jersey, Pennsylvania, and New England, represent a spatial response not to wage rates, but to fashion and market fragmentation, which in turn affects the size of production runs and required speed of delivery. The fashion industry, sportswear, and quality women's outerwear, in particular, require the speed of production made possible by nearby sewing shops. Because fashions change quickly, are design- rather than price-competitive, and require close collaboration between creator and producer, the industry must be well informed about rapidly changing market conditions and buyer preferences. It benefits from spatial concentration not only in New York City but also in other recognized fashion capitals of the world: London, Paris, Milan, Rome, and Los Angeles (for fashion sportswear).

Automobile Industry

This industry typifies manufacturing in a highly advanced economy; the varied outputs of a great many manufacturers are assembled to make a single high-value product. In any given industrially developed country of the commercial economies, labor and plant costs are uniformly high, and tooling costs are enormous. The high transportation costs for the finished product affect location the most. Therefore, we would expect the industry to locate in the major markets of the commercial world. This is generally true for final assembly plants, but not for the earlier stages of manufacturing.

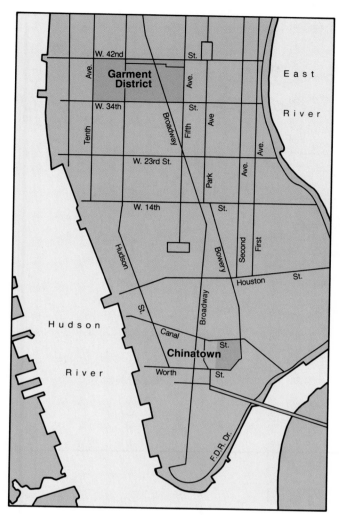

Figure 9.28

The famous midtown Garment District of New York City still contained some 300 factories in 1986 and hundreds of showrooms and wholesale outlets as well as design studios and corporate offices of the apparel trade. But the center of production has shifted to four- and five-story loft buildings in Chinatown and the nearby areas in the Lower East Side where more than 550 unionized factories employed 22,000 workers in 1986, and where an uncounted number of nonunion, low-wage sweatshops were also to be found. Rising rents and renovations of Midtown buildings to contain showrooms rather than workrooms for the "needle trades" have caused the shift.

of the textile industry. Even now, still lower-cost producers have made significant market penetration: China, Bangladesh, Thailand, and others. The trends of the textile industry confirm the observation that in commercial economies "best location" is a transient advantage of specific regions or nations, lost as the changing realities of Weber's locational balances come into play.

Apparel Industry

The conversion of the finished product of the textile industry into clothing is known as the *apparel industry*. Like textile manufacturing, it finds its most profitable production locations in areas of cheap labor. Repetitious,

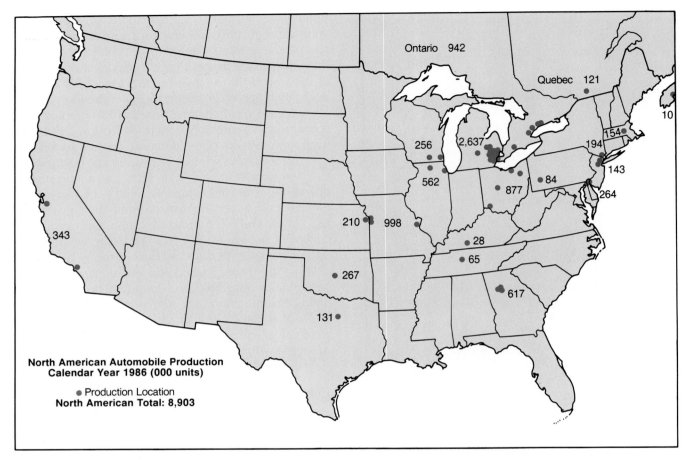

Figure 9.29

The map of automobile production in North America in 1986
shows far more dispersion than concentration, although the older
Michigan-Ontario nucleus is still evident. Further dispersal is
promised as foreign firms locate new assembly plants and
domestic companies build replacement plants away from Detroit.

The industry is so huge that it creates a market itself. That is, the myriad industries that supply the auto
industry locate close to the big plants, resulting in the
development of large industrial complexes or agglomerations. Such agglomerative tendencies are encouraged
by the imposition of "just-in-time" delivery demands
placed upon suppliers by the newest assembly plants.
Location near to those plants enhances the speed and
reliability of delivery of parts for which no backlog supply
is maintained by the assembly operation itself. Cities,
counties, and states bidding for new automotive assembly plants to be located within their boundaries are
simultaneously bidding for the location within the same
area of a multitude of parts suppliers and ancillary services.

Current auto plant locational decisions are not
based upon the same considerations as they were in the
early days of the industry. In the United States, automobile manufacturing got its start in many locations
throughout the Northeast and the Midwest. Essentially,
these were the locations that were already the market

centers of the country in 1900. In the Midwest, however,
the distance between towns was greater than in the
Northeast. There was also flat land and plenty of gravel
for roads. Wood was important to the early auto industry
for body parts, as were the gasoline engines already used
for motorboating on the Great Lakes. These factors, together with the reluctance of eastern bankers to invest
in the fledgling industry and the fact that Michigan was
the home of Henry Ford, led to the first large-scale production of autos in the Detroit area. This was an accident
of history; it could have happened anywhere along the
south shores of the Great Lakes.

Today the earlier centralization of the American
auto industry in the southern Michigan, Ohio, Indiana,
and Wisconsin area is altered by the emergence of a new
generation of car assembly plants. Some have been built
as close to the large markets as possible (Figure 9.29).
Others are being developed, frequently in rural sites well
away from primary metropolitan markets, in response to
monetary and other locational inducements, assessments of quality and wage rates of the labor pool, land

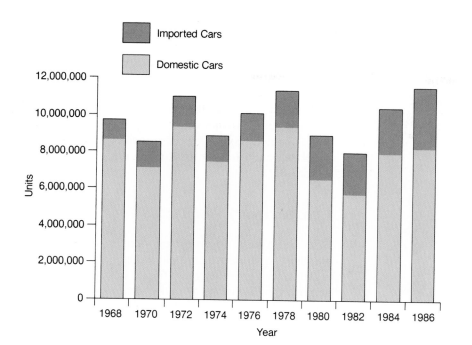

Figure 9.30
United States automobile sales, 1968–1986. Annual sales of new cars in the United States fluctuate. The proportion of those sales accounted for by imports has steadily increased: from 11% in 1968 to 28% in 1986.

prices, rail and expressway access, and other considerations that are somewhat different from the attractions that prevailed in the industry's earlier days. Many of the newest plants are American production sites of foreign automobile companies that increasingly find it advantageous for a variety of economic and political considerations to locate within the country at least part of the assembly of the imports that had, by the late 1970s and early 1980s, taken such a large share of the American market (Figure 9.30).

Imposed Considerations

These few briefly discussed examples of the cost factors affecting the location of individual industries have concentrated on the locational controls cited by Weber and other location economists and geographers. Both theory and observation suggest that in a pure, competitive commercial economy, the costs of material, transportation, labor, and plant should be dominant in locational decisions. Obviously, neither in the United States nor in any other "commercial" nation do the idealized conditions exist. Other constraints—some representing cost considerations, others political or social impositions—also affect, perhaps decisively, the locational decision process. Land use and zoning controls, environmental quality standards, governmental area development inducements, local tax-abatement provisions or developmental bond authorizations, noneconomic pressures on quasi-governmental corporations, and other considerations constitute attractions or repulsions for industry outside of the context and consideration of pure theory. If these noneconomic forces become compelling, the assumptions of the commercial economy classification no longer hold, and the transition to a controlled mixed or wholly planned economy has occurred.

Tertiary Activities

Primary activities are those involved in raw material and basic foodstuff production (mining, agriculture, fishing, forestry); **secondary activities** process raw materials and convert them into demanded products (the manufacturing and construction industries). A separate and essential sector of commercial economies, and the one with which we most often come into contact—**tertiary activities**—consists of those business and labor specializations that provide services to the primary and secondary sectors, to the general community, and to the individual. These include financial and educational institutions, the professions, transport and communication, and the various personal services. Importantly, tertiary activities constitute the vital link between producer and consumer. They are the wholesaling and retailing activities that fulfill the exchange function of advanced economies and provide market availability to satisfy the consumption requirements of individual members of highly interdependent societies. In commercial economies, tertiary activities also provide vitally needed information to manufacturers: the knowledge of market demand without which economically justifiable production decisions are impossible.

The retailing component of tertiary occupations is totally demand-oriented. The spatial distribution of retailing, therefore, is completely controlled by the spatial distribution of effective demand, that is, by wants made meaningful by purchasing power. A considerable amount of theoretical and empirical literature exists reviewing

TABLE 9.3

U.S. Employment Patterns, 1955–1985

	% of nonfarm jobs 1955	% of nonfarm jobs 1985	% increase in jobs 1955–1985
In services:			
Miscellaneous services[1]	12.3	22.4	+ 251
Retail trade	15.0	17.8	+ 129
State and local government	9.4	13.7	+ 103
Finance, insurance	4.5	6.1	+ 158
Wholesale trade	5.8	5.9	+ 97
Transport and utilities	8.2	5.4	+ 28
Federal government	4.3	2.9	+ 31
Total	59.5	74.3	+ 141
In goods production:			
Manufacturing	33.3	19.9	+ 15
Construction	5.6	4.8	+ 64
Mining	1.6	1.0	+ 22
Total	40.5	25.7	+ 22
Total jobs (millions)	50.6	97.7	+ 93

[1]Includes education, health, food and lodgings, etc.
Source: Bureau of Labor Statistics, *Employment and Earnings.*

locational patterns, hierarchies, and land use arrangements of individual stores, store clusters, and communities functioning as retailing centers. Some aspects of this topic are considered in Chapter 11, "Urban Geography."

Wholesaling and retailing are joined in the general tertiary-occupations category by some individuals and businesses that are, in a sense, producers. They produce services sold and purchased as intangible commodities: personal and professional services and financial, administrative, and governmental activities. They, too, represent a mark of contrast between advanced and subsistence societies, and the greater their proliferation, the greater the interdependence of the society of which they are a part.

Their expansion has been great in both the commercial and the planned economies. The decisive element is the development level of the society, not the form of economic administration. Within the United States, nearly three-quarters of nonfarm employment is now accounted for by "services"; manufacturing accounts for

only 20% (Table 9.3). In the mid-1950s the proportion was only 60:40 in favor of services. Since 1955 service jobs have increased by 140%; manufacturing jobs have increased only 15%. Comparable changes are found in other nations. By the mid-1980s, between 60% and 80% of jobs in such commercial economies as Japan, Canada, Australia, and all major western European countries and in the planned economies of the Soviet Union and at least Czechoslovakia in eastern Europe were in the service sector.

Quaternary Activities

In economically advanced economies, whether commercial or planned, many individuals and some entire organizations are engaged in the processing and dissemination of information and in the administration and control of their own or other enterprises. The term **quaternary** is applied to this fourth set of economic activities. It recognizes that in modern societies there are many individuals and groups engaged in business and institutional management, advertising, governmental and corporate planning, and business and financial services for hire or for specific enterprises. They include college professors (and colleges), newspaper reporters, real estate agents, public accountants, bankers, consultants, and even legislators. The list is long. Its diversity and familiarity remind us of the complexity of modern life and of how far we have progressed beyond the dominant farming, herding, and fishing of subsistence economies. As societies advance economically, the share of employment and of national income generated by the primary, secondary, tertiary, and quaternary sectors continually changes. The shift is steadily away from production and processing, and toward the trade, personal, and professional services of the tertiary sector and the information and control activities of the quaternary. The transition is recognized by the now-familiar term "postindustrial."

Quaternary-sector jobs are not spatially tied to resources, affected by the environment, or localized by market. Information, administration, and the "knowledge" activities in their broadest sense are dependent upon communication. Improvements in the technology of communication have freed them from their former concentrations in larger central business districts and in regional and national economic and political capitals. Quaternary activities are also increasingly international in scope and market. Specialization in them, some economists predict, will replace for the most advanced societies their former dependence upon the manufacturing industries now eroded by competition from newly industrializing economies.

Planned Economies

As the name implies, planned economies have a degree of centrally directed control of resources and of key sectors of the economy that permits the attainment of governmentally determined objectives. All advanced economies have elements of planned control, at the very least through the allocation of resources to social welfare programs, mass-transit systems, a defense establishment with associated industries, and the like. However, a clear distinction may be made between those economies categorized as *planned* and those classed as *commercial*. In the former, the preponderant control of resource allocation is centrally administered; in the latter, the allocation of resources results primarily from a response to market forces.

Just as there are variations in the degree of market control within individual commercial economies, so, too, do planned economies differ in the extent to which centralized control is exerted upon the society. The usual distinction made is between socialism and communism. Both derive from the teachings of Karl Marx, but they have taken basically different and mutually exclusive paths to attain the objectives he espoused.

Under *socialism,* planning does not presuppose public ownership of *all* means of production. In fact, according to a declaration adopted by the Socialist International in 1951, planning is "compatible with the existence of private ownership in important fields, for instance in agriculture, handicraft, retail trade, and small and middle-sized industries." What is involved is the concept of the replacement of capitalism by a system of democratically planned production that would achieve socially accepted goals. This aim would be accomplished by public ownership and operation, at various governmental levels, of key industries. The remainder of the economy would be retained in the hands of cooperative organizations and private owners under such public regulation as necessary to reach democratically agreed-upon objectives. Public ownership would not be a goal in itself, but the means of achieving goals.

Under this definition and set of assumptions, many Western nations may be classified as having planned economies on the socialist model. In recent years when under the administration of the Labour party, the government of Great Britain has owned much of the transportation system of the country and has controlled the output of the coal and steel industries as well as the development of most housing. It has determined the locational choices of major industrial investments. Sweden has developed even more central planning direction over the economy. Strong governmental participation in, as well as actual ownership of, major industries, transportation systems, and banks is found in Italy and France. Many economically developing countries have central authorities that control the use of resources and dictate output requirements.

In Marxian theory, *communism* is equated with worker control of the instruments of production, a condition that the theory predicts will evolve with the ultimate "withering away of the state." In practice, communism has emerged as the extreme example of the planned economy. Whether under the name of Marxism-Leninism, Maoism, or some other designation, it is a program of economic organization based on total *state* control of the means and tools of production. Because one of the tools of production is labor, the ideology translates itself into planned direction of all facets of individual and collective social and economic action.

Communism in its idealized form disallows private capitalism, commercial markets, and the informational guidance of the market price system. In their place is state ownership of all enterprise, including agriculture, and of every personal or professional service. Production is divorced from market control; prices are established to achieve predetermined economic or social goals.

Countries in which a Communist party rules are found in Europe, Asia, Africa, and the Americas. In Eastern Europe, communism was imposed by Russian occupation following World War II. In some developing nations outside Europe, militant Communist parties have achieved power through revolution, frequently with the support of disparate groups dissatisfied with the political and social conditions under which they lived. Of course, the forerunner of them all is the Union of Soviet Socialist Republics, which may serve as an example of an advanced planned economy and as a contrast to the theoretical pure commercial and subsistence economies discussed earlier.

The USSR: Planned Economy in Operation

The Communist Revolution of 1917 was followed by a confused period of civil war, a New Economic Policy of limited capitalism, and finally, after the assumption of total authority by Joseph Stalin, the establishment of the principles and mechanisms of the complete state control of economy and society that—with modifications—remain in force today.

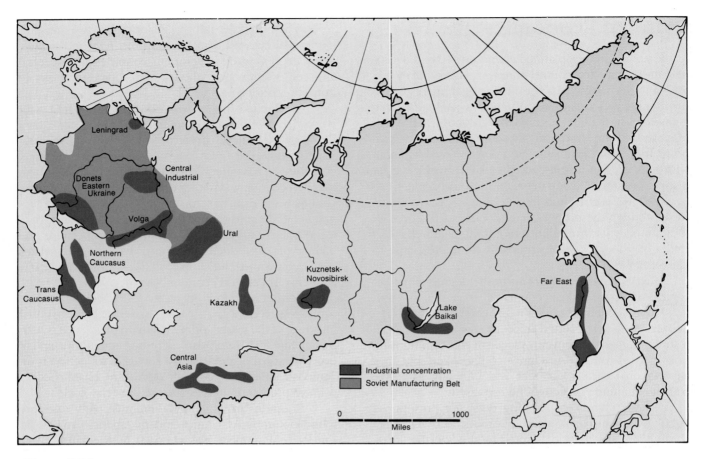

Figure 9.31

Patterns of industrial activity in the Soviet Union. The Volga, the Central Industrial, and the Leningrad concentrations within the Soviet manufacturing belt are not dependent on local raw materials. All other USSR industrial regions have a strong orientation to materials and were developed, by plan, despite their distance from the population centers and markets of European Russia.

Economic development of the then largely peasant society was to proceed, in the form of successive five-year plans, under the direction of the State Planning Agency (*Gosplan*). Beginning with the first plan in 1928, predetermined maximum and minimum production levels for every sector of the economy were forecast, and their achievement was assigned to ministries responsible for the management of designated sectors: heavy industry, agriculture, and so on. Party-established guiding principles of development shaped the details of the plan:

1. **Emphasis upon capital goods, heavy industry, and armaments.**
2. **Full development of the resources of the vast country, no matter how remote from existing European concentrations of population and industry.**
3. **The creation of population centers and industry at the sites of resources.**

4. **Investment capital accumulation by a set of administered prices that would extract from the agricultural and consumer sectors the funds needed for industrial growth.**
5. **The creation of a set of physically separate, independent centers of production of basic commodities, without regard to location of markets, transportation costs, or the other economic constraints accepted in commercial economies.**

In the industrial sector, the application of these policies has resulted in the development of the world's second largest producer of manufactured goods overall and, in many areas of heavy industry—steel, for example—its leading performer. New populous integrated centers of industry have been established in the eastern reaches of the country—in the Urals, in the Kuznetsk Basin of western Siberia, in Kazakhstan, in Central Asia, and in the Soviet Far East (Figure 9.31; see "The Kama

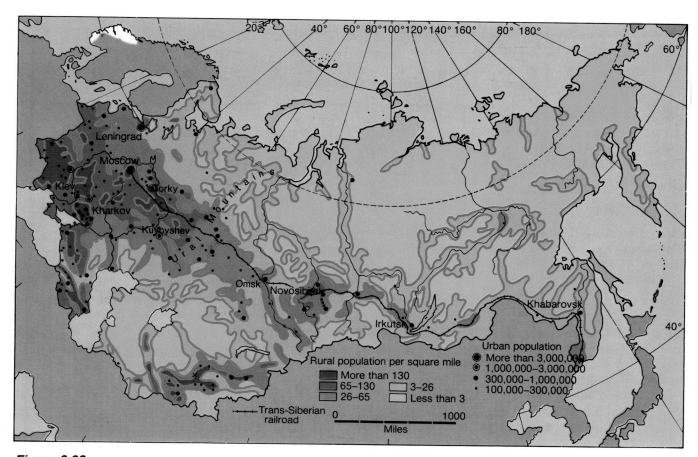

Figure 9.32

Population distribution in the USSR. The majority of the Soviet Union's 284 million people (1987 estimate) live in the European part of the country, west of the Ural Mountains (west of the 60° E longitude line). The Trans-Siberian Railroad is the main connecting link between the Pacific coast in the east through Siberia to the Urals and the industrial center at Moscow. Along it lie the developed areas and higher population densities of Siberia, whose fuller development awaits further transportation improvement and extension.

River Truck Plant" Page 332). Because the primary market for most commodities remains in the heavily settled and industrialized west, the price of delivered goods from the high-cost eastern producers must be discounted to fit within the budget allocations of the receiving establishments (Figure 9.32). The government subsidizes industries in part by ignoring the full cost of transportation services and in part by absorbing any transportation losses in the national industrial budget. The concept of optimal location in the commercial economy sense, though recognized and argued, has characteristically been secondary to the achievement of other planning objectives.

Those objectives are seen clearly enough in broad outline. It is the necessary detailed planning—budget, production goals, material sources, product destination, labor force size, wage rates, and every other facet of plant operation—for some quarter million industrial enterprises that has proved most complicated. An enormous work force is engaged in the planning process carried on through 51 economic ministries—not all, of course, confined to the industrial sector. The planners, lacking the feedback information available in commercial economies through the market price system, must draw up for enterprises, economic sectors, planning regions, and the nation enormous balance sheets with all the supplies for the economy entered on one side and planned uses on the other. Complex and time-consuming bargaining sessions follow, with planners seeking to minimize allocations to enterprises and maximize quotas, and enterprise managers and sector ministries aiming at the opposite goal. Recent administrations, particularly those of Y. V. Andropov and M. S. Gorbachev, have tried to decentralize planning control, give greater autonomy to individual enterprise managers, and introduce a modicum of profit motive and market pricing to the industrial sector to reduce its rigidities and make more efficient an

The Kama River Truck Plant

Six hundred miles east of Moscow in the isolated Tatar Autonomous Republic, the world's largest industrial project, the Kama River truck plant, was constructed during the 1970s. It was as much a symbol of Soviet industrial location philosophies as it was a significant addition to the manufacturing capabilities of the nation.

Kama is big even by Russian standards. The plant alone covers some 40 square miles (100 km²) and is made up of six separate installations, each a giant in its own right: foundry, forge, pressing and stamping, tooling and repair, engines and transmissions, and final assembly. Additional factories set up to produce building materials were added to the plant complex, for the isolation of the assemblage required self-sufficiency in construction.

Certain practical, economic considerations influenced the location of the truck plant. The broad Kama River, a tributary of the Volga, provides cheap, easy transportation of materials and generation of electric power. The largely rural Tatar Republic contained underemployed farm populations that could be turned into factory workers. Proximity to the Ural Mountains and to the Ural-Volga oil and gas fields permitted easy access to raw materials and fuels.

These considerations are comprehensible to industrial locators in commercial economies. Not so

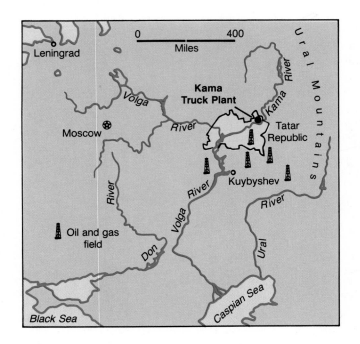

understandable to private entrepreneurs is a decisive Soviet consideration: the determination, at whatever cost, to develop fully every section of the nation, to bring urbanization and industrialization equally to all possible regions, and to develop interior locations less vulnerable to foreign attack than the populous western sections.

The decision to advance those objectives through the Kama River plant entailed investment commitments far beyond the estimated $5 billion cost of the plant itself. The isolated, underdeveloped character of the area demanded the construction of an entire city to house nearly 100,000 workers and their families. As nothing existed

before, not only apartment housing but the full range of urban facilities had to be built: schools, hospitals, stores, restaurants, sports arenas, community centers, and a hotel. To supply food, nine collective farms were organized, and local plants were built to produce vodka, ice cream, bread, and other necessities.

The creation of a single-purpose gigantic company town is fully consistent with Russian developmental and industrialization concepts originated at the start of the Stalinist period. The totally planned economies can envision objectives and allocate resources beyond the interests and capacities of entrepreneurs in commercially oriented societies.

economy officially listing 20 million articles on its "index of products." From the start of 1987, Soviet law was amended to permit profit-making joint ventures between Soviet industries and foreign companies. With up to 49% ownership allowed to foreign firms, the cooperative establishments would be freed from government planning controls, allowed to experiment with Western-style labor contracts, and permitted to compete with government-run enterprises for markets both within the Soviet Union and abroad.

Although a degree of relaxation of planning control in industry might be the result of such joint-venture experimentation, the rigidities and complexities of centralized management continue to exist in agriculture. Governmental control is basic throughout Soviet farming, including ownership of the land, direction of the crop economy, and determination of inputs. There are three main organizational forms in the rural sector: collective farms, state farms, and private holdings. The approximately 500 million acres (200 million hectares) of cultivated land in the "socialized" sector are about equally divided between collective and state farms. The latter are generally much larger than collectives. Some 15 million acres (6 million hectares) are in "private" plots.

Collective farms were created in the late 1920s and the early 1930s in a massive program of forced consolidation of separate landholdings into cooperatives. Individual farm families reside in traditional agricultural villages or in newly constructed centralized "agro-towns," jointly working the allotted communal rent-free land. Brigades of workers are assigned (or, recently, contract to perform) specific tasks during the crop year. Originally they received remuneration only as a share in farm profit, if any was achieved, but increasingly they receive fixed wages supplemented by incentive bonuses based on achievement that are paid from a revolving fund replenished by retained farm income. Crops to be grown and quotas to be met are established by regional agricultural planners. Supervision is under a farm manager, ostensibly elected by the members of the collective but in practice selected by the Communist party.

Collectivization released millions of farm workers for urban employment during the industrialization program of the USSR and so achieved, at great human cost (estimates vary from 8 to 14 million lives lost), one of its primary objectives. A second was also achieved: the generation of investment capital for the industrial sector. By paying low prices for forced delivery of farm products and by disposing of those products in the urban market at much higher prices, the government in effect imposed a massive tax upon the agricultural sector that provided funds for industrial development.

Figure 9.33

A farmers' market in the Soviet Union. For a modest daily fee, collective and state farmers can rent stall space in a government-built and regulated market building, selling commodities produced on their "private" plots. Between one-quarter and one-third of farm family income is earned from the operation of those small properties.

The **state farm** is a government enterprise operated by employees of the state, which provides all inputs and claims all yields. It is a rural factory in every administrative way equivalent to its urban counterpart. The state farm tends to be larger than the collective and more specialized in its output. For ideological reasons, increasing numbers of collectives have been converted to state-farm status in recent years.

Private plots, up to a maximum of about 1 ¼ acres (½ hectare), are labor-intensive holdings granted primarily to collective farm families; state farm workers get about ¾ acre (.3 hectares), and urban workers much less. Collective farmers depend on the plots for home-produced food—including a limited number of animals—and for cash income through sale in urban farmers' markets where state price controls do not apply (Figure 9.33). These carefully tended personal plots, together amounting to less than 5% of the Soviet Union's farmland, yield more than 25% of the value of the gross agricultural output of the country. From them come some 60% of the potatoes, 50% of the fruit, 35% of the eggs and vegetables, and 30% of the meat and milk produced in the USSR. On them are reared a fifth of the sheep, cattle, and hogs.

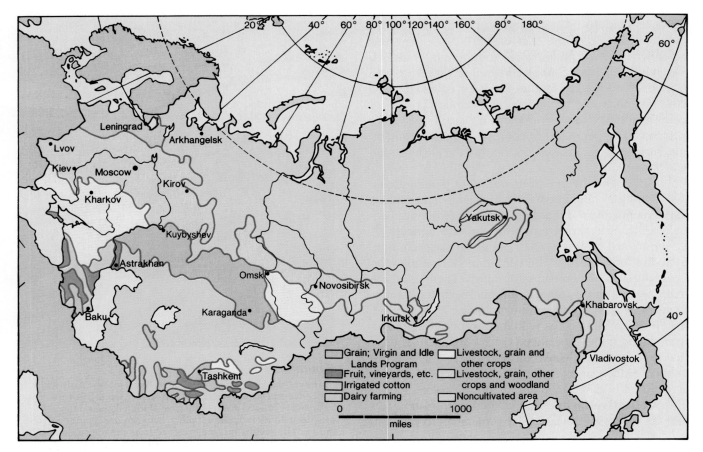

Figure 9.34

The Virgin and Idle Lands program extended grain production, primarily of spring wheat, eastward onto marginal land. Wheat constitutes nearly 90% of total Soviet food-grain production and 50% of all grains grown. Although the Virgin Lands program, begun in the 1950s, added significantly to the established agricultural regions of the nation, variability in weather in the new grain areas greatly affects Soviet wheat supplies and world patterns of wheat trade.

A success of the planned economy in achieving national goals of increased grain production is to be found in the Virgin and Idle Lands program. It spread the pattern of farming from established, primarily European, agricultural regions eastward to the semiarid mid-latitude steppes of the southern Urals, southwestern Siberia, and northern Kazakhstan (Figure 9.34). Launched in 1954, the program directed the cultivation of some 70 million acres (28 million hectares)—ultimately to rise to 115 million acres (47 million hectares)—of new land comparable in physical character to the Dust Bowl area of the United States. Increases in output have been impressive. Grain production grew from an average of 65 million metric tons from 1946–1950, to 205 million in 1976–1980.

Output has been erratic because of climatic conditions, which were largely responsible for a decline in average output to 177 million tons in the 4 years 1981 through 1984, and an increase to over 210 million tons in 1986. The overall improvement in production since inauguration of the Virgin Lands Program has been substantial.

Comparable achievements can be found in all sectors of the Soviet planned economy, for they represent the pursuit of a national objective. That by the standards of commercial economies they may represent a misallocation of resources and unbalanced development is not the point. The planned economy is predicated on the desirability of the achievement of the national priority goals by the assignment of funds and efforts to their fulfillment.

Conclusion

Very few people live in pure subsistence, commercial, or planned economies, although planned economies tend to be more pervasive in the countries where they prevail than do the other two forms in theirs. The trends in the world seem clear. *Subsistence* economies are giving way to planned and commercial organization. *Commercial* economies are introducing more and more government control and planning into their operations. *Planned* economies, seeking ways to overcome at least some of their cumbersome rigidities and inefficiencies, are turning to some of the controls, allocation methods, and assessments of enterprise efficiency characteristic of commercial economies.

It is also clear that economic activities, patterns, and events are the very stuff of life. As they are pivotal to all facets of human existence and experience, so they are central to geographic understanding. World patterns of economic activity join together the physical and cultural sides of the discipline. They are rooted in the realities of climates, soils, mineral deposits, navigable waters, and other natural phenomena. The economies pervasively influence, and are influenced by, the organizational forms of society, religion, custom, and way of life.

One final reminder is needed: an economy in isolation no longer exists. The world network of transportation, communication, trade, and interdependence is too complete to allow totally separated national economies. Events affecting one ramify to affect all. Crop failure in the Soviet Union influences world prices of wheat, patterns of shipping, balances of payments, and even international pacts and agreements. The potato blight in an isolated corner of Europe a century and a half ago still holds its message. Despite differences in language, culture, or ideology, we are inextricably a single people economically.

Key Words

collective farm *333*	intensive commercial	Mediterranean	state farm *333*
commercial	agriculture *309*	agriculture *312*	subsistence
economy *295*	intensive subsistence	nomadic herding *298*	economy *295*
comparative	agriculture *297*	planned economy *295*	tertiary activities *327*
advantage *305*	large-scale wheat	plantation crops *312*	truck farming *309*
extensive commercial	farming *311*	primary activities *327*	variable costs *320*
agriculture *311*	livestock ranching *311*	private plots *333*	von Thünen model *309*
extensive subsistence	livestock-grain	quaternary activities *328*	Weber location
agriculture *297*	farming *311*	secondary activities *327*	model *320*
Green Revolution *304*		shifting cultivation *298*	

For Review

1. What are the distinguishing characteristics of the economic systems labeled *subsistence, commercial,* and *planned?* Are they mutually exclusive, or can they coexist within a single political unit?

2. What are the ecological consequences of the different forms of *extensive subsistence* land use? In what world regions are such systems found? What, in your opinion, are the prospects for these land uses and for the way of life they embody?

3. How is *intensive subsistence* agriculture distinguished from *extensive subsistence* cropping? Why, in your opinion, have such different land use forms developed in separate areas of the warm, moist tropics?

4. What is *comparative advantage?* How is it related to world trade patterns? How does it help us to understand the developmental choices open to underdeveloped nations?

5. Briefly summarize the assumptions and dictates of von Thünen's agricultural model. The model suggests that concentric circles of agricultural specialization develop around a solitary, isolated market. What changes in that proposed pattern would you anticipate if a railroad, a highway, or a navigable waterway were located along a single radius of those circles? Diagram the new pattern that you think would emerge.

6. What role do prices play in the allocation of resources in commercial economies? What role do they play in planned economies? What differences in locational patterns of industry are implicit in the different treatments of costs in the two economic systems?

7. What impact upon industrial and employment patterns in the United States have the wage differentials between this and other countries had? Are those impacts necessarily negative? What do they demonstrate about the economics of industrial location?

8. What kinds of constraints other than pure cost considerations affect industrial location and other forms of economic development in commercial economies? In your opinion, are such constraints appropriate in a market economy? Why or why not?

Suggested Readings

Hartshorn, Truman A. and John W. Alexander. *Economic Geography.* 3d ed. Englewood Cliffs, N.J.: Prentice-Hall, 1988.

Berry, Brian J. L., Edgar Conkling, and D. Michael Ray. *Economic Geography.* Englewood Cliffs, N.J.: Prentice-Hall, 1987.

Gregory, Paul R., and R. C. Stuart. *Soviet Economic Structure and Performance.* 3d ed. New York: Harper & Row, 1986.

Reitsma, H. A., and J. M. G. Kleinpenning. *The Third World in Perspective.* Totowa, N.J.: Rowman and Littlefield, 1985.

Wheeler, James O., and Peter Muller. *Economic Geography.* 2d ed. New York: John Wiley, 1986.

10 Resource Geography

*T*he emergency cooling system of the number 4 reactor at the Chernobyl nuclear power plant was shut off at 2 P.M. on Friday, April 25, as operators prepared to test the turbine generators. But, at the request of a dispatcher at a power distribution station in Kiev, the shutdown of the reactor itself was delayed, and it continued to operate until 1:23 A.M. on April 26, 1986, the moment of the worst nuclear accident in history. Without water to cool them, the nuclear fuel rods overheated, reaching temperatures of approximately 3500° F. The water remaining in the system turned to superheated steam that reacted with the graphite bricks surrounding the fuel assemblies to produce flammable gases. As the gases exploded and an intense graphite fire raged, the reactor core was blown open and partially destroyed. A cloud of radioactive gases and particles was propelled thousands of feet into the air, where southeasterly winds carried it across an entire continent. In Siedlice, Poland; in Bala, Wales; in Allerona, Italy, and hundreds of other towns and cities, rainfall brought down radioactive particles that contaminated plants, soil, and water.

As monitors showed increased levels of radioactive contamination, governments took steps to protect their citizens. The Soviet Union cordoned off a 1000-square-mile area around the power station, evacuated over 100,000 people, and began the enormous task of scraping off the topsoil and carting it away for burial as nuclear waste. The European Common Market banned the importation of fresh vegetables, fruit, and meat from Eastern Europe. Individual countries imposed additional restrictions. Great Britain, for example, banned the slaughter of some 3 million lambs from its sheep farms, and Sweden declared the reindeer, on which Laplanders depend, unfit for human consumption (Figure 9.4).

The accident at Chernobyl does more than raise basic questions about the safety of nuclear energy and the reasons for the differing dependence upon it by the nations of the world. It brings into sharp focus the realization that human exploitation of the earth's resources carries occasional great risks to balance the undoubted great rewards. The cultural and technological advances of humankind have increased our range of resource exploitation, and the danger that our abilities to exploit may exceed the tolerance of the ecosystem to accept errors in judgment, pressures of use, and accidents of technology.

Growing population numbers and economic development have magnified the extent and the intensity of human depletion of the treasures of the earth. Resources of land, of ores, and of energy are finite, but the resource demands of an expanding, economically advancing population appear to be limitless. This imbalance between resource availability and use has emerged as a critical concern in the human–earth-surface equation of geography. Because resources are unevenly distributed in kind, amount, and quality, and do not match uneven distributions of population and demand, a consideration of natural resources falls within the locational-distributional tradition of the discipline of geography.

Although people depend on a wide range of resources contained in the biosphere, the focus in this chapter is on energy and mineral resources. Wildlife resources, freshwater and marine resources, and agricultural resources such as land and soils have already been touched on in Chapter 4. Energy resources are the "master" natural resources, for we use energy to make all other resources, including labor and capital, available. Without the energy resources, all other natural resources would remain in place, unable to be mined, processed, and distributed. When water becomes scarce, we use energy to pump groundwater from greater depths, or to divert rivers and build aqueducts. We increase crop yields in the face of poor soil management by investing energy in fertilizers, herbicides, farm implements, and so on. It will be useful to begin our discussion by defining some commonly employed terms and concepts.

Resource Terminology

Resources are the naturally occurring materials (including people) that a human population, at any given state of economic development and technological awareness, perceives to be necessary and useful to its economic and material well-being. Willing, healthy, and skilled workers constitute a valuable resource, but without access to materials like petroleum or fertile soil, human resources are limited in their effectiveness. In this chapter, we devote our attention to physically occurring resources, or, as they are more commonly called, *natural* resources.

The availability of natural resources is a function of two things: the physical characteristics of the resources (their concentration, depth of burial, grade, and so on) and human economic and technological conditions. The physical processes that govern the formation, distribution, and occurrence of natural resources are determined by physical laws over which people have no direct control. We take what nature gives us. To be considered a

Figure 10.1

The original hardwood forest covering these West Virginia hills was removed by settlers who saw greater resource value in the underlying soils. The soils, in their turn, were stripped away for access to the still more valuable coal deposits. Resources are as a culture perceives them, though their exploitation may consume them and destroy the potential of an area for alternate uses.

resource, however, a given substance must be *understood* to be a resource. This is a cultural, not purely a physical, circumstance. Native Americans may have viewed the resource base of Pennsylvania as composed of forests for shelter and fuel, and as the habitat of the game animals (another resource) on which they depended for food. European settlers viewed the forests as the unwanted covering of the resource that they perceived to be of value: soil for agriculture. Still later, industrialists appraised the underlying coal deposits, ignored or unrecognized as a resource by earlier occupants, as the item of value for exploitation (Figure 10.1).

The interplay between the physical characteristics of the resource and the technology employed to extract and use it determines not only what materials are considered "resources," but also the costs of those resources. Because most raw materials undergo changes in their original form by processing before final consumption takes place, the value of a particular mineral deposit depends on such factors as its accessibility, degree of concentration, and its nearness to markets. The value of low-grade or high-sulfur coal deposits increases as the prices of competing fuels rise, encouraging coal companies to develop the technology to increase the utility of these less desirable deposits.

Natural resources are usually recognized as falling into one of two broad classes: renewable and nonrenewable.

Renewable Resources

Renewable resources are materials that can be consumed and then restored after use. Solar radiation, wood, wind, and water are among the renewable resources. The hydrologic cycle (Figure 4.3) assures that water, no matter how often used or how much abused, will return over and over to the land for further exploitation. Of course, groundwater extracted beyond the replacement rate in arid areas may be as permanently dissipated as if it were a nonrenewable ore. Soils can be continuously used productively and can even be improved by proper management and fertilization practices; they can also be lost by mismanagement that leads to total erosion.

Sometimes the renewal cycle takes hundreds of years. For example, soils depleted of their nutrients can usually regain them if conditions are right, but the process takes a long time. Humans can supplement the process by adding chemical nutrients in the case of soil, or by planting trees in the case of forests, or by cleansing water in the case of that resource.

For practical purposes, a resource cannot be classified as renewable unless the process takes place over a short period of time. Consequently, we should not say that forests are a renewable resource unless humans are planting at least as much wood as is being cut. Even this *sustained-yield* approach to renewable resources is not quite good enough. With rising expectations and growing populations, the renewed resource may not be sufficient to satisfy future demands.

Nonrenewable Resources

In their original forms, **nonrenewable resources** exist in finite amounts. They include the fossil hydrocarbons (coal, crude oil, and natural gas) and the nuclear fuels uranium and thorium. Although the elements of which these resources are composed cannot be destroyed, they can be altered to less useful or available forms, and they are subject to depletion. The energy stored in a unit volume of the fossil fuels may have taken eons to concentrate in usable form; it can be converted to heat in an instant and be effectively lost forever.

Fortunately, many minerals can be *recycled* even though they cannot be *replaced.* If they are not chemically destroyed—that is, if they retain their original chemical composition—they are potentially *reusable.* Aluminum, lead, zinc, and other metallic resources, plus many of the nonmetallics, such as diamonds and petroleum by-products, can be used time and time again. However, many of these materials are used in small amounts in any given manufactured object, so that recouping them is economically unfeasible. In addition, many materials are now being used in manufactured products, so that they are unavailable for recycling unless the product is destroyed. Consequently, the term *reusable resource* must be used carefully. At present, all mineral resources are being mined much faster than they are being recycled.

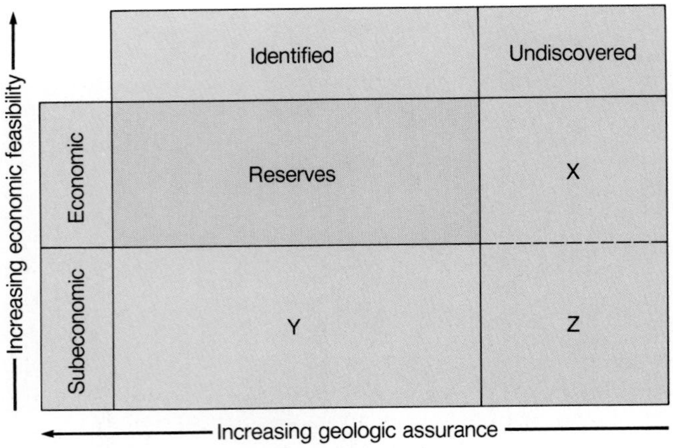

Resource Reserves

Some regions contain many resources, others relatively few. No industrialized country, however, has all the resources it needs to sustain itself. The United States has abundant deposits of many minerals, but it depends on other countries for such items as tin and manganese. The actual or potential scarcity of key nonrenewable resources makes it desirable to predict their availability in the future. We want to know, for example, how much petroleum remains in the earth and how long we will be able to continue using it.

Any answer will be only an estimate, and for a variety of reasons such estimates are difficult to make. Exploration has revealed the existence of certain deposits, but we have no sure way of knowing how many remain undiscovered. Further, our definition of what constitutes a usable resource depends on current *economic* and *technological* conditions. If they change—if, for example, it becomes possible to extract and process ores more efficiently—our estimate of reserves also changes.

Figure 10.2

Usable reserves consist of amounts that have been identified and can be recovered at current prices and with current technology. *X* denotes amounts that would be attractive economically but have not yet been discovered. Identified but not economically attractive amounts are labeled *Y,* and *Z* represents undiscovered amounts that would not now be attractive even if they were discovered.

Finally, the answer depends in part on the rate at which the resource is being used, but it is impossible to predict future rates of use with any certainty. The current rate could drop if a substitute for the resource in question is discovered, or increase if population growth or industrialization places greater demands on it.

A useful way of viewing reserves is illustrated by Figure 10.2. Assume that the large rectangle includes the total stock of a particular resource, all that exists of it in or on the earth. Some deposits of that resource have been discovered; they are shown in the left-hand column as "identified amounts." Deposits that have not been located are called "undiscovered amounts." Deposits that are economically recoverable with current technology are at the top of the diagram, while those labeled "subeconomic" are not attractive for any of a number of reasons (the concentration isn't rich enough, it would require expensive treatment after mining, it isn't accessible, and so on). We can properly term **usable reserves** only the portion of the rectangle indicated by the pink tint. These are the amounts that have been identified and that can be recovered under existing economic and operating conditions. If new deposits of the resource are discovered, the reserve category will shift to the right; improved technology or increased prices for the product can shift the reserve boundary downward. An ore that was not considered a reserve in 1940, for example, may become a reserve in 1990 if ways are found to extract it economically.

The Concept of Energy

People have built their advanced societies by using inanimate energy resources. **Energy**—the ability to do work—exists in two forms: potential and kinetic. *Potential* energy is stored energy; when released, it is in a form that can be harnessed to do work. *Kinetic* energy is the energy of motion; all moving bodies possess kinetic energy.

Assume that a reservoir contains a large amount of stored water. The water is a storehouse of potential energy. When the gates of the dam holding back the water are opened, water rushes out. Potential energy has become kinetic energy, which can be harnessed to do such work as driving a generator of electricity. No energy has been lost, it has simply been converted from one form to another.

Unfortunately, energy conversions are never complete. Not all of the potential energy of the water can be converted into electrical energy. Some potential energy is always converted to heat and dissipated to the surroundings. *Efficiency* is the measure of how well we can convert one form of energy into another without waste.

Resources and Industrialization

There is no necessary correlation between a country's resource base and its technological level or quality of life. Note the prominence of the oil-rich but less developed nations of the Middle East in Figure 10.3. There are countries with a rich resource base whose people are poor, such as Trinidad and Zaire, and equally, there are industrialized countries like West Germany and Japan that compensate for a lack of resources by importing raw materials.

Energy consumption goes hand in hand with industrial production and with increases in per capita income (Figure 10.4). As the graph indicates, there are marked differences in the patterns of energy consumption among countries. The two biggest consumers of energy in the world in 1985 were the United States (24%) and the USSR (19%). By the application of energy, the conversion of materials into commodities and the performance of services far beyond the capabilities of any single individual are made possible. Further, the application of energy can overcome deficiencies in the material world that humans exploit. High-quality iron ore may be depleted, but by massive applications of energy, the iron contained in rocks of very low iron content can be extracted and concentrated for industrial use.

Because of the association of energy and economic development, a basic conflict between societies becomes clear. Nations that can afford high levels of energy production and consumption continue to expand their economies and to increase their levels of living. Those without access to energy, or those unable to afford it, see the gap between their economic prospects and those of the developed nations growing ever greater.

Energy can be extracted in a number of ways. Humans themselves are energy converters, acquiring their fuel from the energy contained in food. Our food is derived from the solar energy stored in plants. In fact, nearly all energy sources are really storehouses for energy originally derived from the sun. Among them are wood, water, the ocean tides, the wind, and the fossil fuels. **Fossil fuels,** the nonrenewable energy resources that occur as sediments or are contained within sedimentary rocks, were formed over millions of years as dead vegetation and other organic matter were transformed into the hydrocarbon compounds of crude oil, natural gas, and coal.

Each of these energy sources has been harnessed to a greater or lesser degree by humans. Preagricultural societies depended chiefly on the energy stored in wild plants and animals for food, although people developed certain tools (such as spears) and customs to exploit the energy base. For example, they added to their own energy resources by using fire for heating, cooking, or clearing forest land.

Sedentary agricultural societies developed the technology to harness increasing amounts of energy. The domestication of plants and animals, the use of wind to power ships and windmills, and of water for waterwheels all expanded the energy base. For most of human history, wood was the predominant source of fuel, and even today at least half the world's people depend largely on fuelwood for cooking and heating.

However, it was the shift from renewable resources to those derived from nonrenewable minerals, chiefly fossil fuels, that sparked the Industrial Revolution, made possible the population increases discussed in Chapter 5, and gave population-supporting capacity to areas far in excess of what would be possible without inanimate energy sources. The enormous increase in individual and national wealth in industrialized nations has been built in large measure on an economic base of coal, oil, and natural gas. They are used to provide heat, to generate electricity, and to run engines.

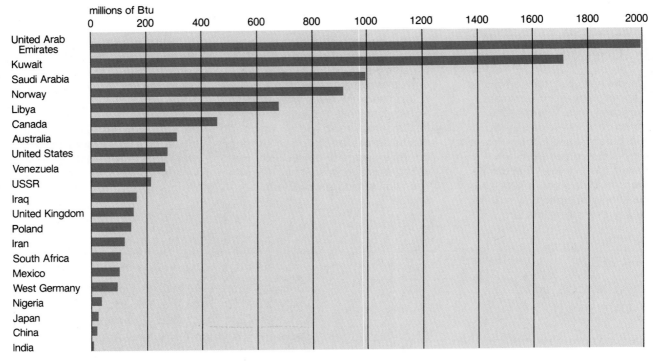

millions of Btu

Figure 10.3

Per capita energy production, 1984 (in millions of BTU).
Per capita energy production reflects the unequal world distribution of energy resources, the application of foreign or domestic capital to their exploitation, the population of producing nations, and the existence of international markets for energy supplies. It has no necessary relationship to national energy consumption or to the degree of economic development

within producing nations. Note the prominence of oil-rich, less-developed nations of the Middle East on this chart, and the differences between it and the record of energy consumption in Figure 10.4.

Source: Data from Energy Information Administration and Population Reference Bureau, Inc.

Figure 10.4

Energy consumption rises with increasing gross national product. Because the internal combustion engine accounts for a large share of national energy consumption, this graph is a statement both of economic development and of the roles of mass transportation, automotive efficiency, and mechanization in different national economies.

Source: Data from World Bank, *World Development Report, 1986.*

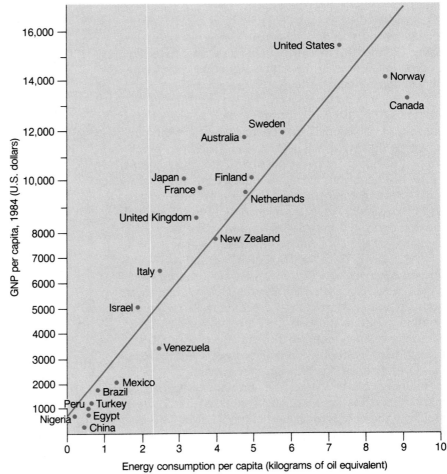

Nonrenewable Resources

Crude oil, natural gas, and coal have formed the basis of industrialization. Figure 10.5 shows past energy-consumption patterns in the United States. Burning wood supplied most energy needs until about 1890; by then coal had risen to prominence. The proportion of energy needs satisfied by burning coal peaked about 1910; from that time on, oil and natural gas were increasingly substituted for coal. The graph shows the absolute dominance of the fossil fuels as energy sources during the last 100 years. In 1986, they accounted for almost 90% of our national energy consumption.

Crude Oil

In 1986, crude oil and its by-products accounted for about 40% of the commercial energy (excluding wood and other traditional fuels) consumed in the world. Some world regions and industrial nations have a far higher dependency (Table 9.1). Figure 10.6 shows the main producers of crude oil.

After it is extracted from the ground, crude oil must be refined. The hydrocarbon compounds are separated and distilled into waxes and tars (for lubricants, asphalt, and many other products) and various fuels. Petroleum rose to importance because of its combustion characteristics and its adaptability as a concentrated energy source for powering moving vehicles. Although there are thousands of oil-based products, fuels such as home heating oil, diesel and jet fuels, and gasoline are the major output of refineries.

The distribution of oil supplies differs markedly from that of the coal deposits on which the urban-industrial markets developed, but the substitution of petroleum for coal did little to alter earlier patterns of manufacturing and population concentration. Because oil is easier and cheaper to transport than coal, it was moved to the existing centers of consumption via intricate and extensive national and international systems of transportation. For many years transfer costs were kept low by high-capacity pipelines and by the construction of large tankers, fostering an enormous world trade in oil.

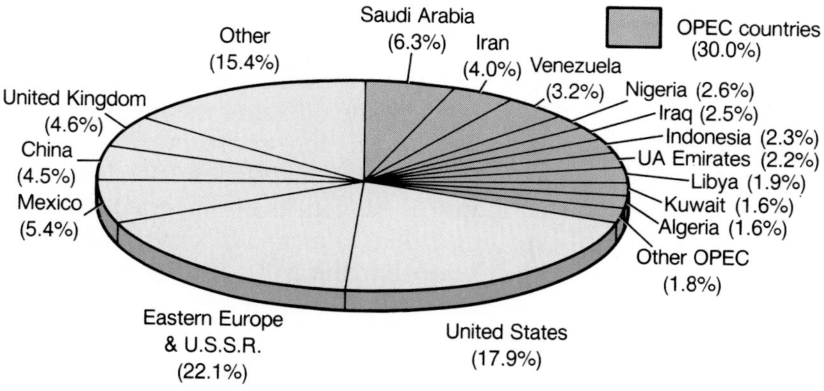

Figure 10.5

Sources of energy in the United States, 1850–1985. The fossil fuels provided 90% of the energy supply in 1985.

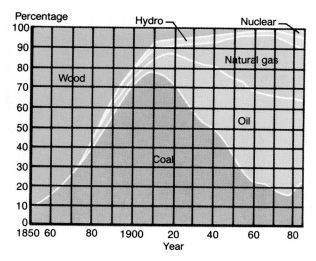

Figure 10.6

Share of total international crude-oil production, 1986. The USSR is the world's largest oil *producer,* followed by the United States. Saudi Arabia and Mexico are the largest oil *exporters.*

Source: British Petroleum Co., *BP Statistical Review of World Energy,* 1987.

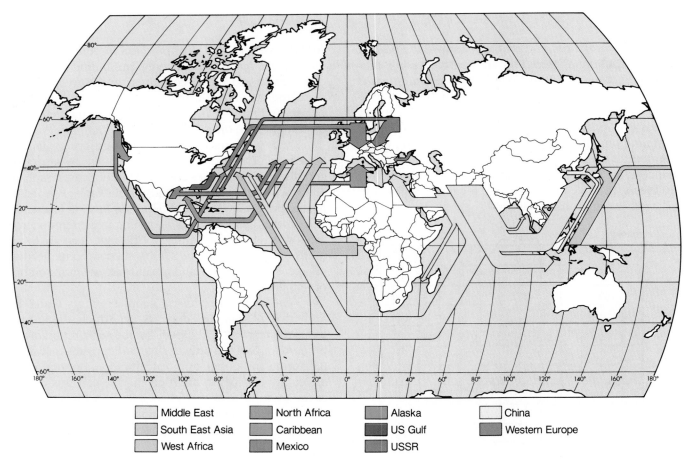

Middle East	North Africa	Alaska	China
South East Asia	Caribbean	US Gulf	Western Europe
West Africa	Mexico	USSR	

Figure 10.7

International crude-oil flow by sea, 1986. Note the dominant position of the Middle East in terms of oil exports. The arrows indicate origin and destination, not specific routes. The line widths are proportional to the volume of movement.

Source: British Petroleum Co., *BP Statistical Review of World Energy*, 1987.

As Figure 10.7 shows, oil from a variety of production centers flows, primarily by water, to the industrially advanced countries. Note that the United States imports oil from a number of regions. The other major importers, Western Europe and Japan, import chiefly Middle Eastern oil. In the United States, where the regions of petroleum production are spatially separated from the regions of consumption, a 250,000-mile network of pipelines carries almost half the oil shipped within the country. The remainder is carried chiefly in tank trucks and by waterborne tankers and barges that move along rivers, lakes, and canals, as well as on the ocean.

The efficiency of pipelines, supertankers, and other modes of transport and the low cost of oil helped to create a world dependence on that fuel even though coal was still generally and cheaply available. The pattern is aptly illustrated by the American reliance on foreign oil. For many years, United States oil production had remained at about the same level, 8 to 9 million barrels per day. Between 1970 and 1977, however, as domestic supplies became much more expensive to extract, consumption

of oil from foreign sources increased dramatically, until almost half the oil consumed nationally was imported. Foreign oil was less expensive than domestic oil because of the large size of the oil fields, the lower costs of exploitation, and newer, more efficient wells.

The dependence of the United States and other advanced industrial economies on imported oil gave the oil-exporting nations tremendous power, reflected in the soaring price of oil in the 1970s. During that decade, oil prices rose dramatically, largely as a result of the strong market position of the Organization of Petroleum Exporting Countries (OPEC). Originally established in 1960 as an intergovernmental study group composed of the countries that then produced almost all the exported oil in the world, the OPEC cartel did not use its united weight until 1973, at the time of the fourth Arab-Israeli war. That and subsequent events—the Iranian revolt against the shah and Iraq's invasion of Iran—increased the power of OPEC as producing nations selectively reduced or halted exports and as importers learned to fear that their supply of oil might be interrupted.

The Strategic Petroleum Reserve

Giant caverns created in salt beds in coastal Texas and Louisiana contain the stockpile necessary to protect the United States in an emergency: approximately one-half billion barrels of crude oil in the government's strategic petroleum reserve (SPR). Created under legislation passed after the Arab oil embargo of 1973–1974, the stockpile is a reserve intended to reduce the impact of any disruption in supply. Oil for the reserve comes primarily from foreign suppliers, chiefly Mexico, Canada, and Venezuela.

At current consumption levels, the stockpile is large enough to replace all oil imports for more than three months. And because the country now imports most of its oil from non-OPEC nations, the SPR could replace imports from members of that cartel for well over a year.

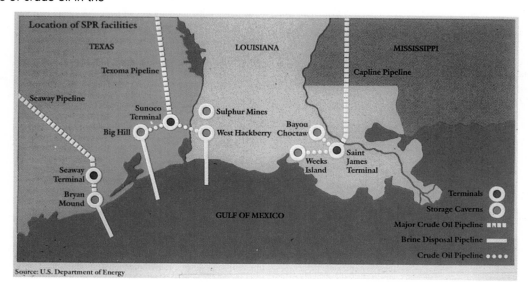

Source: U.S. Department of Energy

Among the side effects of the oil "shocks" of 1973–1974 and 1979–1980 were worldwide recessions, large net trade deficits for oil importers, a reorientation of world capital flows, and a depreciation of the United States dollar against many other currencies. Equally important, the soaring oil prices of the 1970s diminished total energy demand, partly because of the recession, and partly because the high prices fostered conservation. Worldwide, there has been an absolute drop in the demand for oil since 1979. In the United States, by 1986, oil imports had decreased about 50% from 1979 levels. Consequently, domestic oil supplied a higher proportion of United States consumption.

By 1986, oil prices were lower than they had been in a decade. Two factors, diminished demand and new sources of production, had led to an "oil glut" and a collapse in the price of a barrel of crude. OPEC lost control of the export market as major new producers, including Mexico, Great Britain, and Norway, increased oil output. Indeed, Mexico began to export more oil than any member of OPEC except Saudi Arabia. As oil prices declined, the oil producers began lowering their prices, undercutting each other to compete for sales.

The "oil glut" notwithstanding, oil is a scarce resource. Some 670 billion barrels are classified as identified reserves, and another 900 billion are thought to exist in undiscovered reservoirs. If all the oil could be extracted from the earth, and if the current rate of production holds, the known reserves would last only 33 years, and the total, including undiscovered reservoirs, about 75 years. For the United States, the estimate of usable reserves in the mid-1980s was 55 billion barrels; undiscovered reservoirs held perhaps another 83 billion barrels (Figure 10.8). At current rates of production (about 3 billion barrels per year), the life expectancy of known domestic reserves is about 18 years. If assumed undiscovered reservoirs are added, the country has a 45-year supply of oil.

The realization that oil supplies are finite and likely to be depleted in the foreseeable future has sparked renewed interest in another fossil fuel, coal.

Figure 10.8
Estimated United States crude-oil supplies. At 1985 rates of production, the country's usable reserves of oil—some 55 billion barrels—would last for about 18 years.

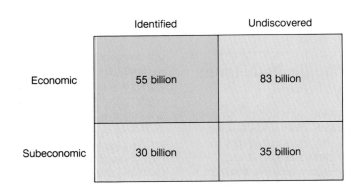

Figure 10.9
International coal production, 1985, in million tons oil equivalent.

Source: British Petroleum Co., *BP Statistical Review of World Energy,* 1986.

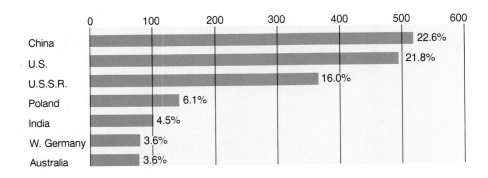

Coal

Coal was the fuel basis of the Industrial Revolution. From 1850 to 1910, the proportion of United States energy supplied by coal rose from 9% to almost 80%. Although the consumption of coal declined as the use of petroleum expanded, coal remained the single most important domestic energy source until 1950 (Figure 10.5).

Although coal is a nonrenewable resource, world supplies are so great that its resource life expectancy may be measured in centuries, not in the tens of years usually cited for oil and natural gas. Of an original estimated world reserve of coal on the order of 10,000 billion (10^{13}) tons, only some 140 billion tons have so far been used. The United States alone possesses over 450 billion tons of coal considered potentially mineable on an economic basis with existing technology. At current consumption levels, these demonstrated reserves would be sufficient to meet the domestic demand for coal for another three centuries.

Worldwide, the most extensive deposits are concentrated in the industrialized middle latitudes of the northern hemisphere (Figure 9.19). As Figure 10.9 indicates, three countries dominate world coal production,

accounting in roughly equal shares for 60% of the coal produced in the world: China, the United States, and the USSR. Note that the fourth largest producer, Poland, was responsible for only 6% of production in 1985, and that remaining producers each accounted for less than 5% of the total world output.

Coal is not a resource of constant quality. It ranges from lignite (barely compacted from the original peat) through bituminous coal (soft coal) to anthracite (hard coal), each **rank** reflecting the degree to which organic material has been transformed. Anthracite has a fixed carbon content of about 90% and contains very little moisture. Conversely, lignite has the highest moisture content and the lowest amount of elemental carbon, and thus the lowest heat value. About half the demonstrated reserve base in the United States is bituminous coal, concentrated primarily in the states east of the Mississippi River.

Besides rank, the **grade** of a coal, which is determined by its content of waste materials (particularly ash and sulfur), helps to determine its quality. Good-quality bituminous coals with the caloric content and the physical properties suitable for producing coke for the steel industry are decreasingly available readily and are increasing in cost. Anthracite, formerly a dominant fuel for

Figure 10.10
Long-distance transportation adds significantly to the cost of the low-sulfur western coals because they are remote from eastern United States markets. To minimize these costs, unit trains carrying only coal engage in a continuous shuttle movement between western strip mines and eastern utility companies.

home heating, is now much more expensive to mine and finds no ready industrial market. The Schuylkill anthracite deposits of eastern Pennsylvania are discussed as a special type of resource region in Chapter 12 on Page 421.

The value of a given coal deposit depends not only on its rank and its grade but also on its accessibility, which depends on the thickness, depth, and continuity of the coal seam and its inclination to the surface. Much coal can be mined relatively cheaply by open-pit (surface) techniques, in which huge shovels strip off surface material and remove the exposed seams. Much, however, is available only by expensive and more dangerous shaft mining, as in Appalachia and most of Europe. In spite of their generally lower heating value, western United States coals are now attractive because of their low sulfur content. They do, however, require expensive transportation to markets or high-cost transmission lines if they are used to generate electricity for distant consumers (Figure 10.10).

The ecological, health, and safety problems associated with the mining and the combustion of coal must also be figured into its cost. The mutilation of the original surface and the acid contamination of lakes and streams associated with the strip mining and the burning of coal are at least partially controlled by environmental protection laws, but these measures add to the energy costs. Eastern United States coals have a relatively high sulfur content, and costly techniques for the removal of sulfur from stack gases (and for the removal of other noxious by-products of burning) are now required by most industrial nations, including the United States.

The cost of moving coal influences its patterns of production and consumption. Coal is bulky and is not as easily transported as nonsolid fuels. As a rule, coal is usually consumed in the general vicinity of the mines. Indeed, the high cost of transporting coal induced the development of major heavy industrial centers directly on coalfields, for example, Pittsburgh, the Ruhr, the English Midlands, and the Donets district of the Ukraine.

One way to reduce transport costs and to help unlock the large western coal reserves in this country would be to use **coal slurry** pipelines. In coal slurry, coal that has been pulverized to a fine dust is mixed with fresh water to form a sludgy liquid that is pumped through pipelines and then separated at the point of discharge by power plants or other industrial users. Mainly because of cost, coal slurry lines have not yet begun to play an important role in coal transport. The only significant slurry pipeline currently operating in this country, the Black Mesa Pipeline, runs 273 miles (435 km) from northeastern Arizona to a power plant in southern Nevada. Environmental and agricultural groups have opposed coal slurries because of their potential for depleting scarce supplies of fresh water in the West, but most opposition comes from the railroads, which fear loss of their business to the pipelines. The railroads currently transport about two-thirds of the coal mined in this country, and coal accounts for almost half the annual rail tonnage.

Natural Gas

Coal is our most abundant fossil fuel, but natural gas has been called the nearly perfect energy resource. It is a highly efficient, versatile fuel that requires little processing and is environmentally benign. Of the fossil fuels, natural gas has the least impact on the environment. It burns cleanly; the chemical products of burned methane are carbon dioxide and water vapor, which are not pollutants, although the former does add to the rising carbon dioxide level of the atmosphere.

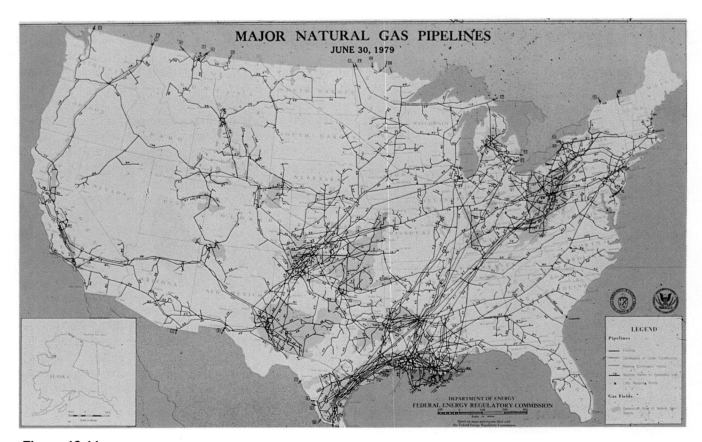

Figure 10.11
Major natural gas pipelines in the United States, 1979.

Courtesy of U.S. Department of Energy and American Petroleum Institute.

As indicated by Figure 10.5, this century has seen an appreciable growth in the proportion of energy supplied by gas. In 1900, it accounted for about 3% of domestic energy supply; by 1980, the figure had risen to 30% and then declined to about 25% of domestic energy consumption by 1985. The trend in the rest of the world has been in the opposite direction. Production increased significantly after the oil shock of 1973–1974 and by 1985 had more than doubled in a number of countries, including the USSR, which replaced the United States as the world's largest producer.

Most gas is used directly for industrial and residential heating. In fact, gas has overtaken both coal and oil as a house-heating fuel, and over 40 million homes in the United States are now heated by gas. A portion is also used in electricity-generating plants, particularly in the Gulf Coast states, and some is chemically processed into products as diverse as motor fuels, plastics, synthetic fibers, and insecticides.

Very large natural gas fields were discovered in Texas and Louisiana as early as 1916. Later, additional large deposits were found in the Kansas–Oklahoma–New Mexico region. At that time, the south-central United States was too sparsely settled to make use of the gas, and in any case, it was oil, not gas, that was being sought.

Many wells that produced only gas were capped. Gas found in conjunction with oil was vented or burned at the wellhead as an unwanted by-product of the oil industry. The situation changed only in the 1930s, when pipelines were built to link the southern gas wells with customers in Chicago, Minneapolis, and other northern cities.

Like oil, natural gas flows easily and cheaply by pipeline. Unlike oil, however, gas does not move freely in international trade. Transoceanic shipment involves costly equipment for liquefaction and for vessels that can contain the liquid under appropriate temperature conditions. **Liquefied natural gas (LNG)** is extremely hazardous because the mixture of methane and air is explosive. Although this country has imported some LNG, chiefly from Algeria, most gas is transferred by pipeline. In the United States, the pipeline system is over a million miles long (Figure 10.11).

Like other fossil fuels, natural gas is nonrenewable; its supply is finite. Estimates of reserves are difficult to make because they depend on what customers are willing to spend for the fuel, and they have risen as the price of gas has increased. Worldwide, two regions contain about two-thirds of the proven gas reserves: the USSR (43%) and the Middle East (25%). The remaining one-third is divided roughly equally among North America, Western

Europe, Africa, Asia, and Latin America, each of which has from 5% to 9% of the total. The gas in these reserves would last about 60 more years at current production rates, but developing countries, particularly in south and southeast Asia, may well have undiscovered deposits that could add significantly to the life expectancy of world reserves if they were developed.

In the United States, the Texas–Louisiana and Kansas–Oklahoma–New Mexico regions account for about 90% of the domestic natural gas output, but there are thought to be gas deposits beneath almost all states except such northern ones as Wisconsin and Vermont. In addition, many offshore areas are known to contain gas. Potential Alaskan reserves are estimated to be at least twice as large as today's proven reserves in the rest of the country, containing enough gas to heat all the houses in the nation for a decade (Figure 10.12). Estimates of United States reserves in the mid-1980s indicated that there is enough gas to last anywhere from 25 to 50 years, if conventional drilling techniques are used. But if the technology necessary to stimulate gas flow in "tight" sand and shale formations, or to extract gas from deep deposits such as those in the Rocky Mountains is developed, gas reserves may be sufficient for the next 90 or so years. Of course, these less accessible supplies will be more costly to develop and hence more expensive.

Synthetic Fuels

The rising energy prices of the 1970s stimulated interest in developing substitutes for oil and natural gas from underused resources. **Synthetic fuels** ("synfuels") can be developed from coal, oil shale, heavy oils including tar sands, and animal or vegetable wastes. This variety of eligible synthetic-fuel resource bases and technologies has made it difficult to determine which synfuel projects have the greatest potential. Furthest advanced are the technologies for converting coal into clean secondary fuels. Two broad categories of coal conversion techniques are gasification and liquefaction.

Coal Gasification and Liquefaction

Coal gasification is a partial combustion process whereby crushed coal is burned under high pressure in the presence of steam or oxygen to produce either a high- or a low-heating-value gas. With additional processing, the product can be converted into a gas of the quality of methane, into liquid hydrocarbons, or into methyl alcohol (Figure 10.13). The sulfur in the coal combines with hydrogen to form hydrogen sulfide, which is removed in a purification step, a technique that permits the use of high-sulfur coal. Although gasification techniques have been employed for several decades, technical problems and inefficiencies characterize most energy conversion processes. As a result, the product has generally been too expensive to compete with natural gas.

Figure 10.12

Drilling rig at Prudhoe Bay on Alaska's North Slope. Alaskan oil is sent southward by pipeline to the port of Valdez for shipment to the "lower" states. A by-product of petroleum extraction is natural gas, for which no outlet to market yet exists. Alaska's potential gas reserves are more than twice the proven reserves of the rest of the country, but lack of market requires the petroleum companies to spend millions of dollars to reinject the gas back into the reservoir from which it came.

Figure 10.13

A coal-gasification plant under construction in Illinois. The plant is intended to convert Illinois high-sulphur coal to a clean, low-BTU gas in an environmentally sound manner.

Figure 10.14
An oil-shale-retorting plant in rugged terrain near Parachute in northwestern Colorado. Pioneers learned about "the rock that burns" from Ute Indians. Here, in the some 1500 square miles (4000 km²) of the Piceance Basin, is one of the richest hydrocarbon deposits in the world. Retorting plants are simply oil-shale "ovens" that cook the rock at high temperatures to drive out the kerogen.

Liquefaction processes are used to convert coal into liquid products comparable to petroleum. One type of liquefaction involves heating coal at relatively low temperatures and under pressure to extract the volatile (easily vaporized) matter. Volatiles are recovered as liquids and gases; the liquids are treated for the removal of organic sulfur, nitrogen, and oxygen. The resulting product, a synthetic crude oil, can be refined into gasoline and fuel oils comparable in quality to petroleum-based products.

Liquefaction processes are not as advanced as gasification techniques, and they appear to be more complex and costly. The market potential for liquid products is conditioned by the national and international factors affecting the supply of and the demand for crude oil and refined products. Both liquefaction and gasification techniques will become economically feasible only as the costs of petroleum and natural gas increase to the point where coal conversion becomes competitive.

Oil Shale

A similar situation colors the prospects of the extraction of oil from **oil shale,** a tremendous potential reserve of hydrocarbon energy. In fact, the rocks involved are not shales but calcium and magnesium carbonates, more similar to limestone than to shale, and the hydrocarbon is not oil but a waxy, tarlike substance called *kerogen* that adheres to the grains of carbonate material. The crushed rock is heated to a temperature high enough (over 900° F) to decompose the kerogen, releasing a liquid oil product, *shale oil* (Figure 10.14). The richest oil shale deposits in the United States are in the Green River area of Colorado, Utah, and Wyoming. They contain astronomical potential quantities of oil, enough to supply the oil needs of the United States for another century. As oil prices eased in the 1980s, interest in synfuels waned and many projects were abandoned. Oil shale research and development efforts continue at only a few sites in Colorado.

Cost and environmental damage are serious considerations in the use of oil shale. If the shale is mined by surface extraction, vast contour disruption, landscape scars, and waste heaps will result despite the best reclamation efforts. The waste heaps would pose the danger of both air and water pollution. Ecologically more promising are techniques for burning the rock in place in underground mines and pumping the oil to the surface. In either case, the inputs of energy, equipment, and money assure high product prices.

Tar Sands

Another potential source of synthetic fuels is the **tar sands,** sandstones with pore spaces containing viscous to solid petroleum that cannot be recovered by conventional methods. Global tar sand resources are thought to be many times larger than conventional oil resources, containing more than 2 trillion barrels of oil, most of it in Canada and Venezuela. Most of the tar sands in the United States are in Utah. Although there is no commercial production of tar-sand oil in the United States, oil is produced from similar deposits in Canada (Figure 10.15).

Two major obstacles limit the use of tar sands. First, extraction of the heavy oils by mining or pumping through wells is both difficult and expensive. Second, because the oil is thicker than molasses, costly processing and upgrading into higher-quality crude oil are required if the oil is to find a ready market.

Because they exist in vast amounts, coal, oil shale, and tar sands could be plentiful sources of gaseous and liquid fuels. Unlike nuclear energy or most renewable energy sources (such as solar or hydroelectric power), they could provide the gasoline, jet, and other fuels on which industrialized societies depend. The lowering of oil and gas prices in the 1980s and the decline in federal government support put a temporary halt to research and

The Athabasca Tar Sands

Beneath the wilderness bogs of northern Alberta, Canada, lie a potential but expensive 1000 billion barrels of petroleum in the world's largest known single deposit. Although only slightly smaller than the proven Middle East oil reserves, the Athabasca tar sands have for over 100 years defied all efforts to exploit their potential at an economical cost.

Actually the tar sands are misnamed, for they are really saturated not with tar but with bitumen, a viscous high-carbon petroleum thicker than molasses, which contains both sulfur and traces of vanadium. Each grain of tar sand is coated with a thin film of water surrounded by bitumen. This physical separation of the petroleum and the sand makes the extraction of the oil, though difficult, at least possible.

Deposits that can be mined from the surface lie as much as 200 feet below overlying sand and shale topped by some 20 feet of muskeg (thick peat accumulations), which must be bulldozed away during the months of winter freeze; in summer, heavy equipment simply disappears into the unstable bog. With the overburden of sand and shale removed, huge excavators dig the sticky tar sands, whose abrasive qualities drastically shorten the useful life of all mining equipment.

In the refinery, the tar sand is mixed with hot water, steam, and a bit of sodium hydroxide. The resultant mixture is aerated and transferred to holding vessels, where the bitumen rises to the surface to be skimmed off for further processing. The sand-free bitumen is next heated and separated ("cracked") into solid coke and both a light gas and a heavy gas oil. Sulfur is extracted from the two individually treated oils, and a high-quality, light crude oil is produced by a recombination of the original gas oils. The cost of these operations makes crude oil from the tar sands among the world's most expensive to produce.

But the cost is more than monetary. The delicate muskeg ecosystem is destructively disturbed by the vast open-pit mining operations, refineries, and road network. The plan is to return the cleansed sands to the pits, to mix them with muskeg, to fertilize them, and to plant them. The reestablishment of the subarctic vegetative cover, however, is extremely difficult, and, even if successful, it involves a great many years in areas of a short growing season and a harsh climate. Even if revegetation is successful, the inevitable sulfur dioxide emissions to the air and the necessary deposit of alkaline processing water in immense holding ponds assure serious ecological disturbance in a largely pristine area.

The price of satisfying energy needs is high and varied.

Figure 10.15

A giant bucket-wheel reclaimer about to drop freshly mined tar sands on a conveyor belt at a syncrude plant in northern Alberta, Canada. The size of the bucket wheeler is indicated by the fact that each bucket can hold 6 people. Production of synthetic oil from the tar sands involves 4 steps: removal of overburden, mining and transporting it to extraction units, adding steam and hot water to separate the bitumen from the tailings residue, and refining the bitumen into coke and distillates.

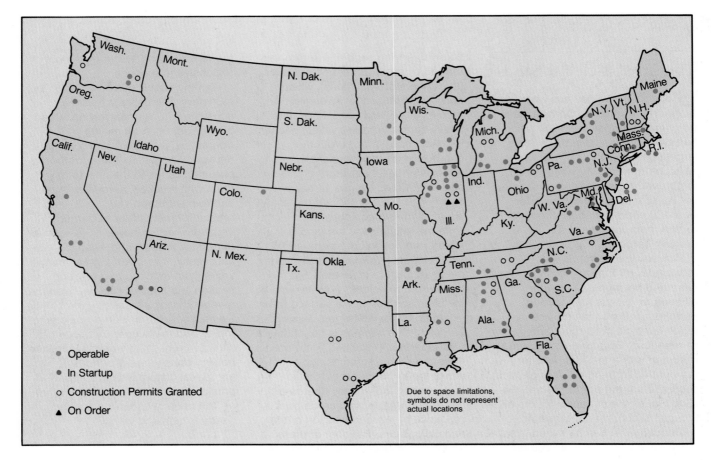

Figure 10.16

The location and status of nuclear reactor units in the United States, January 1, 1986. Construction of some of these plants has been halted, and no new orders have been placed since 1978.

Source: *Annual Energy Review*, 1985, Energy Information Administration.

development of ways to make synfuel products economically competitive with oil imports. The prospect of depletion of oil and natural gas resources within the next century, however, indicates that at some point their prices are bound to rise. When that occurs, countries are likely to try to turn to coal, oil shale, and the tar sands as sources of fuel. In the meantime, technologies employing nuclear energy as a source of nonfuel power are already being developed.

Nuclear Energy

Proponents herald nuclear power as a major long-term solution to the energy crisis. Assuming that the technical problems can be solved, they contend nuclear fuels could provide a cheap and virtually inexhaustible source of energy. Other commentators, pointing to the dangers inherent in any system dependent on the use of radioactive fuels, oppose the use of nuclear power because of the risk it poses to future generations. They argue that it involves technological, political, social, and environmental problems for which we have no solutions.

Nuclear Fission

Basically, energy can be created from the atom in two ways: nuclear fission and nuclear fusion. **Nuclear fission** for power production involves the controlled "splitting" of an atomic nucleus of uranium-235, the only naturally occurring fissile isotope. When U-235 atoms are split, about one-thousandth of the original mass is converted to heat. The released heat is transferred through a heat exchanger to create steam, which drives turbines to generate electricity.

Within a single gram of uranium-235 is the energy equivalent of nearly 14 barrels of oil. To tap that energy, 374 commercial nuclear power plants had already been built in 26 countries around the world by early 1986. These power stations produced about 16% of the world's electricity. In that same year, 100 plants were in operation in the United States alone (Figure 10.16). A large number were under construction, both in the United States and abroad, but increasing opposition by environmental groups and many scientists resulted first in stringent regulations (leading to higher costs) and then in a growing number of plant cancellations and a virtual moratorium on new orders.

Figure 10.17
The Cherobyl nuclear plant near Kiev in Ukraine, USSR, after the accident in April 1986.

Some countries are much more dependent than others on nuclear power. It provides only 16% of the electric power needs of the United States, but two European countries (France and Belgium) receive over half their electricity from nuclear power. Some countries have decided to reject the nuclear option altogether. Sweden, which receives over 40% of its electricity from nuclear power plants, will phase out all its reactors by 2010; Denmark and Austria have also decided against the use of nuclear energy. The Soviet Union, on the other hand, remains committed to nuclear power generation in the area west of the Urals and continues to expand its installed capacity.

The Fast Breeder Reactor

Uranium-235 is a raw material that is itself in short supply. The economically exploitable reserves are expected to be depleted by about the year 2025. The anticipated shortage of U-235 for conventional fission plants has spurred interest in the **fast breeder reactor** as a "bridge" between the fission plants of today and the nuclear fusion plants of the long-term future. Breeder reactors could stretch uranium supplies to more than 700 years.

If a core of fissionable U-235 is wrapped in a blanket of U-238 (a more abundant but nonfissionable isotope), the U-235 both generates electricity and converts its wrapping to fissionable plutonium. This is the breeder reactor, so named because it can "breed" its own fuel;

once made, plutonium can replace U-235 as the breeder's core to fuel chain reactions. In theory, a breeder could produce more plutonium fuel than it consumes.

France, Great Britain, the USSR, and Japan all operate demonstration fast breeder reactors. Funding for this country's first demonstration project was halted in 1982. The project had faced continuous opposition from those who argued that the breeder reactor is unnecessary, uneconomical, and unsafe. Indeed, problems of ensuring complete safety in any fission reactor have not been solved (Figure 10.17). Equally troublesome is the matter of the disposal of the great amount of radioactive waste produced by either fission method of power production. With half-lives in the hundreds and thousands of years, radioactive wastes—deadly contaminants all—represent a potential for ecological and human disaster on an unimaginable scale.

Nuclear Fusion

Nuclear fusion represents the same kind of energy generation that is carried on in the sun and the stars. The process involves forcing two atoms of deuterium to fuse into a single atom of helium, with a consequent release of large amounts of energy, about 400 times as much energy as fission of the same weight of uranium. The process is accomplished in an uncontrolled fashion in the

TABLE 10.1
World Mineral Reserves*

Mineral	Years remaining
Mercury	21
Silver	22
Zinc	27
Lead	29
Copper	42
Tin	43
Manganese	111
Iron ore	195
Bauxite	257

*Approximate number of years the world's identified reserves of selected minerals will last, based on the expected rate of consumption between the years 1983 and 2000. Such figures are only suggestive because reserve totals and consumption rates fluctuate over time. Resource estimates for all the minerals listed (except mercury and tin) are increased considerably if seabed deposits are included.
Source: Data from *Minerals Handbook, 1984–85*, Phillip Crowson, ed., Macmillan, London, 1984.

hydrogen bomb. The problem, still not solved for commercial use despite 25 years of research, is to control fusion for a usable release of its energy. The technical problems involve the heating of deuterium to 180 million °F (82.2 million °C), sustaining the temperature long enough to create more energy than is required to produce the fusion reaction, and containing a substance of that temperature.

If the developmental problems are solved, human energy requirements would presumably be satisfied for millions of years. One cubic kilometer of ocean water, the source of deuterium atoms, contains as much potential energy as that available from the world's entire known oil reserves. Fusion also offers the advantages of the total absence of radioactive wastes and greater inherent safety in energy generation than fission plants.

Nonfuel Mineral Resources

The mineral resources already discussed provide the energy that enables people to do their work. Equally important to our economic well-being are the **nonfuel minerals,** for they can be processed into steel, aluminum, and other metals, and into glass, cement, and other products. Our buildings, tools, and weapons are chiefly mineral in origin.

Natural processes produce minerals so slowly that they fall into the category of nonrenewable resources. Usable mineral deposits are finite in the number and size of reserves, with predictable rates and dates of effective exhaustion. Table 10.1 gives one estimate of the "years remaining" for important metals. It should be taken as suggestive rather than definitive, as new discoveries and new recovery techniques may amend the predictions.

Although human societies began to use metals as early as 3500 B.C., world demand remained small until the Industrial Revolution of the 18th and 19th centuries. It was not until after World War II that increasing shortages and rising prices—and in the United States, increasing dependence on foreign sources—began to impress themselves on the general consciousness. Worldwide technological development has established ways of life in which minerals are the essential constituent. That industrialization has proceeded so rapidly and so cheaply is the direct result of the earlier ready availability of rich and accessible deposits of the requisite materials. Economies grew fat by skimming the cream.

All workable mineral deposits are the result of geologic processes that have concentrated one or more desired materials in particular locations. Although additional techniques of concentration may be necessary after mining, ultimately the result of the use of any mineral resource is to disperse the material used. In the case of metals, the dispersion occurs as fabricated materials rust, corrode, or abrade, or are included as fractional alloys in other metals.

Our successes in exploiting metallic minerals have been achieved not only at the expense of the usable world reserves but at increasing monetary costs as the high-grade deposits are depleted. The quick national assets represented by the richest, most accessible ores are soon exhausted. Costs increase as more advanced energy-consuming technologies must be applied to extract the desired materials from ever greater depths or from new deposits of smaller mineral content. The consequent environmental costs increase greatly.

The Distribution of Resources

Because usable mineral deposits are the result of geologic accident, it follows that the larger the nation, the more probable it is that such accidents will have occurred within the national territory. And in fact, the USSR, Canada, China, the United States, Brazil, and Australia possess abundant and diverse mineral resources. As Table 10.2 indicates, these nations are the leading mining countries. They produce the bulk of the metals (such as iron, manganese, nickel, and tungsten) and nonmetals (such as potash and sulfur). However, the table also reflects the fact that many developing countries are important sources of one or more critical raw materials.

TABLE 10.2

Leading Producers of Critical Raw Materials (Percent of World Total). The USSR is a major producer of eight of the nine minerals listed, the United States of five.

Bauxite	%	Copper	%	Iron ore	%
Australia	29.8	United States	18.8	USSR	28.1
Guinea	14.1	Chile	13.2	Brazil	11.6
Jamaica	13.6	USSR	11.6	Australia	10.0
Brazil	6.2	Canada	8.8	United States	8.6
USSR	5.4	Zambia	7.2	China	8.1
	69.1		59.6		66.4

Lead	%	Manganese	%	Tin	%
United States	13.3	USSR	39.8	Malaysia	23.7
USSR	13.1	South Africa	21.4	USSR	14.2
Australia	11.7	Brazil	8.0	Indonesia	13.8
Canada	9.9	China	6.8	Thailand	12.6
Peru	5.6	India	6.3	Bolivia	11.8
	53.6		82.3		76.1

Uranium	%	Industrial diamonds	%	Phosophate rock (for fertilizer)	%
United States	37.8	USSR	29.3	United States	36.8
Canada	18.0	Zaire	24.8	USSR	21.2
South Africa	12.6	South Africa	21.0	Morocco	13.5
Namibia	9.7	Botswana	14.5	China	7.9
Niger	9.7	Ghana	3.1	Tunisia	3.2
	87.8		92.7		82.6

Source: U.S. Bureau of Mines, *Mineral Commodity Summaries*, 1983.

Not even the large nations have domestic reserves of all needed minerals. Some, like the United States, which were bountifully supplied by nature, have spent much of their assets and depend on foreign sources. Although the United States was virtually self-sufficient in mineral supplies in the 1940s and 1950s, it is not today. Because of its past history of use of domestic reserves and its continually expanding economy, the United States now depends on a number of countries for supplies of such materials as iron ore and mercury, as Figure 10.18 demonstrates.

Conservation and Replacement

The increasing costs and the declining availability of metals encourage the search for substitutes. The fact that industrial chemists and metallurgists have been so successful in the search for new materials that substitute for the traditional resources has tended to allay fears of possible resource depletion. But it must be understood that for some minerals, no adequate replacements have been found. Other substitutes are frequently synthetics, often employing increasingly scarce and costly hydrocarbons in their production. Many, in their use or disposal, constitute environmental hazards, and all have their own high and increasing price tags.

Clearly, then, both conservation in the use and the recycling of minerals to as great an extent as possible constitute desirable approaches to the problem of mineral resource depletion. By **conservation** we mean resource use with as little waste as possible. The term implies a deliberate policy of trying to assure society as a whole as much future benefit from the exploitation of natural resources as is now experienced. Conservation is a human-centered concern; it seeks to maximize human benefit by maintaining the supply and the quality of natural resources at levels sufficient to meet present and future needs. Fuel-efficiency requirements for cars sold in the United States have resulted in vehicle size and weight reductions, achieving at a stroke conservation of both fuel and minerals. Such gains are not easily made and are offset by rising populations even when per capita consumption remains constant.

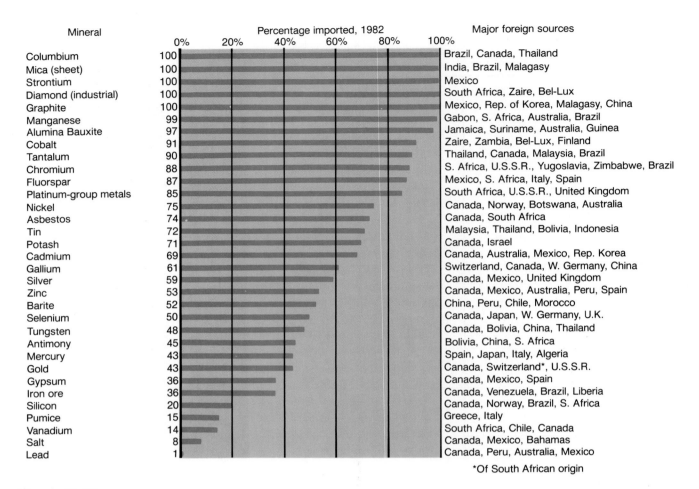

Figure 10.18

Imports supply a significant percentage of the minerals and metals consumed in the United States. Reliance on foreign suppliers for most of the listed materials continues to increase, posing problems of national security and balance of payments.

Source: Data from U.S. Bureau of Mines, *Mineral Commodity Summaries*, 1983.

Much more promising is the **recycling** of already extracted and used minerals. Old cars can then be made into new ones; old aluminum cans may, at 10% of the energy input of new aluminum production, be made into new cans; and so on. At present in the United States, about 40,000 pounds of new raw materials are consumed per capita per year (Figure 10.19), and of this amount, about half becomes solid waste. Only about 1000–2000 pounds are recycled.

Copper: A Case Study

Table 10.1 indicates that the world reserves of copper will last only another 40 or so years, based on current rates of consumption and assuming that no new extractable reserves appear. Copper is a relatively scarce mineral, and its importance to industrialized societies is evidenced by the fact that more copper is mined annually than any other nonferrous metal except aluminum. Two properties make copper desirable: it conducts both heat and electricity extremely well, and it can be hammered or drawn into thin films or wires. More than half of all copper produced is used in electrical equipment and supplies,

mainly as wire. Most of the remainder is used in construction, in industrial and farm machinery, for transportation, and for plumbing pipes. In alloys with other metals, copper is used to make bronze and brass.

The leading mining regions are in western North America; in western South America (Peru and Chile); in the copperbelt of southern Africa (Zambia and Zaire); and in the USSR. Although the United States is a leading copper producer and has very large proved reserves (about 17% of the world's total), for commercial and economic reasons it still imports copper to meet part of its demands. It is interesting to note that copper mining and refining have side benefits, yielding such resources as arsenic, plutonium, gold, silver, and molybdenum.

The threatened scarcity and the ultimate depletion of copper supplies have had several effects that suggest how societies will cope with shortages of other raw materials. First, the grade of mined ores has decreased steadily. Those with the highest percentage of copper (2% and above) were mined early. Now, ore of 0.5% grade is the average. Thus 1000 tons of rock must be mined and

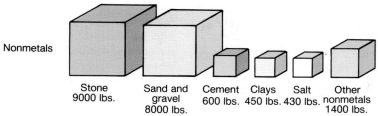

Approximately 38,000 pounds of new mineral materials are required annually for each U.S. citizen

Nonmetals

Stone 9000 lbs. Sand and gravel 8000 lbs. Cement 600 lbs. Clays 450 lbs. Salt 430 lbs. Other nonmetals 1400 lbs.

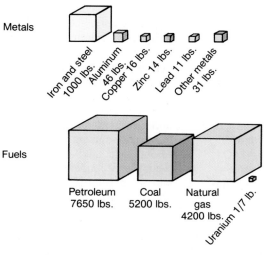

Metals

Iron and steel 1000 lbs. Aluminum 46 lbs. Copper 16 lbs. Zinc 14 lbs. Lead 11 lbs. Other metals 31 lbs.

Fuels

Petroleum 7650 lbs. Coal 5200 lbs. Natural gas 4200 lbs. Uranium 1/7 lb.

Figure 10.19

The per capita consumption of new minerals in the United States (in pounds). More conservative mineral use, the substitution of alternatives for scarce minerals, and recycling are seen as devices for reducing some demands for new minerals.

Source: Data from U.S. Bureau of Mines, *Mineral Facts and Problems.*

processed to yield 5 tons of copper—or, in more practical terms, 3 tons of rock are necessary to equip one automobile with the copper used in its radiator and its various electrical components. The remaining 2985 tons is waste, generated at the mine, the concentrator, and the smelter.

Second, the recovery of copper by recycling has increased. In 1982, old scrap provided a half-million tons of copper in the United States. New scrap, from fabricating operations, yielded almost 700,000 tons of the mineral. The recycling of old scrap is expected to supply about one-third of the copper in the United States by the year 2000.

Developing countries that produce copper have moved to gain control over their own resources and have banded together in a cartel. Copper mines in Chile, Peru, Zambia, and Zaire, once privately owned, have all been nationalized. Those countries are all members of the Intergovernmental Council of Copper Exporting Countries (CIPEC), which exports roughly half of the world's traded copper. Should world demand for copper increase, and should serious shortages appear imminent, the power of such an organization to affect prices would likely grow.

Finally, price rises have spurred the search for substitutes. In many of its applications, copper is being replaced by other, less expensive materials. Aluminum is replacing copper in some electrical applications and in heat exchangers; plastics are supplanting copper in

plumbing pipes and building materials; steel can be used in shell casings and coinage. In this regard, it is significant that the per capita consumption of copper in the United States has decreased since 1950, from a high of 19 pounds to less than 16 pounds per person in 1982.

Renewable Resources

The problems posed by the use of nuclear energy, the threatened depletion of some finite fossil fuels, and the desire to be less dependent on foreign sources of energy have increased interest in the *renewable* resources. One of the advantages of such resources is their ubiquity. Most places on earth have an abundance of sunlight, rich plant growth, strong wind, or heavy rainfall. The most common renewable source of energy is plant matter.

Biomass

Over half the people of the world depend on wood and other forms of **biomass**—plant material and animal waste used as fuel—for their daily energy requirements. In Burkina Faso and Malawi, biomass supplies 94% of total energy consumption; in India and Costa Rica, the figure is 42%. In contrast, energy from the conversion of wood, grasses, and other organic matter is insignificant in the developed world.

Forest Resources in the Tropics

Forests are generally considered a renewable resource, but that designation is proper only if measures are taken to conserve them. The opposite situation occurs in the developing countries, where there is a rising demand for, but decreasing supply of, wood. About 2.5 billion people depend on fuelwood as their principal source of energy; yet over half these people are cutting wood faster than it can grow back. Deforestation rates have exceeded reforestation rates by 10–30 times in recent years. In tropical America, Africa, and Asia, some 76,000 acres (31,000 hectares) of trees are cleared daily, the equivalent, on a yearly basis, of a loss of forested area the size of Pennsylvania. The world's tropical forests have already shrunk from an estimated 5 million square miles (13 million km²) to 3.5 million (9 million km²). At present, only 1 acre is replanted for every 10 acres cleared. The ratio of tree clearing to planting in Africa is an alarming 29:1.

Three major forces are driving deforestation, but behind them all is population pressure. The first is the clearance of forests to create cropland, as more families try to survive on marginal land. Unfortunately, much of the clearance occurs in regions with soils that aren't fertile enough to sustain crop production in the long run. The other two causes of deforestation are timber harvesting for commercial purposes and fuelwood gathering. According to the UN Food and Agriculture Organization (FAO), nearly 1.2 billion people are meeting their fuel needs only by overcutting and depleting their forests. As the supplies of wood decrease, people are forced to spend more time searching for fuelwood. The FAO reports that in parts of Tanzania in East Africa, over 250 workdays are needed to fill the yearly firewood requirements of a single household.

The environmental consequences of deforestation are severe. In a vicious cycle, forest clearance increases soil erosion and river siltation, which in turn leaves the area vulnerable to flooding and drought, leading in turn to future shortages of food, wood, and fuel. Furthermore, the forests are one component in an intricate ecosystem that has developed over eons. The trees, vines, flowering plants, animals, and insects depend on one another for survival. The destruction of the habitat by clearance will lead to the extinction of millions of plant and animal species that exist nowhere else. The tropical forests are known for their diversity. One stand of rain forest in Kalimantan (formerly Borneo) contains over 700 species of tree, as many as exist in all of North America. Forty-three species of ant inhabit a single tree species in Peru, dependent on it for food and shelter and providing in return protection from other insects. The tropical forests also yield an abundance of industrial products (oils, gums, latexes, and turpentines) and chemical products used to manufacture medicines. Indeed, half of all modern drugs, including strychnine, quinine, and ipecac, come from the tropical forests. A single flower, the Madagascar periwinkle, produces two drugs used to treat leukemia and Hodgkin's disease.

There are two major source of biomass: (1) crops grown explicitly for their energy value, and (2) wastes from nonenergy processes. These wastes include crop residues, animal wastes, garbage, and human sewage. Biomass can be transformed into fuel in many ways, including direct combustion, gasification, and anaerobic digestion. Further, conversion processes can be designed to produce, in addition to electricity, solid (wood and charcoal), liquid (oils and alcohols), and gaseous (methane and hydrogen) fuels that can be easily stored and transported.

Energy Crops

Energy crops are plants grown specifically as sources of energy; they include not only trees but also grasses and legumes, grain and sugar crops, and oil-bearing plants like sunflowers. The great majority of the energy that is produced from biomass, however, comes from *wood.* In 1850 the United States obtained about 90% of its energy needs from wood. Although wood now contributes only 3–4% to the energy mix of the country as a whole, the percentage varies by region, with wood fuel providing 16% and 13% of the energy used in Maine and Vermont, respectively. In developing countries, wood is a key source of energy, used for space heating, cooking, water heating, and lighting.

A second biomass contribution to energy systems is *alcohol,* which can be produced from a variety of plants. Brazil, which is poorly endowed with fossil fuels, has embarked on an effort to develop its indigenous energy supplies in order to reduce the country's dependence on imported oil. By 1986, about 9 million of its cars were running on a 20% gasoline-alcohol mixture (the alcohol is made from sugarcane and cassava), and another 1.2 million cars ran on straight alcohol. Overall, alcohol met over 40% of the country's fuel needs.

Figure 10.20

An anaerobic digester on a small farm in China. Animal and vegetable wastes are significant sources of fuel in countries such as India and China. The smaller unit at the left is the "input" tube. Human, animal, and vegetable wastes, inserted here, pass downward to an underground pit containing water and decaying organic material. The large unit in the middle is the gas-collection chamber. The copper tube leads to a community kitchen that serves ten families.

Figure 10.21

Power station at the Akosombo Dam on the Volta River in Ghana. The dam is part of a major hydroelectric complex that supplies much of Ghana's electricity. Africa's hydropower potential is thought to be 4 to 5 times greater than that of the United States.

Waste

Waste, including crop residues and animal and human refuse, represents the second broad category of organic fuels. New York City, San Francisco, Miami, and a host of other American cities have refuse-based power plants that burn garbage to produce steam and electricity. In the United States, household trash amounts to about one-half ton per person per year. Because burning reduces the volume of waste by about 90%, it significantly eases the pressure on municipal landfills.

In rural areas, particularly, energy can also be obtained by fermenting human and animal wastes, in a process known as *anaerobic digestion,* to produce methane gas (also called *biogas*). A number of countries, including India, South Korea, and Thailand, have national biogas programs, but the largest effort to generate substantial quantities of methane gas for rural households has been undertaken in China. There, millions of backyard fermentation tanks (biodigesters) on communes and farms supply as many as 35 million people with fuel for cooking, lighting, and heating (Figure 10.20). The technology has been kept intentionally simple. A stone fermentation tank constructed in a pit is fed with wastes, which can include straw and other crop residues in addition to manure. These are left to ferment under pressure, producing methane gas that is later drawn through a hose into the farm kitchen. After the gas is spent, the remaining waste is pumped out and used in the fields for fertilizer.

Hydroelectric Power

Biomass, particularly wood, is the most commonly used source of renewable energy. The second most common is **hydropower,** which supplies roughly one-fourth of the world's electricity (or about 7% of total commercial energy production). In the United States, where biomass does not make a major contribution to energy production, hydropower is the *largest* renewable energy resource.

Hydropower is generated when water falls from one level to another, either naturally or over a dam. The falling water can then be used to turn waterwheels, as it was in ancient Egypt, or modern turbine blades, powering a generator to produce electricity (Figure 10.21). The amount of electricity produced is a function of the height of the water above the turbines and the volume of water flowing through them.

Tied as it is to a source of water, hydroelectric power production is diffuse. In the United States, it is generated at more than 1900 sites in 47 states. Still, over 60% of the nation's developed capacity is concentrated in just three areas: the Pacific states (Washington, Oregon, California), the multistate Tennessee River valley area of the Southeast, and New York (Figure 9.21). That pattern is a result both of the location of the resource base and the role that agencies like the Tennessee Valley Authority (TVA) have played in hydropower development.

Figure 10.22

The Suntree apartment complex in Davis, California, is oriented toward the sun. Each of the ninety-five units uses solar-heated water for bathing, water and space heating. The City of Davis promotes conservation and the use of solar energy through building codes that specify, among other things, building orientation, insulation, and window areas.

Transmitting hydroelectricity over long distances is costly, so it is generally consumed in the region where it is produced. This fact helps to account for variations in the pattern of consumption and for the energy mix used in specific regions. Thus, while hydropower provides the United States as a whole with one-seventh of its electricity, it supplies all of Idaho's needs and about 90% of the electricity used in Oregon and Washington.

Consumption patterns vary widely around the world. Canada gets about 70% of its electricity from hydropower, a share roughly comparable to that found in such Western European countries as Switzerland and Austria. Almost 40 countries in Asia, Africa, and South America obtain more than half their electricity from hydropower; indeed, it supplies more than 90% of the electricity in a half-dozen of these nations (Bhutan, Ghana, Laos, Uganda, Zaire, and Zambia).

Solar Power

Each year the earth intercepts solar energy that is equivalent to many thousands of times the energy people currently use. Inexhaustible and nonpolluting, **solar energy** is the ultimate origin of most forms of utilized energy: fossil fuels and plant life, waterpower, and wind power. It is, however, the direct capture of solar energy that has excited recent interest.

The technology for using solar energy for domestic uses such as hot-water heating, and space heating and cooling is well known. In the United States, both passive and active solar heating technologies have secured a permanent foothold in the marketplace. Well over one-half million homes use solar radiation for water and space heating (Figure 10.22). Japan is reported to have some 4 million solar hot-water systems, and in Israel, half the homes heat water with solar energy. Present technology

Figure 10.23
Solar One, the world's largest solar
electric-power plant. Located in the desert
area near Barstow, California, the facility
uses solar energy to produce steam to
generate electricity. The 1818 large
heliostats, arranged in concentric circles,
focus the sun's rays on water-filled tubes
in a boiler on top of the central tower.

cannot maintain usable heat levels for more than few days at a time, so that solar heating must be supplemented by heat from conventional sources during cloudy periods or in cool areas.

A second type of solar-energy use under development involves converting concentrated solar energy into thermal energy to generate electricity. Research efforts are focusing on a variety of thermal-electric systems, including power towers, parabolic troughs, parabolic dishes, and solar ponds. Most involve the concentration of the sun's rays onto a collecting system. In a power tower system, large mirrors called *heliostats* focus solar rays on a boiler in a central tower, developing heat levels (over 1000° F) sufficient to convert the water to steam, which drives a turbine generator on the ground. An experimental solar-thermal power plant of this type, the Solar One facility in the Mojave Desert of southern California, produces 10 megawatts of electricity, enough to satisfy the electrical needs of about 10,000 people (Figure 10.23). That plant had massive government subsidies, but if one or more of the thermal-electric systems prove to

be economically feasible, they could be important in any area of the world that has a high incidence of solar radiation.

Solar One generates electricity indirectly by first converting light to heat, but electricity can also be generated directly from solar rays by **photovoltaic** (PV) **cells.** Silicon solar cells have been used to provide energy for spacecraft as well as for some terrestrial needs—for example, powering lighthouses, navigation buoys, and mountaintop radio relays. Some 6000 houses in the United States are outfitted with PV systems, and silicon cells are being increasingly used in small-scale applications, such as wristwatches and pocket calculators.

The harnessing of solar rays is seen by many as the best hope of satisfying a large proportion of future energy needs with minimal environmental damage and maximum conservation of earth resources. Considerable research and development will be necessary, however, if PV power systems are to become economically and technically viable. The costs of the energy cells are still high, and their efficiency is relatively low.

Figure 10.24

The tidal power station on the Rance estuary in Brittany, France. Tidal power originates in the gravitational energy of the earth-sun-moon system. The back-and-forth movement of oceans under the influence of tides represents kinetic energy. Tidal stations capture some of the kinetic energy before it is converted into heat energy and lost.

Other Renewable Energy Resources

In addition to biomass, hydropower, and solar power, there are a number of renewable sources of energy that might be exploited. In the last decade, research has focused primarily on three of these: the ocean, the wind, and geothermal energy. Although none appears able to make a major contribution to the world's energy needs, each might have limited, often localized, potential if appropriate technology can be developed.

Energy from the Ocean

Four forms of energy exist in or at the edges of oceans: tidal power, thermal power, waves, and currents. In theory, all four could be used to generate electricity, although their potential contributions vary significantly. All are still at experimental stages; the technology for using tidal power is furthest advanced.

Tidal power harnesses electricity from the ebb and flow of water as ocean tides rise and fall. If a dam is built across an estuary or bay with a high tidal range (the *tidal range* is the difference between high and low tides), electricity can be generated as water moves both into and out of the bay over reversible turbines. Fewer than 50 sites with suitable tidal ranges (at least 15 feet) and geographic features have been identified around the world, and only three in the United States.

The world's first tidal power plant was completed in 1966 on the Rance River in France, using the tidal flows of the English Channel (Figure 10.24). Canada and the USSR also operate experimental tidal plants, which should help provide answers to concerns of environmentalists about the possibility that such facilities might alter ocean currents, increase coastal flooding, and disrupt salt marshes where fish and birds breed.

Wind Power

Although **wind power** was used for centuries to pump water, grind grain, and drive machinery, its contribution to the energy supply virtually disappeared over a century ago, when windmills were replaced first by steam and later by the fossil fuels. The fuel shortages and escalating electrical rates of the 1970s stimulated interest in exploring again the potential of this renewable resource.

Windmills offer many advantages as sources of electrical power. They can turn turbines directly, do not use any fuel, and can be built and erected rather quickly.

Figure 10.25
A "wind park" in the Tehachapi Mountains Wind Resource Area
of California. The wind turbine generators harness wind power to
produce electricity. California's wind parks produce enough
electricity for 200,000 households, or about 1% of the state's
population.

They need only strong, steady winds to operate, and these
exist at many sites—in the United States, for example,
along the ocean coasts, in mountainous areas, and on the
Great Plains. Furthermore, wind turbine generators
(WTGs) avoid the negative ecological effects of other
energy sources. They don't pollute the air or water and
don't deplete scarce natural resources.

Much of the world's wind-power generating capacity (and over 90% of the U.S. total) is in California,
which has some 80 "wind parks" (areas containing clusters of small WTGs) like that pictured in Figure 10.25.
Over 9000 WTGs have been hooked into the state's electric power grid. California's dominance is due less to favorable wind energy conditions than to nongeographic
factors. These include the state's commitment to developing renewable resources, favorable tax credits, and the
support of utility companies. Because the wind parks have
only recently been built, and various experimental windmill designs are still being evaluated, it is not yet clear
whether WTGs can deliver electricity at costs competitive with other energy sources.

Geothermal Power

People have always been fascinated by volcanoes, gey-
sers, and hot springs, all of which are manifestations of
geothermal (literally, "earth-heat") **energy.** There are
several methods of deriving energy from the earth's heat
as it is captured in hot water and steam trapped a mile
or more beneath the earth's surface. Deep wells can tap
these reservoirs and use the heat energy either to generate electricity or for direct heat applications like heating
houses or drying crops. New Zealand, Japan, Iceland,
Mexico, and the United States are among the countries
already using geothermal energy (Figure 10.26). Reyk-
javík, the capital of Iceland, is almost entirely heated by
geothermal steam.

Conventional methods of tapping geothermal energy depend on hot water reservoirs beneath the earth's
surface. A potentially important new technology mines
the earth's interior heat instead. In the "hot dry rock"

(a)

Figure 10.26

(a) Schematic cross section of a geothermal plant. Magma radiates heat through the solid rock above it, heating water from underground sources and surface runoff. Drilled wells tap the steam and bring it to the surface, where it is piped to the power plant. (Courtesy of Pacific Gas and Electric Co. and American Petroleum Institute.) (b) Electric generating plant powered by geothermal steam supplies electricity to Manila in the Philippines.

(b)

Figure 10.27

Typical newspaper articles about price increases during the oil embargo of 1979.

project near Los Alamos, New Mexico, a well was drilled into the magma (molten rock) 2.5 miles (4 km) below the earth's surface; then millions of gallons of water were pumped down the well into the fractured, naturally heated rock. Superheated to over 350° F (177° C), the fluid rises back up through a second well to the surface, where it drives turbines in a small power station. If the experimental project proves successful, it will be followed by installation of a commercial electric generating station.

The advantage of the hot dry rock method of using geothermal energy is that it does not depend on the fortuitous availability of hot water reservoirs. It could be used anywhere that magma has intruded into the earth's crust and is heating subsurface rock to high temperatures. In addition, this method is free of the pollution problems that sometimes accompany geothermal energy development based on natural underground water. These include hydrogen sulfide ("rotten egg") emissions, pollution of groundwater by excess steam concentrate, and land subsidence due to the withdrawal of underground fluids.

Resource Management

During the 1970s, it was forcefully and unexpectedly brought home to the American people just how dependent they were on energy resources that were uncertain in quantity and availability. Long lines at service stations, wildly escalating prices, shortages of natural gas during frigid winters, balance-of-payment difficulties associated with oil imports, and daily news stories about the "energy crisis" all contributed to a national awareness of energy shortages (Figure 10.27). Studies like *The Limits to Growth* predicted the collapse of industrialized societies if their growth continued unchecked. The federal government established the U.S. Department of Energy, passed legislation mandating greater vehicular fuel efficiencies, and promoted energy conservation.

By the mid-1980s, the energy outlook seemed to have changed dramatically. Instead of shortages, there was an "oil glut." Gasoline at the pump was half the price it had been a few years earlier. The power of OPEC to curtail production and affect world oil prices had diminished. Domestically, government and the private sector all but withdrew from synthetic fuels development, and tax credits promoting conservation or the use of renewable resources disappeared, as did nightly news reports on the energy crisis.

Most observers agree that the situation in the late 1980s is at best a temporary respite from a crisis (or a series of crises) that is likely to emerge again as supplies of oil, natural gas, and other critical resources inevitably shrink. Nations such as the United States are approaching the end of a period in which resources were cheap, readily available, and lavishly used. The earth was until recently viewed as an almost inexhaustible storehouse of resources for humans to exploit and, simultaneously, as a vast repository for the waste products of society. That view has been replaced by the realization that many resources can be depleted, even "renewable" ones like forests, that many have life spans measured only in decades, and that the air, water, and soil—also resources—cannot absorb massive amounts of pollutants and yet retain their life-supporting abilities.

The wise management of resources entails three strategies: conservation, reuse, and substitution. Opportunities to reduce energy consumption by *conservation* are many and varied. Estimates are that the United States could decrease the amount of energy it now consumes by anywhere from 10% to 40%. Nearly everything can be made more energy-efficient. The nation uses a significant portion of the world's oil output as gasoline for cars and trucks. Doubling their fuel efficiency by reducing auto weight and using more efficient engines and tires would save some 8% of the world's total annual oil output. Industries have an enormous potential for saving energy by

Two Views of the Future

W hen he was president, Jimmy Carter asked 13 federal agencies to participate in an exhaustive study of likely long-term changes in the world's population, natural resources, and environment. In 1980 the Council on Environmental Quality and the Department of State gave him their analysis, *The Global 2000 Report to the President*. The book warned:

If present trends continue, the world in 2000 will be more crowded, more polluted, less stable ecologically and more vulnerable to disruption than the world we live in now. Serious stresses involving population, resources and environment are clearly visible ahead. Despite greater material output, the world's people will be poorer in many ways than they are today.

Alarmed by the pessimism evident in "Global 2000," Herman Kahn, then head of the Hudson Institute, and Julian Simon, a senior fellow at the Heritage Foundation, asked a number of experts to undertake their own study of the future. Modeling their language on the earlier report, the contributors to *The Resourceful Earth* predicted:

If present trends continue, the world in 2000 will be less crowded (though more populated), less polluted, more stable ecologically, and less vulnerable to resource-supply disruption than the world we live in now. Stresses involving population, resources, and environment will be less in the future than now. The world's people will be richer in most ways than they are today. The outlook for food and other necessities of life will be better. Life for most people on earth will be less precarious economically than it is now.

---

using more efficient equipment and processes. The Japanese steel industry, for example, uses one-third less energy to produce a ton of steel than does that industry in most other countries. Energy used for heating, cooling, and lighting homes and office buildings could be reduced by half if they were properly constructed, and expansion of mass transit lines would add further energy savings. Other conservation measures include experimentation with rate reductions during off-peak hours, cost penalties rather than rewards for high-volume energy users, and the use of cogeneration facilities (using industrial boiler heat for in-plant electrical generation).

The *reuse* of materials that it has taken energy to produce also reduces resource consumption. Instead of being buried in landfills, waste can be burned or decomposed and fermented to provide energy. Recycling of paper, aluminum, steel, and other materials can be greatly increased to yield back the energy invested in them (Figure 10.28). It has been estimated that throwing away an aluminum soft drink container wastes as much energy as filling the can half full of gasoline and pouring it on the ground.

Finally, the *substitution* of other energy sources for gas and oil, and the substitution of other materials for nonfuel minerals in short supply can be more actively pursued. If the technology to exploit them economically can be developed, coal and oil shale could supply fuel needs well into the future. The renewable energy resources—biomass, solar and geothermal power, and so on—are virtually infinite in their amount and variety. While no single renewable source is likely to be as important as oil or gas, collectively they could make a significant contribution to energy needs. No matter what alternate sources are developed, however, they will be exploited only at economic costs reflecting the conversion of energy from a cheap resource to be used prodigally to a scarce resource to be used as conservatively as possible.

Conclusion

Energy and nonfuel mineral resources are but two types of earth materials that humans have learned to use and come to depend on for maintenance of both traditional and modern societies. They have been made the primary focus of this chapter because they clearly summarize fundamental concerns of resource evaluation and use in general. These concerns include spatial concentration; assessment of size, value, and use potential of reserves; economic and political consequences of reserve location and depletion; considerations of resource cost and substitution possibilities; and the relationship between growing resource demand and the finite supply of the earth's raw materials.

Energy has become in the late 20th century a symbol of national wealth and economic development. It figures importantly in national foreign trade accounts and in household budgets; its availability (or shortage) and cost can trigger changes in the style and quality of life. The problem that all countries are now facing or will soon face is best viewed not as a resource problem but as a cultural problem. Technology provides options but not answers; it is simply a tool to achieve societal goals. It is the way people think and act that will determine the shape of the future, and for that reason the conservation ethic appears mandated for all the world's economies regardless of their stage of development or apparent current access to resources.

Figure 10.28

The approximately 140 million tons of refuse thrown away by Americans each year contain an estimated 11.5 million tons of steel and iron; 870,000 tons of aluminum; some 435,000 tons of other metals, principally copper; and more than 13 million tons of glass. Burnable organic matter, including paper, equals some 60 million tons—the equivalent of 150 million barrels of crude oil in energy content. The photographs show one method of recycling industrial scrap to fuel steam-power boilers. Solid waste is collected at the plant (upper left) and shredded (upper right), and the nonburnable materials are removed (lower left). The remaining scrap is stored in a huge silo (lower right), later to be burned with coal to produce steam power.

Key Words

For Review

1. What is the basic distinction between a *renewable* and a *nonrenewable* resource? Does the use of nonrenewable bauxite pose the same problem of resource depletion as the consumption of nonrenewable petroleum? Why or why not?

2. With what indices is energy production positively correlated? Why is there not always a correlation between energy production and consumption?

3. Briefly summarize the effects of the oil shocks of 1973–1974 and 1979–1980.

4. Why has the proportion of United States energy supplied by coal increased since 1961? What ecological and social problems are associated with the use of coal?

5. Review some of the techniques by which synthetic fuels might be produced. Why are synfuels not yet economically competitive with the other fossil fuels?

6. What are the different methods of generating nuclear energy? Why is there public opposition to nuclear power?

7. What in general are the leading mining countries? What role do developing countries play in the production of critical raw materials? How have producing countries reacted to the threatened scarcity of copper?

8. Which are the most widely used ways of using renewable resources to generate energy? What are some of the renewable sources still under development?

9. Discuss three ways of reducing demands on resources.

Suggested Readings

Cuff, David J., and William J. Young. *The United States Energy Atlas,* 2d rev. ed. New York: Macmillan, 1986.

Gates, David M. *Energy and Ecology.* Sunderland, Mass.: Sinauer Associates, 1985.

Glassner, Martin I., ed. *Global Resources: Challenges of Interdependence.* New York: Praeger Publishers, for the Foreign Policy Association, 1983.

Miller, G. Tyler. *Environmental Science: An Introduction.* Belmont, Calif.: Wadsworth, 1986.

Owen, Oliver S. *Natural Resource Conservation,* 3d ed. New York: Macmillan, 1980.

Sawyer, Stephen W. *Renewable Energy: Progress, Prospects.* Washington, D.C.: Association of American Geographers, 1986.

World Resources 1987, A Report by the World Resources Institute and the International Institute for Environment and Development. New York: Basic Books, 1987.

11 Urban Geography

*T*wenty-five years ago, when Mexico City contained some 2.5 million people, it was described as a beautiful and impressive city, its bright and colorful older streets lined with mansions, stately palaces, and lovely churches. Nearby, new residential districts accommodated residents in modern houses and apartment buildings set amid trees and flower gardens. The 200-foot-wide Paseo de la Reforma, the chief avenue, was said to be one of the most beautiful boulevards in the world, shaded by a double row of trees and lined with luxurious residences.

What will Mexico City be like in the year 2000? The generally accepted forecast is that Mexico City will be the largest city in the world, with a population of about 31 million people. People are flowing into Mexico City at an incredible rate. The city's housing and its sanitary and transportation systems have already been overwhelmed (Figure 11.1). By 2000, the smog problem, which is currently among the worst in the world, will make breathing an exhausting experience. The hard-pressed government is building housing, but it can accommodate only a small percentage of the people. A result is one of the largest slums in the world. Mexico City is but one example of a post–World War II trend toward unbelievably large cities. There is no indication that the trend will come to a halt in the next 20 or 30 years.

Figure 11.2 presents evidence that the growth of major metropolitan areas has been astounding in this century. Today there are over 250 metropolitan areas that have a population in excess of 1 million people, though at the beginning of the century there were only 13. Nine metropolises (Beijing, Buenos Aires, Cairo, Los Angeles, Mexico City, New York, São Paulo, Shanghai, and Tokyo) have over 10 million people. In 1900, there were none of that size. Of course, as we saw in Chapter 5, the world's population has greatly increased, so that we would also expect the urban component to increase. But the fact remains that urbanization and metropolitanization have increased more rapidly than the growth of total population.

Figure 11.1
Sprawling Mexico City.

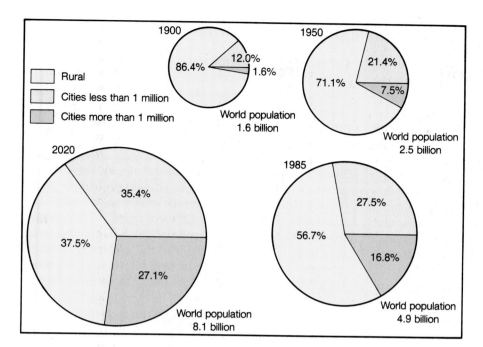

Figure 11.2
Patterns of world urbanization.

From Elaine M. Murphy, *World Population: Toward the Next Century,* Population Reference Bureau, Inc., May 1985.

The amount of urban growth differs from continent to continent and from country to country, but all countries have one thing in common: the proportion of the people living in cities is rising.

Table 11.1 shows world urban population by region. Note that the most industrialized parts of the world, North America and Western Europe, are the most urbanized in terms of percentage of people living in cities, while Asia and Subsaharan Africa have the lowest proportions of urban population. What is evident is that the industrialization process has also been one of urbanization. As the world, especially the Third World, continues to industrialize, we can expect further large increases in urbanization. It is interesting to note in Table 11.1 that even though East Asia (China, Japan, and Korea) and South Asia (India, Pakistan, and Bangladesh) have relatively low proportions of people in urban regions, the absolute number of people in urban areas there is among the highest in the world. Given the huge populations in Asia and the relatively heavy emphasis on agriculture (except in Japan and the Republic of Korea), it sometimes escapes us that there are many large cities throughout parts of the world where subsistence agriculture still engages most people.

In order to understand trends in urbanization, it is first necessary to look at the reasons for the existence of cities. Since nearly all cities were at one time villages, the principles involved are those that concern the founding of villages. As we will see, these principles have as much to do with the location of the settlements as they do with the forces that make them grow. In this chapter, then, our

TABLE 11.1
Estimated Urban Share of Total Population, 1950 and 1986, with Projections to 2000

Region	1950	1986	2000
North America	64%	74%	78%
Europe	56	73	79
Soviet Union	39	65[a]	74
East Asia	43	70	79
South Asia	15	24	35
China	12	32	40
Latin America	41	65	77
Africa	15	30	42
Oceania	61	65	73
World	29	43	48

[a]1987

Source: Lester R. Brown and Jodi L. Jacobson, *The Future of Urbanization.* Worldwatch Paper 77 (Washington, D.C.: Worldwatch Institute, 1987), p. 8. Original data from Population Reference Bureau, Inc.

first objective is to consider the major factors responsible for the size and the location of cities. The second goal is to identify the nature of land use patterns within cities, and third, we will attempt to differentiate cities around the world by a review of the factors that help to explain their special nature.

The Functions of Urban Areas

Except for the occasional recluse or hermit, people gather together to form couples, families, groups, organizations, towns, and so forth. This desire to be near one another, however, is more than a function of the need to socialize. Our human support systems are based on the flow of information, goods and services, and cooperation among people who are located at places convenient to one another. Unless individuals can produce all that they need themselves, and relatively few can, they must depend on shipments of food and supplies to their home place or convenient outlet centers. Nonsubsistence groups establish stores, places of worship, repair centers, and production sites as close to their home places as possible and reasonable. The result is the establishment of towns. These may grow to the size of a Mexico City (about 20 million people) or a Tokyo metropolitan area (about 18 million people).

Whether they are villages, towns, or cities, urban settlements exist for the efficient performance of functions required by the society that creates them, functions that cannot be adequately carried out in dispersed locations. They reflect the saving of time, energy, and money that the agglomeration of people and activities implies. The more accessible the producer to the consumer, the worker to the workplace, the citizen to the town hall, the worshiper to the church, or the lawyer or doctor to the client, the more efficient is the performance of their separate activities, and the more effective is the integration of urban functions.

Urban areas provide all or some of the following types of functions: retailing, wholesaling, manufacturing, business service, entertainment, religious service, political and official administration, military defensive needs, social service, public service including sanitation and police service, transportation and communication service, meeting place activity, visitor service, and places for its residents to live. Because all urban functions and people cannot be located at a single point, cities themselves must take up space, and land uses and populations must have room within them. Because interconnection is essential, the nature of the transportation system will have an enormous bearing on the total number of services that can be performed and the efficiency with which the functions can be carried out. The totality of people and functions of a city constitutes a distinctive cultural landscape whose similarities and differences from place to place are the subjects for urban geographic analysis.

Some Definitions

Urban areas are not of a single type, structure, or size. What they have in common is that they are nucleated, nonagricultural settlements. At one end of the size scale, urban areas are small towns with perhaps a single main street of shops; at the opposite end, they are complex, multifunctional metropolitan areas, supercities, or a "megalopolis" (Figure 11.3). The word *urban* often is used in place of such terms as town, city, suburb, and metropolitan area, but it is a general term, and it is not used to specify a particular type of settlement. Although the terms designating the different types of urban settlements, like *city,* are employed in common speech, everyone does not use them in the same way. What is recognized as a city by a resident of rural Vermont or West Virginia might not at all be afforded that name and status by an inhabitant of California or New Jersey. It is necessary in this chapter to agree on the meanings of terms commonly employed but varyingly interpreted.

The words **city** and town denote nucleated settlements, multifunctional in character, including an established central business district and both residential and nonresidential land uses. **Towns** have smaller size and less functional complexity than cities, but still have a nuclear business concentration. **Suburb** implies a subsidiary area, a functionally specialized segment of a large urban complex. It may be dominantly or exclusively residential, industrial, or commercial, but by the specialization of its land uses and functions, suburbs are dependent on urban areas outside their boundaries. Suburbs, however, can be independent political entities. For large cities having many suburbs, it is common to call that part of the urban area contained within the official boundaries of the main city around which the suburbs have been built the **central city.**

Some or all of these urban types may be spatially and functionally associated into larger landscape units. The **urbanized area** refers to a continuously built-up landscape defined by building and population densities with no reference to the political boundaries that limit the legal city of which it is the extension. It may be viewed as the physical city and may contain a central city and many contiguous cities, towns, suburbs, and other urban tracts. A **metropolitan area,** on the other hand, refers to a large-scale functional entity, perhaps containing several urbanized areas, discontinuously built up but nonetheless operating as a coherent and integrated economic whole. Figure 11.4 shows these areas in a hypothetical county.

Figure 11.3

The differences in size, density, and land use complexity between
New York City and a small town are immediately apparent.
Clearly, one is a city and one is a town, but both are urban areas.

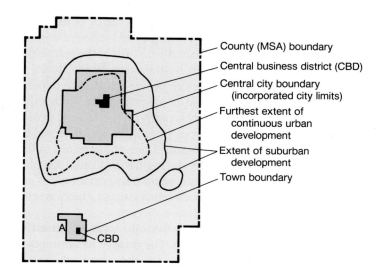

County (MSA) boundary

Central business district (CBD)

Central city boundary
(incorporated city limits)

Furthest extent of
continuous urban
development

Extent of suburban
development

Town boundary

A ⯀

CBD

Figure 11.4

A hypothetical spatial arrangement of urban units within a Metropolitan Statistical Area. Sometimes official limits of the central city are very extensive and contain areas commonly considered suburban. Older eastern U.S. cities more often have restricted limits and contain only part of the high-density land uses associated with them.

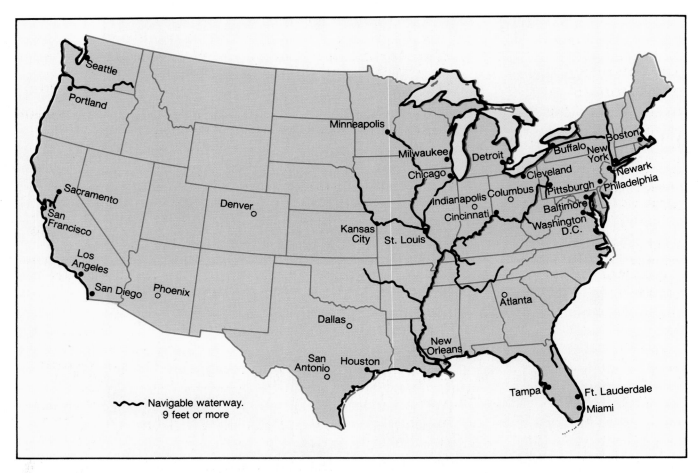

Figure 11.5

The 35 most populous Standard Metropolitan Statistical Areas in the United States (1980 Census). Notice the association of principal cities and navigable water; before the advent of the railroads in the middle of the 19th century, all major cities were associated with waterways. Only seven of these metropolitan areas are not on navigable water: Atlanta, Columbus, Dallas, Denver, Indianapolis, Phoenix, and San Antonio.

The Location of Urban Settlements

Cities are functionally connected to other cities and to rural areas. In fact, the reason for the existence of a city is not only to provide services for itself, but for others outside of the city. The city is a consumer of food, a processor of materials, and an accumulator and dispenser of goods and services, but it must depend on the "countryside" that established it for its supplies and as a market for its activities. In order to perform adequately the tasks that support it and to add new functions as demanded by the larger economy, the city must be efficiently located. That efficiency may be marked by spatial or transportation centrality to the area served. It may derive from location at points whose physical characteristics permit the performance of assigned functions. Or placement may be related to the resources, productive regions, and transportation network of the nation, so that the effective performance of a wide array of activities is possible without reference to the requirements of the immediate local region.

In discussing urban settlement location, geographers frequently differentiate between site and situation. The **site** is the exact location of the settlement and can be described in terms of latitude and longitude or in terms of the physical characteristics of the site. For example, the site of Philadelphia is an area bordering and west of the Delaware River north of the junction with the Schuylkill River in southeastern Pennsylvania. The description can be more or less exhaustive depending on the purpose it is meant to serve. In the Philadelphia case, the fact that the city is partly on the Atlantic Coastal Plain, partly on a piedmont (foothills), and is served by navigable rivers is important if one is interested in the development of the city during the Industrial Revolution. As Figure 11.5 suggests, water transportation was an important localizing factor when the major American cities were established.

If site suggests absolute location, **situation** indicates relative location. The relative location places a settlement in relation to the physical and human

The Definition of *Metropolitan* in the United States

Definitions of various types of urban areas must be clear if proper accounting is to be made by governmental authorities. The United States Bureau of the Census has refined and redefined the concept of "metropolitan" from time to time to summarize the realities of the changing population, physical size, and functions in urban regions.

Until 1983, the *Standard Metropolitan Statistical Area* (SMSA) was recognized. It was made up of one or more functionally integrated counties focusing upon a central city of at least 50,000 inhabitants. Now, the minimum size requirement for central cities has been dropped, and central-city status is determined by other qualities—whether, for example, a city is an employment center surrounded by bedroom-community suburbs. Automatically, the number of central cities (and metropolitan areas) increased. The statistical structure of urban America has been altered as individual communities exercised their rights to withdraw from former metropolitan affiliations or opted to join neighboring cities in new ones.

In the mid-1980s, old and new metropolitan areas have been redefined into *Metropolitan Statistical Areas* (MSAs are economically integrated urbanized areas in one or more contiguous counties), *Primary Metropolitan Statistical Areas* (PMSAs are those counties that are part of MSAs that have less than 50% resident workers working in a different county), and *Consolidated Metropolitan Statistical Areas* (an MSA becomes a CMSA if it contains 1 million or more people and is composed of PSMAs).

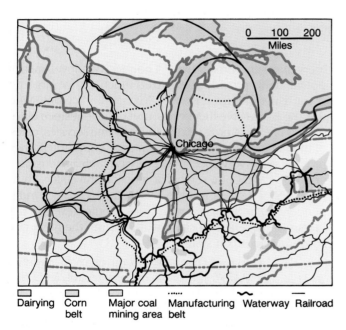

Dairying | Corn belt | Major coal mining area | Manufacturing belt | Waterway | Railroad

Figure 11.6
The situation of Chicago helps to suggest the reasons for its functional diversity and size.

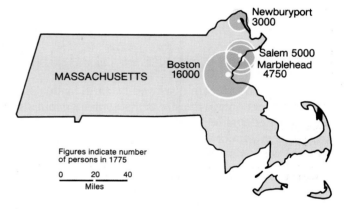

Figure 11.7
Maritime activities, situational advantages, and diversity of functions led to the early rise of urban agglomerations in eastern Massachusetts. Their development and success encouraged continued urbanization despite changing economic conditions.

characteristics of the surrounding areas. Very often it is important to know what kinds of possibilities and activities exist in the area near a settlement, such as the distribution of raw materials, market areas, agricultural regions, mountains, and oceans. The site of central Chicago is 41°52′ N, 87°40′ W, on a lake plain, but more important is its situation close to the deepest penetration of the Great Lakes system into the interior of the country, astride the Great Lakes–Mississippi waterways, and near the western margin of the manufacturing belt, the northern boundary of the corn belt, and the southeastern reaches of a major dairy region. References to railroads, coal deposits, and ore fields would amplify its situational characteristics (Figure 11.6). From this description of Chicago's situation, implications relating to market, to raw materials, and to transportation centrality can be drawn.

The site or situation that originally gave rise to an urban unit may not long remain the essential ingredient for its growth and development. Agglomerations, originally successful for whatever reason may, by their success, attract people and activities totally unrelated to the initial localizing forces (Figure 11.7). By what has been called a process of "circular and cumulative causation," a successful urban unit may acquire new populations and functions attracted by the already existing markets, labor force, and urban facilities.

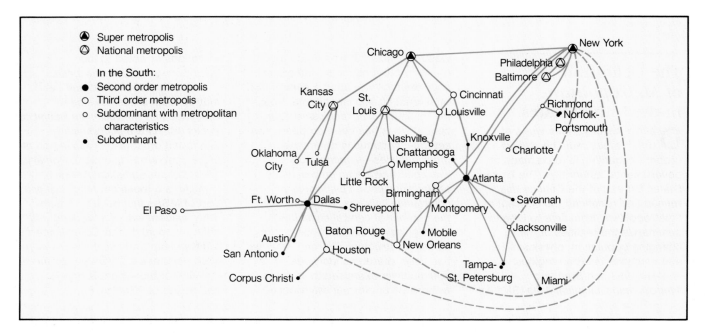

Figure 11.8
Hierarchical relationships among southern U.S. cities.

Systems of Urban Settlements

The functional structure displayed by an individual city is reflected not only in the size of that settlement but also in its location and in its relationship with other units in the larger system of which it is part.

The Urban Hierarchy

Perhaps the most effective way to recognize the way systems of cities are organized is to consider the **urban hierarchy.** Cities can be divided into size classes on the basis of their functional complexity and importance. One can measure the numbers and kinds of services each city or metropolitan area provides. The hierarchy is then like a pyramid; the few large and complex cities are at the top and the many smaller cities are at the bottom. There are always more smaller cities than larger ones. One can envisage, say, a seven-level hierarchy where the complexity of cities increases as one rises in the pyramid.

When a spatial dimension is added to the hierarchy as in Figure 11.8, it becomes clear that a spatial system of metropolitan centers, large cities, small cities, and towns exists. Goods, services, communication lines, and people flow up and down the hierarchy according to the location of the urban areas. The few high-level metropolitan areas provide specialized functions for large regions while the smaller cities serve smaller regions. Note that the cities serve the area around them, but since cities of the same level provide roughly the same services, cities of the same size tend not to serve each other unless they provide some very specialized service such as housing a political capital of a region or a major university. Thus, the cities of a given level in the hierarchy are not independent but interrelated with cities of other levels in the hierarchy. Together all cities at all levels in the hierarchy constitute an urban system.

Central Places

An effective way to see how cities and towns are interrelated is to consider urban settlements exclusively as centers for the distribution of economic goods and services. This special case, called the marketing principle, was developed by the German geographer Walter Christaller, who laid the foundations for **central place theory.**

In 1933, Christaller attempted to explain the size and the location of settlements. Although he was not totally successful, he performed a valuable service by developing a framework for understanding much about urban interdependence. Christaller first decided that his

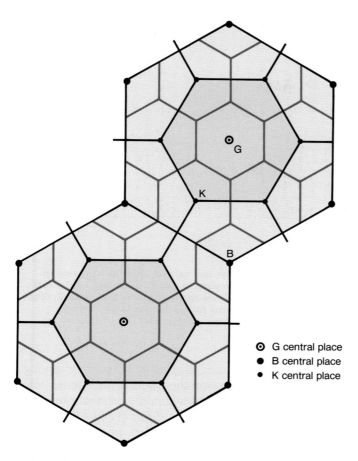

- ⊙ G central place
- ● B central place
- ● K central place

Figure 11.9

The two G central places are the largest on this diagram of one of Christaller's models. The B central places offer fewer goods and services for sale and serve only the areas of the intermediate-sized hexagons. The many K central places, which are considerably smaller and more closely spaced, serve still smaller hinterlands. The goods offered in the K places are also offered in the B and G places; the latter offer considerably more and higher-order goods. Notice that the places of the same size are equally spaced.

theory could best be developed in rather idealized circumstances. He assumed that towns that provide the surrounding countryside with such fundamental goods as groceries and clothing would develop in a plain where farmers specialized in commercial agricultural production. He assumed also that the farm population would be dispersed in an even pattern and that the characteristics of the people would be uniform; that is, they would possess similar tastes, demands, and incomes. Christaller knew full well that such assumptions are hardly realistic, but he recognized that his theory would have to be infinitely more complicated if such assumptions were not used.

Christaller further assumed that each kind of product or service supplied to the dispersed population had its own **threshold,** which is the minimum market (number of consumers or volume of purchasing power) needed to support the supply of a product. For example, a shoe store can prosper in an area only if there is sufficient demand for the shoes and money to buy shoes.

There will be stores offering products with high thresholds (requiring high purchasing power) only if there is a large market area. High-threshold goods, such as diamonds or fur coats, are either expensive or not in great demand. On the other hand, bread, a low-threshold product, can be marketed from many locations because demand is high and the product is relatively inexpensive. A smaller hinterland (market area) can support an enterprise selling bread and other low-order goods.

Christaller assumed that the business people of the agricultural plain would divide the area into noncompeting markets where each entrepreneur would have exclusive rights to the sale of a particular product. This idea follows from a set of assumptions about the efficient use of transportation by consumers—they will travel to the nearest opportunity (store)—and the efficient division of market areas by those supplying goods and services.

When all of the assumptions and definitions are considered simultaneously, the result is a series of hexagonal market areas that cover the entire plain (Figure 11.9). The hexagonal market area is a necessary postulate; the hexagon is the most compact geometric form into which the entire hypothetical plain can be subdivided, leaving no unserved hinterland and no areas unassigned to a closest central place. The largest hexagonal market areas have as their centers central places that supply goods and services of the highest and lowest orders, that is, all goods and services supportable by the demand and the purchasing power of the entire area. Contained within or at the edge of the largest market areas are central places serving a smaller population and offering fewer goods and services to their respectively smaller hinterlands.

Christaller reached two important conclusions. First, towns of the same size will be evenly spaced, and larger towns will be farther apart than smaller ones. This means that there will be many more small than large towns. In the region under consideration, the ratio of the number of small towns to towns of the next larger size will be constant. In Figure 11.9 the ratio in the completed network would be 3 to 1. This distinct, steplike series of towns in size classes differentiated by both size and function is called a *hierarchy of central places.*

Second, the system of towns is interdependent. If one town were eliminated, the entire system would have to readjust. Consumers need a variety of products, each of which has a different threshold. The towns containing many goods and services become regional retailing centers, and the small central places serve just the people immediately in their vicinity. The higher the threshold of a desired product, the farther, on average, the consumer must travel to purchase it.

These conclusions have been shown to be generally valid in widely differing areas within the commercial world. When varying incomes, cultures, landscapes, and transportation systems are taken into consideration, the results of Christaller hold up rather well. They are particularly applicable to agricultural areas, especially with regard to the size and spacing of cities and towns (Figure 11.10). One has to stretch things a bit to see the model operating in highly industrialized areas, where cities are more than just retailing centers. However, if we combine a Christaller-type approach with the ideas that help us understand industrial location and transportation alignments (see Chapter 9), we have a fairly good understanding of the location of the majority of cities and towns.

The Economic Base

When one or more urban settlements within a well-linked system increases its productivity, perhaps because of an increase in demand for the special goods or services that it produces, all members of the system are likely to benefit. The concept of the **economic base** shows how settlements are affected by changes in economic conditions. In the discussion that follows, we concentrate on change within one city, and from our knowledge of central place theory, we can deduce how changes in one city affect the well-being of those living in other cities with which it is linked.

Part of the employed population of an urban unit is engaged in the production of goods or the performance of services for areas and people outside the city itself. They are workers engaged in "export" activities, whose efforts result in money flowing into the community. Collectively, they constitute the **basic sector** of the city's total economic structure. Other workers support themselves by producing things for residents of the urban unit itself. Their efforts, necessary to the well-being and the successful operation of the city, do not generate new money for it but comprise a **service** or **nonbasic sector** of its economy. These people are responsible for the internal functioning of the urban unit. They are crucial to the continued operation of its stores, professional offices, city government, local transit, and school systems.

The total economic structure of a city equals the sum of its basic and nonbasic activities. In actuality, it is the rare urbanite who can be classified as belonging entirely to one sector or another. Some part of the work of most people involves financial interaction with residents of other areas. Doctors, for example, may have mainly local patients and thus are members of the nonbasic sector, but the moment they provide a service to someone from outside the community, they bring new money into the city and become part of the basic sector.

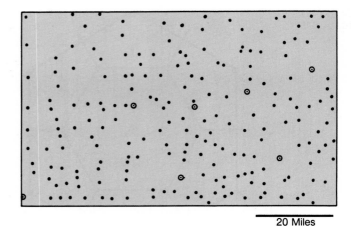

20 Miles

Figure 11.10
The pattern of hamlets, towns, and cities in a portion of Indiana. This map represents an area 82 miles by 51 miles just north of Indianapolis. Cities containing more than 10,000 people are circled. Notice that the pattern is remarkably even and includes a number of linear arrangements that correspond to highways and railroads.

Variations in basic employment structure among urban units characterize the specific functional role played by individual cities within the national economy and provide a means for a classification and comparison of cities. Some settlements, particularly smaller ones, may be termed unifunctional; their basic sector is composed entirely of a single activity type. They are the small farm-market towns, the coal-mining or lumbering towns, the county seats or university towns without other basic activities. Most cities, however, perform many export functions, and the larger the urban unit, the more multifunctional it becomes. Nonetheless, even in cities with a diversified economic base, one or a very small number of export activities tends to dominate the structure of the community and to identify its operational purpose within a system of cities.

Such functional specialization permits the classification of cities into categories: manufacturing, retailing, wholesaling, transportation, government, and so on. Such specialization may also evoke images when the city is named: Detroit, Michigan, or Tokyo, Japan, as manufacturing centers; Tulsa, Oklahoma, for oil production; Nice, France, as a resort; Ottawa, Canada, in public administration; and so on. Certain large regional, national, or world capitals—as befits major multifunctional concentrations—call up a whole series of mental associations, such as New York with banking, the stock exchange, entertainment, the fashion industry, port activities, and so on.

Assuming it were possible to divide with complete accuracy the employed population of a city into totally separate basic and nonbasic components, a ratio between the two employment groups could be established.

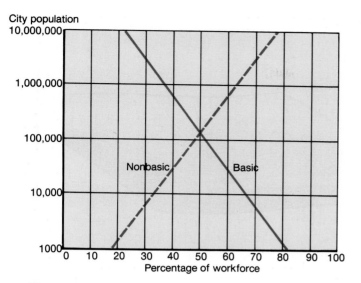

Figure 11.11

A generalized representation of the proportion of the work force engaged in basic and nonbasic activities by city size. As settlements become larger, a greater proportion of their work force is employed in nonbasic activities. Larger settlements are therefore more self-contained.

With some exception for high-income communities, this basic/nonbasic ratio is roughly similar for cities of similar size irrespective of their functional specializations. Further, as a city increases in size, the number of nonbasic personnel grows faster than the number of new basic workers. Thus, in cities with a population of 1 million, the ratio is about 2 nonbasic workers for every basic worker; the addition of 10 new basic employees implies the expansion of the labor force by 30 (10 basic, 20 nonbasic) and an increase in total population equal to the added workers plus their dependents. A **multiplier effect** thus exists, associated with economic growth. The term *multiplier effect* implies the addition of nonbasic workers and dependents to a city's total employment and population as a supplement of new basic employment; the size of the effect is determined by the city's basic/nonbasic ratio. These changing numerical relationships are shown in Figure 11.11.

The changing numerical relationships are understandable when we consider how settlements add functions and grow in population. A new industry selling services to other communities requires new workers, who thus increase the basic work force. These new employees, in turn, demand certain goods and services, such as clothing, food, and medical assistance, which are provided locally. Those who perform such services must themselves have services available to them. For example, a grocery clerk must also buy groceries. The more nonbasic workers a city has, the more nonbasic workers are needed to support them, and the application of the multiplier effect becomes obvious.

We have also seen that the growth of cities may be self-generating—"circular and cumulative" in a way related not to the development of basic industry but to the attraction of what would be classified as *service* in an economic base study. Banking and legal services, a sizable market, a diversified labor force, extensive public services, and the like may attract or generate additions to the labor force not classically basic by definition. The multiplier effect in larger metropolitan centers is discerned in many guises.

In much the same way as settlements grow in size and complexity, so do they decline. When the demand for the goods and services of an urban unit falls, obviously there is a need for fewer workers, and thus both the basic and the services components of a settlement system are affected. There is, however, a resistance to decline that impedes the process and delays its impact. Whereas cities can grow rapidly as migrants respond quickly to the need for more workers, under conditions of decline those that have developed roots in the community are hesitant to leave or may be financially unable to move to another locale. One reason for the chronic poverty in Appalachia, for example, is the unwillingness of many residents to leave their families and the land that they know and love.

Inside the City

An understanding of the nature of cities is incomplete without a knowledge of their internal characteristics. So far, we have explored the location, the size, the spacing, and the growth and decline tendencies of cities. Now we look into the city itself in order to understand better how land uses are distributed, how social areas are formed, and how institutional controls such as zoning regulations affect its structure. In our search for understanding, we consider not only landscape evidence of urban regularities but also bodies of theory that document and support our observations of urban structural patterns. This discussion will primarily relate to cities in the United States, although most cities of the world have been formed in a similar manner. It is a common observation that a recurring pattern of land use arrangements and population densities exists within older central cities and their suburban fringes. It is deduced that those regularities have been created by the circumstances of differential accessibility, the impersonal operations of a free market in land, and the collective consequences of individual residential locational decisions. The whole urban structure is, of course, also being modified by land use controls, obsolescence, transportation alterations, and governmental redevelopment programs.

Figure 11.12
Generalized pattern of land values.
The land value surface reflects demand for
accessibility within a hypothetical urban
area.

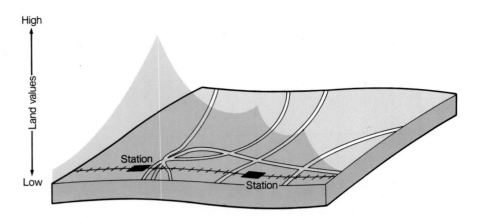

The Competitive Bidding for Land

For its effective operation, the city requires close spatial association of the functions that it houses and of the populations carrying out those functions. As long as these functions were few and the population small, pedestrian movement and pack-animal haulage were sufficient for the effective integration of the urban community. With the addition of large-scale manufacturing and the accelerated urbanization of the economy during the 19th century, however, functions and populations—and therefore city area—grew beyond the interaction capabilities of pedestrian movement alone. Increasingly efficient, and costly, mass transit systems were installed. Even with their introduction, however, only land within walking distance of the mass transit routes or terminals could successfully be incorporated into the expanding urban structure. Usable land, therefore, was a scarce commodity, and by its scarcity, it assumed high market value and demanded intensive, high-density use. The industrial city of the mass transit era, because of its limited supply of usable land, was compact, was characterized by high residential and structural densities, and showed a sharp break on its margins between urban and nonurban uses. The older central cities of, particularly, the northeastern United States and southeastern Canada were of that vintage and pattern.

Within the city, parcels of land were allocated among alternate potential users on the basis of the relative ability of those users to outbid their competitors for a chosen site. There was, in gross generalization, a continuous open auction in land in which users would locate, relocate, or be displaced in accordance with "rent-paying" ability. The attractiveness of a parcel, and therefore the price that it could command at auction, was a function of its accessibility. Ideally, the most desirable and efficient location for all the functions and the people of a city would be at the single point at which the maximum possible interchange could be achieved. Such total coalescence of activity is obviously impossible.

Because uses must therefore arrange themselves spatially, the attractiveness of a parcel is rated by its relative accessibility to all other land uses of the city. Store owners wish to locate where they can easily be reached by potential customers; factories need a convenient assembly of their workers; residents desire easy connection with jobs, stores, and schools; and so forth. Within the older central city, the radiating mass transit lines established the elements of the urban land use structure by freezing in the landscape a clear-cut pattern of differential accessibility. The convergence of that system on the city core gave that location the highest accessibility, the highest desirability, and, hence, the highest land values of the entire built-up area. Similarly, transit junction points were accessible to larger segments of the city than locations along single traffic routes; these latter were more desirable than parcels lying between the radiating lines (Figure 11.12).

Society deems certain functions desirable without regard to their economic competitiveness. Schools, parks, and public buildings are assigned space without being participants in the auction for land. Other uses, through the process of that auction, are assigned spaces by market forces. The merchants with the highest-order goods and the largest threshold requirements bid most for, and occupy, parcels within the **central business district (CBD),** which is located at the convergence of mass transit lines. The successful bidders for slightly less accessible CBD parcels are the developers of the tall office buildings of major cities, the principal hotels, and similar land uses.

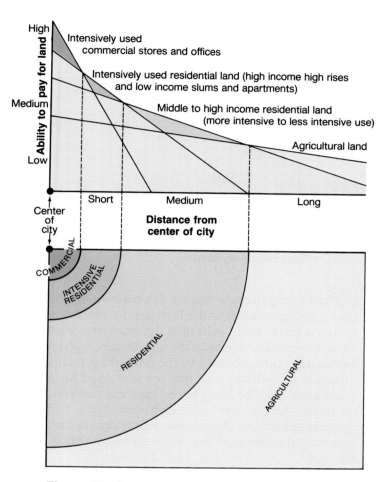

Figure 11.13
The location of various land uses in an idealized city where the highest bidder gets the most accessible land.

Figure 11.14
A bird's-eye view of Toronto showing clusters of tall buildings in the central business district and at important intersections.

Comparable, but lower-order, commercial aggregations develop at the outlying intersections—transfer points—of the mass transit system. Industry takes control of parcels adjacent to essential cargo routes: rail lines, waterfronts, rivers, or canals. Strings of stores, light industries, and high-density apartment structures can afford and benefit from location along high-volume transit routes. The least accessible locations within the city are left for the least competitive users: low-density residences. A diagrammatic summary of this repetitive allocation of space among competitors for urban sites is shown in Figure 11.13.

Land Values and Population Density

Theoretically, the open land auction should yield two separate although related distance-decay patterns, one related to land values and the other to population density (as distance increases away from the CBD, the population density decreases). If one views the land value surface of the central city as a topographic map (Figure 11.14), with hills representing high valuations and depresions showing low prices, a series of peaks, ridges, and valleys would reflect the differentials in accessibility marked by the pattern of mass transit lines, their intersections, and the unserved interstitial areas. Dominating these local variations, however, is an overall decline of valuations with increasing distance away from the peak value intersection, the most accessible and costly parcel of the central business district. As we would expect in a distance-decay pattern, the drop in valuation is precipitous within a short linear distance from that point, and then the valuation declines at a lesser rate to the margins of that built-up area.

Figure 11.15
A generalized population density curve. As distance from the area of multistory apartment buildings increases, the population density declines.

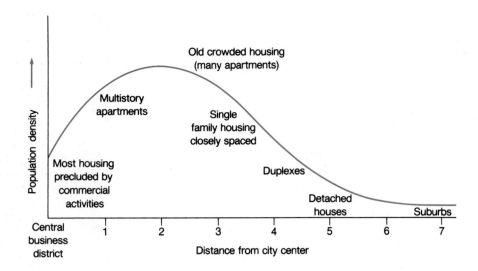

Figure 11.16
Population density of Des Moines, Iowa, 1960 to 1980. The scale on the left is in natural logs. For example, ln 9 = 8100 people per square mile, ln 8 = 3000, ln 7 = 1100, ln 6 = 400, ln 5 = 150.

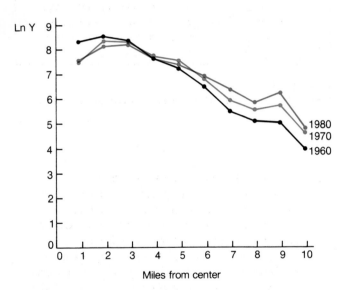

With one important variation, the population density pattern of the central city shows a comparable distance-decay arrangement (Figure 11.15). The exception is the tendency to form a hollow *at the center,* the CBD, which represents the inability of all but the most costly apartment houses to compete for space against alternative occupants desirous of supremely accessible parcels. Yet accessibility is attractive to a number of residential users and brings its penalty in high land prices. The result is the high-density residential occupance of parcels *near to the center* of the city—by those who are too poor to afford a long-distance journey to work; who are consigned by their poverty to the high-density, obsolescent slum tenements near the heart of the inner city; or who are self-selected occupants of the high-density, high-rent apartments made necessary by the price of land. Other urbanites, if financially able, may opt to trade off higher commuting costs for lower-priced land and may reside on larger parcels away from high-accessibility, high-congestion locations. Residential density declines with increasing distance from the city center as this option is exercised.

As a city grows in population, the peak densities no longer increase, and the pattern of population distribution becomes more uniform. Secondary centers begin to compete with the CBD for customers and industry, and the residential areas become less associated with the city center and more dependent on high-speed transportation arteries. Peak densities in the inner city decline, and peripheral areas increase in population concentration.

The validity of these generalizations may be seen on a time series graph of population density patterns for Des Moines, Iowa, over a 30-year period (Figure 11.16). The peak density was 2 miles from the CBD in 1960, but by 1980 it was at 3 miles. As the city expanded, density decreased close to the center, but beyond 3 miles from the center, population density increased.

Models of Urban Land-Use Structure

Generalized models of urban growth and land use patterns have been proposed to summarize the observable results of these organizing forces and controls. The starting point of all of them is the distinctive central business district that every older central city has. The **core** of this area is characterized by intensive land development: tall buildings, many stores and offices, and crowded streets. Just outside the core is the **frame** of the CBD;

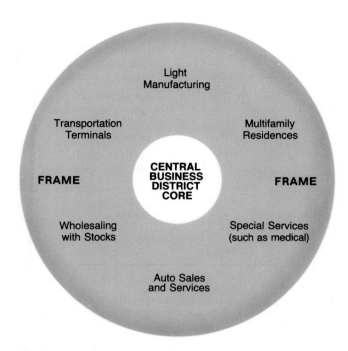

Figure 11.17

The core-frame concept. Commercial land is used intensively at the core of the central business district; the frame contains less intensive commercial uses. Some of the activities of the frame are related to each other and to the core.

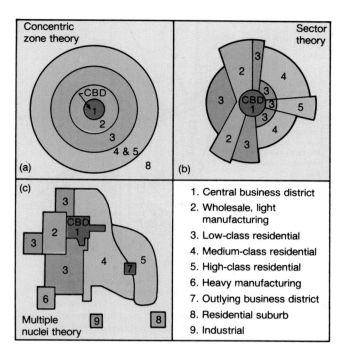

Figure 11.18

Three classic models of the internal structure of cities.

this is an area of wholesaling activities, transportation terminals, warehouses, new-car dealers, furniture stores, and even light industries (Figure 11.17). Just beyond the central business district is the beginning of residential land uses.

The land use models depicted on Figure 11.18 diverge in their summaries of patterns outside the CBD. The **concentric zone model** (Figure 11.18a) was developed to explain the observed sociological patterning of American cities. The model recognizes five concentric circles of mostly residential diversity at increasing distance in all directions from the core.

1. **The high-density CBD (core) with wholesaling, warehousing, light industry, and transport depots at its margins (frame).**
2. **A zone in transition marked by the deterioration of old residential structures abandoned, as the city expanded, by the former more wealthy occupants and now containing high-density, low-income slums, rooming houses, and perhaps ethnic ghettos.**
3. **A zone of "independent workingmen's homes" occupied by industrial workers, perhaps second-generation Americans able to afford modest, older homes on small lots.**

4. **A zone of better residences, single-family homes, or high-rent apartments occupied by those wealthy enough to exercise choice in housing location and to afford the longer, more costly journey to CBD employment.**
5. **A commuters' zone of low-density, isolated residential suburbs, just beginning to emerge when this model was proposed in the 1920s.**

The model is dynamic; it imagines the continuous expansion of inner zones at the expense of the next outer developed circles and suggests a ceaseless process of invasion, succession, and population segregation by income level.

The **sector model** (Figure 11.18b) also concerns itself with patterns of housing and wealth, but it arrives at the conclusion that high-rent residential areas dominate expansion patterns and grow outward from the center of the city along major arterials, new housing for the wealthy being added as an outward extension of existing high-rent axes as the city grows. Middle-income housing sectors lie adjacent to the high-rent areas, and low-income residents occupy the remaining sectors of growth. There tends to be a "filtering-down" process as older areas are abandoned by the outward movement of

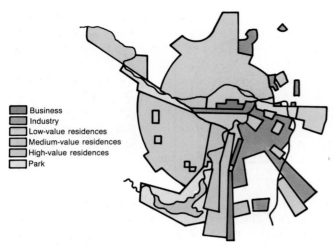

Figure 11.19

The land use pattern of Calgary, Alberta, in 1961. The circular arrangement of uses suggested by the concentric zone theory (Figure 11.18a) might result if a city developed on a flat surface. In reality, hills, rivers, railroads, and highways affect land uses in uneven ways. Physical and cultural barriers and the evolution of cities over time tend to result in a sectoral pattern of similar land uses. Calgary's central business district is the focus for many of the sectors.

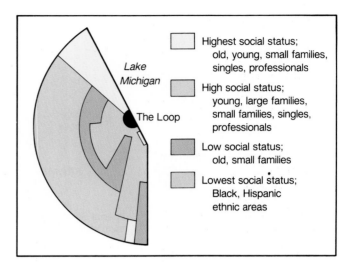

Figure 11.20

A diagrammatic representation of the major social areas of the Chicago region. The central business district is known as the "Loop."

their original inhabitants, with the lowest-income populations (closest to the center of the city and farthest from the current location of the wealthy) the dubious beneficiaries of the least desirable vacated areas. The accordance of the sector model with the actual pattern of Calgary, Canada, is suggested in Figure 11.19.

These "single-node" models of growth and patterning are countered by a **multiple-nuclei model** (Figure 11.18c), which postulates that large cities develop by peripheral spread not from one but from several nodes of growth. Individual nuclei of special function—commercial, industrial, port, residential—are originally developed in response to the benefits accruing from the spatial association of like activities. Peripheral expansion of the separate nuclei eventually leads to coalescence and the juxtaposition of incompatible land uses along the lines of juncture. The urban land use pattern, therefore, is not regularly structured from a single center in a sequence of circles or a series of sectors, but based on separate expanding clusters of contrasting activities.

Social Areas of Cities

Although too simplistic to be fully satisfactory, these classical models of American city structure receive some confirmation from modern interpretations of social segregation within urban areas. The more complex cities are economically and socially, the stronger is the tendency for city residents to segregate themselves into groups based on *social status, family status,* and *ethnicity.* In a large metropolitan region, this territorial behavior may be a defense against the unknown or the unwanted. Most people feel more secure when they are near those with whom they can easily identify. In traditional societies, these groups are the families and tribes. In modern society, people group according to income or occupation (social status); and language or race (ethnic characteristics). Many of these groupings are fostered by the size and the value of the available housing. Land developers, especially in cities, produce homes of similar quality in specific areas. Of course, as time elapses, there is a change in the quality of houses, and new groups may replace old groups. In any case, neighborhoods of similar social characteristics evolve.

Social Status

The social status of an individual or a family is determined by income, education, occupation, and home value. In the United States, high income, a college education, a professional or managerial position, and high home value constitute high status. High home value can mean an expensive rented apartment as well as a large house with extensive grounds.

A good housing indicator of social status is persons per room. A low number of persons per room tends to indicate high status. Low status characterizes people with low-income jobs living in low-value housing. There are many levels of status, and people tend to filter out into neighborhoods where most of the heads of households are of similar rank.

In most cities, people of similar social status are grouped in sectors whose points are in the innermost urban residential areas. The pattern in Chicago is illustrated in Figure 11.20. If the number of people within a

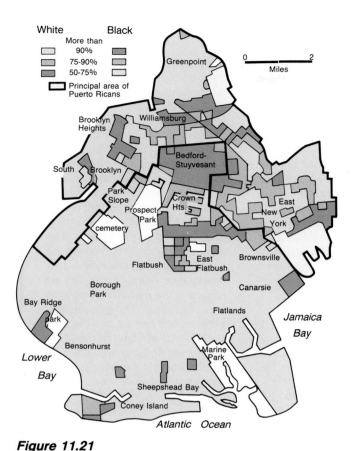

Figure 11.21

Black and Puerto Rican areas of Brooklyn, New York, 1970.
The main black area is in the Bedford-Stuyvesant district, not far
from Brooklyn's commercial center, which is just north of
Prospect Park. In most U.S. cities, the black ghetto is in the most
densely populated area close to the CBD.

given social group increases, they tend to move away from
the central city along an arterial connecting them with
the old neighborhood. Major transport routes leading to
the city center are the usual migration routes out from
the center. Social-status patterning agrees with sector
theory.

Family Status

As the distance from the origin of each sector increases,
the average age of the head of the household declines,
or the size of the family increases, or both. Within a particular sector—say, that of high status—older people
whose children do not live with them or young professionals without families tend to live close to the city
center. Between these are the older families who lived
at the outskirts of the city in an earlier period. The young
families seek space for child rearing, and older people

covet more the accessibility to the cultural and business
life of the city. Where inner-city life is unpleasant, there
is a tendency for older people to migrate to the suburbs
or to retirement communities.

Within the lower-status sectors, the same pattern
tends to emerge. Transients and single people are housed
in the inner city, and families, if they find it possible or
desirable, live farther from the center. The arrangement
that emerges is a concentric-circle patterning according
to family status.

Ethnicity

For some groups ethnicity is a more important residential locational determinant than social or family status.
Areas of homogeneous ethnic identification appear in the
social geography of cities as separate clusters or nuclei,
reminiscent of the multiple-nuclei concept of urban
structure. For some ethnic groups, cultural segregation
is both sought and vigorously defended, even in the face
of pressures for neighborhood change exerted by potential competitors for housing space. The durability of
"Little Italys" and "Chinatowns," and of Polish, Greek,
Armenian, and other ethnic neighborhoods in many
American cities is evidence of the persistence of self-
maintained segregation.

Certain ethnic or racial groups, especially blacks,
have had segregation in nuclear communities forced on
them. Every city has one or more black areas, which in
many respects may be considered cities within a city.
Figure 11.21 illustrates the concentration of blacks and
of Puerto Ricans in certain portions of Brooklyn, New
York. The barriers to movement outside the area have
always been high. Whites have consistently blocked
blacks from gaining the social status that would allow
them a greater choice in neighborhood selection. In most
American cities, the poorest residents are the blacks, who
are relegated to the lowest-quality housing in the least
desirable areas of the city. Similar restrictions have been
placed on Hispanics and other non-English-speaking
minorities.

Of the three patterns, family status has undergone
the most widespread change in recent years. Today, the
suburbs house large numbers of singles and childless
couples, and areas near the central business district have
become popular for young professionals (see "Gentrification"). Much of this is a result of changes in family
structure and the advent of large numbers of new jobs
for professionals in the suburbs and the central business
districts, but not in between.

Gentrification

As urban governments and state and federal authorities attempt to find ways to revitalize the ailing older American cities, a potentially significant process is occurring: **gentrification,** or the movement of middle-class people to deteriorated portions of the inner city. This movement does not yet counterbalance the exodus to the suburbs that has taken place since the end of World War II, but it marks an interesting reversal to what had seemed to be the inevitable decline of the cities. According to an estimate of the Urban Land Institute, 70% of all sizable American cities are experiencing a significant renewal of deteriorated areas. Gentrification is especially noticeable in the major cities of the North and the East, from Boston down the Atlantic Coast to Charleston, South Carolina, and Savannah, Georgia.

During the early years of suburbanization, there was also a major movement of low-income nonwhites into the central cities. That migration stream has now slowed to a trickle, and attention is centered on the movement of young, affluent, tax-paying professionals into the neighborhoods close to the city center. These formerly depressed areas are being rehabilitated and made attractive by the new residents. The prices of houses and apartments in former slums have soared. New, trendy restaurants and boutiques open daily.

The reasons for the upsurge of interest in urban housing reflect to some extent the recent changes in American family structure and in the employment structure of central business districts. Just a decade ago the suburbs, with their green spaces, were a powerful attraction for young married couples who considered single-family houses with ample yards ideal places to raise children. Now that the proportion of single people (whether never or formerly married) and childless couples in the American population has increased, the attraction of nearby jobs and social and recreational activities has become a more important residential location factor. Many of the new jobs in the central business district are designed for professionals in the fields of banking, insurance, and financial services. These jobs are replacing the manufacturing jobs of an early period. The gentrification process has been a positive force in the renewal of some of the depressed housing areas in neighborhoods surrounding the central business district.

Institutional Controls

Most governments have instituted innumerable laws to control all aspects of urban life, including the rules for using streets, the provision of sanitary services, and the use of land. In this section we touch only upon land use. Institutional or governmental controls have strongly influenced the land-use arrangements and growth patterns of most cities in the world. Cities have adopted land-use plans and enacted subdivision control regulations and zoning ordinances to realize those plans. They have adopted building, health, and safety codes to assure legally acceptable urban development and maintenance. All such controls are based on broad applications of the police powers of municipalities and their rights to assure public health, safety, and well-being even when private-property rights are infringed. These nonmarket controls on land use are designed to minimize incompatibilities (residences adjacent to heavy industry, for example), provide for the creation in appropriate locations of public uses (the transportation system, waste disposal, government buildings, parks) and private uses (colleges, shopping centers, housing) needed for and conducive to a balanced, orderly community. In theory, such careful planning should preclude the emergence of slums, so often the result of undesirable adjacent uses, and should stabilize neighborhoods by reducing market-induced pressures for land use change.

Such controls in the United States, particularly zoning ordinances and subdivision control regulations specifying acre or larger residential building lots and large house-floor areas, have been adopted as devices to exclude from upper-income areas lower-income populations or those who would choose to build or occupy other forms of residences: apartments, special housing for the aged, and so forth. Bitter court battles have been waged, with mixed results, over "exclusionary" zoning practices that in the view of some serve to separate rather than to unify the total urban structure and to maintain or increase diseconomies of land use development. In addition, it is well known that some real estate agents "steer" people of certain racial and ethnic groups into neighborhoods that the agent thinks are appropriate.

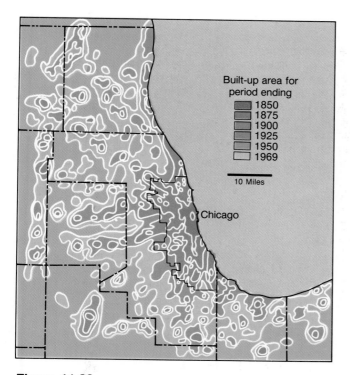

Built-up area for
period ending
■ 1850
■ 1875
■ 1900
■ 1925
■ 1950
□ 1969

10 Miles

Chicago

Figure 11.22

Urban sprawl. In Chicago, as in most large U.S. cities, the size
of the urbanized areas has increased dramatically during the last
40 years.

Suburbanization in the United States

The 20 years before World War II saw the creation of a
technological, physical, and institutional structure that
resulted, after that war, in a sudden and massive alteration of past urban forms. The improvement of the automobile increased its reliability, range, and convenience,
freeing its owner from dependence on fixed-route public
transit for access to home, work, or shopping. The new
transport flexibility opened up vast new acreages of nonurban land to urban development. The acceptance of a
maximum 40-hour work week guaranteed millions of
Americans the time for a commuting journey not possible when the workdays of 10 or more hours were
common. Finally, to stimulate the economy by the ripple
effect associated with a potentially prosperous construction industry, the Federal Housing Administration was
established as part of the New Deal programs under President Franklin D. Roosevelt. It guaranteed creditors the
security of their mortgage loans, thus reducing down-
payment requirements and lengthening mortgage repay-
ment periods. In addition, veterans of World War II were
granted generous terms on new housing.

Demands for housing, pent up by years of economic depression and wartime restrictions, were loosed
in a flood after 1945, and a massive suburbanization of
people and urban functions altered the existing pattern
of urban America. Between 1950 and 1970, the two most
prominent patterns of population growth were the metropolitanization of people and, within metropolitan
areas, their suburbanization. During the 1960s the interstate highway system was substantially completed, allowing sites 20 and 30 miles from workplaces to be within
commuting distance. Growth patterns for the Chicago
area are shown in Figure 11.22. The high energy prices
of the 1970s slowed the rush to the suburbs, but in the
1980s suburbanization proceeded apace, although the
tendency is as much for "filling in" as it is for continued
sprawl.

In the last 25 years, more and more industries of all
types (first manufacturers, then service industries) moved
to the suburbs. Their moves reflected the economies that
began to accrue from modern single-story facilities with
plenty of parking space for employees. No longer did industries need to locate near rail facilities; the freeways
presented new opportunities for lower-cost, more flexible truck transportation. Service industries took advantage of the large, well-educated labor force now living
in the suburbs. In addition to the land that was developed around freeway intersections in the 1960s, and along
major commercial routeways intersecting freeways in the
1970s, was the residential land that filled the interstitial
areas between the major routeways. Now the major shopping areas of the metropolitan areas, besides the central
business district, are the shopping centers at the freeway
intersections, the freeway frontage roads, and the major
connecting highways.

In time, in the United States, an established social
and functional pattern of suburban land use emerged,
giving evidence of a lower-density, more extensive rep-
etition of the models of land use developed to describe
the structure of the central city. Multiple nuclei of specialized land uses were created, expanded, and co-
alesced. Sectors of high-income residential use continued
their outward extension beyond the central-city limits,
usurping the most scenic and the most desirable suburban areas and segregating them by price and zoning
restrictions. Middle-, lower-middle-, and lower-income
groups found their own income-segregated portions of
the fringe; and ethnic minorities, frequently relegated to
the older industrial suburbs, found theirs.

With increasing suburban sprawl and the rising
costs implicit in the ever-greater spatial separation of the
functional segments of the fringe, the limits of feasible
expansion were reached, the supply of developable land

TABLE 11.2

Metropolitan Area Population Change, 1970–1980. Since 1970, the large, older eastern central cities have lost population; in some cases, their suburbs have also lost population. In the South and the West, however, the pattern is generally much like the 1950–1970 pattern: large increases are noted in the metropolitan areas, especially in the suburbs.

Metropolitan area	Percentage change in population, 1970 to 1980		Metropolitan area	Percentage change in population, 1970 to 1980	
	Central city	Suburbs		Central city	Suburbs
North and East			South and West		
New York	−10.4	−1.4	Los Angeles	+5.5	+7.2
Chicago	−10.8	+13.6	Houston	+29.2	+71.2
Philadelphia	−13.4	+5.4	Dallas	+7.1	+47.9
Detroit	−20.5	+7.8	San Diego	+25.5	+49.4
Baltimore	−13.1	+19.1	San Antonio	+20.1	+22.4
Indianapolis	−4.9	+24.5	Phoenix	+30.9	+92.1
Washington	−15.7	+12.5	San Francisco	−5.1	+10.0
Milwaukee	−11.3	+10.8	Memphis	+3.6	+26.9
Cleveland	−23.6	+0.9	San Jose	+38.4	+8.8
Columbus	+4.6	+10.6	New Orleans	−6.1	+38.9
Boston	−12.2	−2.6	Jacksonville	+7.3	+67.2

Source of data: U.S. Bureau of the Census.

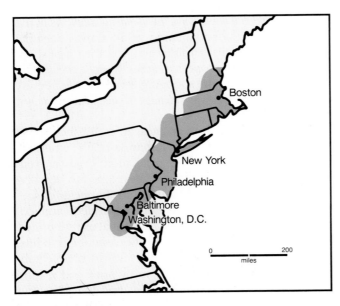

Figure 11.23

The northeastern U.S. Boston-to-Washington urban region. See also Figure 12.18.

was reduced (with corresponding increases in its price), and the intensity of land development grew. Changing life-styles and cost constraints have resulted in a proliferation of suburban apartment complexes and the disappearance of open land. The maturation and the coalescence of urban land uses have resulted in the emergence of coherent metropolitan-area cities that are suburban in traditional name only, containing the business districts and the mix of land uses that the designation *city* implies.

In recent years, suburban areas have expanded to the point where metropolitan areas are coalescing. Within these suburban areas new major centers of growth are appearing. Office buildings and huge shopping centers are being developed in such places as Oak Brook, Illinois (in the western Chicago suburbs), Meadowlands, New Jersey (west of New York), King of Prussia, Pennsylvania (northwest of Philadelphia), and Costa Mesa, California (south of Los Angeles). In fact, in the 1980s more office space has been created in the suburbs than in the central cities of America. The Boston-to-Washington corridor is now a continuous, functionally urban region (a megalopolis) with many new centers that compete with the business districts of Boston, Providence, New York, Philadelphia, Baltimore, and Washington (see Table 11.2). One summary of the new pattern is shown in Figure 11.23. A further analysis of this urban corridor is offered in "Megalopolis" (Page 423).

The Self-Sufficient Suburbs

The suburbanization of the American population and of the commerce and industry dependent on its purchasing power and labor has led to more than the simple physical separation of the suburbanite from the central city. It has created a metropolitan area, many of whose inhabitants have no connection with the core city, feel no ties to it, and find satisfaction of all their needs within the peripheral zone.

A New York Times "suburban poll," conducted in the summer of 1978, revealed a surprising lack of interest in New York City on the part of those usually assumed to be intimately involved in the economic, cultural, and social life of the "functional city," or the metropolitan area dominated by New York. The poll discovered that in 80% of suburban households, the principal wage earner did not work in New York City. The concept of the commuter zone obviously no longer has validity if *commuting* implies, as it formerly did, a daily journey to work in the central city.

Just as employment ties have been broken, now that the majority of suburbanites work in or near their outlying areas of residence, the cultural and service ties that were traditionally thought to bind the metropolitan area into a single unit dominated by the core city have also been severed. The *Times* survey found that one-half of the suburbanites polled made fewer than five nonbusiness visits to New York City each year; one-quarter said that they never went there. Even the presumed concentration in the city of high-order goods and services—those with the largest service areas—did not constitute an attraction for most suburbanites. Only 20% of the respondents thought that they would journey to New York City to see a medical specialist, and only 7% or 8% to see a lawyer or an accountant or to make a major purchase. Even needs at the upper reaches of the central-place hierarchy were satisfied within the fringe.

The suburbs, as the *Times* survey documented, have outgrown their former role as bedroom communities and have emerged as a chain of independent, multinucleated urban developments. Together, they are largely self-sufficient and divorced from the central city and have little feeling of subordination to or dependence on that city.

Central City Change

While the process of suburban spread continued, many central cities of the United States, especially in the Northeast and the Midwest, lost population. In the 1970s, there was a decided shift of population to what are called the "Sunbelt" cities. This trend was partly a result of the lower living costs (especially lower fuel costs) in the South and the West, but it is chiefly a function of manufacturers' seeking nonunionized workers or new plants, unencumbered by high social and transportation costs. Perhaps as important as that trend was the movement of people out of metropolitan areas to nearby nonmetropolitan districts. In a sense, this latter trend represents a further sprawl of the metropolitan districts.

The structure, the economic base, and the financial stability of the central city have been grievously damaged by the process of suburbanization. In earlier periods of growth, as new settlement areas or functional nodes developed beyond the political margins of the city, annexation absorbed new growth within the corporate boundaries of the expanding older city. The additional tax base and employment centers became part of the municipal whole. But in the states that recognized the right of the separate incorporation of the new growth areas—particularly in the eastern part of the United States—the ability of the city to continue to expand was restricted. Lower suburban residential densities made septic tanks and private wells adequate substitutes for expensive city-sponsored sewers and water mains; lower structural densities and lower crime rates meant less felt need for high-efficiency city fire and police protection. Where possible, suburbanites opted for a separation from the central city and for aloofness from the costs, the deterioration, and the adversities associated with it. Their homes, jobs, shopping, schools, and recreation all existed outside the confines of the city from which they had divorced themselves, and the close-knit fabric of urban life was rent.

The redistribution of population caused by suburbanization resulted not only in the spatial but also in the political segregation of the social groups of the metropolitan area. The upwardly mobile resident of the city—younger, wealthier, and better educated—took advantage of the automobile and the freeway to leave the central city. The poorer, older, least advantaged urbanites

TABLE 11.3
Race and Income for Selected Metropolitan Areas, 1980. Central cities and their suburbs display significant differences in racial composition and household income.

Metropolitan area	Racial composition (%)			Median annual household income	Metropolitan area	Racial composition (%)			Median annual household income
	White	Black	Hispanic[a]			White	Black	Hispanic[a]	
North and East					*South and West*				
New York	60.7	25.2	19.9	$13,855	Los Angeles	61.2	17.0	27.5	15,746
Suburbs	89.0	7.6	0.4	23,740	Suburbs	71.9	9.6	28.9	19,410
Chicago	49.6	39.8	14.0	15,301	Houston	61.3	27.6	17.6	18,474
Suburbs	90.8	5.6	3.9	24,811	Suburbs	86.3	6.7	10.9	24,847
Philadelphia	58.2	37.8	3.7	13,169	Dallas	61.4	29.4	12.3	16,227
Suburbs	89.8	8.1	1.7	20,904	Suburbs	92.2	3.9	5.3	21,301
Detroit	34.3	63.0	0.2	13,981	San Diego	75.4	8.9	14.8	16,409
Suburbs	94.1	4.2	0.1	24,038	Suburbs	85.9	2.6	14.6	17,808
Baltimore	43.9	54.8	1.0	12,811	San Antonio	78.6	7.3	53.6	13,775
Suburbs	89.2	9.0	1.0	22,272	Suburbs	88.1	5.2	23.4	19,452
Indianapolis	77.1	21.8	0.9	17,279	Phoenix	84.0	4.9	15.1	17,419
Suburbs	98.3	1.0	0.5	26,854	Suburbs	89.4	1.4	11.2	18,085
Washington	26.8	70.3	2.7	16,211	San Francisco	58.3	12.7	12.2	15,867
Suburbs	78.5	16.7	3.1	25,728	Suburbs	81.1	6.5	10.6	22,375
Milwaukee	73.3	23.1	4.1	16,061	Memphis	51.6	47.6	—	14,040
Suburbs	98.4	0.5	1.0	23,805	Suburbs	77.8	21.0	1.0	18,753
Cleveland	53.6	43.8	2.9	12,277	San Jose	73.9	4.6	22.0	22,886
Suburbs	91.8	7.1	0.6	22,024	Suburbs	83.1	2.1	13.0	23,832
Columbus	76.2	22.1	0.8	14,834	New Orleans	46.2	55.3	—	11,834
Suburbs	97.4	1.9	0.5	20,604	Suburbs	85.4	12.6	—	19,678
Boston	69.9	22.4	6.4	12,500	Jacksonville	73.0	25.4	—	14,426
Suburbs	96.6	1.6	1.5	20,469	Suburbs	88.3	10.7	—	16,084

Source: U.S. Bureau of the Census.
[a]As used by the Bureau of the Census, "Hispanic" race has no genetic connotation.

were left behind. As indicated by Table 11.3, the central cities and the suburbs are becoming increasingly differentiated. Large areas within the cities now contain only the poor and minority groups, a population little able to pay the rising costs of the social services that their numbers, neighborhoods, and condition require.

The services needed to support the poor include welfare payments, social workers, extra police and fire protection, health delivery systems, and subsidized housing. Central cities by themselves are unable to raise the taxes needed to support such an array and intensity of social services now that they have lost the tax bases represented by suburbanized commerce, industry, and

upper-income residential uses. Lost, too, are the job opportunities formerly a part of the central-city structure. Increasingly, the poor and the minorities are trapped in a central city without the possibility of nearby employment and isolated by distance, immobility, and unfamiliarity from the few remaining low-skill jobs, which are now largely in the suburbs.

The abandonment of the central city by people and functions has nearly destroyed the traditional active, open auction of urban land, which led to the replacement of obsolescent uses and inefficient structures in a continuing process of urban modernization. In the vacuum left by the departure of private investors, the federal government, particularly since the landmark Housing Act of 1949, has initiated urban renewal programs with or

Figure 11.24
Some elaborate governmental projects for urban redevelopment, slum clearance, and rehousing of the poor have been failures. The demolition of the Wayne Miner public housing project in Kansas City, Missouri in 1987 was a recognition that public high-rise developments do not always meet the housing and social needs of their inhabitants.

without provisions for a partnership with private housing and redevelopment investment. Under a wide array of programs instituted and funded since the late 1940s, slum areas have been cleared; public housing has been built; cleared land has been conveyed at subsidized cost to private developers for middle-income housing construction; cultural complexes and industrial parks have been created; and city centers have been reconstructed. With the continuing erosion of the urban economic base and the disadvantageous restructuring of the central-city population base, the hard-fought governmental battle to maintain or revive the central city is frequently judged to be a losing one (Figure 11.24).

There are two trends however, that do give a new flavor to the central city. One is gentrification (see Page 386) and the other is the vitality of the CBD. New office buildings are springing up in most large central cities. These represent a shift in employment emphasis from manufacturing to more service-oriented activities such as headquarters for industries, financial services, and business services in general. The number of jobs in the central city has not increased, but the change favors the professionals over blue-collar workers. Many of the workers in these office buildings now live within the gentrified neighborhoods of the central city.

The Making of World Metropolitan Regions

Urban settlements have been increasing in size and number for the last 200 to 300 years. Until the end of World War II, one could easily distinguish a city by its distinctive central business district, the high-density housing close to the CBD, and the lower densities that ended fairly abruptly at the outer ends of the public transport system. Now that about 75% or more of the people of Western Europe, Japan, Australia, and North America live in urban settlements (about 600 to 700 million people), it would seem to follow that many cities would coalesce with one another. In fact, this has been the case, and the urban region can no longer be described in simple terms. We now must talk of multiple

TABLE 11.4

Largest Urban Agglomerations: 1980 and Expected in 2000

Urban region	Country	Population in millions 1980	2000
1. Mexico City	Mexico	14	32
2. Tokyo–Yokohama	Japan	20	26
3. São Paulo	Brazil	13	26
4. New York–NE New Jersey	U.S.A.	18	22
5. Calcutta	India	10	20
6. Rio de Janiero	Brazil	10	19
7. Shanghai	China	12	19
8. Bombay	India	9	19
9. Beijing	China	10	19
10. Seoul	Korea	9	19
11. Jakarta	Indonesia	7	17
12. Cairo	Egypt	8	16
13. Karachi	Pakistan	6	16
14. Los Angeles–Long Beach	U.S.A.	11	15
15. Buenos Aires	Argentina	10	14
16. Tehran	Iran	6	14
17. New Delhi–(old) Delhi	India	6	13
18. London	United Kingdom	11	13
19. Manila	Philippines	6	13
20. Osaka–Kobe	Japan	10	13
21. Paris	France	10	12
22. Lima	Peru	5	12
23. Rhine–Ruhr	West Germany	10	11
24. Bangkok	Thailand	4	11
25. Baghdad	Iraq	5	11
26. Moscow	USSR	8	11
27. Madras	India	5	10
28. Bogotá	Colombia	4	10

Source: From Table 6 in *Global Review of Human Settlements, Statistical Annex,* United Nations: Department of Economic and Social Affairs, 1986, pp. 77–87.

centers and greatly varying population densities not necessarily associated directly with business districts. Table 11.4 provides a list of the current and predicted largest metropolitan areas in the world.

The coalescence of many cities into great metropolitan (megalopolitan) masses is more notable than ever before. The major **megalopolis** of the United States is the continuous urban string that stretches from north of Boston (southern New Hampshire) to south of Washington, D.C. (northern Virginia) (see "The Self-Sufficient Suburbs," Page 389). Other megapolises include the southern Lake Michigan regions stretching from north of Milwaukee through Chicago and across northwestern Indiana and southwest Michigan; and Los Angeles, including the myriad suburbs and cities extending from Santa Barbara to San Diego and even including the 1 million people living in Tijuana in Mexico. Outside the United States, megapolises include Toronto and the many industrial cities in southern Ontario; the north of England, including Liverpool, Manchester, Leeds, Sheffield, and the several dozen industrial towns in between; the Ruhr of West Germany, with a dozen large connected cities like Essen and Düsseldorf; and the Tokyo region of Japan.

Population Pressure in Cairo, Egypt

C airo is a vast, sprawling metropolis—hot, crowded, choked by automobiles, bothered by noxious fumes from traffic and by sand that blows from the Sahara. The city is representative of many of the principal cities of Third World countries where population growth far outstrips economic development. A steady stream of migrants arrives in Cairo daily because it is the place where opportunities are perceived to be available, where people think that life will be better and brighter than in the crowded countryside. Cairo is the symbol of modern Egypt, a place where young people are willing to undergo deprivation for the chance to "make it." In addition, the population of the country was never more than 12 million people during its 7000-year history, but now better health care has led to a big drop in infant mortality, and the result is a country of 51 million people and an expected year 2000 population of 70 or 80 million.

In the last 30 years, the population of Cairo has increased sixfold. Three decades ago, the estimated population was 2.3 million; now some 14 million reside in the greater metropolitan area. Real opportunities continue to be low relative to the size of the metropolis. The poor, of which there are millions, crowd into warrens of slums, often without water or sewage systems. Yet the city continues to grow onto valued farmland, thus decreasing the food available for the country's 51 million people.

The sewer and water systems, modernized in the 1930s, were intended to serve only 2 million people. Now it is not uncommon for burst sewer mains to dump raw sewage into the city streets. When such conditions prevail in the hot summer, the chances of the spread of disease increase many times over. It is also not unusual to find people living on little boats on the river, under the arches of bridges, and in makeshift, poorly constructed buildings that, on occasion, cave in.

One's first impression when arriving in central Cairo is one of opulence and prosperity, a stark contrast to what lies outside the center. High-rise apartments, regional headquarters buildings of multinational corporations, and modern hotels stand amid clogged streets. This part of the city is clearly Western-oriented, but the fancy restaurants, plush apartments, and high-priced cars stand only a short distance from the horrible slums that are home to masses of underemployed people.

The story of Cairo in Egypt can be repeated for city after city in the Third World. Bombay and Calcutta in India, São Paulo in Brazil, Kinshasa in Zaire, Lagos in Nigeria, Bangkok in Thailand, Jakarta in Indonesia, Karachi in Pakistan, and Mexico City are already housing millions more than their service structures are designed to accommodate, and one can foresee only worsening conditions in the future.

Figure 11.25
Open sewer in a slum in Nigeria.

Figure 11.26
Modern office buildings in Lagos, Nigeria.

Were it not for the mountains between Tokyo and Nagoya, and between Nagoya and Osaka–Kobe–Kyoto, there is no doubt that these huge metropolises would by now be one megalopolitan area.

Large metropolises are being created in developing countries as well, and wherever the automobile or the modern transport system is an integral part of the growth, the form of the metropolis takes on Western characteristics. On the other hand, in places like Bombay (India), Lagos (Nigeria), Jakarta (Indonesia), Kinshasa (Zaire), and Cairo (Egypt), where modern roads have not yet made a significant impact and the public transport system is limited, the result has been overcrowded cities centered on a single major business district in the old tradition. In such societies, the impact of urbanization and the responses to it differ from the patterns and problems observable in the cities of the United States. The situation in Cairo (see "Population Pressure in Cairo, Egypt," Page 393) is indicative of the reasons for and the results of urbanization in Third World cities.

The developing countries, emerging from formerly dominantly subsistence economies, have experienced disproportionate population concentrations, particularly in their national and regional capitals. Lacking or relatively undeveloped is the substructure of maturing, functionally complex smaller and medium-sized centers characteristic of more advanced and diversified economies. The term used to describe the huge, one-to-a-country (or to a large region, as is the case in India and Brazil) metropolises, is **primate city.** More specifically, the primate city is one that has a population much greater than twice the population of the second largest city. Greater Cairo contains 30% of Egypt's total population; 34% of all Panamanians live in Panama City; and Baghdad contains 24% of the Iraqi populace. Vast numbers of surplus, low-income rural populations have been attracted to these developed seats of wealth and political centrality in the hope of finding a job (Figure 11.25). Although attention may be lavished on creating urban cores on the skyscraper model of Western cities (Figure 11.26), most of the new urban multitudes have little choice but to pack themselves into squatter shanty communities on the fringes of the city, isolated from the sanitary facilities, the public utilities, and the job opportunities that are found only at the center. Such impoverished squatter districts are found around most major cities in Africa, Asia, and Latin America.

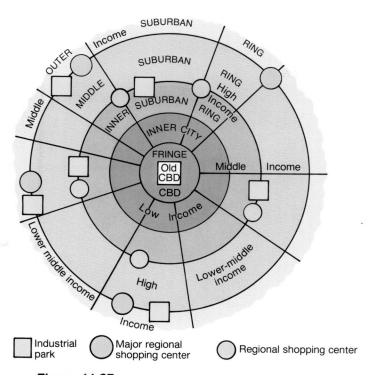

Industrial park

Major regional shopping center

Regional shopping center

Figure 11.27

A diagrammatic representation of the present-day large U.S. city. Note that aspects of the concentric-zone, sector, and multiple-nuclei patterns are evident.

Contrasts among Metropolitan Areas

Figures 11.27, 11.28, and 11.29 show typical land use patterns for three large cities in different parts of the world. The first is a diagrammatic representation of a typical United States city, the second a European city and the third an Asian city. Each reflects the way recent developments have altered the traditional city. In the case of the American city, already discussed, the effect of freeways and the use of the automobile is clearly evident in the development of new centers of commercial activity in suburban districts.

For the European city, the effect of institutional controls to contain cities within specified limits has had the effect of building up housing densities to heights not common in the United States. Single-family homes are more the exception than the rule. Usually a *green belt* rings the city as a place for relaxation and a way to protect rural areas and the environment. The historic core is now mainly a tourist and shopping district usually contained within the walls of the old city. High-income people are the only residents. Renewal mainly takes place in the area

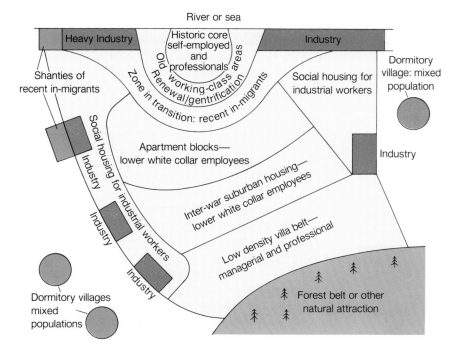

Figure 11.28

A diagrammatic representation of the Western European city.

Figure 11.29
A diagrammatic representation of a colonial-based South Asian city.

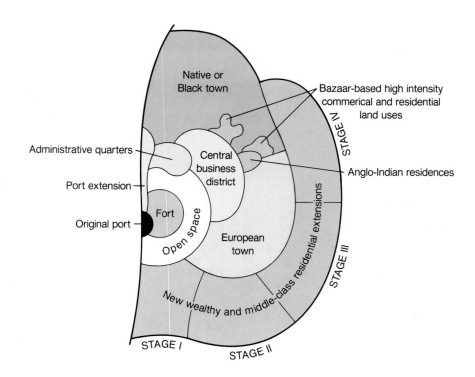

outside of the old city where the upwardly mobile professionals are replacing the working class. Many housing districts, especially near industry, contain public housing (usually apartments) built for the industrial workers.

Many of the large cities of Asia were founded and developed by European colonialists. For example, the British built Calcutta and Bombay in India, the French developed Ho Chi Minh City (Saigon) in Vietnam, the Dutch had as their main outpost Jakarta in Indonesia, and many colonial countries established Shanghai. These and many other cities in Asia, and Africa as well, have certain important similarities. They stand in marked contrast to the North American and European cities in that large areas on the periphery contain slum housing. Since the transportation systems are not well developed, high-income residents usually live close to the city center, where the amenities of the city are easily available. Colonists usually built a fort, left an open area around the fort, and then developed administrative and commercial services outside of the open area. A European town and a native town developed side by side. In Africa, the native section is typically subdivided into ethnic areas. In recent years, the well-to-do have been spreading to outer areas away from the slums, usually where new roads have been built.

The socialist city is somewhat different from the others in form. The typical city is completely designed by planners who have attempted to develop livable, efficient land-use arrangements. Cities of the USSR and to a certain extent, Eastern Europe, are characterized by huge blocks of apartments that are as self-contained as possible. Shops, services, and some green space are available in each of the "super-blocks," which are ringed by boulevards. The center of the city is reserved for museums and public buildings, and is used for parades and other civic activities.

Conclusion

The city is the essential functional node in the chain of linkages and interdependencies that marks any society or economy advanced beyond the level of subsistence. The more complex and functionally integrated the society, the greater the degree of specialization and exchange that it demands for its maintenance, and the more urbanized it becomes. Although cities are among the oldest marks of civilization, only in this century have they become the home of the majority of the people in industrialized countries and the inhospitable place of refuge for uncounted millions in modernizing subsistence economies.

In all their incarnations, ancient and modern, recurring themes and regularities are evident within the cities. All perform functions—have an economic base—generating the income necessary to support themselves and to sustain their contained population. None exists in a vacuum; each is part of a larger urban and nonurban society and economy with which it has essential reciprocal connections. Each has a more or less orderly internal arrangement of land uses, social groups, and economic functions partially planned and controlled, and partially determined by individual decisions, competition for sites, and operational efficiencies. All, large or small, ancient or modern, have or have had maladjustments of land use, recurring and serious social problems and conflicts, environmental inadequacies, and concerns of administration and orderly development. Yet cities remain, though flawed, the capstone of our cultures, the organizing focuses of modern societies and economies.

Key Words

basic sector *378*
central business district (CBD) *380*
central city *372*
central place theory *376*
city *372*
concentric zone model *383*

core *382*
economic base *378*
frame *382*
gentrification *386*
megalopolis *392*
metropolitan area *372*
multiple-nuclei model *384*

multiplier effect *379*
nonbasic sector *378*
primate city *394*
sector model *383*
service sector *378*
site *374*
situation *374*
suburb *372*

threshold *377*
town *372*
urban hierarchy *376*
urbanized area *372*

For Review

1. Consider the city or town in which you live or attend school or with which you are most familiar. In a brief paragraph, discuss that community's site and situation. Point out the connection, if any, between its site and situation and the basic functions that it earlier performed or now performs.

2. Describe the multiplier effect as it relates to the population growth of urban units.

3. What area does a central place serve, and what kinds of functions does it perform? If an urban system were composed solely of central places, what summary statements could you make about the spatial distribution and the urban size hierarchy of that system?

4. Is there a hierarchy of retailing activities in the community with which you are most familiar? Of how many and of what kinds of levels is that hierarchy composed? What localizing forces affect the distributional pattern of retailing within that community?

5. Briefly describe the urban land use patterns predicted by the concentric-circle, the sector, and the multiple-nuclei models of urban development. Which one, if any, best corresponds to the growth and land use pattern of the community most familiar to you? How well does Figure 11.27 depict the land use patterns in the metropolitan area with which you are most familiar?

6. In what ways do social status, family status, and ethnicity affect the residential choices of households? What expected distributional patterns of urban social areas are associated with each? Does the social geography of your community conform to the predicted pattern?

7. How has suburbanization damaged the economic base and the financial stability of the central city?

8. Why are metropolitan areas in developing countries expected to grow larger than many Western metropolises by the year 2000? What do you expect the population density profile for Mexico City to look like in the year 2000?

9. Why are Third World primate cities overburdened? What can be done to alleviate the difficulties?

Suggested Readings

Association of American Geographers Comparative Metropolitan Analysis Project. *A Comparative Atlas of America's Great Cities: Twenty Metropolitan Regions,* vol. 3. Minneapolis: University of Minnesota Press, 1976.

Brunn, Stanley D., and Jack F. Williams. *Cities of the World: World Regional Development.* New York: Harper & Row, 1983.

Cadwallader, Martin. *Analytical Urban Geography.* Englewood Cliffs, N.J.: Prentice-Hall, 1985.

Christian, Charles M., and Robert A. Harper. *Modern Metropolitan Systems.* Columbus, Ohio: Charles E. Merrill, 1982.

Johnston, R. J. *The American Urban System: A Geographical Perspective.* New York: St. Martin's Press, 1982.

Muller, Peter O. *Contemporary Suburban America.* Englewood Cliffs, N.J.: Prentice-Hall, 1981.

Yeates, Maurice, and Barry Garner. *The North American City,* 3d ed. New York: Harper & Row, 1980.

Part Four

The Area Analysis Tradition

Julius Caesar began his account of his transalpine campaigns by observing that all of Gaul was divided into three parts. With that spatial summary, he gave to every schoolchild an example of geography in action.

Caesar's report to the Romans demanded that he convey to an uninformed audience a workable mental picture of place. He was able to achieve that aim by aggregating spatial data, by selecting and emphasizing what was important to his purpose, and by submerging or ignoring what was not. He was pursuing the geographic tradition of area analysis, a tradition that is at the heart of the discipline and focuses on the recognition of spatial uniformities and the elucidation of their significance.

The tradition of area analysis is commonly associated with the term *regional geography*, the study of particular portions of the earth's surface. As did Caesar, the regional geographer attempts to view a particular area and to summarize what is spatially significant about it. One cannot, of course, possibly know everything about a region, nor would "everything" contribute to our understanding of its essential nature. Regional geographers, however, approach a preselected earth space—a continent, a nation, or some other division—with the intent of making it as fully understood as possible in as many facets of its nature as they deem practicable. It is from this school of area analysis that "regional geographies" of, for example, Africa, the United States, or the Pacific Northwest emanate.

Because the scope of their inquiries is so broad, regional geographers must become thoroughly versed in all of the topical subfields of the discipline, such as those discussed in this book. Only in that way can area specialists select those phenomena that give insight into the essential unity and diversity of their region of study. In their approach, and based upon their topical knowledge, regional geographers frequently seek to delimit and study regions defined by one or a limited number of criteria.

The three topical traditions already discussed have often misleadingly been contrasted to the tradition of area analysis. In actuality, practitioners of both topical and regional geography inevitably work within the tradition of area analysis. Both seek an organized view of earth space. The one asks what are the regional units that evolve from the consideration of a particular set of preselected phenomena; the other asks how, in the study of a region, its varied content may best be clarified.

Chapter 12 is devoted to the area-analysis tradition. Its introductory pages explore the nature of regions and methods of regional analysis. The body of the chapter consists of a series of regional vignettes, each based upon a theme introduced in one of the preceding chapters of the book. Since those chapters were topically organized, the separate studies that follow demonstrate the tradition of area analysis in the context of topical (sometimes called "systematic") geography. They are examples of the ultimate regionalizing objectives of geographers who seek an understanding of their data's spatial expression. The step from these limited topical studies to the broader, composite understandings sought by the area analyst is both short and intellectually satisfying.

12 The Regional Concept

The Nature of Regions
The Structure of This Chapter
Part I. Regions in the Earth-Science Tradition
   Landforms as Regions
   Dynamic Regions in Weather and Climate
   Ecosystems as Regions
Part II. Regions in the Culture-Environment
Tradition
   Population as Regional Focus
   Language as Region
   Political Regions
   Mental Regions
Part III. Regions in the Locational Tradition
   Economic Regions
   Regions of Natural Resources
   Urban Regions
Conclusion
Key Words
For Review
Suggested Readings

The questions geographers ask, we saw in the Introduction to this book, ultimately focus on matters of location and character of place. We asked how things are distributed over the surface of the earth; how physical and cultural features of areas are alike or different from place to place; how the varying content of different places came about; and what all these differences and similarities mean for people.

The Nature of Regions

In the earlier chapters of this book we examined some of the physical and cultural content of areas and some spatially important aspects of human behavior. We looked at the physical earth processes that lead to differences from place to place in the environment. We studied ways in which humans organize their actions in earth space—through political institutions, by the economic systems and practices they develop, and by the cultural and social processes influencing their spatial behavior and interaction. Population and settlement patterns and areal differences in human use and misuse of earth resources all were considered as part of the mission of geography.

For each of the topics we studied—from landforms to cities—we found spatial regularities. We discovered that things are not irrationally distributed over the surface of the earth, but reflect an underlying spatial order based upon understandable physical and cultural processes. We found, in short, that although no two places are *exactly* the same, it is possible and useful to recognize segments of the total world that are internally similar in some important characteristic and distinct in that feature from surrounding areas.

These regions of significant uniformity of content are the geographer's equivalent of the historian's "eras" or "ages": brief summary names for specific areas that are different in some important way from adjacent or distant territories. The **region,** then, is a device of areal generalization. It is an attempt to separate into recognizable component parts the otherwise overwhelming diversity and complexity of the earth's surface.

All of us have a general idea of the meaning of region, and all of us refer to regions in everyday speech and action. We visit "the old neighborhood" or "go downtown"; we plan to vacation or retire in the "Sunbelt," or we speculate on the effects of weather conditions in the "Northeast" or the "corn belt" on grain surpluses or next year's food prices. In each instance we have mental images of the areas mentioned. Those images are based upon place characteristics and areal generalizations that seem useful to us and recognizable to

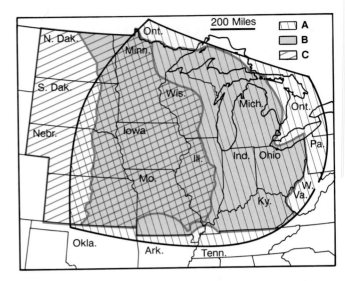

Figure 12.1

The Middle West as seen by different professional geographers. Agreement upon the need to recognize spatial order and to define regional units does not imply unanimity in the selection of boundary criteria. Each of the sources concurs in the significance of the Middle West as a regional entity in the spatial structure of the United States and agrees upon its core area. These sources differ, however, in their assessments of its limiting characteristics.

(a) from John H. Garland, ed., *The North American Midwest*, New York: John Wiley, 1955; (b) from John R. Borchert and Jane McGuigan, *Geography of the New World*, Chicago: Rand McNally, 1961; (c) from Otis P. Starkey and J. Lewis Robinson, *The Anglo-American Realm*, New York: McGraw-Hill, 1969.

our listeners. We have, in short, engaged in an informal place classification to pass along quite complex spatial, organizational, or content ideas. We have applied the **regional concept** to bring order to the immense diversity of the earth's surface.

What we do informally as individuals, geography attempts to do formally as a discipline—define and explain regions (Figure 12.1). The purpose is clear: to make the infinitely varying world around us understandable through spatial summaries. That world is only rarely subdivided into neat, unmistakable "packages" of uniformity. Neither the environment nor human areal actions present us with a compartmentalized order, any more than the sweep of human history has predetermined "eras," or all plant specimens come labeled in nature with species names. We all must classify to understand, and the geographer classifies in regional terms.

Regions are spatial expressions of ideas or summaries useful to the analysis of the problem at hand. Although as many possible regions exist as there are physical, cultural, or organizational attributes of area, the geographer selects for study those areal variables that contribute to the understanding of a specific topic or areal problem. All other variables are disregarded as irrelevant. In our reference to the "corn belt" we delimit a

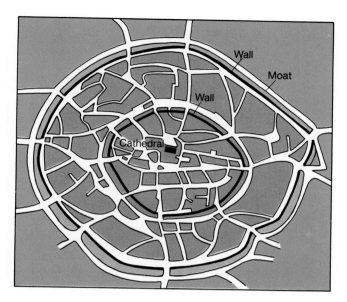

Figure 12.2

Aachen in 1649. The acceptance of regional extent implies the recognition of regional boundaries. At some defined point, *urban* is replaced by *nonurban,* the Middle West ends and the Plains begin, or the rain forest ceases and the savanna emerges. Regional boundaries are, of course, rarely as precisely and visibly marked as were the limits of the walled medieval city. Its sprawling modern counterpart may be more difficult to define, but the boundary significance of the concept of *urban* remains.

portion of the United States showing common characteristics of farming economy and marketing practices. We dismiss—at that level of generalization—differences within the region based upon slope, soil type, state borders, or characteristics of population. The boundaries of the corn belt are assumed to be marked where the region's internal unifying characteristics change so materially that different agricultural economies become dominant and different regional summaries are required. It is the content of the region that suggests its definition and determines the basis of its delimitation.

Although regions may vary greatly, they all share certain common characteristics related to earth space:

1. Regions have *location,* often expressed in the regional name selected, such as the Middle West, the Near East, North Africa, and the like. This form of regional name underscores the importance of *relative location* (see Introduction, p. 3).
2. Regions have *spatial extent.* They are recognitions of the fact that there are differences from place to place in the physical and cultural attributes of area, and that those differences are important.

3. Regions have *boundaries* based upon the areal spread of the features selected for study. Since regions are the recognition of the features defining them, their boundaries are drawn where those features no longer occur or dominate (Figure 12.2).
4. Regions, as we saw in the Introduction (pp. 14–15), may be either *formal* or *functional.*

Formal regions are areas of essential uniformity throughout in one or a limited combination of physical or cultural features. In previous chapters we encountered such formal physical regions as the humid subtropical climate zone and the Sahel region of Africa, as well as formal (homogeneous) cultural regions in which standardized characteristics of language, religion, ethnicity, or livelihood existed. The endpaper maps of topographical regions and national units show other formal regional patterns. Whatever the basis of its definition, the formal region is the largest area over which a valid generalization of attribute uniformity may be made. Whatever is stated about one part of it holds true for its remainder.

The **functional region,** in contrast, is a spatial system defined by the interactions and connections that give it a dynamic, organizational basis. Its boundaries remain constant only as long as the interchanges establishing it remain unaltered. The commuting region shown in Figure I.9 retained its shape and size only as long as the road pattern and residential neighborhoods on which it was based remained as shown.

5. Regions are *hierarchically arranged.* Although regions vary in scale, type, and degree of generalization, none stands alone as the ultimate key to areal understanding. Each defines only a part of spatial reality. On a formal regional scale of size progression, the Delmarva Peninsula of the eastern United States may be seen as part of the Atlantic Coastal Plain, which is in turn a portion of the eastern North American humid continental climatic region, a hierarchy that changes the basis of regional recognition as the level and purpose of generalization alter (Figure 12.3). The central business district of Chicago is one land-use complex in the functional regional hierarchy that describes the spatial influences of the city of Chicago and of the metropolitan region of which it is the core. Each recognized regional entity in such progressions may stand alone and, at the same time, exist as a part of a larger, equally valid, territorial unit.

These generalizations about the nature of regions and the regional concept are meant to instill firmly the understanding that regions are human intellectual creations designed to serve a purpose. Regions focus our attention upon spatial uniformities; they bring clarity to the seeming confusion of the observable physical and cultural features of the world we inhabit. Regions provide the framework for the purposeful organization of spatial data.

The Structure of This Chapter

The remainder of the chapter contains examples of how geographers have organized regionally their observations about the physical and cultural world they examine. Each study or vignette explores a different aspect of regional reality. Each organizes its data in ways appropriate to its subject matter and objective, but in each some or all of the common characteristics of regional delimitation and structure may be recognized. Each of the regional examples is based upon the content of one of the earlier chapters of this book, and the page numbers preceding the different selections refer to the material within those chapters most closely associated with, or explained further by, the regional case study. As a further aid to visualizing the application of the regional concept, the examples that follow are introduced by reference to the "traditions" of geography that unite their subject matter and approach.

Part I. Regions in the Earth-Science Tradition

The simplest of all regions to define, and generally the easiest to recognize, is the formal region based upon a single readily apparent component or characteristic. The island is land, not water, and its unmistakable boundary is naturally given where the one element passes to the other. The terminal moraine may mark the transition from the rich, black soils of recent formation to the particolored clays of earlier generation. The dense forest may break dramatically upon the glade or the open prairies. The nature of change is singular and apparent.

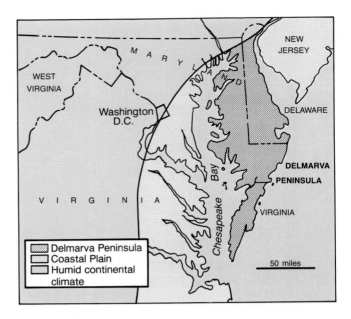

Figure 12.3
One possible nesting of regions within a *regional hierarchy* defined by differing criteria. Each regional unit has internal coherence; the recognition of each adds comprehension of the nature of its subdivisions.

The physical geographer, although concerned with all the earth sciences that explain the natural environment, deals at the outset with *single factor* formal regions. Many of the earth features of concern to physical geography, of course, do not exist in simple, clearly defined units. They must be arbitrarily "regionalized" by the application of boundary definitions. A stated amount of received precipitation, the presence of certain important soil characteristics, the dominance in nature of particular plant associations—all must be decided upon as regional limits, and all such limits are subject to change through time or by purpose of the regional geographer.

Landforms as Regions *(See Stream Landscapes in Humid Areas, Page 74)*

The landform region exists in a more sharply defined fashion than such transitional physical features as soil, climate, or vegetation. For these latter areas, the boundaries depend upon definitional decisions made (and defended) by the researcher. The landform region, on the

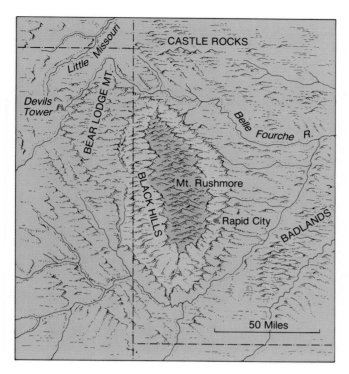

Figure 12.4
The Black Hills physiographic province.

other hand, arises—visibly and apparently unarguably—from nature itself, independent of human influence and unaffected by time on the human scale. Landforms constitute basic, naturally defined regions of physical geographic concern. The existence of major landform regions—mountains, lowlands, plateaus—is unquestioned in popular recognition or scientific definition. Their influence on climates, vegetational patterns, even upon the primary economies of subsistence populations has been noted in earlier portions of this book. The following discussion of a distinct landform region, describing its constitution and its relationships to other physical and cultural features of the landscape, is adapted from a classic study by Wallace W. Atwood.

The Black Hills Province.* The Black Hills rise abruptly from the surrounding plains (Figure 12.4). The break in topography at the margin of this province is obvious to the most casual observer who visits that part of our country. Thus the boundaries of this area, based on contrasts in topography, are readily determined.

*Adapted from *The Physiographic Provinces of North America* by Wallace W. Atwood, © Copyright, 1940, by Ginn and Company. Used by permission of Silver, Burdett & Ginn, Inc.

One who looks more deeply into the study of the natural environment may recognize that in the neighboring plains the rock formations lie in a nearly horizontal position. They are sandstones, shales, conglomerates, and limestones. In the foothills those same sedimentary formations are bent upward and at places stand in a nearly vertical position. Precisely where the change in topography occurs, we find a notable change in the geologic structure and thus discover an explanation for the variation in relief.

The Black Hills are due to a distinct upwarping, or doming, of the crustal portion of the earth. Subsequent removal by stream erosion of the higher portions of that dome and the dissection of the core rocks have produced the present relief features. As erosion has proceeded, more and more of a complex series of ancient metamorphic rocks has been uncovered. Associated with the very old rocks of the core and, at places, with the sedimentary strata, there are a number of later intrusions which have cooled and formed solid rock. They have produced minor domes about the northern margin of the Black Hills.

With the elevation of this part of our country there came an increase in rainfall in the area, and with the increase in elevation and rainfall came contrasts in relief, in soils, and in vegetation.

As we pass from the neighboring plains, where the surface is monotonously level, and climb into the Black Hills area, we enter a landscape having great variety in the relief. In the foothill belt, at the southwest, south, and east, there are hogback ridges interrupted in places by *water gaps,* or gateways, that have been cut by streams radiating from the core of the Hills. Between the ridges there are roughly concentric valley lowlands. On the west side of the range, where the sedimentary mantle has not been removed, there is a plateau-like surface; hogback ridges are absent. Here erosion has not proceeded far enough to produce the landforms common to the east margin. In the heart of the range we find deep canyons, rugged intercanyon

Figure 12.5
The "Needles" of the Black Hills result from erosion along vertical cracks and crevices in granite.

ridges, bold mountain forms, craggy knobs, and other picturesque features (Figure 12.5). The range has passed through several periods of mountain growth and several stages, or cycles, of erosion.

The rainfall of the Black Hills area is somewhat greater than that of the brown, seared, semiarid plains regions, and evergreen trees survive among the hills. We leave a land of sagebrush and grasses to enter one of forests. The dark-colored evergreen trees suggested to early settlers the name Black Hills. As we enter the area, we pass from a land of cattle ranches and some seminomadic shepherds to a land where forestry, mining, general farming, and recreational activities give character to the life of the people. In color and form, in topography, climate, vegetation, and economic opportunities, the Black Hills stand out conspicuously as a distinct geographic unit.

Dynamic Regions in Weather and Climate *(See Air Masses, Page 102)*

The unmistakable clarity and durability of the Black Hills landform region and the precision with which its boundaries may be drawn are rarely echoed in other types of formal physical regions. Most of the natural environment, despite its appearance of permanence and certainty, is dynamic in nature. Vegetations, soils, and climates change through time by natural process or by the action of humans. Boundaries shift, perhaps abruptly, as witnessed by the recent southward migration of the Sahara. The core characteristics of whole provinces change as marshes are drained or forests are replaced by cultivated fields.

That complex of physical conditions that we recognize locally and briefly as weather and summarize as climate displays particularly clearly the temporary nature of much of the natural environment that surrounds us. Yet even in the turbulent change of the atmosphere, distinct regional entities exist with definable boundaries and internally consistent horizontal and vertical properties.

"Air masses" and the consequences of their encounters constitute a major portion of contemporary weather analysis and prediction. Air masses further fulfill all the criteria of *multifactor formal regions,* though their dynamic quality and their patterns of movement obviously mark them as being of a nature distinctly different from such stable physical entities as landform regions, as the following extract from *Climatology and the World's Climates* by George R. Rumney makes clear.

Air Masses. * An air mass is a portion of the atmosphere having a uniform horizontal distribution of certain physical characteristics, especially of temperature and humidity. These qualities are acquired when a mass of air stagnates or moves very slowly over a large and relatively unvaried surface of land or sea. Under these circumstances surface air gradually takes on properties of temperature and moisture approaching those of the underlying surface, and there then follows a steady, progressive transmission of properties to greater heights, resulting finally in a fairly clearly marked vertical transition of characteristics. Those parts of the earth where air masses acquire their distinguishing qualities are called source regions.

The height to which an air mass is modified depends upon the length of time it remains in its source region and also upon the difference between the initial properties of the air when it first arrived and those of the underlying surface. If, for example, an invading flow of air is cooler than the surface beneath as it comes to virtual rest over a source region, it is warmed from below, and convective currents are formed, rapidly bearing aloft new characteristics of temperature and moisture to considerable heights. If, on the other hand, it is warmer than the surface of the source region, cooling of its surface layers takes place, vertical thermal currents do not develop, and the air is modified only in its lower portions. The process of modification may be accomplished in just a few days of slow horizontal drift, although it often takes longer, sometimes several weeks.

Radiation, convection, turbulence, and advection are the chief means of bringing it about.

The prerequisite conditions for these developments are very slowly migrating, outward spreading, and diverging air and a very extensive surface beneath that is fairly uniform in nature. Light winds and relatively high barometric pressure characteristically prevail. Hence, most masses form within the great semipermanent anticyclonic regions of the general circulation, where calms, light variable winds, and overall subsidence of the atmosphere are typical.

Four major types of source regions are recognized: continental polar, maritime polar, continental tropical, and maritime tropical. Polar air masses are continental when they develop over land or ice surfaces in high latitudes; these are cold and dry. They are maritime when they form over the oceans in high latitudes. An air mass from these sources is cold and moist. Similarly, tropical air is continental when it originates along the Tropics of Cancer and Capricorn over northern Africa and northern Australia and is therefore warm and dry. It is maritime when it forms along the Tropics over the oceans, where it develops as a mass of warm, moist air. A single air mass usually covers thousands of square miles of the earth's surface when fully formed.

An air mass is recognizable chiefly because of the uniformity of its primary properties—temperature and humidity—and the vertical distribution of these. Secondary qualities, such as cloud types, precipitation, and visibility, are also taken into account. These qualities are retained for a remarkably long time, often for several weeks, after an air mass has traveled far from its source region, and they are thus the means of distinguishing it from other masses of air.

The principal air masses of North America, their source regions, and seasonal movements are shown in Figure 12.6.

* Copyright George R. Rumney, 1968 (Macmillan, New York). Adapted from pp. 64–65 with permission of the author.

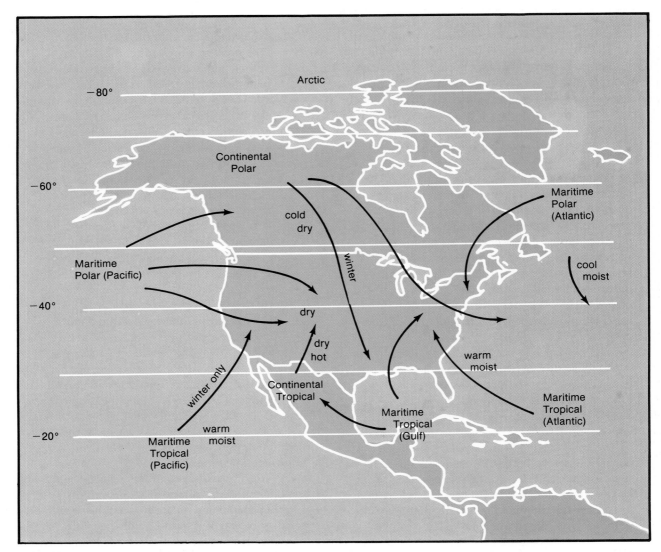

Figure 12.6
**Air masses of North America, their source regions, and
their paths of movement.**

Source: After Haynes, U.S. Department of Commerce.

Air masses, intangible and ephemeral as they may appear, adhere to the requirements for recognition as special-type geographic regions. They have location—at their origin and on their paths of movement to ultimate dissipation. They may be assigned regional names indicative of their origin locations, their characteristics, or both. They have extent, a surface area over which their properties are present and dominant. And they have definable boundaries—called *fronts*, as we saw in Chapter 3— where different sets of temperature and humidity properties are encountered.

Ecosystems as Regions *(See Ecosystems, Page 124)*

A traditional though oversimplified definition of geography as "the study of the areal variation of the surface of the earth" suggested that the discipline centered on the classification of areas and on the subdivision of the earth into its constituent regional parts. Considerations of organization and function were secondary and even unnecessary to the implied main purpose of regional study: the definition of cores and boundaries of areas uniform in physical or cultural properties.

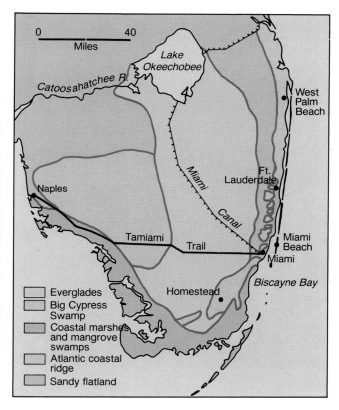

Figure 12.7

The Everglades is part of a complex of ecosystems stretching southward in Florida from Lake Okeechobee to the sea. Drainage and water-control systems have altered its natural condition.

Newer research approaches stress the need for the study of spatial relationships from the standpoint of *systems analysis,* which emphasizes the organization, structure, and functional dynamics within an area and provides for the quantification of the linkages between the things in space. The *ecosystem* or *biome,* introduced in Chapter 4, provides a systems-analytic concept of great flexibility. It brings together in a single framework environment, humans, and the biological realm, permitting an analysis of the relationships between these components of area. Since that relationship is structured, structure rather than spatial uniformity attracts attention and leads to new understanding of the region. The ecosystem concept, particularly, provides a point of view for investigating the complex consequences of human impact upon the natural environment, the theme of Chapter 4.

Note how, in the following description drawn from an article by William J. Schneider, the ecosystem concept is employed in the recognition and analysis of regions and subregions of varying size, complexity, and nature.

Introduced, too, is the concept of *ecotone,* or zone of ecological stress, in this case induced by human pressures upon the natural system.

The Everglades.* The Everglades is a river. Like the Hudson or the Mississippi, it is a channel through which water drains from higher to lower ground as it moves to the sea. It extends in a broad, sweeping arc from the southern end of Lake Okeechobee in central Florida to the tidal estuaries of the Gulf Coast and Florida Bay. As much as 70 miles wide, but generally averaging 40, it is a large, shallow slough that weaves tortuously through acres of saw grass and past "islands" of trees, its waters, even in the wet season, rarely deeper than 2 feet. But, again like the Hudson or the Mississippi, the Everglades bears the significant imprint of civilization.

Since the close of the Pleistocene, 10,000 years ago, the Everglades has been the natural drainage course for the periodically abundant overflow of Lake Okeechobee. As the lake filled during the wet summer months or as hurricane winds blew and literally scooped the water out of the lake basin, excess water spilled over the lake's southern rim. This overflow, together with rainfall collected en route, drained slowly southward between Big Cypress Swamp and the sandy flatlands to the west and the Atlantic coastal ridge to the east, sliding finally into the brackish water of the coastal marshes (Figure 12.7).

Water has always been the key factor in the life of the Everglades. Three-fourths of the annual average 55 inches (140 cm) of rainfall occurs in the wet season, June through October, when water levels rise to cover 90% of the land area of the Everglades. In normal dry seasons in the past, water covered no more than 10% of the land surface. Throughout much of the Everglades, and prior to recent engineering activities, this seasonal rain cycle caused

* From William J. Schneider, "Water and the Everglades." With permission from *Natural History,* November, 1966, pp. 32–40. Copyright American Museum of Natural History, 1966.

Figure 12.8
Open water, saw grasses, and tree islands all visible in this scene constitute separate *biomes* of the Everglades, which is home also of a teeming animal life including the great egret, shown in the foreground.

fluctuations in water levels that averaged 3 feet. Both occasional severe flooding and prolonged drought accompanied by fire imposed periodic stress upon the ecosystem. It may be that randomly occurring ecologic trauma is vital to the character of the Everglades.

Three dominant biological communities—open water, saw grass, and woody vegetation—reflect small, but consistent, differences in the surface elevation of peat soils that cover the Everglades (Figure 12.8). The open-water areas occur at the lower soil elevations; inundated much of the year, they contain both sparse, scattered marsh grasses and a mat of algae. The saw-grass communities develop on a soil base only a few inches higher than that in the surrounding open glades. The soil base is thickest under the tree islands. The few inches' difference in soil depth apparently governs the species composition of these three communities.

Today, the Everglades is no longer precisely a natural river. Much of it has been altered by an extensive program of water management, including drainage, canalization, and the building of locks and dams. Large withdrawals of groundwater for municipal and industrial use have depleted the underlying aquifer and permitted the landward penetration of sea water through the aquifer and through the surface canals. Thousands of individual water-supply wells have been contaminated by encroaching saline water; large biotic changes have taken place in the former freshwater marshes south of Miami. Mangroves—indicators of salinity—have extended their habitat inland, and fires rage across areas that were formerly much wetter. The ecotone—the zone of stress between dissimilar adjacent ecosystems—is altering as a consequence of these human-induced modifications of the Everglades ecosystem.

The organization, structure, and functional dynamics of the Everglades ecosystems are thus undergoing change. The structured relationships of its components—in nature affected and formed by stress—are being subjected to distortions by humans in ways not yet fully comprehended.

Part II. Regions in the Culture–Environment Tradition

The earth-science tradition of geography imposes certain distinctive limits upon area analysis. However defined, the regions that may be drawn are based upon nature and do not result from human action. The culture–environment tradition, however, introduces to regional geography the infinite variations of human occupation and organization of space. There is a corresponding multiplication of recognized regional types and of regional boundary decisions.

Despite the differing interests of physical and cultural geographers, one element of study is common to their concerns: that of process. The "becoming" of an ecosystem, of a cultural landscape, or of the pattern of exchanges in an economic system is an important open or implied part of nearly all geographic study. Evidence of the past as an aid to understanding the present is involved in much geographical investigation, for present-day distributional patterns or qualities of regions mark a merely temporary stage in a continuing process of change.

Population as Regional Focus *(See World Population Distribution, Page 179)*

In no phase of geography are process and change more basic to regional understanding than in population studies. The human condition is dynamic and patterns of settlement are ever-changing. Although these spatial distributions are related to the ways that people use the physical environment in which they are located, they are also conditioned by the purposes, patterns, and solutions of those who went before. In the following extract taken from the work of Glenn Trewartha, a dean of American population geographers, notice how population regionalization—used as a focal theme—ties together a number of threads of regional description and understanding. The aspirations of colonist-conquerors, past and contemporary transportation patterns, physical geographic conditions, political separatism, and the history and practice of agriculture and rural landholdings are all introduced to give understanding to population from a regional perspective.

Population Patterns of Latin America.[*] A distinctive feature of the spatial arrangement of population in Latin America is its strongly nucleated character; the pattern is one of striking clusters. Most of the population clusters remain distinct and are separated from other clusters by sparsely occupied territory. Such a pattern of isolated nodes of settlement is common in many pioneer regions; indeed, it was characteristic of early settlement in both Europe and eastern North America. In those regions, as population expanded, the scantily occupied areas between individual clusters gradually filled in with settlers, and the nodes merged. But in Latin America such an evolution generally did not occur, and so the nucleated pattern persists. Expectably, the individual population clusters show considerable variations in density.

The origin of the nucleated pattern of settlement is partly to be sought in the gold and missionary fever that imbued the Spanish colonists. Their settlements were characteristically located with some care, since only areas with precious metals for exploitation and large Indian populations to be Christianized and to provide laborers could satisfy their dual hungers. A clustered pattern was also fostered by the isolation and localism that prevailed in the separate territories and settlement areas of Latin America.

Almost invariably each of the distinct population clusters has a conspicuous urban nucleus. To an unusual degree the economic, political, and social life within a regional cluster centers on a single large primate city, which is also the focus of the local lines of transport.

The prevailing nucleated pattern of population distribution also bears a relationship to political boundaries. In some countries . . . a single population cluster represents the core area of the nation. In more instances, however, a population cluster forms the core of a major political subdivision of a nation-state, so that a country may contain more than one cluster. A

[*] From Glenn T. Trewartha, *The Less Developed Realm: A Geography of Its Population.* Copyright © 1972 by John Wiley and Sons, Inc. Reprinted from pp. 43–46.

consequence of this simple population distribution pattern and its relation to administrative subdivisions is that political boundaries ordinarily fall within the sparsely occupied territory separating individual clusters. In Latin America few national or provincial boundaries pass through nodes of relatively dense settlement (Figure 12.9).

Another feature arising out of the cluster pattern of population arrangement is that the *total national territory* of a country is often very different from the *effective national territory,* since the latter includes only those populated parts that contribute to the nation's economic support.

A further consequence of the nucleated pattern is found in the nature of the transport routes and systems. Overland routes between population clusters are usually poorly developed, while a more efficient network ordinarily exists *within* each cluster, with each such regional network joined by an overland route to the nearest port. Thus, the chief lines of transport connecting individual population clusters are often sea lanes rather than land routes. Gradually, as highways are developed and improved, intercluster overland traffic tends to increase.

. . . The spectacular rates at which population numbers are currently soaring in Latin America are not matched by equivalent changes in their spatial redistributions. Any population map of Latin America reveals extensive areas of unused and underutilized land. Part of such land is highland and plagued by steep slopes, but by far the larger share of it is characterized by a moist tropical climate, either tropical wet or tropical wet and dry. Such a climatic environment, with its associated wild vegetation, soils, and drainage, admittedly presents many discouraging elements to the new settler of virgin lands. . . . [T]ropical climate alone is scarcely a sufficient explanation for the abundance of near-empty lands south of the Rio Grande. Cultural factors are involved as much as, if not more than, physical ones. One of the former is the unfortunate land-holding system that has been fastened on the continent, under which vast areas of potentially cultivable land are held out of active use by a small number of absentee landlords, who not only themselves make ineffective use of the land, but at the same time refuse to permit its cultivation by small

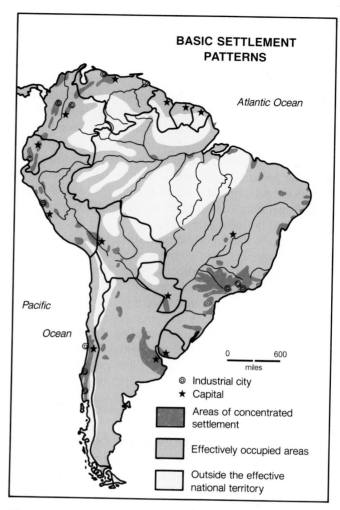

Figure 12.9

South America—basic settlement patterns. Population clusters focused on urban cores and separated by sparsely populated rural areas were the traditional pattern in Latin American countries.

operators. Because of the land-holding system, peasant proprietors are unable to secure their own lands, a situation that discourages new rural settlement. . . .

The changes now in progress in the spatial distributions accompanying the vast increase in numbers of people do not appear to involve any large-scale push of rural settlement into virgin territory. Only to a rather limited extent is new agricultural settlement taking place. Intercluster regions are not filling rapidly. The overwhelming tendency is for people to continue to pile up in the old centers of settlement in and around their cities, rather than to expand the frontier into new pioneer-settlement areas. . . .

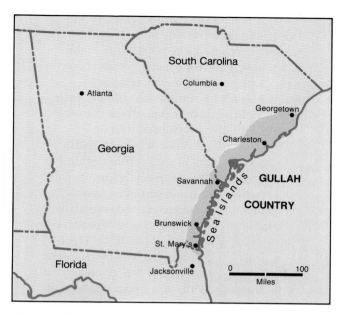

Figure 12.10
Gullah speakers are concentrated on the Sea Islands and the
coastal mainland of South Carolina and Georgia. The isolation
that promoted their linguistic distinction is now being eroded.

Language as Region (See Language, Page 213)

The great culture realms of the world outlined in Chapter
6 are historically based composites of peoples. They are
not closely identified with nation, language, religion, or
technology, but with all these and more in varying combinations. Culture realms are therefore *multifactor regions* that obscure more than they clarify the distinctions
between peoples that are so fundamental to the human
mosaic of the earth. Basic to cultural geography is the
recognition of small regions of single-factor homogeneity that give character to their areas of occurrence
and that collectively provide a needed balance to the
sweeping generalizations of the culture realm.

Language provides an example of such small-area
variation, one partially explored in Chapter 6. The language families shown on Figure 6.18 conceal the identity of and the distinctions among the different official
tongues of nation-states. These, in turn, ignore or sub-
merge the language forms of minority populations, who
may base their own sense of proud identity upon their
regional linguistic separateness. In scale and recognition
even below these ethnically identified regional languages are those local speech variants frequently denied
status as identifiable languages and cited as proof of the
ignorance and the cultural deprivation of their speakers.

Yet such a limited-area, limited-population tongue contains all the elements of the classic culturally based region. Its area is defined; its boundaries are easily drawn;
it represents homogeneity and majority behavior among
its members; and it summarizes, by a single cultural trait,
a collection of areally distinctive outlooks.

Gullah as Language. Isolation is a key
element in the retention or the creation of
distinctive and even externally unintelligible
languages. The isolation of the ancestors of the
some quarter-million present-day speakers of
Gullah—themselves called Gullahs—was nearly
complete. Held by the hundreds as slaves on the
offshore islands and in the nearly equally remote
low country along the southeastern United States
coast from South Carolina to the Florida border
(Figure 12.10), the blacks retained the speech
patterns of the African languages—Ewe, Fanti,
Bambara, Twi, Wolof, Ibo, Malinke, Yoruba,
Efik—native to the slave groups. Forced to use
English words for minimal communication with
their white overseers, but modifying, distorting,
and interjecting African-based substitute words
into that unfamiliar language, the Gullahs kept
intonations and word and idea order in their
spoken common speech that made it
unintelligible to white masters or to more
completely integrated mainland blacks. Because
the language was not understood, its speakers
were considered ignorant, unable to master the
niceties of English. Because ignorance was
ascribed to the Gullahs, they learned to be
ashamed of themselves, of their culture, and of
their tongue, which even they themselves did
not recognize as a highly structured and
sophisticated separate language.

In common with many linguistic minorities,
the Gullahs are losing their former sense of
inferiority and gaining pride in their cultural
heritage and in the distinctive tongue that
represents it. Out of economic necessity,
standard English is being taught to their
schoolchildren, but an increasing scholarly and
popular interest in the structure of their
language and in the nature of their culture has
caused Gullah to be rendered as a written
language, studied as a second language, and
translated into English.

In both the written and the spoken versions, Gullah betrays its African syntax patterns, particularly in its employment of terminal locator words: "Where you goin' at?" The same African origins are revealed by the absence in English translation of distinctive tenses: "I be tired" conveys the concept that "I have been tired for a period of time." Though tenses exist in the African root languages, they are noted more by inflection than by special words and structures.

"He en gut no morratater fer mak no pie wid" may be poor English, but it is good Gullah. Its translation—"He has no more sweet potatoes for making pie"—renders it intelligible to ears attuned to English but loses the musical lilt of original speech and, more importantly, obscures the cultural identity of the speaker, a member of a regionally compact group of distinctive Americans whose territorial extent is clearly marked by its linguistic dominance.

Political Regions   *(See Boundaries: The Limits of the State, Page 246)*

The most rigorously defined formal cultural region is the nation-state. Its boundaries are presumably carefully surveyed and are, perhaps, marked by fences and guard posts. There is no question of an arbitrarily divided transition zone or of lessening toward the borders of the basic quality of the regional core. This rigidity of a nation's boundaries, its unmistakable placement in space, and the trappings—flag, anthem, government, army—that are uniquely its own give to the nation-state an appearance of permanence and immutability not common in other, more fluid cultural regions. But its stability is often more imagined than real. Political boundaries are not necessarily permanent. They are subject to change, sometimes violent change, as a result of internal and external pressures. The Indian subcontinent illustrates the point.

Political Regions in the Indian Subcontinent.
The history of the subcontinent since about 400 B.C. has been one of the alternating creation and dissolution of empires, of the extension of central control based upon the Ganges Basin, and of resistance to that centralization by the marginal territories of the peninsula. British India, created largely unintentionally by 1858, was only the last, though perhaps the most successful, attempt to bring under unified control the vast territory of incredibly complex and often implacably opposed racial, religious, and linguistic groupings.

A common desire for independence and freedom from British rule united the subcontinent's disparate populations at the end of World War II. That common desire, however, was countered by the mutual religious antipathies felt by Muslims and Hindus, each dominant in separate regions of the colony and each unwilling to be affiliated with or subordinated to the other. When the British surrendered control of the subcontinent in 1947, they recognized these apparently irreconcilable religious differences and partitioned the subcontinent into the second and seventh most populous nations on earth. The independent nation of India was created out of the largely Hindu areas constituting the bulk of the former colony. Separate sovereignty was granted to most of the Muslim-majority area under the name of Pakistan. Even so, the partition left boundaries, notably in the Vale of Kashmir, dangerously undefined or in dispute.

An estimated 1 million persons died in the religious riots that accompanied the partition decision. In perhaps the largest short-term mass migration in history, some 10 million Hindus moved from Pakistan to India, and 7.5 million Muslims left India for Pakistan, "The Land of the Pure."

Unfortunately, the purity resided only in common religious belief, not in spatial coherence or in shared language, ethnicity, customs, food, or economy. During its 23 years of existence as originally conceived, Pakistan was a sorely divided nation. The partition decision created an eastern and a western component separated by more than 1000 miles (1600 km) of foreign territory and united only by a common belief in Mohammed (Figure 12.11). West Pakistan, as large as Texas and Oklahoma combined, held 55 million largely light-skinned Punjabis with Urdu language and strong Middle Eastern cultural ties. Some 70 million Bengali-speakers, making up East Pakistan, were crammed into an Iowa-sized portion of the delta of the Ganges and Brahmaputra rivers. The western segment of the nation was part of the semiarid world of western Asia; the eastern portion of Pakistan was joined to humid, rice-producing Southeast Asia.

Beyond the affinity of religion, little else united the awkwardly separated nation. East Pakistan felt itself exploited by a dominating western minority that sought to impose its language and its economic development, administrative objectives, and military control.

Figure 12.11

The sequence of political change on the Indian subcontinent. British India was transformed in 1947 to the nations of India and Pakistan, the latter a Muslim state with a western and an eastern component. In 1970 Pakistan itself was torn by civil war based on ethnic and political contrasts, and the eastern segment became the new nation of Bangladesh.

Rightly or wrongly, East Pakistanis saw themselves as aggrieved and abused; they complained of a per capita income level far below that of their western compatriots, claimed discrimination in the allocation of investment capital, found disparities in the pricing of imported foods, and asserted that their exports of raw materials—particularly jute—were supporting a national economy in which they did not share proportionally. They argued that their demands for regional autonomy, voiced since nationhood, had been denied.

When, in November of 1970, East Pakistan was struck by a cyclone and tidal wave that took an estimated 500,000 lives (Figure I.1), the limit of eastern patience was reached. Resentful over what they saw as a totally inadequate West Pakistani effort of aid in the natural disaster that had befallen them, the East Pakistanis were further incensed by the refusal of the central government to convene on schedule a national assembly to which they had won an absolute majority of delegates. Civil war resulted, and the separate nation-state of Bangladesh was created. The sequence of political change in the subcontinent is traced in Figure 12.11. That further boundary change through war or forced annexation will occur seems implicit in the structure of divergent national claims and cultural affiliations and diversities of the subcontinent.

The nation-state so ingrained in our consciousness and so firmly defined, as displayed on the map inside the front cover, is both a recent and an ephemeral creation of the cultural-regional landscape. It rests upon a claim, more or less effectively enforced, of a monopoly of power and allegiance resident in a government and superior to the communal, linguistic, ethnic, or religious affiliations that preceded it or that claim loyalties overriding it. As the violent recent history of the Indian subcontinent demonstrates, nationalism may be sought, but its maintenance is not assured by the initiating motivations.

Mental Regions *(See Mental Maps, Page 273)*

The regional units so far used as examples, and the methods of regionalization they demonstrate, have a concrete reality. They are formal or functional regions of specified, measurable content. They have boundaries drawn by some objective measures of change or alteration of content, and they have location upon an accurately measured global grid.

Individuals and whole cultures may operate—and operate successfully—with a much less formalized and less precise picture of the nature of the world and of the structure of its parts. The mental maps discussed in Chapter 8 represent personal views of regions and regionalization. The private world views they represent are, as we also saw, colored by the culture of which their holders are members.

Primitive societies, particularly, have distinctive world views by which they categorize what is familiar, and satisfactorily account for what is not. The Yurok Indians of the Klamath River area of northern California were no exception. Their geographic concepts were reported by T. T. Waterman, from whose paper, "Yurok Geography," the following summary is drawn.

The Yurok World View.* The Yurok imagines himself to be living on a flat extent of landscape, which is roughly circular and surrounded by ocean. By going far enough up the river, it is believed, "you come to salt water again." In other words, the Klamath River is considered, in a sense, to bisect the world. This whole earth-mass, with its forests and mountains, its rivers and sea cliffs, is regarded as slowly rising and falling, with a gigantic but imperceptible rhythm, on the heaving, primeval flood. The vast size of the "earth" causes you not to notice this quiet heaving and settling. This earth, therefore, is not merely surrounded by the ocean but floats upon it. At about the central point of this "world" lies a place which the Yurok call qe'nek, on the southern bank of the Klamath, a few miles below the point where the Trinity River comes in from the south. In the Indian concept, this point seems to be accepted as the center of the world.

At this locality also the sky was made. Above the solid sky there is a sky-country, wo'noiyik, about the topography of which the Yurok's ideas are almost as definite as are his ideas of southern Mendocino County, for instance. Downstream from qe'nek, at a place called qe'nek-pul ("qe'nek-downstream"), is an invisible ladder leading up to the sky-country. The ladder is still thought to be there, though no one to my knowledge has been up it recently. The sky-vault is a very definite item in the Yurok's cosmic scheme. The structure consisting of the sky dome and the flat expanse of landscape and waters that it encloses is known to the Yurok as ki-we'sona (literally "that which exists"). This sky, then, together with its flooring of landscape, constitutes "our world." I used to be puzzled at the Yurok confusing earth and sky, telling me, for example, that a certain gigantic redwood tree "held up the world." Their ideas are of course perfectly logical, for the sky is as much a part of the "world" in their sense as the ground is.

The Yurok believe that passing under the sky edge and voyaging still outward you come again to solid land. This is not our world, and mortals ordinarily do not go there; but it is good, solid land. What are breakers over here are just little ripples over there. Yonder lie several regions. To the north (in our sense) lies pu'lekūk, downstream at the north end of creation. South of pu'lekūk lies tsī'k-tsīk-ol ("money lives") where the dentalium-shell, medium of exchange, has its mythical abode. Again, to the south there is a place called kowe'tsik, the mythical home of the salmon, where also all have a "house." About due west of the mouth of the Klamath lies rkrgr', where lives the culture-hero wo'xpa-ku-mä ("across-the-ocean that widower").

Still to the south of rkrgr' there lies a broad sea, kiolaaopa'a, which is half pitch—an Algonkian myth idea, by the way. All of these solid lands just mentioned lie on the margin, the absolute rim of things. Beyond them the Yurok does not go even in imagination. In the opposite direction, he names a place pe'tskuk ("up-river-at"), which is the upper "end" of the river but still in this world. He does not seem to concern himself much with the topography there.

The Yurok's conception of the world he lives in may be summed up in the accompanying diagram (Figure 12.12).

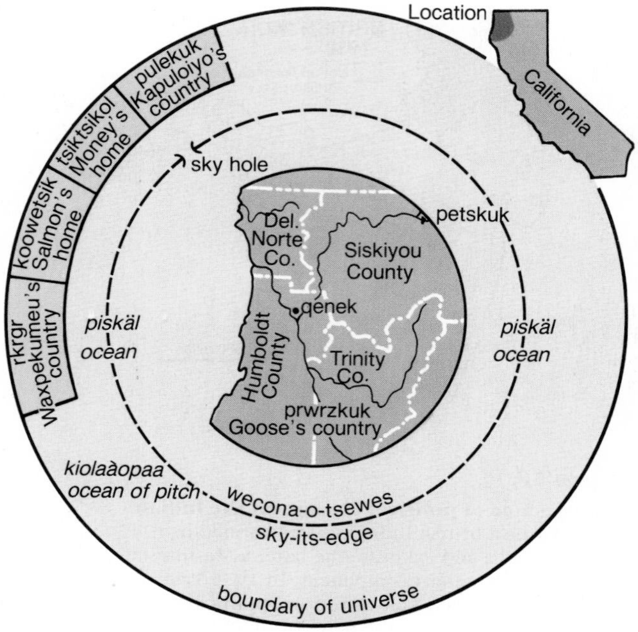

Figure 12.12
The world view of the Yuroks as pieced together by T. T. Waterman during his anthropological study of the tribe. Qe'nek, in the center of the diagram, marks the center of the world in Indian belief.

Redrawn with permission from T. T. Waterman, "Yurok Geography," *University of California Publications in American Archaeology and Ethnology*, vol. 16, 1920.

* *University of California Publications in American Archaeology and Ethnology*, Vol. 16, No. 5 (1920). Selection from pp. 189–193.

Figure 12.13
A deserted farmhouse in Green County, Tennessee, is a mute reminder of the economic and social changes in Appalachia. In the 1950s and 1960s, large numbers of people left this area for the industrial cities to the north.

Part III. Regions in the Locational Tradition

While location, as we have seen, is a primary attribute of all regions, regionalization in the locational tradition of geography implies far more than a named delimitation of earth space. The central concern is with the distribution of human activities and of the resources upon which those activities are based.

In this sense, world regionalization of agriculture and of the soils and climates with which it is related is within the locational tradition. For practical and accepted reasons, however, such underlying physical patterns have been included under the earth-science tradition. But the point is made: the locational tradition emphasizes the "doing" in human affairs, and "doing" is not an abstract thing but an interrelation of life and the resources on which life depends.

The locational tradition, therefore, encourages the recognition and the definition of a far wider array of regional types than do the earth-science or culture–environment traditions. Any single pattern of economic activity or of resource distribution invites the recognition of definable *formal regions.* The interchange of commodities, the control of urban market areas, the flows of capital, or the collection and distribution activities of ports are just a few examples of the infinite number of analytically useful *functional regions* that one may recognize.

Economic Regions   *(See Word Games, Page 315)*

Economic regionalization is among the most frequent, familiar, and useful employments of the regional method. Through economic regions the geographer identifies activities and resources, maps the limits of their occurrence or use, and examines the interrelationships and flows that are part of the complexities of the contemporary world.

The economic region, examples of which were explored in Chapter 9, should be seen as potentially more than a device for recording what *is* in either a formal or a functional sense. It has increasingly become a device for examining what might or should be. The concept of the economic region as a tool for planning and a framework for the manipulation of the people, resources, and economic structure of a composite region first took root in the United States during the Great Depression years of the 1930s. The key element in the planning region is the public recognition of a major territorial unit in which economic change or decline is seen as the cause of a variety of interrelated problems including, for example, population out-migration, regional isolation, cultural deprivation, underdevelopment, and poverty.

Appalachia. Until the early 1960s, "Appalachia" was for most a loose reference to the complex physiographic province of the eastern United States associated with the Appalachian mountain chain. If thought of at all, it was apt to be visualized as rural, isolated, and tree-covered; as an area of coal mining, hillbillies, and folksongs (Figure 12.13).

During the 1950s, however, the economic stagnation and the functional decline of the area became increasingly noticeable in the national context of economic growth, rising personal incomes, and growing concern with the elimination of the poverty and the deprivation of every group of citizens. Less dramatically but just as decisively as the Dust Bowl or the Tennessee Valley of an earlier era, Appalachia became simultaneously a popularly recognized economic and cultural region and a governmentally determined planning region.

Evidences of poverty, underdevelopment, and social crisis were obvious to a nation committed to recognize and eradicate such conditions within its own borders. By 1960, per capita income within Appalachia was $1400 when the national average was $1900. In the decade of the 1950s, mine employment fell 60% and farm jobs declined 52%; the rest of the country lost only 1% of mining jobs and 35% of agricultural employment. Rail employment fell with the drop in coal mining. Massive out-migration occurred among the young adults, with such cities as Chicago, Detroit, Dayton, Cleveland, and Gary the targets. Even with these departures, unemployment among those who remained in Appalachia averaged 50% higher than the national rate. Because of the departures, the remnant population—only 47% of whom lived, in 1960, in or near cities, as against 70% for the entire United States—was distorted in age structure. The very young and the old were disproportionally represented; the productive working-age groups were, at least temporarily, emigrants.

When these and other socioeconomic indicators were plotted by counties and by state economic areas, an elongated but regionally coherent and clearly bounded Appalachia as newly understood was revealed by maps (Figure 12.14). It extended through 13 states from Mississippi to New York, covered some 195,000 square miles (505,000 km²), and contained 18 million people, 93% of them white.

By 1963, awareness of the problems of the area at federal and state levels passed beyond recognition of a multifactor economic region to the establishment of a planning region. A joint federal–state Appalachian Regional Commission was created to develop a program designed to meet the perceived needs of the entire area. The approach chosen was one of limited investment in a restricted number of highly localized developments, with the expectation that these would spark economic growth supported by private funds.

In outline, the plan was (1) to ignore those areas of poverty and unemployment that were in isolated, inaccessible "hollows" throughout the region; (2) to designate "growth centers" where development potential was greatest and concentrate all spending for economic expansion there; the regional growth potential in targeted expenditure was deemed sufficient to overcome charges of aiding the prosperous and depriving the poor; and (3) to create a new network of roads so that the isolated jobless could commute to the new jobs expected to form in and near the favored growth centers; road construction would

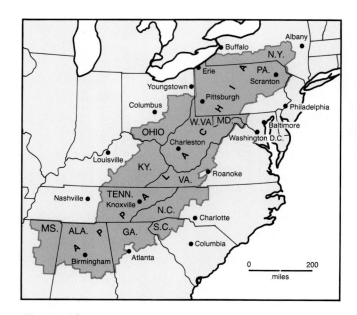

Figure 12.14

The boundary criteria of "Appalachia" as defined by the Appalachian Regional Commission were social and economic conditions, colored by political considerations, not topography.

also, of course, open inaccessible areas to tourism and strengthen the economic base of the entire planning region.

In the years since the Appalachian Regional Commission was established, and in ways not anticipated, the economic prospects of Appalachia have altered. National energy crises have revived mining and transportation employment; new industrial jobs have multiplied as manufacturing has relocated to, or has been newly developed in, the Appalachia portion of the Sunbelt. New employment opportunities have exceeded local labor pools, and out-migrants have returned home from cities outside the region, bringing again a more balanced population pyramid.

By the end of 1986, the Appalachian Regional Commission had over the years committed $5 billion to the area, and another $10 billion had come from other sources. The percentage of people living in poverty within the region declined from 31% in 1960 to 14% in 1980, and per capita income, which had been 79% of the national average in 1960, rose to 85% by 1986. Population flows stabilized, with the number of people coming into Appalachia about equaling those leaving. Some 2000 miles (3200 km) of road had been laid by the end of 1986, though their construction did not give the total regional

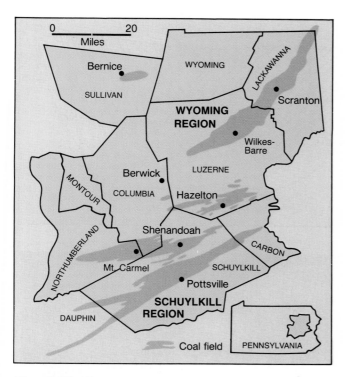

Figure 12.15
The anthracite regions of northeastern Pennsylvania are sharply defined by the geologic events that created them.

access that had been hoped. Smaller towns and their hinterlands have been by-passed and remain as isolated as ever. Some basic social services have, for the first time, reached essentially everyone in the region. Each of the 397 counties of the commission territory, for example, has been supplied with a clinic or a doctor.

Despite these evidences of progress, Appalachia still remained economically distressed in the late 1980s. The mining and manufacturing recession of the earlier years of the decade was sorely felt within a region so dependent upon coal mining and with an industrial base of primary and fabricated metals, wood products, textiles, and apparel—all eroded by imports replacing domestic products. Unemployment rates continued to be between 5% and 10% above the national average, and a third of the mining and manufacturing jobs disappeared between 1979 and 1987.

Regions of Natural Resources
(See Coal, Page 346)

The unevenly distributed resources upon which people depend for existence are logical topics of interest within the locational tradition of geography. Resource regions are mapped, and raw material qualities and quantities are

discussed. Areal relationships to industrial concentrations and the impacts of material extraction upon alternate uses of areas are typical interests in resource geography and in the definition of resource regions.

Those regions, however, are usually treated as if they were expressions of observable surface phenomena; as if, somehow, an oilfield were as exposed and two-dimensional as a soil region or a manufacturing district. What is ignored is that most mineral resources are three-dimensional regions beneath the ground. In addition to the characteristics of an area that may form the basis for regional delimitations and descriptions of surface phenomena, regions beneath the surface add their own particularities to the problem of regional definition. They have, for example, upper and lower boundaries in addition to the circumferential bounds of surface features. They may have an internal topography divorced from the visible landscape. Subsurface relationships—for example, mineral distribution and accessibility in relation to its enclosing rock or to groundwater amounts and movement—may be critical in understanding these specialized, but real, regions. An illustration drawn from the Schuylkill field of the anthracite region of northeastern Pennsylvania helps illustrate the nature of regions beneath the surface.

The Schuylkill Anthracite Region. Nothing in the wild surface terrain of the anthracite country suggested the existence of an equally rugged subterranean topography of coal beds and interstratified rock, slate, and fire clays forming a total vertical depth of 3000 feet (900 m) at greatest development. Yet the creation of the surface landscape was an essential determinant of the areal extent of the Schuylkill district, of the contortions of its bedding, and of the nature of its coal content. A county history of the area reports, "The physical features of the anthracite country are wild. Its area exhibits an extraordinary series of parallel ridges and deep valleys, like long, rolling lines of surf which break upon a flat shore." Both the surface and the subsurface topographies reflect the strong folding of strata after the coal seams were deposited; the anthracite (hard) coal resulted from metamorphic carbonization of the original bituminous beds. Subsequent river and glacial erosion removed as much as 95% of the original anthracite deposits and gave to those that remained discontinuous existence in sharply bounded fields like the Schuylkill (Figure 12.15), a discrete areal entity of 181 square miles (470 km²).

Figure 12.16
The deep folding of the Schuykill coal seams made them costly to exploit. The Mammoth Seam runs as deep as 1500–2000 feet below the surface.

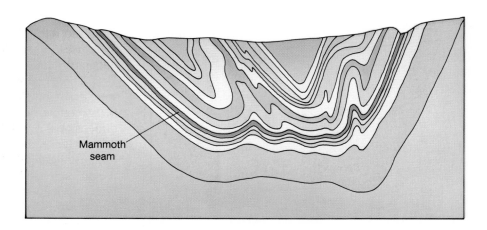

Mammoth seam

Figure 12.17
The Schuylkill Navigation Canal, like its counterpart Lehigh Canal shown here, provided outlet to market for the anthracite region's coal resources after 1825.

The irregular topography of the underground Schuylkill region means that the interbedded coal seams, the most steeply inclined of all the anthracite regions (Figure 12.16), outcrop visibly at the surface on hillsides and along stream valleys. The outcrops made the presence of coal known as early as 1770, but not until 1795 did Schuylkill anthracite find its first use by local blacksmiths. Reviled as "stone coal" or "black stone" that would not ignite, anthracite found no ready commercial market, although it was used in wire and rolling mills located along the Schuylkill River before 1815 and to generate steam in the same area by 1830.

The resources of the subterranean Schuylkill region affected human patterns of surface regions only after the Schuylkill Navigation Canal was completed in 1825 (Figure 12.17), providing a passage to rapidly expanding external markets for the fuel and for the output of industry newly located atop the region. Growing demand induced a boom in coal exploration, an exhaustion of the easily available outcrop coal, and the beginning of the more arduous and dangerous underground mining.

The early methods of mining were simple: merely quarrying the coal from exposed outcrops, usually driving on a slight incline to

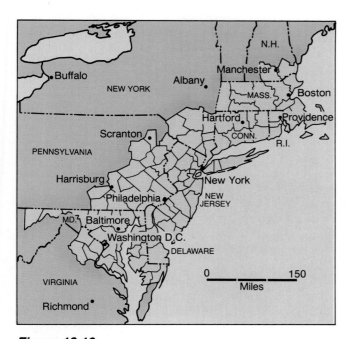

Figure 12.18

Megalopolis in 1960. The region was then composed of counties that, by United States census definition, were "urban" in population characteristics. Much of the area is still distinctly "rural" in land use.

permit natural drainage. Deep shafts were unnecessary and, indeed, not thought of, since the presence of anthracite at depth was not suspected. Later, when it was no longer possible to secure coal from a given outcrop, a small pit was sunk to a depth of 30–40 feet (9–12 m); when the coal and the water that accumulated in the pit could no longer safely be brought to the surface by windlass, the pit was abandoned and a new one was started. Shaft mining, in which a vertical opening from the surface provides penetration to one or several coal beds, eventually became a necessity; with it came awareness of the complex interrelationships between seam thickness, the nature of interstratified rock and clays, the presence of gases, and the movement of subsurface water.

The subterranean Schuylkill region has a three-dimensional pattern of use. The configuration and the variable thickness of the seams demand concentration of mining activities. Minable coal is not uniformly available along any possible vertical or horizontal cross section because of the interstratification and the extreme folding of the beds. Mining is concentrated further by the location of shafts and the construction of passages, in their turn determined by both patterns of ownership and thickness of seam. In general, no seam less than 2 feet (.6 m) is worked, and the absolute thickness—50 feet (15 m)—is found only in the Mammoth seam of the Schuylkill region.

Friable interstratified rock increases the danger of coal extraction and raises the costs of cave-in prevention. Although the Schuylkill mines are not gassy, the possibility of gas release from the collapse of coal pillars left as mine supports makes necessary systems of ventilation even more elaborate than those minimally required to provide adequate air to miners. Water is ever present in the anthracite workings, and constant pumping or draining is necessary for mine operation. The collapse of strata underlying a river may result in sudden disastrous flooding.

The Schuylkill subterranean anthracite region presents a pattern of complexity in distribution of physical and cultural features and of interrelationships between phenomena every bit as great and inviting of geographical analysis as any purely surficial region.

Urban Regions (See Suburbanization in the United States, Page 387)

Urban geography represents a climax stage in the locational tradition of geography. Modern integrated, interdependent society on a world basis is urban-centered. Cities are the indispensable functional focuses of production, exchange, and administration. They exist individually as essential elements in interlocked hierarchical systems of cities. Internally, they display complex but repetitive spatial patterns of land uses and functions.

Because of the many-sided character of urbanism, cities are particularly good subjects for regional study. They are themselves, of course, formal regions. In the aggregate, their distributions give substance to formal regions of urban concentration. Cities are also the cores of functional regions of varying types and hierarchical orders. Their internal diversity of functional, land use, and socioeconomic patterns invite regional analysis. The employment of that approach in both its formal and functional modes is clearly displayed by Jean Gottmann, who, in examining the data and landscapes of the eastern United States at mid-century, recognized and analyzed Megalopolis. The following is taken from his study.

Megalopolis.[*] The northeastern seaboard of the United States is today the site of a remarkable development—an almost continuous stretch of urban and suburban areas from southern New Hampshire to northern Virginia and from the Atlantic shore to the Appalachian foothills (Figure 12.18). The processes of urbanization, rooted deep in the American past, have worked

* Jean Gottmann, *Megalopolis: The Urbanized Northeastern Seaboard of the United States.* Copyright © 1961 by the Twentieth Century Fund, Inc.

Figure 12.19
The "Pine Barrens" of New Jersey is still a largely undisturbed natural enclave in the heart of Megalopolis. Its preservation is a subject of dispute between environmentalists and developers.

steadily here, endowing the region with unique ways of life and of land use. No other section of the United States has such a large concentration of population, with such a high average density, spread over such a large area. And no other section has a comparable role within the nation or a comparable importance in the world. Here has been developed a kind of supremacy, in politics, in economics, and possibly even in cultural activities, seldom before attained by an area of this size.

Great, then, is the importance of this section of the United States and of the processes now at work within it. And yet it is difficult to single this area out from surrounding areas, for its limits cut across established historical divisions, such as New England and the Middle Atlantic states, and across political entities, since it includes some states entirely and others only partially. A special name is needed, therefore, to identify this special geographical area.

This particular type of region is new, but it is the result of age-old processes, such as the growth of cities, the division of labor within a civilized society, the development of world resources. The name applied to it should, therefore, be new as a place name but old as a symbol of the long tradition of human aspirations and endeavor underlying the situations and problems now found here. Hence the choice of the term *Megalopolis*, used in this study.

As one follows the main highways or railroads between Boston and Washington, D.C., one hardly loses sight of built-up areas, tightly woven residential communities, or powerful concentrations of manufacturing plants. Flying this same route one discovers, on the other hand, that behind the ribbons of densely occupied land along the principal arteries of traffic, and in between the clusters of suburbs around the old urban centers, there still remain large areas covered with woods and brush alternating with some carefully cultivated patches of farmland (Figure 12.19). These green spaces, however, when inspected at closer range, appear stuffed with a loose but immense scattering of buildings, most of them residential but some of industrial character. That is, many of these sections that look rural actually function largely as suburbs in the orbit of some city's downtown. Even the farms, which occupy the larger tilled patches, are seldom worked by people whose only occupation and income are properly agricultural.

Thus the old distinctions between rural and urban do not apply here any more. Even a quick look at the vast area of Megalopolis reveals a revolution in land use. Most of the people living in the so-called rural areas, and still classified as "rural population" by recent censuses, have very little, if anything, to do with agriculture. In terms of their interests and work they are what used to be classified as "city folks," but their way of life and the landscapes around their residences do not fit the old meaning of urban.

In this area, then, we must abandon the idea of the city as a tightly settled and organized unit in which people, activities, and riches are crowded into a very small area clearly separated from its nonurban surroundings. Every city in this region spreads out far and wide around its original nucleus; it grows amidst an irregularly colloidal mixture of rural and suburban landscapes; it melts on broad fronts with other mixtures, of somewhat similar though different texture, belonging to the suburban neighborhoods of other cities.

Thus an almost continuous system of deeply interwoven urban and suburban areas, with a total population of about 37 million people in 1960, has been erected along the Northeastern Atlantic seaboard. It straddles state boundaries, stretches across wide estuaries and bays, and encompasses many regional differences. In fact, the landscapes of Megalopolis offer such variety that the average observer may well doubt the unity of the region. And it may seem to him that the main urban nuclei of the seaboard are little related to one another. Six of its great cities would be great individual metropolises in their own right if they were located elsewhere. This region indeed reminds one of Aristotle's saying that cities such as Babylon had "the compass of a nation rather than a city."

The description of *Megalopolis* in 1960 represents, as does any regional study, a captured moment on the continuum of areal change. When that moment is lucidly described and the threads of areal character clearly delineated, the regional study serves as both a summary of the present and an augury of a future whose roots may be discerned in it. In the years since Gottmann described it, *Megalopolis* has continued to develop along the lines he summarized. Urbanization has continued, physically as well as functionally, to encroach upon the rural landscapes without regard for state boundaries or even the metropolitan cores that dominated in 1960. As then, new growth centers—becoming equivalents of central cities in their own right—are linked to the expanding transportation corridors, though now the lines of importance are increasingly expressways rather than railroads.

In the south, the corridors around Washington, D.C. are the Capital Beltway in Fairfax County and an extension from it westward to Loudoun County and Dulles Airport in Virginia and, to the north into Montgomery County, Maryland, the I-270 corridor. The Virginia suburbs, focusing on Tysons Corners, have specialized in defense-related industries, but vast office building complexes and commercial centers are rapidly converting rural land to general urban uses. The Maryland suburbs specialize in health, space, and communications interests in their own new office, industrial, and commercial "parks" and complexes. Still farther to the north, in the "Princeton Corridor", a 26-mile stretch along Route 1 from New Brunswick to Trenton, New Jersey, huge corporate parks provide office and research space to companies relocating from the New York area and to new technology firms attracted by proximity to Princeton University. Morristown, New Jersey, and White Plains, in Westchester County, New York, are other similar concentrations of commercial, industrial, and office developments north and west of New York City. To the east, Stamford, Connecticut, with 150,000 daily in-commuters, has become the headquarters center for major national corporations and has emerged as a truly major central city new on the scene in its present form since Megalopolis was first described. And still farther to the north, the I-495 beltway around Boston has joined Route 128 as a magnet for electronics, computer, electrical, and other technology-related companies.

Conclusion

The region is a mental construct, a created entity whose sole function is the purposeful organization of spatial data. The scheme of that organization, the selection of data to be analyzed, and the region resulting from these decisions are reflections of the intellectual problem posed.

This chapter has not attempted to explore all aspects of regionalism and of the regional method. It has tried, by example, only to document its basic theme: the geographer's regions are arbitrary but deliberately conceived devices for the isolation of things, patterns, interrelations, and flows that invite geographic analysis. In this sense, all geographers are regional geographers, and the regional examples of this chapter may logically complete our survey of the four traditions of geography.

Key Words

formal region *405*   functional region *405*   region *404*   regional concept *404*

For Review

1. What do geographers seek to achieve when they recognize or define regions? On what basis are regional boundaries drawn? Are regions concrete entities whose dimensions and characteristics are agreed upon by all who study the same general segment of earth space? Ask three fellow students who are not participants in this course for their definition of the "South." If their answers differed, what implicit or explicit criteria of regional delimitation were they employing?

2. What are the spatial elements or the identifying qualities shared by all regions?

3. What is the identifying characteristic of a formal region? How are its boundaries determined? Name three different examples of formal regions drawn from any earlier chapters of this book. How was each defined and what was the purpose of its recognition?

4. How are functional regions defined? What is the nature of their bounding criteria? Give three or four examples of functional regions that were defined earlier in the text.

5. The ecosystem was suggested as a viable device for regional delimitation. What regional geographic concepts are suggested in ecosystem recognition? Is an ecosystem identical to a formal region? Why or why not?

6. National, linguistic, historical, planning, and other regions have been recognized in this chapter. With what other regional entities do you have acquaintance in your daily affairs? Are fire protection districts, police or voting precincts, or zoning districts regional units identifiable with geographers' regions as discussed in this chapter or elsewhere in this book? How influenced are you in your private life by your, or others', regional delimitations?

Suggested Readings

Freeman, T. W. *A Hundred Years of Geography.* Ch. 2, "The Regional Approach." Chicago: Aldine, 1961.

Johnston, R. J. *Geography and Geographers,* 2d ed. London: Methuen, 1983.

Minshull, Roger. *Regional Geography.* Chicago: Aldine, 1967.

Murphey, Rhoads. *The Scope of Geography,* 3d ed. Ch. 2, "The Region." New York and London: Methuen, 1982.

Whittlesey, Derwent. "The Regional Concept and the Regional Method." In *American Geography: Inventory and Prospect.* Preston James and Clarence F. Jones, eds. Syracuse, N.Y.: Syracuse University Press (for the Association of American Geographers), 1954.

Appendix

1987 WORLD POPULATION DATA

Region or Country	Population Estimate mid-1987 (millions)	Crude Birth Rate	Crude Death Rate	Natural Increase (annual, %)	Population "Doubling Time" in Years (at current rate)	Population Projected to 2000 (millions)	Population Projected to 2020 (millions)	Infant Mortality Rate[a]	Total Fertility Rate[b]	% Population Under Age 15/65+	Life Expectancy at Birth (years)	Urban Population (%)	Population with Access to Safe Water Supply (%, 1983)	Energy Consumption Per Capita, 1984 (gigajoules)/ Percent Change Since 1970	Per Capita GNP, 1985 (US$)
WORLD	5,026	28	10	1.7	40	6,158	7,992	81	3.6	33/6	63	43	—	55/6	$2,880
MORE DEVELOPED	1,191	15	10	0.5	128	1,263	1,330	16	2.0	22/11	73	72	—	—/—	9,930
LESS DEVELOPED	3,836	32	11	2.1	33	4,896	6,662	90	4.2	37/4	59	34	—	—/—	660
LESS DEVELOPED (Excl. China)	2,774	36	12	2.4	29	3,696	5,301	97	4.8	40/4	57	35	51	—/—	810
AFRICA	601	44	16	2.8	24	880	1,479	113	6.3	45/3	51	30	41	12/41	710
NORTHERN AFRICA	135	39	11	2.7	25	187	279	93	5.6	42/4	58	42	73	—/—	1,150
Algeria	23.5	42	10	3.2	22	33.7	49.4	81	6.4	46/4	60	43	90	21/95	2,530
Egypt	51.9	37	11	2.6	26	71.2	103.0	93	5.3	40/4	59	46	75	19/142	680
Libya	3.8	39	9	3.0	23	5.6	8.4	90	5.6	45/3	61	76	98	111/463	7,500
Morocco	24.4	36	10	2.5	27	33.3	48.5	90	4.8	42/4	60	43	—	9/69	610
Sudan	23.5	45	16	2.8	24	33.4	55.6	112	6.5	45/3	49	20	48	2/−46	330
Tunisia	7.6	32	7	2.5	27	9.9	13.6	78	4.5	40/4	62	53	—	20/137	1,220
WESTERN AFRICA	187	46	18	2.8	25	278	480	127	6.6	45/2	48	28	34	—/—	580
Benin	4.3	51	20	3.0	23	7.1	14.2	115	7.1	49/5	45	39	20	1/−14	270
Burkina Faso	7.3	48	20	2.8	25	10.5	18.2	146	6.5	44/3	46	8	30	1/125	140
Cape Verde	0.3	35	8	2.6	26	0.5	0.7	76.5	5.1	46/6	60	27	—	6/—	430
Cote d'Ivoire	10.8	46	15	3.0	23	17.3	32.1	105	6.7	46/3	52	43	20	6/5	620

Region or Country	Population Estimate mid-1987 (millions)	Crude Birth Rate	Crude Death Rate	Natural Increase (annual, %)	Population "Doubling Time" in Years (at current rate)	Population Projected to 2000 (millions)	Population Projected to 2020 (millions)	Infant Mortality Rate[a]	Total Fertility Rate[b]	% Population Under Age 15/65+	Life Expectancy at Birth (years)	Urban Population (%)	Population with Access to Safe Water Supply (%, 1983)	Energy Consumption Per Capita, 1984 (gigajoules)/Percent Change Since 1970	Per Capita GNP, 1985 (US$)
Gambia	0.8	49	28	2.1	33	1.1	1.8	169	6.4	46/3	36	21	—	3/49	230
Ghana	13.9	42	14	2.8	25	20.5	33.9	94	5.8	47/3	54	31	43	2/−55	390
Guinea	6.4	47	23	2.4	29	8.9	14.3	153	6.2	43/3	41	22	17	2/−28	320
Guinea-Bissau	0.9	41	21	2.0	35	1.2	1.8	138	5.4	44/4	44	27	33	1/−40	170
Liberia	2.4	48	16	3.2	22	3.6	6.7	127	6.9	47/3	50	40	40	10/−26	470
Mali	8.4	51	22	2.9	24	12.3	21.1	175	6.7	46/3	43	18	14	1/60	140
Mauritania	2.0	50	20	3.0	23	3.0	5.2	132	6.9	46/3	45	35	—	5/2	410
Niger	7.0	51	22	2.9	24	10.6	18.9	141	7.1	47/3	44	16	34	2/133	200
Nigeria	108.6	46	18	2.8	25	160.9	273.6	124	6.6	45/2	49	28	37	7/383	760
Senegal	7.1	46	18	2.8	24	10.6	19.7	131	6.7	46/3	45	36	44	5/32	370
Sierra Leone	3.9	47	29	1.8	38	5.4	8.9	176	6.2	41/3	35	28	23	2/−44	370
Togo	3.2	47	15	3.1	22	4.9	8.9	117	6.6	44/3	53	22	37	2/6	250
EASTERN AFRICA	**179**	**48**	**18**	**3.0**	**23**	**273**	**496**	**117**	**6.9**	**47/3**	**48**	**16**	**—**	**—/—**	**250**
Burundi	5.0	47	18	2.9	24	7.2	11.0	119	6.4	44/4	48	5	26	z/51	240
Comoros	0.4	47	14	3.3	21	0.7	1.3	91	7.0	47/3	52	23	—	2/—	280
Djibouti	0.3	43	18	2.5	28	0.4	0.7	132	6.5	46/3	47	74	—	8/34	—
Ethiopia	46.0	46	23	2.3	30	66.5	111.2	152	6.7	46/4	41	10	—	z/−34	110
Kenya	22.4	52	13	3.9	18	38.3	79.2	76	8.0	51/2	54	16	28	3/−8	290
Madagascar	10.6	44	16	2.8	25	15.6	25.8	63	6.1	44/3	50	22	23	2/−20	250
Malawi	7.4	53	21	3.2	22	11.4	21.1	157	7.0	47/3	46	12	51	1/−10	170
Mauritius	1.1	19	7	1.2	57	1.3	1.6	25.1	2.3	32/5	68	42	95	8/−18	1,070
Mozambique	14.7	45	19	2.6	27	21.1	35.7	147	6.1	46/3	46	13	13	3/−31	—
Reunion	0.6	25	6	1.9	36	0.7	0.8	12	2.7	33/5	70	60	—	—/—	—
Rwanda	6.8	53	16	3.7	19	11.0	21.8	122	8.5	48/2	48	6	60	1/280	290
Seychelles	0.1	27	7	1.9	36	0.1	0.1	17.9	3.5	37/6	70	37	—	—/—	—
Somalia	7.7	48	23	2.5	28	10.4	18.6	150	7.1	45/4	41	34	31	4/228	270
Tanzania	23.5	50	15	3.5	20	36.6	68.6	111	7.1	48/3	52	18	46	1/−24	270
Uganda	15.9	50	16	3.4	20	24.7	46.4	108	7.0	48/3	50	10	16	1/−64	—
Zambia	7.1	50	15	3.5	20	11.6	22.0	84	7.0	47/3	52	43	—	10/−11	400
Zimbabwe	9.4	47	12	3.5	20	15.1	29.4	76	6.5	50/3	57	24	52	13/−46	650

Region or Country	Population Estimate mid-1987 (millions)	Crude Birth Rate	Crude Death Rate	Natural Increase (annual, %)	Population "Doubling Time" in Years (at current rate)	Population Projected to 2000 (millions)	Population Projected to 2020 (millions)	Infant Mortality Rate[a]	Total Fertility Rate[b]	% Population Under Age 15/65+	Life Expectancy at Birth (years)	Urban Population (%)	Population with Access to Safe Water Supply (%, 1983)	Energy Consumption Per Capita, 1984 (gigajoules)/ Percent Change Since 1970	Per Capita GNP, 1985 (US$)
MIDDLE AFRICA	61	44	17	2.8	25	89	149	114	6.1	45/3	49	35	22	—/—	430
Angola	8.0	47	22	2.5	28	11.5	19.1	143	6.4	45/3	43	25	28	4/−25	—
Cameroon	10.3	43	16	2.7	26	14.5	23.5	103	5.9	43/4	51	42	26	13/600	810
Central African Republic	2.7	44	19	2.5	28	3.8	5.9	142	5.9	43/4	43	42	16	1/−26	270
Chad	4.6	43	23	2.0	35	6.3	9.5	143	5.9	44/2	43	27	26	1/12	—
Congo	2.1	47	13	3.4	21	3.2	5.9	112	6.8	46/3	55	48	29	3/−29	1,020
Equatorial Guinea	0.3	38	20	1.8	38	0.4	0.7	130	5.6	41/4	45	60	—	3/−24	—
Gabon	1.2	34	18	1.6	43	1.6	2.4	112	4.5	35/6	49	41	—	33/139	3,340
Sao Tome and Principe	0.1	36	9	2.7	25	0.1	0.3	61.7	5.4	41/6	65	35	—	—/—	310
Zaire	31.8	45	15	3.1	23	47.6	81.6	103	6.1	46/3	51	34	19	2/2	170
SOUTHERN AFRICA	39	34	10	2.4	29	52	76	76	4.8	39/4	55	53	—	88/20	1,880
Botswana	1.2	48	14	3.4	20	1.8	3.4	70	6.7	48/4	58	22	65	—/—	840
Lesotho	1.6	41	15	2.6	27	2.3	3.6	106	5.8	42/4	50	17	14	—/—	480
Namibia	1.3	44	11	3.3	21	2.0	3.7	110	6.4	45/3	49	51	—	—/—	—
South Africa	34.3	33	10	2.3	30	45.3	63.5	72	4.6	38/4	56	56	—	—/—	2,010
Swaziland	0.7	47	16	3.1	22	1.0	1.9	124	6.5	46/3	50	26	—	—/—	650
ASIA	2,930	28	10	1.9	37	3,598	4,584	86	3.7	34/5	61	32	—	20/55	970
ASIA (Excl. China)	1,868	33	11	2.2	32	2,398	3,223	95	4.4	38/4	58	32	50	—/—	1,400
WESTERN ASIA	121	37	10	2.7	26	171	265	83	5.5	41/4	62	55	70	—/—	3,130
Bahrain	0.4	32	5	2.8	25	0.7	1.0	32	4.6	41/2	67	81	—	388/402	9,560
Cyprus	0.7	20	9	1.1	63	0.8	0.9	12	2.5	25/11	74	53	—	56/56	3,790
Gaza	0.6	47	8	4.0	18	0.8	1.4	59	7.4	50/3	64	79	—	—/—	—
Iraq	17.0	46	13	3.3	21	26.5	43.8	80	7.2	49/4	62	68	73	15/−13	—
Israel	4.4	23	7	1.7	41	5.3	6.6	12.3	3.1	33/9	75	90	—	69/10	4,920
Jordan	3.7	45	8	3.7	19	5.7	9.9	54	7.4	51/3	67	60	89	29/274	1,560
Kuwait	1.9	34	3	3.2	22	2.7	4.0	19.0	4.4	40/1	72	80	89	195/37	14,270
Lebanon	3.3	30	8	2.2	32	4.1	5.8	52	3.8	38/5	65	80	92	25/26	—
Oman	1.3	47	14	3.3	21	2.0	3.2	117	7.1	44/3	52	9	—	391/8,434	7,080
Qatar	0.3	34	4	3.0	23	0.5	0.7	42	5.6	34/2	69	86	—	637/65	15,980
Saudi Arabia	14.8	39	7	3.1	22	23.2	43.2	79	6.9	37/2	63	72	93	104/667	8,860

Region or Country	Population Estimate mid-1987 (millions)	Crude Birth Rate	Crude Death Rate	Natural Increase (annual, %)	Population "Doubling Time" in Years (at current rate)	Population Projected to 2000 (millions)	Population Projected to 2020 (millions)	Infant Mortality Rate[a]	Total Fertility Rate[b]	% Population Under Age 15/65+	Life Expectancy at Birth (years)	Urban Population (%)	Population with Access to Safe Water Supply (%, 1983)	Energy Consumption Per Capita, 1984 (gigajoules)/Percent Change Since 1970	Per Capita GNP, 1985 (US$)
Syria	11.3	47	9	3.8	18	17.8	29.0	59	7.2	49/4	63	49	71	27/141	1,630
Turkey	51.4	30	9	2.1	33	65.4	86.7	92	4.0	36/4	62	46	63	26/84	1,130
United Arab Emirates	1.4	30	4	2.6	27	1.9	2.6	38	5.9	30/1	68	81	93	223/210	19,120
Yemen, North	6.5	53	19	3.4	20	10.0	19.5	137	7.8	49/3	47	15	31	6/1,243	520
Yemen, South	2.4	47	17	3.0	23	3.6	6.3	135	7.3	48/3	48	40	50	25/241	540
SOUTHERN ASIA	**1,112**	**36**	**13**	**2.3**	**30**	**1,448**	**1,988**	**110**	**4.8**	**40/4**	**54**	**25**	**50**	**—/—**	**250**
Afghanistan	14.2	48	22	2.6	27	24.5	39.1	182	7.6	46/4	39	16	10	2/24	—
Bangladesh	107.1	44	17	2.7	26	144.9	201.5	140	6.2	44/4	50	13	42	2/—	150
Bhutan	1.5	38	18	2.0	34	1.9	2.5	142	5.5	40/3	46	5	17	z/—	160
India	800.3	33	12	2.1	33	1,013.3	1,310.0	101	4.3	38/4	55	25	54	7/71	250
Iran	50.4	45	13	3.2	21	73.9	130.6	113	6.3	44/3	57	51	—	39/40	—
Maldives	0.2	48	10	3.8	18	0.3	0.6	68	7.1	45/2	51	26	—	—/—	290
Nepal	17.8	42	17	2.5	28	24.4	37.4	112	6.1	41/3	52	7	15	z/14	160
Pakistan	104.6	44	15	2.9	24	145.3	242.2	125	6.6	45/4	50	28	39	6/30	380
Sri Lanka	16.3	25	7	1.8	38	19.4	24.2	29.8	3.7	35/4	70	22	36	4/−1	370
SOUTHEAST ASIA	**421**	**32**	**9**	**2.3**	**30**	**544**	**720**	**74**	**4.2**	**39/3**	**60**	**25**	**43**	**—/—**	**690**
Brunei	0.2	30	4	2.6	26	0.3	0.4	12	3.6	38/3	62	64	—	—/—	17,580
Burma	38.8	34	13	2.1	33	49.8	67.5	103	4.4	39/4	53	24	25	2/40	190
East Timor	0.7	48	23	2.5	28	0.9	1.1	183	5.8	35/3	40	12	—	—/—	—
Indonesia	174.9	31	10	2.1	33	219.8	284.2	88	4.2	40/3	58	22	33	8/126	530
Kampuchea	6.5	39	18	2.1	33	8.5	12.0	160	4.7	35/3	43	11	—	z/−92	—
Laos	3.8	41	16	2.5	28	5.0	6.9	122	5.8	43/3	50	16	21	1/−72	—
Malaysia	16.1	31	7	2.4	28	20.2	26.3	30	3.9	39/4	67	32	80	26/57	2,050
Philippines	61.5	35	7	2.8	25	85.5	115.4	50	4.7	41/3	65	40	54	9/17	600
Singapore	2.6	17	5	1.1	61	2.9	3.1	9.3	1.6	24/5	71	100	100	196/429	7,420
Thailand	53.6	29	8	2.1	33	65.5	81.6	57	3.5	36/3	63	17	65	12/125	830
Vietnam	62.2	34	8	2.6	27	86.0	121.2	55	4.5	40/4	63	19	—	4/−59	—

Region or Country	Population Estimate mid-1987 (millions)	Crude Birth Rate	Crude Death Rate	Natural Increase (annual, %)	Population "Doubling Time" in Years (at current rate)	Population Projected to 2000 (millions)	Population Projected to 2020 (millions)	Infant Mortality Rate[a]	Total Fertility Rate[b]	% Population Under Age 15/65+	Life Expectancy at Birth (years)	Urban Population (%)	Population with Access to Safe Water Supply (%, 1983)	Energy Consumption Per Capita, 1984 (gigajoules)/Percent Change Since 1970	Per Capita GNP, 1985 (US$)
EAST ASIA	**1,275**	**20**	**7**	**1.3**	**55**	**1,435**	**1,611**	**55**	**2.4**	**28/5**	**67**	**39**	**—**	**—/—**	**1,490**
China	1,062.0	21	8	1.3	53	1,200.0	1,361.0	61	2.4	28/5	66	32	—	19/87	310
Hong Kong	5.6	14	5	0.9	77	6.4	7.0	7.5	1.6	24/7	75	92	99	—/—	6,220
Japan	122.2	12	6	0.6	124	126.6	122.9	5.5	1.8	22/10	77	76	—	111/18	11,330
Korea, North	21.4	30	5	2.5	28	28.4	38.5	33	4.0	39/4	65	64	—	78/29	—
Korea, South	42.1	20	6	1.4	51	48.0	52.0	30	2.1	31/4	67	65	—	44/131	2,180
Macao	0.4	23	6	1.7	41	0.6	0.7	12	3.7	34/8	68	97	—	—/—	—
Mongolia	2.0	37	11	2.6	26	2.8	4.4	53	5.1	42/3	62	51	—	47/85	—
Taiwan	19.6	17	5	1.2	59	22.3	24.4	8.9	1.8	30/5	73	67	—	—/—	—
NORTH AMERICA	**270**	**15**	**9**	**0.7**	**101**	**296**	**326**	**10**	**1.8**	**22/12**	**75**	**74**	**—**	**—/—**	**16,150**
Canada	25.9	15	7	0.8	91	27.9	29.3	7.9	1.7	22/10	76	76	—	288/11	13,670
United States	243.8	16	9	0.7	102	268.0	296.6	10.5	1.8	22/12	75	74	—	280/−12	16,400
LATIN AMERICA	**421**	**30**	**8**	**2.2**	**31**	**537**	**712**	**58**	**3.7**	**38/4**	**66**	**67**	**70**	**—/—**	**1,700**
CENTRAL AMERICA	**109**	**33**	**7**	**2.6**	**27**	**143**	**197**	**53**	**4.3**	**42/4**	**66**	**63**	**70**	**—/—**	**1,830**
Belize	0.2	33	6	2.7	26	0.2	0.3	27	4.5	46/5	70	52	—	—/—	1,130
Costa Rica	2.8	31	4	2.7	25	3.7	5.0	18.9	3.5	35/4	74	48	93	13/22	1,290
El Salvador	5.3	36	10	2.6	27	7.2	10.3	65	4.7	46/4	66	43	51	5/−10	710
Guatemala	8.4	41	9	3.2	22	12.2	19.7	71	5.8	46/3	60	39	51	6/6	1,240
Honduras	4.7	39	8	3.1	22	7.0	12.0	69	5.6	47/3	63	40	69	7/−3	730
Mexico	81.9	31	7	2.5	28	104.5	137.8	50	4.0	42/4	67	70	74	50/65	2,080
Nicaragua	3.5	43	9	3.4	20	5.1	7.7	69	5.7	47/3	61	53	53	8/−11	850
Panama	2.3	27	5	2.2	32	2.9	3.7	25	3.3	38/4	72	51	62	22/−6	2,020
CARIBBEAN	**32**	**26**	**8**	**1.8**	**38**	**38**	**48**	**55**	**3.1**	**34/7**	**67**	**54**	**—**	**—/—**	**—**
Antigua and Barbuda	0.1	15	5	1.0	71	0.1	0.1	10	1.7	27/6	72	34	—	—/—	2,030
Bahamas	0.2	23	6	1.8	39	0.3	0.4	26.7	2.9	38/4	70	75	—	—/—	7,150
Barbados	0.3	17	8	0.9	78	0.4	0.5	13.1	2.0	30/11	73	32	—	36/42	4,680
Cuba	10.3	18	6	1.2	58	11.4	12.5	16.5	1.8	27/8	73	71	—	43/47	—

Region or Country	Population Estimate mid-1987 (millions)	Crude Birth Rate	Crude Death Rate	Natural Increase (annual, %)	Population "Doubling Time" in Years (at current rate)	Population Projected to 2000 (millions)	Population Projected to 2020 (millions)	Infant Mortality Rate[a]	Total Fertility Rate[b]	% Population Under Age 15/65+	Life Expectancy at Birth (years)	Urban Population (%)	Population with Access to Safe Water Supply (%, 1983)	Energy Consumption Per Capita, 1984 (gigajoules)/Percent Change Since 1970	Per Capita GNP, 1985 (US$)
Dominica	0.1	22	5	1.7	41	0.1	0.2	13	3.0	40/7	75	—	—	—/—	1,160
Dominican Republic	6.5	33	8	2.5	28	8.4	11.5	70	4.0	41/3	63	52	60	14/55	810
Grenada	0.1	26	7	1.9	37	0.1	0.1	22	3.1	39/7	72	—	—	—/—	970
Guadeloupe	0.3	20	7	1.3	52	0.4	0.5	14	2.4	31/7	72	46	—	—/—	—
Haiti	6.2	36	13	2.3	30	7.7	10.1	107	4.9	39/6	53	26	33	2/43	350
Jamaica	2.5	26	5	2.0	34	3.0	4.1	20	3.1	36/7	73	54	86	35/−2	940
Martinique	0.3	17	7	1.1	64	0.4	0.4	13	2.0	28/8	74	71	—	—/—	—
Netherlands Antilles	0.2	19	6	1.4	51	0.2	0.2	10	2.1	30/7	75	50	—	—/—	6,110
Peurto Rico	3.3	19	7	1.3	54	3.7	4.1	15.7	2.2	30/9	75	67	—	—/—	4,850
St. Kitts-Nevis	0.05	26	11	1.6	45	0.1	0.1	27.8	2.8	37/9	67	45	—	—/—	1,520
Saint Lucia	0.1	30	6	2.5	28	0.2	0.2	17.6	4.1	44/6	71	40	—	—/—	1,210
St. Vincent and the Grenadines	0.1	26	7	2.0	35	0.1	0.2	26.5	3.2	43/6	69	—	—	—/—	840
Trinidad and Tobago	1.3	27	7	2.0	34	1.6	2.2	20	3.2	34/6	70	34	99	148/15	6,010
TROPICAL SOUTH AMERICA	**232**	**30**	**8**	**2.2**	**31**	**300**	**401**	**64**	**3.7**	**37/4**	**65**	**68**	**72**	**—/—**	**1,590**
Bolivia	6.5	40	14	2.6	27	9.2	13.2	127	5.1	43/3	53	48	43	10/47	470
Brazil	141.5	29	8	2.1	33	179.5	233.8	63	3.5	36/4	65	71	76	19/51	1,640
Colombia	29.9	28	7	2.1	33	38.0	49.3	48	3.1	36/3	65	65	81	25/41	1,320
Ecuador	10.0	35	8	2.8	25	13.6	19.7	66	4.7	42/4	65	51	59	20/140	1,160
Guyana	0.8	26	6	2.0	35	0.8	1.1	36	3.0	38/4	68	32	80	18/−41	570
Paraguay	4.3	36	7	2.9	24	6.0	8.8	45	4.9	41/4	65	43	25	7/54	940
Peru	20.7	35	10	2.5	28	28.0	38.6	94	4.8	41/4	60	69	52	19/5	960
Suriname	0.4	27	7	2.1	33	0.5	0.6	33	3.3	37/4	69	66	—	49/−28	2,570
Venezuela	18.3	32	6	2.7	26	24.7	35.4	38	3.9	40/3	69	76	—	96/46	3,110
TEMPERATE SOUTH AMERICA	**47**	**23**	**8**	**1.5**	**46**	**55**	**67**	**31**	**3.0**	**31/8**	**69**	**84**	**70**	**—/—**	**1,920**
Argentina	31.5	24	8	1.6	44	37.2	45.6	35.3	3.3	31/9	70	84	63	50/9	2,130
Chile	12.4	22	6	1.6	44	14.8	17.7	19.5	2.4	32/6	68	83	85	27/−20	1,440
Uruguay	3.1	18	10	0.8	90	3.2	3.5	30.4	2.5	27/11	71	85	79	18/−29	1,660

Region or Country	Population Estimate mid-1987 (millions)	Crude Birth Rate	Crude Death Rate	Natural Increase (annual, %)	Population "Doubling Time" in Years (at current rate)	Population Projected to 2000 (millions)	Population Projected to 2020 (millions)	Infant Mortality Rate[a]	Total Fertility Rate[b]	% Population Under Age 15/65+	Life Expectancy at Birth (years)	Urban Population (%)	Population with Access to Safe Water Supply (%, 1983)	Energy Consumption Per Capita, 1984 (gigajoules)/Percent Change Since 1970	Per Capita GNP, 1985 (US$)
EUROPE	**495**	**13**	**10**	**0.3**	**272**	**507**	**502**	**13**	**1.8**	**21/13**	**74**	**73**	**—**	**124/13**	**7,280**
NORTHERN EUROPE	**84**	**13**	**11**	**0.2**	**402**	**85**	**84**	**9**	**1.8**	**20/15**	**75**	**85**	**—**	**—/—**	**9,200**
Denmark	5.1	11	11	−0.1	(—)	5.0	4.7	7.9	1.4	18/15	75	84	—	133/−16	11,240
Finland	4.9	13	10	0.3	224	5.0	4.9	6.5	1.7	19/13	75	60	—	146/23	10,870
Iceland	0.2	16	7	0.9	76	0.3	0.3	5.7	1.9	26/10	77	89	—	150/22	10,720
Ireland	3.5	18	9	0.8	85	4.2	5.0	8.9	2.5	30/11	73	56	—	96/27	4,840
Norway	4.2	12	11	0.2	408	4.3	4.3	8.3	1.7	20/16	76	70	—	193/44	13,890
Sweden	8.4	12	11	0.1	636	8.3	8.0	6.8	1.7	18/17	77	83	—	137/−19	11,890
United Kingdom	56.8	13	12	0.2	462	57.3	56.6	9.4	1.8	19/15	74	90	—	138/−3	8,390
WESTERN EUROPE	**156**	**12**	**11**	**0.1**	**620**	**155**	**146**	**9**	**1.5**	**18/14**	**75**	**79**	**—**	**—/—**	**10,270**
Austria	7.6	12	12	0.0	(—)	7.5	7.0	11.2	1.5	18/14	74	55	—	117/22	9,150
Belgium	9.9	12	11	0.0	1,732	9.8	9.0	9.4	1.5	19/14	73	95	—	145/−12	8,450
France	55.6	14	10	0.4	178	57.3	58.4	8.1	1.8	21/13	75	73	—	114/1	9,550
Germany, West	61.0	10	12	−0.2	(—)	58.4	50.5	9.5	1.3	15/15	74	85	—	163/8	10,940
Luxembourg	0.4	11	11	0.0	3,465	0.4	0.3	9.0	1.4	17/13	73	78	—	328/−34	13,380
Netherlands	14.6	12	9	0.4	182	15.0	14.6	8.0	1.5	20/12	76	89	—	172/28	9,180
Switzerland	6.6	12	9	0.2	301	6.5	6.1	6.9	1.5	18/14	76	57	—	108/14	16,380
EASTERN EUROPE	**113**	**16**	**12**	**0.4**	**174**	**118**	**125**	**18**	**2.1**	**24/11**	**71**	**63**	**—**	**—/—**	**—**
Bulgaria	9.0	13	12	0.1	578	9.4	9.5	15.8	2.0	22/11	72	66	—	166/50	-
Czechoslovakia	15.6	15	12	0.3	257	16.2	16.8	14.0	2.1	24/11	71	74	—	183/16	—
Germany, East	16.7	14	14	0.0	3,465	16.8	16.8	9.6	1.8	19/14	73	77	—	223/24	—
Hungary	10.6	12	14	−0.2	(—)	10.7	10.4	20.4	1.8	22/12	70	56	—	112/36	1,940
Poland	37.8	18	10	0.8	88	40.8	44.6	18.5	2.3	25/9	71	60	—	133/27	2,120
Romania	22.9	16	11	0.5	141	24.5	26.6	25.6	2.3	25/9	71	53	—	134/52	—
SOUTHERN EUROPE	**144**	**12**	**9**	**0.3**	**201**	**149**	**148**	**17**	**1.7**	**23/12**	**74**	**69**	**—**	**—/—**	**4,630**
Albania	3.1	26	6	2.0	34	3.8	4.6	43	3.3	35/5	71	34	—	39/117	—
Greece	10.0	12	9	0.2	289	10.2	10.1	14.0	1.8	22/13	74	70	—	66/103	3,550
Italy	57.4	10	10	0.1	1,155	57.5	53.5	10.9	1.4	21/13	75	72	—	90/16	6,520

Region or Country	Population Estimate mid-1987 (millions)	Crude Birth Rate	Crude Death Rate	Natural Increase (annual, %)	Population "Doubling Time" in Years (at current rate)	Population Projected to 2000 (millions)	Population Projected to 2020 (millions)	Infant Mortality Rate[a]	Total Fertility Rate[b]	% Population Under Age 15/65+	Life Expectancy at Birth (years)	Urban Population (%)	Population with Access to Safe Water Supply (%, 1983)	Energy Consumption Per Capita, 1984 (gigajoules)/ Percent Change Since 1970	Per Capita GNP, 1985 (US$)
Malta	0.4	16	8	0.8	89	0.4	0.5	11.7	2.0	24/9	73	85	—	45/62	3,300
Portugal	10.3	12	10	0.3	257	10.9	11.4	16.7	1.9	24/12	73	30	—	38/74	1,970
Spain	39.0	13	8	0.5	147	41.0	41.6	10.5	1.8	23/13	76	91	—	65/50	4,360
Yugoslavia	23.4	16	9	0.7	102	25.0	25.9	28.8	2.1	25/9	71	46	—	73/77	2,070
USSR	**284**	**19**	**11**	**0.9**	**79**	**312**	**355**	**26**	**2.5**	**26/9**	**69**	**65**	**—**	**176/45**	**7,400**
OCEANIA	**25**	**20**	**8**	**1.2**	**59**	**29**	**35**	**36**	**2.6**	**28/8**	**72**	**71**	**—**	**135/25**	**8,320**
Australia	16.2	16	8	0.8	86	17.9	20.0	9.9	2.0	24/10	76	86	—	180/27	10,840
Fiji	0.7	28	5	2.3	31	0.9	1.3	18.5	3.2	37/4	67	37	—	12/−12	1,700
French Polynesia	0.2	28	4	2.4	29	0.3	0.3	23	3.6	37/3	71	57	—	—/—	7,840
New Caledonia	0.2	24	6	1.8	38	0.2	0.2	36	3.1	36/4	68	60	—	—/—	5,760
New Zealand	3.3	16	8	0.8	92	3.6	3.8	10.8	1.9	24/11	74	84	—	113/38	7,310
Papua-New Guinea	3.6	36	12	2.4	29	4.8	6.8	100	5.3	42/2	54	13	16	9/93	710
Solomon Islands	0.3	42	6	3.6	19	0.5	0.8	43	6.6	48/3	68	9	—	8/26	510
Vanuatu	0.2	39	5	3.3	21	0.2	0.3	39	5.8	45/3	68	18	—	—/—	—
Western Samoa	0.2	31	7	2.4	29	0.2	0.4	51	5.0	38/3	65	21	—	—/—	660

[a]Infant deaths per 1,000 live births
[b]Average number of children born to a woman during her lifetime
A dash (—) indicates that no estimate is available
A (z) signifies an amount which rounds to zero

Glossary

A

absorbing barrier
A natural, cultural, transportational, or distance blockage preventing the diffusion of an innovation. *See also* interrupting barrier

accessibility
The relative ease with which a destination may be reached from other locations.

acculturation
Cultural modification or change resulting from one culture group or individual adopting traits of a more advanced or dominant society; cultural development through "borrowing."

acid rain
Precipitation that is unusually acidic; created when oxides of sulfur and nitrogen change chemically as they dissolve in water vapor in the atmosphere and return to earth as acidic rain, snow, or fog.

activity space
The area within which people move freely on their rounds of regular activity.

adaptation
A presumed modification of heritable traits through response to environmental stimuli.

agglomeration economies
The savings that result from the spatial groupings of activities, such as industries or retail stores.

air mass
A large body of air with little horizontal variation in temperature, pressure, and humidity.

air pressure
The weight of the atmosphere as measured at a point on the earth's surface.

alluvial fan
A fan-shaped accumulation of alluvium deposited by a stream at the base of a hill or mountain.

alluvium
Sediment carried by a stream and deposited in a floodplain or delta.

anaerobic digestion
The process by which organic waste is decomposed in an oxygen-free environment to produce methane gas (biogas).

antecedent boundary
A boundary line established before the area in question is well populated.

anticline
A folded rock layer bowed upward in an archlike manner.

aquifer
Underground porous and permeable rock that is capable of holding groundwater, especially rock that supplies economically significant quantities of water to wells and springs.

Arctic haze
Air pollution resulting from the transport by air currents of combustion-based pollutants to the area north of the Arctic Circle.

arithmetic density
See crude density.

arroyo
A steep-sided, flat-bottomed gully, usually dry, carved out of desert land by rapidly flowing water.

artificial boundary
(*syn:* geometric boundary) A boundary without obvious physical geographic basis; often a section of a parallel of latitude or a meridian of longitude.

assimilation
The social process of merging into a composite culture, losing separate ethnic or social identity and becoming culturally homogenized.

asthenosphere
A partially molten, plastic layer above the core and lower mantle of the earth.

atmosphere
The gaseous mass surrounding the earth.

atoll
A near circular low coral reef formed in shallow water enclosing a central lagoon; most common in the central and western Pacific Ocean.

B

barchan
A crescent-shaped sand dune; the horns of the crescent point downwind.

basic sector
Those products of an urban unit that are exported outside the city itself, earning income for the community.

bench mark
A surveyor's mark indicating the position and elevation of some stationary object; used as a reference point in surveying and mapping.

biochore
A major structural subdivision of natural vegetation.

biocide
A chemical used to kill plant and animal pests and disease organisms.

biological magnification
The accumulation of a chemical in the fatty tissue of an organism and its concentration at progressively higher levels in the food chain.

biomass
Living matter, plant and animal, in any form.

biomass fuels
Combustible and/or fermentable material of plant or animal origin, such as wood, corn cobs, human and animal wastes.

biome
See ecosystem.

biosphere
(*syn:* ecosphere) The thin film of air, water, and earth within which we live, including the atmosphere, surrounding and subsurface waters, and the upper reaches of the earth's crust.

birth rate
The ratio of the number of live births during one year to the total population, usually at the midpoint of the same year, expressed as the number of births per year per 1000 population.

boundary
A line separating one political unit from another.

boundary definition
A general agreement between two states about the allocation of territory between them.

boundary delimitation
The plotting of a boundary line on maps or aerial photographs.

boundary demarcation
The actual marking of a boundary line on the ground; the final stage in boundary development.

butte
A small, flat-topped, isolated hill with steep sides, common in dry-climate regions.

C

carcinogen
A substance that produces or incites cancerous growth.

carrying capacity
The numbers of any population that can be adequately supported by the available resources upon which that population subsists; for humans, the numbers supportable by the known and used resources—usually agricultural—of an area. *See also* homeostatic plateau.

cartogram
A map that has been simplified to present a single idea in a diagrammatic way; the base is not normally true to scale.

central business district (CBD)
The center or "downtown" of an urban unit, where retail stores, offices, and cultural activities are concentrated and where land values are high.

central city
That part of the urban area contained within the boundaries of the main city around which suburbs have developed.

central place
A nodal point for the distribution of goods and services to a surrounding hinterland population.

central place theory
A deductive theory formulated by Walter Christaller (1893–1969) to explain the size and distribution of settlements through reference to competitive supply of goods and services to dispersed rural populations.

channelization
The modification of a stream channel; specifically, the straightening of meanders or dredging of the stream channel to deepen it.

channelized migration
The tendency for migration to flow between areas that are socially and economically allied by past migration patterns, by economic and trade connections, or by some other affinity.

chemical weathering
The decomposition of earth materials due to chemical reactions that include oxidation, hydration, and carbonation.

chlorofluorocarbons (CFCs)
A family of synthetic chemicals that have significant commercial applications but whose emissions are contributing to the depletion of the ozone layer.

circumpolar vortex
High-altitude winds circling the poles from west to east.

city
A multifunctional nucleated settlement with a central business district and both residential and nonresidential land uses.

climate
A summary of weather conditions in a place or region over a period of time.

coal gasification
A process by which crushed coal is burned in the presence of steam or oxygen to produce a synthetic gas.

coal liquefaction
A process whereby coal is heated to produce a variety of liquid products that can be used as fuels.

coal slurry
A mixture of finely ground coal and water that is moved by pipeline.

cogeneration
The simultaneous use of a single fuel for the generation of electricity and low-grade central heat.

cohort
A population group unified by a specific common characteristic, such as age.

collective farm
In the Soviet planned economy, the term refers to the cooperative operation of an agricultural enterprise under state control of production and market, but without full status or support as a state enterprise.

commercial economy
The production of goods and services for exchange in competitive markets where price and availability are determined by supply and demand forces.

commercial energy
Commercially traded fuels such as coal, oil, or natural gas and excluding wood, vegetable or animal wastes, or other biomass.

compact state
A state whose territory is nearly circular.

comparative advantage
A region's profit potential for a productive activity compared to alternate areas of production of the same good or to alternate uses of the region's resources.

complementarity
A term used to describe the actual or potential relationship between two units that each produce different goods or services for which the other has a demand, resulting in an exchange between the units.

concentric zone
One of a series of circular belts of land uses around the central business district, each belt housing distinct functions.

conformal map projection
One on which the shapes of small areas are accurately portrayed.

conic map projection
One based on the projection of the grid system onto a cone.

consequent boundary
(*syn:* ethnographic boundary) A boundary line that coincides with some cultural divide, such as religion or language.

conservation
The wise use or preservation of natural resources so as to maintain supplies and qualities at levels sufficient to meet present and future needs.

continental drift
The hypothesis that an original single land mass (Pangaea) broke apart and that the continents have moved very slowly over the asthenosphere to their present locations.

contour interval
The vertical distance separating two adjacent contour lines.

contour line
A map line along which all points are of equal elevation above or below a datum plane, usually mean sea level.

convectional precipitation
Rain produced when heated, moisture-laden air rises and then cools below the dew point.

coral reef
A rocklike landform in shallow tropical water composed chiefly of compacted coral and other organic material.

core
In urban geography, that part of the central business district characterized by intensive land development.

Coriolis effect
A fictitious force used to describe motion relative to a rotating earth; specifically, the force that tends to deflect a moving object or fluid to the right (clockwise) in the northern hemisphere and to the left (counterclockwise) in the southern hemisphere.

creole
A language developed from a pidgin to become the native tongue of a society.

critical distance
The distance beyond which cost, effort, and/or means play an overriding role in the willingness of people to travel.

crude birth rate (CBR)
See birth rate

crude death rate (CDR)
See death rate

crude density
(*syn:* arithmetic density). The number of persons per unit area of land.

crude oil
A mixture of hydrocarbons that exists in a liquid state in underground reservoirs; petroleum as it occurs naturally, as it comes from an oil well, or after extraneous substances have been removed.

cultural integration
The observation that all aspects of a culture are interconnected; no part can be altered without impact upon other culture traits.

cultural lag
The retention of established culture traits despite changing circumstances rendering them inappropriate.

cultural landscape
The surface of the earth as modified by human action, including housing types, settlement patterns, and agricultural land use.

culture
The totality of learned behaviors and attitudes transmitted within a society to succeeding generations by imitation, instruction, and example.

culture complex
An integrated assemblage of culture traits structuring the behaviors and values of a society.

culture hearth
A region within which an advanced, distinct set of culture traits developed and from which there was diffusion of distinctive technologies and ways of life.

culture realm
A major world area having sufficient distinctiveness to be perceived as set apart from other realms in its cultural characteristics and complexes.

culture system
A generalization suggesting shared, identifying traits uniting two or more culture complexes.

culture trait
A single distinguishing feature of regular occurrence within a culture, such as the use of chopsticks or the observance of a particular caste system.

cyclone
A type of atmospheric disturbance in which masses of air circulate rapidly about a region of low atmospheric pressure.

cyclonic precipitation
See frontal precipitation.

cylindrical projection
Any of several map projections based on the projection of the globe grid onto a cylinder.

D

data base
In cartography, a digital record of geographic information.

DDT
A chlorinated hydrocarbon that is among the most persistent of the biocides in general use.

death rate
(*syn:* mortality rate) A mortality index usually calculated as the number of deaths per year per 1000 population.

decomposers
Microorganisms and bacteria that feed on dead organisms, causing their chemical disintegration.

deforestation
The clearing of land through total removal of forest cover.

delta
A triangular-shaped deposit of mud, silt, or gravel created by a stream where it flows into a body of standing water.

demographic momentum
(*syn:* population momentum) The tendency for population growth to continue despite stringent family planning programs because of a relatively high concentration of people in the childbearing years.

demographic transition
A model of the effect of economic development on population growth. A first stage involves both high birth and death rates; the second phase displays high birth rates and falling mortality rates and population increases. Phase three shows reduction in population growth as birth rates decline to the level of death rates. The final, fourth, stage implies again a population stable in size but larger in numbers than at the start of the transition cycle.

demography
The scientific study of population, with particular emphasis upon quantitative aspects.

density of population
(*syn:* crude density) A measurement of the numbers of persons per unit area of land within predetermined limits, usually political or census boundaries. *See also* physiological density.

dependency ratio
The number of dependents, old or young, that each 100 persons in the productive years must support.

deposition
The process by which silt, sand, and rock particles accumulate and create landforms such as stream deltas and talus slopes.

desertification
Extension of the desert landscape as a result of overgrazing, destruction of the forests, or other human-induced changes.

developable surface
A geometric surface such as a cylinder or cone that may be spread out flat wthout tearing or stretching.

dew point
The temperature at which air becomes saturated with water vapor.

dialect
A regional variation of a more widely spoken language.

diastrophism
The earth force that folds, faults, twists, and compresses rock.

dibble
Any small hand tool or stick to make a hole for planting.

diffusion
The gradual spread from an origin area of new ideas or practices.

directional bias
The tendency for people to have greater knowledge of places in some directions than in others as a result of barriers to the flow of information.

distance decay
The exponential decline of an activity or function with increasing distance from its point of origin.

distance-affected diffusion
The process by which contacts between people and the resulting diffusion of things or ideas is more likely to occur over short rather than long distances.

domestication
The successful transformation of plant or animal species from a wild state to a condition of dependency upon human management, usually with distinct physical change from wild forebears.

doubling time
The time period required for any beginning total, experiencing a compounding growth, to double in size.

dune
A wavelike desert landform created by wind-blown sand.

E

ecology
The scientific study of how living creatures affect each other and what determines their distribution and abundance.

economic rent
The net surplus yielded by any factor of production (frequently land) after all costs, including interest and all income from possible alternative uses, have been subtracted.

ecosphere
See biosphere.

ecosystem
(*syn:* biome) A population of organisms existing together in a particular area, together with the energy, air, water, soil, and chemicals upon which it depends.

ecotone
The zone of stress between dissimilar, adjacent ecosystems.

ecumene
Permanently inhabited areas of the earth.

electromagnetic spectrum
The entire range of radiation, including the shortest as well as the longer wavelengths.

el Niño
The periodic (every 3 to 7 or 8 years) buildup of warm water along the west coast of South America, replacing the cold Humboldt current off the Peruvian coast. El Niño is associated with both a fall in plankton levels (and decreased fish supply) and with short-term, widespread weather modification.

elongated state
A state whose territory is long and narrow.

energy
The ability to do work. *See also* kinetic energy, potential energy.

energy crop
Any plant grown specifically as a source of energy, including trees for burning and oil-bearing plants.

energy efficiency
The ratio of the output of useful energy from a conversion process to the total energy inputs.

environment
Surroundings; the totality of things that in any way may affect an organism, including both physical and cultural conditions; a region characterized by a certain set of physical conditions.

environmental determinism
The theory that the physical environment, particularly climate, molds human behavior.

environmental perception
The way people observe and interpret, and the ideas they have about, near or distant places.

environmental pollution
See pollution.

equal-area projection
See equivalent map projection.

equator
An imaginary line that encircles the globe halfway between the North and South Poles.

equidistant map projection
One on which true distances in all directions can be measured from one or two central points.

equinox
Day and night of equal length.

equivalent map projection
One on which the areas of regions are represented in correct or constant proportions to earth reality; also called *equal-area*.

erosion
The wearing away and removal of rock and soil particles from exposed surfaces by agents such as moving water, wind, or ice.

estuary
The lower course or mouth of a river where tides cause fresh water and salt water from the sea to mix.

ethnic separatism
Desire of a culturally distinctive group within a larger, politically dominant culture for regional autonomy.

ethnicity
Social status afforded to, usually, a minority group within a national population. Recognition is based primarily upon culture traits such as religion, distinctive customs, or native or ancestral national origin.

ethnocentrism
The belief that one's own ethnic group is superior to all others.

ethnographic boundary
See consequent boundary.

European Economic Community (EEC)
An economic association established in 1957 of a number of western European states that promotes free trade among member countries; often called the Common Market.

eutrophication
The increase of nutrients in a body of water; the nutrients stimulate the growth of algae, whose decomposition decreases the dissolved oxygen content of the water.

exclusive economic zone (EEZ)
As proposed in the Law of the Sea Convention, a zone of exploitation extending 200 nautical miles seaward from a coastal state in which it has exclusive mineral and fishing rights.

expansion diffusion
Spread of ideas, behaviors, or articles from one culture to others through contact and exchange of information; the dispersion leaves the phenomenon intact or intensified in its area of origin. *See also* relocation diffusion.

extensive agriculture
Crop or livestock system in which land quality or extent is more important than capital or labor inputs in determining output.

extrusive rock
Rock solidified from molten material that has issued out onto the earth's surface.

F

false-color image
A remotely sensed image whose colors do not appear natural to the human eye.

fast breeder reactor
A nuclear reactor that uses uranium-235 to release energy from the more abundant uranium-238.

fault
A break or fracture in rock produced by stress or the movement of lithospheric plates.

fault escarpment
A steep slope formed by the vertical movement of the earth along a fault.

filtering
In urban geography, a process whereby individuals of a lower income group replace residents of a portion of an urban area who are of a higher income group.

fjord
A glacial trough the lower end of which is filled with sea water.

floodplain
A valley area bordering a stream that is subject to inundation by flooding.

folding
The buckling of rock layers under pressure of moving lithospheric plates.

food chain
A sequence of nutritional energy transfers in an ecosystem accomplished when organisms at one trophic level feed upon those at a lower level.

food web
The total flow of energy in an ecosystem.

footloose
A descriptive term applied to manufacturing activities for which the cost of transporting material or product is not important in determining location of production.

formal region
An earth area throughout which a single feature or limited combination of features is of such uniformity that it can serve as the basis for an areal generalization.

forward-thrust capital
A capital city deliberately sited in a state's frontier zone.

fossil fuels
Hydrocarbon compounds of crude oil, natural gas, and coal that are derived from the accumulation of plant and animal remains in ancient sedimentary rocks.

fragmented state
A state whose territory contains isolated parts, separated and discontinuous.

frame
In urban geography, that part of the central business district characterized by such low-intensity land uses as warehouses and automobile dealers.

friction of distance
A measurement indicating the effect of distance upon the extent of interaction between two points. Generally, the greater the distance, the less the interaction or exchange or the greater the cost of achieving the exchange.

front
The line or zone of separation between two air masses of different temperatures and humidities.

frontal precipitation
(*syn:* cyclonic precipitation) Rain or snow produced when moist air of one air mass is forced to rise over the edge of another air mass.

frontier
That portion of a country adjacent to its boundaries and fronting another political unit; as a **frontier zone,** a belt lying between two nation-states or between settled and uninhabited or sparsely settled areas.

functional region
An earth area recognized as an operational unit based upon defined organizational criteria.

G

genetic drift
A chance modification of gene composition occurring in an isolated population and becoming accentuated through inbreeding.

gentrification
The movement into the inner portions of American cities of middle-class people who replace low-income populations and rehabilitate structures.

geometric boundary
see artificial boundary.

geomorphology
The scientific study of landform origins and evolutions and their processes.

geothermal power
Energy generated when hot water or steam is extracted from reservoirs in the earth's crust and fed to steam turbines at electric generating plants.

gerrymander
To divide an area into voting districts in such a way as to give one political party an unfair advantage in elections, to fragment voting blocks, or to achieve other nondemocratic objectives.

glacial till
Deposits of rocks, silt, and sand left by a glacier after it has receded.

glacial trough
A deep, U-shaped valley or trench formed by glacial erosion.

glacier
A huge mass of slowly moving land ice.

globe properties
Characteristics of the grid system of longitude and latitude on a globe; see Page 26.

gradation
The process responsible for the gradual erosional reduction of the land surface.

grade (of coal)
A classification of coals based on their content of waste materials.

graphic scale
A graduated line included in a map legend by means of which distances on the map may be measured in terms of ground distances.

gravity model
A mathematical prediction of the interaction between two bodies as a function of their size and of the distance separating them.

Green Revolution
Term suggesting the great increases in food production, primarily in subtropical areas, accomplished through the introduction of very high-yielding grain crops, particularly wheat and rice.

greenhouse effect
Heating of the earth's surface as short-wave solar energy passes through the atmosphere, which is transparent to it but opaque to reradiated long-wave terrestrial energy. Also refers to increasing the opacity of the atmosphere through addition to it of increased amounts of carbon dioxide from burning of fossil fuels.

grid system
The set of imaginary lines of latitude and longitude that intersect at right angles to form a reference system for locating points on the earth.

gross national product (GNP)
Tht total value of all goods and services produced by a nation per year.

groundwater
Subsurface water that accumulates below the water table in the pores and cracks of rock and soil.

growth center
An urban center that, if its economy were stimulated, would serve to generate economic expansion in surrounding areas.

H

half-life
The time required for one-half of the atomic nuclei of an isotope to decay.

hazardous waste
Discarded solid, liquid, or gaseous material that may pose a substantial threat to human health or the environment when it is improperly disposed of or stored.

heliostat
A mirror that reflects sunlight onto a central receiver.

herbicide
A chemical that kills plants, especially weeds. *See also* biocide.

hierarchical diffusion
The process by which contacts between people and the resulting diffusion of things or ideas occurs first among those at the same level of a hierarchy and then among elements at a lower level of the hierarchy; e.g., small-town residents acquire ideas or articles after they are common in large cities.

hierarchical migration
The tendency for individuals to move from small places to larger ones.

hierarchy of central places
The steplike series of urban units in classes differentiated by both size and function.

high-level waste (HLW)
Nuclear waste with a relatively high level of radioactivity.

hinterland
The market area or region served by an urban unit.

homeostatic plateau
The equilibrium level of population that can be supported adequately by available resources. Equivalent to *carrying capacity.*

humid continental climate
A climate of east coasts and continental interiors of middle latitudes, displaying large annual temperature ranges resulting from cold winters and hot summers. Precipitation at all seasons.

humid subtropical climate
A climate of the east coast of continents in lower middle latitudes, characterized by hot summers with convectional precipitation and by cool winters with cyclonic precipitation.

hurricane
A severe tropical cyclone with winds exceeding 75 mph (120 kmph) originating in the tropical region of the Atlantic Ocean, Caribbean Sea, or Gulf of Mexico.

hydroelectric power
The kinetic energy of moving water converted into electrical power by a power plant whose turbines are driven by flowing water.

hydrologic cycle
The system by which water is continuously circulated through the biosphere.

hydrosphere
All water at or near the earth's surface that is not chemically bound in rocks; includes the oceans, surface waters, groundwater, and water held in the atmosphere.

I

icebox effect
The tendency for certain kinds of air pollutants to lower temperatures on earth by reflecting incoming sunlight back into space, preventing it from reaching the earth.

iconography
In political geography, a term denoting the study of symbols that unite a nation.

ideological subsystem
The complex of ideas, beliefs, knowledge, and means of their communication that characterize a culture.

igneous rock
Rock formed as molten earth materials cool and harden either above or below the earth's surface.

Industrial Revolution
The term applied to the rapid economic and social changes in agriculture and manufacturing that followed the introduction of the factory system to the textile industry of England in the last quarter of the 18th century.

infant mortality rate
A refinement of the death rate to specify the ratio of deaths of infants age one year or less per 1000 live births.

infrared
Electromagnetic radiation having wavelengths greater than those of visible light.

innovation
Introduction into an area of new ideas, practices, or objects; an alteration of custom or culture that originates within the social group itself.

insolation
The solar radiation received at the earth's surface.

intensive agriculture
The application of large amounts of capital and/or labor per unit of cultivated land to increase output.

interaction model
See gravity model.

International Date Line
By international agreement, the designated line where each new day begins, generally following the 180th meridian.

interrupting barrier
An obstacle temporarily halting or deflecting the path of diffusion of an innovation. *See also* absorbing barrier.

intrusive rock
Rock resulting from the hardening of magma beneath the earth's surface.

isochrone
A line connecting points equidistant in travel time from a common origin.

isoline
A map line connecting points of constant value.

isotropic plain
A hypothetical portion of the earth's surface where, it is assumed, the land is everywhere the same and the characteristics of the inhabitants are everywhere similar.

J

jet stream
A high-speed wind near the boundary between the troposphere and the stratosphere generally moving west to east at speeds of 200 mph (320 kmph) or more.

K

karst
A limestone region marked by sinkholes, caverns, and underground streams.

kerogen
A waxy, organic material occurring in oil shales that can be converted into crude oil by distillation.

kinetic energy
The energy that results from the motion of a particle or body. *See also* energy.

L

Landsat satellite
One of a series of continuously orbiting satellites that carry scanning instruments to measure reflected light in both the visible and near-infrared portions of the spectrum.

lapse rate
The rate of change of temperature with altitude in the troposphere; the average lapse rate is about 3.5° F per 1000 feet (6.4° C per 1000 m).

large-scale map
A representation of a small land area, usually with a representative fraction of 1:75,000 or less.

latitude
A measure of distance north or south of the equator, given in degrees.

lava
Molten material that has emerged onto the earth's surface.

Law of the Sea Convention
A code of sea law approved by the United Nations in 1982 that authorizes, among other provisions, territorial waters extending 12 nautical miles from shore and 200-nautical-mile-wide exclusive economic zones.

levee
In agriculture, a continuous embankment surrounding areas to be flooded. *See also* natural levee.

linguistic family
A group of languages thought to have a common origin.

liquefied natural gas (LNG)
Methane gas that has been liquefied by refrigeration for storage or transportation.

lithosphere
Solid shell of rocks resting on the asthenosphere.

locational rent
The additional value of land above its physical properties that is related to the location of the parcel.

loess
A deposit of wind-blown silt.

longitude
A measure of distance east or west of the prime meridian, given in degrees.

longshore current
A current that moves roughly parallel to the shore and transports the sand that forms beaches and sand spits.

low-level waste (LLW)
Low-level hazardous waste, produced principally by nuclear power plants and industries.

M

magma
Underground molten material.

malnutrition
Food intake insufficient in quantity or deficient in quality to sustain life at optimal conditions of health.

Malthusianism
Doctrine proposed by Thomas R. Malthus (1766–1834), that unless checked by self-control, war, or natural disaster, population will inevitably increase faster than will the food supplies needed to sustain it.

map projection
A method of transferring the grid system from the earth's curved surface to the flat surface of a map.

map scale
The ratio between length or distance on a map and the corresponding measurement on the earth. Scale may be represented verbally, graphically, or as a fraction.

marginal cost
The additional cost incurred to produce an additional unit of output.

marine west coast climate
A regional climate found on the west coast of continents in upper mid-latitudes, rainy all seasons with relatively cool summers and relatively mild winters.

mechanical weathering
The physical disintegration of earth materials, commonly by frost action, root action, or the development of salt crystals.

Mediterranean climate
A climate of lower mid-latitudes characterized by mild, wet winters and hot, dry, sunny summers.

megalopolis
A large, sprawled urban complex with contained open, nonurban land, created through the spread and joining of separate metropolitan areas; the name applied to the continuous functionally urban area of coastal northeastern United States from Maine to Virginia.

megawatt
A unit of power equal to 1 million watts (1000 kilowatts) of electricity.

mental map
A map drawn to represent the mental image(s) a person has of an area.

Mercator projection
A true conformal cylindrical projection first published in 1569, useful for navigation.

meridian
A north–south line of longitude; on the globe, all meridians are of equal length and converge at the poles.

mesa
An extensive, flat-topped elevated tableland with horizontal strata, a resistant cap rock, and one or more steep sides; a large butte.

metamorphic rock
Rock transformed from igneous and sedimentary rocks by earth forces that generate heat, pressure, or chemical reaction.

metropolitan area
A large functional entity, perhaps containing several urbanized areas, discontinuously built up but operating as a coherent economic whole.

microdistrict
The basic residential planning unit, usually a superblock, characteristic of new urban developments in planned economies.

migration
The movement of people or other organisms from one region to another.

migration field
An area that sends major migration flows to a given place or the area that receives major flows from a place.

mineral
A natural inorganic substance that has a definite chemical composition and characteristic crystal structure, hardness, and density.

monoculture
Agricultural system dominated by a single crop.

monsoon
A wind system that reverses direction seasonally, producing wet and dry seasons; used especially to describe the wind system of South, Southeast, and East Asia.

moraine
Any of several types of landforms composed of debris transported and deposited by a glacier.

mortality rate
See death rate.

mountain breeze
The downward flow of heavy, cool air at night from mountainsides to lower valley locations.

multiple-nuclei model
The idea that large cities develop by peripheral spread not from one but from several nodes of growth, and that therefore there are many origin points of the various land use types in an urban area.

multiplier effect
The expected addition of nonbasic workers and dependents to a city's total employment and population that accompanies new basic employment.

N

nation
A culturally distinctive group of people occupying a particular region and bound together by a sense of unity arising from shared beliefs and customs.

nationalism
A sense of unity binding the people of a state together; devotion to the interests of a particular nation; an identification with the state and an acceptance of national goals.

nation-state
A state whose territory is identical to that occupied by a particular nation.

natural boundary
A boundary line based on recognizable physiographic features, such as mountains, rivers, or deserts.

natural gas
A mixture of hydrocarbons and small quantities of nonhydrocarbons existing in a gaseous state or in solution with crude oil in natural reservoirs.

natural hazard
A process or event in the physical environment that has consequences harmful to humans.

natural increase
The growth of a population through excess of births over deaths, excluding the effects of immigration or emigration.

natural levee
An embankment on the sides of a meandering river formed by deposition of silt during floods.

natural resource
A physically occurring item that a population perceives to be necessary and useful to its maintenance and well-being.

neo-Malthusianism
The advocacy of population control programs to preserve and improve general national prosperity and well-being.

niche
The place an organism or species occupies in an ecosystem.

nomadic herding
Migratory but controlled movement of livestock solely dependent upon natural forage.

nonbasic sector
Those economic activities of an urban unit that service the resident population.

nonfuel mineral resources
Mineral used for purposes other than providing a source of energy.

nonrenewable resource
A natural resource that exists in a finite amount, such as fossil fuels and other minerals.

North and South Poles
The end points of the axis about which the earth spins.

North Atlantic drift
The massive movement of warm water in the Atlantic Ocean from the Caribbean Sea and Gulf of Mexico in a northeasterly direction to the British Isles and Scandinavia.

nuclear fission
The controlled splitting of an atom to release energy.

nuclear fusion
The combining of two atoms of deuterium into a single atom of helium in order to release energy.

nuclear power
Electricity generated by a power plant whose turbines are driven by steam produced by the fissioning of nuclear fuel in a reactor.

nutrient
A mineral or other element that an organism requires for normal growth and development.

O

oil shale
Sedimentary rock containing solid organic material (kerogen) that can be extracted and converted into a crude oil by distillation.

organic
Derived from living organisms; plant or animal life.

Organization of Petroleum Exporting Countries (OPEC)
An international cartel composed of 13 nations that aims to pursue common oil-marketing and pricing policies.

orographic precipitation
Rain or snow caused when warm, moisture-laden air is forced to rise over hills or mountains in its path and is thereby cooled.

orthophotomap
An aerial photograph to which a grid system and certain map symbols have been added.

out-sourcing
Producing parts or products abroad for domestic use or sale.

outwash plain
A gently sloping area in front of a glacier composed of neatly stratified glacial till carried out of the glacier by meltwater streams.

overburden
Soil and rock of little or no value that overlies a deposit of economic value, such as coal.

overpopulation
A value judgment that the resources of an area are insufficient to sustain adequately its present population numbers.

oxbow lake
A crescent-shaped lake contained in an abandoned meander of a river.

ozone
A gas molecule consisting of 3 atoms of oxygen (O_3) formed when diatomic oxygen (O_2) is exposed to ultraviolet radiation. As a damaging component of photochemical smog formed at the earth's surface, it is a faintly blue, poisonous agent with a pungent odor.

ozone layer
A layer of ozone in the high atmosphere that protects life on earth by absorbing ultraviolet radiation from the sun.

P

Paleolithic period
Old Stone Age; the earliest stage of human culture, it was characterized by hunting and gathering economies and by use of fire and worked tools.

Pangaea
The name given to the supercontinent that is thought to have existed 200 million years ago.

parallel of latitude
An east-west line indicating the distance north or south of the equator.

PCBs
Polychlorinated biphenyls, compounds containing chlorine that can be biologically magnified in the food chain.

peak value intersection
The most accessible and costly parcel of land in the central business district and, therefore, in the entire urbanized area.

permafrost
Permanently frozen subsoil.

pesticide
A chemical that kills insects, rodents, fungi, weeds, and other pests. *See also* biocide.

petroleum
A general term applied to oil and oil products in all forms, such as crude oil and unfinished oils.

photochemical smog
A form of polluted air produced by the interaction of hydrocarbons and oxides of nitrogen in the presence of sunlight.

photovoltaic cell
A device that converts solar energy directly into electrical energy. *See also* solar power.

physiological density
The number of persons per unit area of agricultural land. *See also* density of population.

pidgin
An auxiliary language derived, with reduction of vocabulary and simplification of structure, from other languages. Not a native tongue, it is employed to provide a mutually intelligible vehicle for limited transactions of trade or administration.

pixel
An extremely small sensed unit of a digital image.

place utility
The perceived attractiveness of a place in its social, economic, or environmental attributes.

planar projection
Any of several map projections based on the projection of the globe grid onto a plane.

planned economy
Production of goods and services, usually consumed or distributed by a governmental agency, in quantities and at prices determined by governmental program.

plantation
A large agricultural holding, frequently foreign-owned, devoted to the production of a single export crop.

plate tectonics
The theory that the lithosphere is divided into plates that slide or drift very slowly over the asthenosphere.

playa
A temporary lake or lake bed found in a desert environment.

Pleistocene
The geological epoch dating from 2 million to 11,000 years ago during which four stages of continental glaciation occurred.

pollution
The introduction into the biosphere of materials that, because of their quantity, chemical nature, or temperature, have a negative impact on the ecosystem or that cannot be readily disposed of by natural recycling processes.

population-density-affected diffusion
The process by which contacts between people and the resulting diffusion of things or ideas is more likely to occur in areas where population density is great rather than sparse.

population momentum
See demographic momentum.

population projection
A report of future size, age, and sex composition of a population based upon assumptions applied to current data.

population pyramid
A graphic depiction of the age and sex composition of a (usually national) population.

potential energy
The energy stored in a particle or body. *See also* energy.

precipitation
All moisture, solid and liquid, that falls to the earth's surface from the atmosphere.

pressure gradient force
Differences in air pressure between areas that induce air to flow from areas of high to areas of low pressure.

primary activities
Those parts of the economy involved in making natural resources available for use or further processing; includes mining, agriculture, forestry, fishing or hunting, grazing.

primate city
The largest city of a country, with a population more than twice (in central place theory, three times) the size of the next largest city; usually the capital city and a center of wealth and power enabling it to increase its relative importance.

prime meridian
An imaginary line passing through the Royal Observatory at Greenwich, England, serving by agreement as the zero degree line of longitude.

private plot
In Soviet and Eastern Bloc economies, small garden parcels allotted to collective farmers and urban workers.

projection
See map projection.

prorupt state
A state that possesses one or more narrow extensions of territory.

pull factor
A characteristic of a region that acts as an attractive force, drawing migrants from other regions.

push factor
A characteristic of a region that contributes to the dissatisfaction of residents.

Q

quaternary activity
That employment concerned with research, with the gathering or disseminating of information, and with administration, including administration of the other economic activity levels.

R

race
A subset of human population whose members share certain distinctive, inherited biological characteristics.

radar
A device for detecting distant objects by analysis of very high-frequency radio waves beamed at and reflected from their surfaces.

rank (of coal)
A classification of coals based on their age and energy content; those of higher rank are more mature and richer in energy.

rate
The frequency of occurrence of an event during a specified time period.

recycling
The reuse of resources after they have passed through some form of treatment (e.g., melting down glass bottles to produce new bottles).

reflection
The process of returning to outer space some of the earth's received insolation.

region
In geography, the term applied to an earth area that displays a distinctive grouping of physical or cultural phenomena or is functionally united as a single organizational unit.

regional autonomy
A measure of self-governance for a region within a state.

relative humidity
A measure of the relative dampness of the atmosphere; the ratio between the amount of water vapor in the air and the maximum amount that it could hold at the same temperature, given in percent.

relict boundary
A former boundary line that is still discernable and marked by some cultural landscape feature.

relocation diffusion
The transfer of ideas, behaviors, or articles from one place to another through the migration of those possessing the feature transported; also, spatial relocation in which a phenomenon leaves an area of origin as it is transported to a new location. *See also:* expansion diffusion.

remote sensing
Any of several techniques of obtaining images of an area without having the sensor in direct physical contact with it, as by air photography or satellite sensors.

renewable resource
A naturally occurring material that can be consumed and restored after use, either because it flows continuously (as solar radiation or wind) or is renewed within a short period of time (as biomass). *See also* sustained yield.

representative fraction (RF)
The scale of a map expressed as a ratio of a unit of distance on the map to distance measured in the same unit on the ground, e.g., 1:250,000.

reradiation
A process by which the earth returns solar energy to space; some of the short-wave solar energy that is absorbed into the land and water is returned to the atmosphere in the form of long-wave terrestrial radiation.

resource
See natural resource.

return migration
The stream of migrants who subsequently decide to return to their point of origin.

rhumb line
A line of constant compass bearing; it cuts all meridians at the same angle.

Richter scale
A logarithmic scale used to express the magnitude of an earthquake.

S

Sahel
The semiarid zone between the Sahara and the savanna area to the south in West Africa; a district of recurring drought, famine, and environmental degradation.

salinization
The process by which soil becomes saturated with salt, rendering the land unsuitable for agriculture; occurs when land that has poor drainage is improperly irrigated.

sandbar
An offshore shoal of sand created by the backwash of waves.

sanitary landfill
Disposal of solid wastes by spreading them in layers covered with enough soil or ashes to control odors, rats, and flies.

savanna
A tropical grassland characterized by widely dispersed trees and experiencing pronounced yearly wet and dry seasons.

scale
See map scale.

secondary activities
Those parts of the economy involved in the processing of raw materials derived from primary activities; includes manufacturing, construction, power generation.

sector theory
A model used to describe wedge-shaped sectors of different land uses radiating outward from the central business district.

sedimentary rock
Rock formed from particles of gravel, sand, silt, and clay that were eroded from already existing rocks.

seismic waves
Vibrations within the earth set off by earthquakes.

shaded relief
A method of representing the three-dimensional quality of an area by use of continuous graded tone to simulate the appearance of sunlight and shadows.

shale oil
The crude oil resulting from the distillation of kerogen in oil shales.

shifting cultivation
(*syn:* slash-and-burn agriculture, swidden agriculture) Crop production of forest clearings kept in cultivation until their quickly declining fertility is lost. Cleared plots are then abandoned and new sites are prepared.

sinkhole
A deep surface depression formed when ground collapses into a subterranean cavern.

site
The place where something is located; the immediate surroundings.

situation
The location of something in relation to the physical and human characteristics of a larger region.

slash-and-burn agriculture
See shifting cultivation.

small-scale map
A representation of a large land area on which small features (highways, buildings) cannot be shown true to scale.

sociological subsystem
The totality of expected and accepted patterns of interpersonal relations common to a culture or subculture.

soil depletion
The loss of some or all of the vital nutrients from soil.

soil horizon
A layer of soil distinguished from other soil zones by color, texture, and other characteristics resulting from soil-forming processes.

solar power
The radiant energy generated by the sun; sun's energy captured and directly converted for human use. *See also* photovoltaic cell

solid waste
Unwanted materials generated in production or consumption processes that are solid rather than liquid or gaseous in form.

solstice
The date when the sun's direct rays are farthest north or south of the equator and, therefore, the hours of daylight in either hemisphere are most or fewest. Solstice days are 21–22 June and 21–22 December.

source region
In climatology, an area where an air mass forms.

spatial diffusion
The outward spread of a substance, concept, or population from its point of origin.

spatial interaction
The movement (e.g., of people, goods, information) between different places.

spatial search
The process by which individuals evaluate the alternative locations to which they might move.

spring wheat
Wheat sown in spring for ripening during the summer or autumn.

standard parallel
The tangent circle, usually a parallel of latitude, in a conic projection; along the standard line the scale is as stated on the map.

state
An independent political unit occupying a defined, permanently populated territory and having full sovereign control over its internal and foreign affairs.

state farm
Under Soviet and other planned economies, a government agricultural enterprise operated with paid employees.

steppe
Name applied to treeless mid-latitude grasslands.

strategic petroleum reserve (SPR)
Petroleum stocks maintained by the Federal Government for use during periods of major supply interruption.

stratosphere
The layer of the atmosphere that lies above the troposphere and extends outward to about 35 miles (56 km).

stream load
Eroded material carried by a stream in one of three ways, depending on the size and composition of the particles: (1) in dissolved form, (2) suspended by the water, (3) rolled along the stream bed.

subduction
The process by which one lithospheric plate is forced down into the asthenosphere as a result of collision with another plate.

subsequent boundary
A boundary line that is established after the area in question has been settled, and that considers the cultural characteristics of the bounded area.

subsidence
Settling or sinking of a portion of the land surface, sometimes as a result of the extraction of fluids such as oil or water from underground deposits.

subsistence agriculture
Any of several farm economies in which most crops are grown for food, nearly exclusively for local (producer) consumption.

subsistence economy
A system in which goods and services are created for the use of producers or their immediate families. Market exchanges are limited and of minor importance.

suburb
A functionally specialized segment of a large urban complex located outside the boundaries of the central city.

Superfund Law
The popular reference to the Comprehensive Environmental Response, Compensation and Liability Act, the primary goals of which are to identify and analyze all uncontrolled and abandoned hazardous waste sites in the U.S. and to assure that remedial action is taken at sites that pose grave danger to public health and the environment.

superimposed boundary
A boundary line placed over, and ignoring, an existing cultural pattern.

sustained yield
The practice of balancing harvesting with growth of new stocks so as to avoid depletion of the resource and ensure a perpetual supply. *See also* renewable resource.

swidden agriculture
See shifting cultivation.

syncline
A folded rock layer bowed downward in a troughlike manner.

syncretism
The development of a new form of, for example, religion or music, through the fusion of distinctive parental elements.

synthetic fuel (synfuel)
Crude oil and natural gas substitutes that can be synthesized from a variety of nonoil and nongas sources, including coal, tar sands, and wood.

systems analysis
An approach to the study of large systems through (1) segregation of the entire system into its component parts, (2) investigation of the interactions between system elements, and (3) study of inputs, outputs, flows, interactions, and boundaries within the system.

T

talus slope
A landform composed of rock particles that have accumulated at the base of a cliff, hill, or mountain.

tar sands
Sand and sandstone impregnated with heavy oil.

technological subsystem
The complex of material objects together with the techniques of their use by means of which people carry out their productive activities.

technology
An integrated system of knowledge and skills developed within a culture to carry out successfully purposeful and productive tasks.

temperature inversion
The condition caused by rapid reradiation in which air at lower altitudes is cooler than air aloft.

territoriality
Persistent attachment of most animals to a specific area; the behavior associated with the defense of the home territory.

tertiary activities
Those parts of the economy that fulfill the exchange function and that provide market availability of commodities; includes wholesale and retail trade and associated transportational, governmental, and informational services.

thermal pollution
The introduction of heated water into the environment, with consequent adverse effects on plants and animals.

thermal scanner
A remote sensing device that detects the energy (heat) radiated by objects on earth.

Third World
Those nations that, in contrast to the industrialized countries, typically have low per capita incomes, high population growth rates, and the majority of their labor forces engaged in agriculture.

threshold
In economic geography, the minimum market needed to support the supply of a product or service.

tidal power
The kinetic energy of ocean tides as they rise and fall harnessed to generate electricity as the water flows over reversible turbines.

topographic map
One that portrays the surface features of a relatively small area, often in great detail.

tornado
A small, violent storm characterized by a funnel-shaped cloud of whirling winds that can form beneath a cumulonimbus cloud in proximity to a cold front with wind speeds to 300 mph (480 kmph).

total fertility rate
The average number of children that would be born to each woman if, during her child-bearing years, she bore children at the current year's rate for women that age.

town
A nucleated settlement that contains a central business district but that is smaller and less functionally complex than a city.

trophic level
A nutritional transfer stage in an ecosystem.

tropical rain forest
Tree cover composed of tall, high-crowned evergreen deciduous species, associated with the continuously wet tropical lowlands.

tropical rain forest climate
The continuously warm, frost-free climate of tropical (and equatorial) lowlands, with abundant moisture year round.

troposphere
The atmospheric layer closest to the earth, extending outward about 7 to 8 miles (11 to 13 km) at the poles to about 16 miles (26 km) at the equator.

truck farming
The production of fruits and vegetables for market.

tsunami
A seismic sea wave generated by an earthquake or volcanic eruption.

tundra
The treeless area lying between the tree line of Arctic regions and the permanently ice-covered zone.

typhoon
Name given to hurricanes occurring in the western Pacific Ocean region.

U

Unigov
The unified government of a city and all or parts of its metropolitan region.

urban hierarchy
A ranking of cities based on their size and functional complexity.

urbanization
Transformation of a population from rural to urban status; the process of city formation and expansion.

urbanized area
A continuous built up landscape defined by building and population densities with no reference to the political boundaries of the city; may contain a central city and many contiguous cities, towns, suburbs, and unincorporated areas.

usable reserve
That portion of a resource that has been identified and is economically extractable with current technology.

V

valley breeze
The flow of air up mountain slopes during the day.

variable costs
In economic geography, costs of production inputs that display place to place differences.

verbal scale
A statement of the relationship between units of measure on a map and distance on the ground, as "one inch represents one mile."

volcanism
The earth force that transports heated material to or toward the surface of the earth.

von Thünen model
Model developed by Johann H. von Thünen (1783–1850) to explain the forces that control the prices of agricultural commodities and how those variable prices affect patterns of agricultural land use.

W

washes
The dry, braided channels in deserts that remain after the rush of rainfall runoff water.

water table
The upper limit of the saturated zone and therefore of groundwater.

weather
The state of the atmosphere at a given time and place.

weathering
Mechanical and chemical processes that fragment and decompose rock materials.

Weber model
Analytical model devised by Alfred Weber (1868–1958) to explain the principles governing the optimum location of industrial establishments.

wetland
A lowland area that is saturated with moisture, such as a marsh or tidal flat.

wind power
The kinetic energy of wind converted into mechanical energy by wind turbines that drive generators to produce electricity.

winter wheat
Wheat sown in fall for ripening the following spring or summer.

Z

zero population growth (ZPG)
A term suggesting a population in equilibrium, with birth rates and death rates nearly identical.

zoning
Designating by ordinance areas in a municipality for particular types of land use.

Credits

CHAPTER 7

Figure 7.8 © Michael Siluk.
Figure 7.15 AP/Wide World Photos.
Figure 7.18 Courtesy of United States Coast Guard.
Figure 7.23 AP/Wide World Photos.

CHAPTER 8

Figure 8.16 © William E. Ferguson.
Figure 8.17 United Nations/Photo by J. K. Isaac.

CHAPTER 9

Figure 9.1 © John Maher/EKM-Nepenthe.
Figure 9.2 © Kurt Thorson/EKM-Nepenthe.
Figure 9.4 © Wedigo Ferchland/Bruce Coleman, Inc.
Figure 9.5 © Thomas J. Bassett.
Figure 9.6 © V. Crowl/Visuals Unlimited.
Figure 9.7 © V. Crowl/Visuals Unlimited.
Figure 9.10 AP/Wide World Photos.
Figure 9.11 © Jean Fellmann.
Figure 9.15 © J. D. Cunningham/Visuals Unlimited.
Figure 9.25 AP/Wide World Photos.
Figure 9.33 © Jonathan Wright/Bruce Coleman, Inc.

CHAPTER 10

Figure 10.1 © Grant Heilman/Grant Heilman Photography.
Figure 10.10 © BLM/Visuals Unlimited.
Figure 10.12 Courtesy of Exxon Corporation and American Petroleum Institute.
Figure 10.13 © Frank Hoffman/Illustrator's Stock Photos.
Figure 10.14 Courtesy of Tosco Corporation and American Petroleum Institute.
Figure 10.15 Courtesy of Gulf Oil Company and American Petroleum Institute.
Figure 10.17 AP/Wide World Photos.
Figure 10.20 © Harold W. Scott.
Figure 10.21 © Robert Frerck/Odyssey Productions.
Figure 10.22 Courtesy of the City of Davis, CA.
Figure 10.23 Courtesy of Southern California Edison Company.
Figure 10.24 Courtesy of American Petroleum Institute.
Figure 10.25 Courtesy of Southern California Edison Company and American Petroleum Institute.
Figure 10.26b Courtesy of Union Oil Company of California and American Petroleum Institute.
Figure 10.27 Courtesy of Olympia Resource Management and American Petroleum Institute.
Figure 10.28 (all) Courtesy of GM Truck and Bus Group Public Relations.

CHAPTER 11

Figure 11.1 © Elizabeth L. Hovinen.
Figure 11.3a © B. F. Molnia/Terraphotographics/BPS.
Figure 11.3b © Carl May/Terraphotographics/BPS.
Figure 11.14 © John A. Jakle.
Figure 11.24 AP/Wide World Photos.
Figure 11.25 © Robin L. Ferguson/William E. Ferguson.
Figure 11.26 © Robert Frerck/Odyssey Productions.
Page 393 United Nations.

CHAPTER 12

Figure 12.5 © B. F. Molnia/Terraphotographics/BPS.
Figure 12.8 © William H. Allen, Jr.
Figure 12.13 © Ben Mathes.
Figure 12.17 © Gary and Elizabeth Hovinen.
Figure 12.19 © Elizabeth L. Hovinen.

ILLUSTRATION CREDITS

INTRODUCTION

Figure I.8 From Howard J. Nelson and William A. V. Clark, *The Los Angeles Metropolitan Experience*, p. 49. Association of American Geographers, Washington, D.C., 1976. Reprinted by permission.
Figure I.9 Prepared by the Chicago Area Transportation Study.
Figure I.12 From John R. Borchert, "America's Changing Metropolitan Regions," *Annals of the Association of American Geographers*, Vol. 62, No. 2, p. 358. Association of American Geographers, Washington, D.C., 1972. Reprinted by permission.

CHAPTER 1

Figure 1.9b Copyright 1977, Brooks-Roberts; with permission.
Figure 1.11 Copyright The University of Chicago Department of Geography.
Figure 1.16 Reproduced with the permission of the federal office of topography from 25.11.1987. [Bundesamt fur Landestopographie, Wabern, Switzerland]
Figure 1.23b From Daniel D. Arreola, "Urban Mexican Americans" in *Focus*, Vol. 34, No. 3, January/February 1984. Copyright © 1984 American Geographical Society. Reprinted by permission.
Figure 1.27 Reprinted with permission from *The Economist*, London.
Figure 1.30 From Map Supplement to the *Annals of the Association of American Geographers*, Vol. 76, No. 4. Association of American Geographers, Washington, D.C., 1986. Reprinted by permission.

CHAPTER 2

Figures 2.2a, 2.7b, 2.18, and 2.30a From Montgomery, Carla W., *Physical Geology*. © 1987 Wm. C. Brown Publishers, Dubuque, Iowa. All Rights Reserved. Reprinted by permission.
Figure 2.3 Source: Exxon Corporation, 1983.
Figure 2.4 Source: American Petroleum Institute.
Figure 2.5 A portion of the "World Ocean Floor" Panorama by Bruce C. Heezen and Marie Tharp, 1977, copyright © 1977 Marie Tharp.
Figure 2.6 From Barazangi and Dorman, *Bulletin of Seismological Society of America*, 1969. Reprinted by permission.
Figure 2.11b Copyright © 1986 by The New York Times Company. Reprinted by permission.
Figure 2.34 From Charles B. Hunt, *Geology of Soils: Their Evolution, Classification, and Uses*. Copyright © W. H. Freeman and Company.

CHAPTER 4

Figure 4.3 From Montgomery, Carla W., *Physical Geology*. © 1987 Wm. C. Brown Publishers, Dubuque, Iowa. All Rights Reserved. Reprinted by permission.
Figure 4.5 Redrawn from "Channelization: Shortcut to Nowhere" by Raymond V. Corning, in *Virginia Wildlife*, February 1975. Copyright © *Virginia Wildlife*. Reprinted by permission.
Figure 4.10 From O'Donnell, Frank, "Acid Rain Controls," *Sierra*, November/December 1983, p. 32. Reprinted by permission.
Figure 4.25 Reprinted with permission from *The Economist*, London.
Figure 4.27 Artwork by Richard Farrell. Copyright © 1983 Richard Farrell.

CHAPTER 5

Figure 5.3 Data from United Nations *Demographic Yearbook*, 1981.
Figure 5.7 Data from United Nations *Demographic Yearbook*, Population Reference Bureau, and the Censuses of Mexico, the United States, and West Germany.
Figure 5.26 Redrawn with permission from J. Norwine and A. Gonzalez (eds.), *The Third World*. © 1987 by Allen and Unwin.

CHAPTER 6

Figure 6.9 Adapted from Ernest S. Easterly, III, "Global Patterns of Legal Systems: Notes Toward a New Geojurisprudence," in *Geographical Review*, Vol. 67. © 1977 by the American Geographical Society. Reprinted by permission.
Figure 6.14 Redrawn with permission from Gerald F. Pyle and K. David Patterson, "Influenza Diffusion in European History: Patterns and Paradigms" in *Ecology of Disease*, Vol. 2, No. 3, 1984, pp. 173–184. Pergamon Journals, Ltd., Oxford, England.
Figure 6.19 Redrawn with Kurath, Hans, *A World Geography of the Eastern United States*. Copyright © 1949 by The University of Michigan Press. Reprinted by permission.
Figure 6.24 From Wilbur Zelinski, "An Approach to the Religious Geography of the United States: Patterns of Church Membership in 1952," *Annals of the Association of American Geographers*, Vol. 51, No. 2, p. 174. Association of American Geographers, Washington, D.C., 1961. Reprinted by permission.
Figure 6.28 Fig. 21-e from the Appendix, "Black Metropolis 1961" in *Black Metropolis*/ Vol. 1, by St. Claire Drake and Horace R. Cayton. © 1962 by St. Claire Drake and Horace R. Cayton. Reprinted by permission of Harper & Row, Publishers, Inc.
Figure 6.29 From Samuel C. Kincheloe, *The American City and Its Church*. Copyright © 1938 Friendship Press, Inc., New York.

CHAPTER 7

Figure 7.1 From Edward F. Bergman, *Modern Political Geography*. Copyright © 1975 Wm. C. Brown Publishers, Dubuque, Iowa. Reprinted by permission of Edward F. Bergman.
Figure 7.7 From *World Regional Geography: A Question of Place* by Paul Ward English, with James Andrew Miller. Reprinted with permission of Harper & Row, Publishers, Inc.
Figure 7.12 Source: *Oxford Atlas for New Zealand*. Copyright © 1960 Oxford University Press, Oxford, England.
Figure 7.16 Redrawn from *Tension Areas of the World*, by D. Gordon Bennett, editor, p. 310. © 1982 by Park Press.

CHAPTER 8

Figure 8.2 Redrawn from Robert A. Murdie, "Cultural Differences in Consumer Travel" in *Economic Geography*, Vol. 41, No. 3, July 1965. Copyright © 1965 *Economic Geography*. Reprinted by permission.
Figure 8.4 Source: *Metropolitan Toronto and Region Transportation Study*, by Maurice Yeates. The Queen's Printer, Toronto, 1966, fig. 42.

Figure 8.6   From Cole, J. P. and P. Whysall, "Places in the News, A Study of Geographical Information," Bulletin of Quantitative Data for Geographers, 1968, no. 17. Reprinted by permission.

Figure 8.9   From Paul F. Mattingly, "Disadoption of Milk Cows by Farmers in Illinois: 1930–1978," *The Professional Geographer,* Vol. 36, No. 1, p. 45. Association of American Geographers, Washington, D.C., 1984. Reprinted by permission.

Figures 8.11 and 8.12   Redrawn from Yoshio Sugiura, "Diffusion of Rotary Clubs in Japan, 1920–1940: A Case of Non-Profit Motivated Innovation Diffusion Under a Decentralized Decision-Making Structure" in *Economic Geography,* Vol. 62, No. 2, 1986. Copyright © 1986 *Economic Geography.* Reprinted by permission.

Figure 8.13   From Herbert A. Whitney, "Preferred Locations in North America: Canadians, Clues, and Conjectures, *Journal of Geography,* Vol. 83, No. 5, 1984, p. 222. With permission of the National Council for Geographic Education.

Figure 8.14   From Stanley Milgram, "Two Views of Paris" in *Journal of Nonverbal Behavior.* Copyright © 1976 Human Sciences Press, New York. Reprinted by permission.

Figure 8.19   From C. C. Roseman, *Association of American Geographers Resource Paper 77-2,* p. 5. Association of American Geographers, Washington, D.C., 1977. Reprinted by permission.

Figure 8.20   Data from T. Hagerstrand, *Innovation Diffusion as a Spatial Process.* Copyright © 1967 The University of Chicago Press.

Figure 8.21   From J. A. Jakle, S. Brunn, and C. C. Roseman, *Human Spatial Behavior: A Social Geography.* Copyright © 1976, reissued 1985 by Waveland Press, Inc., Prospect Heights, Illinois, p. 167. Reprinted with permission.

Figure 8.22   From C. C. Roseman, "Channelization of Migration Flows From the Rural South to the Industrial Midwest," *Proceedings of the Association of American Geographers,* Vol. 3, p. 142. Association of American Geographers, Washington, D.C., 1971. Reprinted by permission.

Box, Page 281   After Howard Raiffa, Harvard University.

CHAPTER 9

Figure 9.8   Data from Robert E. Huke, "The Green Revolution," *Journal of Geography,* Vol. 84, No. 6, 1985, p. 248 and p. 251. With permission of the National Council for Geographic Education.

Figure 9.9   From *World Atlas for Students, 1980.* Copyright © 1980 Hammond Incorporated, Maplewood, New Jersey.

Figure 9.12b   From Bernd Andreae: *Farming, Development and Space.* Translated from the German by H. F. Gregor. Walter de Gruyter, Berlin, New York. 1981.

Figure 9.22   Data from "A Survey of Hydroelectric Power in 100 Developing Countries", World Bank Energy Department, *Energy Department Paper No. 17.* Copyright © 1984 The International Bank for Reconstruction and Development/The World Bank.

Figures 9.29 and 9.30   Reprinted with permission from *Automotive News.* Copyright, 1986.

CHAPTER 10

Figures 10.2 and 10.8   From D. J. Cuff and W. J. Young, *The United States Energy Atlas.* Copyright © 1986 Macmillan Publishing Co., New York.

Page 345   Source: Unocal Corporation, 1984.

Figure 10.25   Source: Southern California Edison Company, 1933.

Figure 10.26a   Sources: Pacific Gas and Electric Company and American Petroleum Institute.

CHAPTER 11

Figure 11.7   Data from *Atlas of Early American History.* Princeton University Press, 1976.

Figure 11.8   Redrawn by permission from R. B. Vance and S. Smith, "Metropolitan Dominance and Integration" in *The Urban South,* R. B. Vance and N. S. Demeruth, Eds., Copyright © 1954 University of North Carolina Press, Chapel Hill, North Carolina.

Figure 11.9   From A. Getis and J. Getis, "Christaller's Central Place Theory," *Journal of Geography,* 1966. With permission of the National Council for Geographic Education.

Figure 11.16   From Arthur Getis, "Urban Population Spacing Analysis" in *Urban Geography,* Vol. 6, 1985. Copyright © 1985 V. H. Winston and Son, Inc., Silver Springs, Maryland.

Figure 11.17   Redrawn from E. Horwood and R. Boyce, *Studies of the Central Business District and Urban Freeway Development.* Copyright © 1959 The University of Washington Press, Seattle, Washington. By permission.

Figure 11.19   Redrawn from P. J. Smith, "Calgary" A Study in Urban Patterns" in *Economic Geography,* Vol. 38, 1962. Copyright © 1962 *Economic Geography.* Reprinted by permission.

Figure 11.20   Redrawn from Philip Rees, "The Factorial Ecology of Metropolitan Chicago". M. A. Thesis, University of Chicago, 1968.

Figure 11.21   From E. F. Bergman and T. W. Pohl, *A Geography of the New York Metropolitan Region.* Copyright © 1975 by Kendall/Hunt Publishing Company, Dubuque, Iowa. Used with permission.

Figure 11.22   Redrawn with permission from B. J. L. Berry, Chicago: *Transformation of an Urban System.*

Figure 11.27   Fig. 10.27, p. 385 from *The North American City,* 3d ed. by Maurice Yeates and Barry Garner. Copyright © 1980 Maurice Yeates and Barry Garner. Reprinted by permission of Harper & Row, Publishers, Inc.

Figure 11.28   Redrawn from Paul White, *The West European City: A Social Geography.* Copyright © Longman Group, Ltd., Essex, England.

Figure 11.29   Fig. 9.7, p. 338 from *Cities of the World: World Regional Development* by S. D. Brunn and J. F. Williams. Copyright © 1983 S. D. Brunn and J. F. Williams. Reprinted by permission of Harper & Row, Publishers, Inc.

CHAPTER 12

Figure 12.1   From John H. Garland, ed., *The North American Midwest,* 1955, p. 15, fig. 4. John Wiley and Sons, Inc.; John R. Borchert and Jane McGuigan, *Geography of the New World,* 1961, p. 117. Rand McNally and Company; and Otis P. Starkey and J. Lewis Robinson, *The Anglo-American Realm,* 1969, p. 202, fig. 10.1. McGraw-Hill Book Company.

Figure 12.9   From Preston E. James, Introduction to Latin America. Copyright © 1954 Odyssey Press. Reprinted by permission of Macmillan Publishing Co., Inc.

Excerpt, Page 407 "The Black Hills Province"   Excerpted from the *Physiographic Provinces of North America* by Wallace W. Atwood. Copyright © 1940 Ginn and Company.

Excerpt, Page 409 "Air Masses"   Excerpted from *Climatology and the World's Climates* by George R. Rumney. Copyright © 1968 George R. Rumney. Reprinted with permission of Macmillin Publishing Company, Inc.

Excerpt, Page 413 "Population Patterns of Latin America"   From Glenn T. Trewartha, *The Less Developed Realm: A Geography of Its Population.* Copyright © 1972 John Wiley & Sons, Inc.

END-PAPER MAPS   From Richard E. Murphy, Map Supplement to the *Annals of the Association of American Geographers,* Vol. 58, No. 1. Association of American Geographers, Washington, D.C., 1968. Reprinted by permission.

APPENDIX   From *1987 World Population Data Sheet.* Population Reference Bureau, Washington, D.C.

Index